Internet of Things

Technology, Communications and Computing

Series editors

More information about this series at http://www.springer.com/series/11636

Giancarlo Fortino · Zhelong Wang
Editors

Advances in Body Area Networks I

Post-Conference Proceedings of BodyNets 2017

Springer

Editors
Giancarlo Fortino
DIMES, Cubo 41C
Università della Calabria
Rende, Cosenza, Italy

Zhelong Wang
Faculty of Electronic Information
and Electrical Engineering
Dalian University of Technology
Dalian, Liaoning, China

ISSN 2199-1073 ISSN 2199-1081 (electronic)
Internet of Things
ISBN 978-3-030-02818-3 ISBN 978-3-030-02819-0 (eBook)
https://doi.org/10.1007/978-3-030-02819-0

Library of Congress Control Number: 2018958486

This Springer imprint is published by the registered company Springer Nature Switzerland AG
The registered company address is: Gewerbestrasse 11, 6330 Cham, Switzerland

Preface

We are delighted to introduce the post-proceedings of the 2017 European Alliance for Innovation (EAI) International Conference on Body Area Networks (BODYNETS). BodyNets 2017 aims to provide a world-leading and unique opportunity for bringing together researchers and practitioners from diverse disciplines to plan, analyze, design, build, deploy, and experiment with/on body area networks (BANs).

The technical program of BodyNets 2017 consisted of 34 full papers, including 2 invited papers in oral presentation sessions at the main conference tracks. The conference tracks were: Track 1—Wearable Computing and Smart Clothing; Track 2—Embedded Devices and Medical Applications; Track 3—Communications and Networking; Track 4—Systems and Applications—Ambient Intelligence. Each track has regular and short paper presentations. The program will also consist of the four special tracks: ST1—Antenna/Propagation and WiBEC project (APWiBEC); ST2—Cloud-assisted Body Area Networks (CBAN); ST3—Smart Clothing (SC); ST4—Sensors and Algorithms for Human Motion Analysis and Classification (SAHM). Aside from the high-quality technical paper presentations, the technical program also featured two keynote speeches. The two keynote speeches were Prof. Wang Lei from Shenzhen Institutes of Advanced Technology, Chinese Academy of Sciences, and Prof. Ryuji Kohno from Centre on Medical Information and Communication Technology, Yokohama National University in Japan.

Coordination with the steering chair, Prof. Imrich Chlamtac, and steering member, Prof. Giancarlo Fortino, was essential for the success of the conference. We sincerely appreciate their constant support and guidance. It was also a great pleasure to work with such an excellent organizing committee team for their hard work in organizing and supporting the conference. In particular, the Technical Program Committee, led by our TPC Co-Chairs, Dr. Qiong Wang, Dr. Dongyi Chen, Dr. Hassan Ghasemzadeh, Dr. Min Chen, and Dr. Mehmet Yuce, completed the peer-review process of technical papers and made a high-quality technical program. We are also grateful to Conference Managers, Lenka Bilska, for her support and all the authors who submitted their papers to the BodyNets 2017 conference.

We strongly believe that BodyNets conference provides a good forum for all researchers, developers, and practitioners to discuss all science and technology aspects that are relevant to body area networks and their applications. We also expect that the future BodyNets conference will be as successful and stimulating, as indicated by the contributions presented in this volume.

Dalian, China — Prof. Zhelong Wang
Rende, Italy — Prof. Giancarlo Fortino

Contents

Characterization and Identification of Driver Distraction During Naturalistic Driving: An Analysis of ECG Dynamics

Shantanu V. Deshmukh and Omid Dehzangi

Abstract One of the most contributing factors to the accidents on the roadways is distracted driving. While in-vehicle, driver may get distracted by variety of ways such as talking on the cellphone, conversing with the accompanying passengers, texting while driving, etc. In order to reduce potential chances of road-accidents, it is highly essential to characterize and identify distracted situations in real-time. In this paper, we investigate Electrocardiogram (ECG) signals as the physiological measure to characterize driver distraction. We aim to provide an empirical guideline for accurate and in real-time analysis irrespective of the body physic. ECG-based driver distraction identification has significant advantages in practice such as being easy to capture, minimally intrusive, and reliable in biometric patterns. ECG dynamics encompass multiple descriptors that we examine in this investigation for efficient characterization of driver state toward real-time identification of distracted driving. In this effort, six drivers were actively participated in our naturalistic driving experiments, where the distraction is introduced by the cellphone conversation and the conversation with the passenger. Our study mainly focuses on the efficient characterization of distraction by localizing R-R interval based on temporal features as well as spectral features. In addition to this, we further investigated the real-time predictive ability of the extracted features through state of the art predictive algorithms. Our experimental results demonstrated ~83% average predictive accuracy of driver distraction identification in near real-time.

Keywords Driver distraction identification · ECG characterization · R-R interval Temporal and spectral features

S. V. Deshmukh (✉)
University of Michigan Dearborn, Dearborn, MI 48126, USA
e-mail: svdeshmu@umich.edu; svdeshmu@umich.com

O. Dehzangi
Rockefeller Neuroscience Institute, West Virginia University, Morgantown, WV 26505, USA
e-mail: omid.dehzangi@hsc.wvu.edu
URL: http://www.wssplab.net

G. Fortino and Z. Wang (eds.), *Advances in Body Area Networks I*,
Internet of Things, https://doi.org/10.1007/978-3-030-02819-0_1

1 Introduction

Distracted driving is a serious concern toward the safety of driver and passengers. Mainly, driver distraction can be classified into three discrete groups: visual, manual, and cognitive distraction. Visual distraction involves taking eyes of the road, manual distraction contains taking hands of the steering wheel, and cognitive distraction includes taking drivers attention of the driving task.

Among several driver distracting activities, it is observed that the principle contribution in the process of taking drivers attention of is caused by driver conversing either on cellphone or with interactions with accompanying passengers. These activities are technically dangerous because of their involvement in all of the three pre-described distraction cases at the same. According to the National Highway Traffic Safety Administration (NHTSA), distracted driving claimed 3,477 number of lives and 391,000 number of people injured in 2013 [1]. Reducing those numbers is the driving factor for the significant urge of characterizing distraction to avoid potential accidents on the road.

There have been extensive research explorations in order to detect and possibly prevent high levels of driver distraction considering the significance of the task. Majority of studies were focused on monitoring driving behavior and performance related to distraction. In a previous work [2], the authors used steering entropy in order to evaluate driver's secondary workload. They observed that the steering angle has a direct correlation with the amount of effort involved in performing additional tasks. However, the steering entropy is often subjective depending on the individual way of driving. In addition to that, the resulting detection might be rather late as the focus is majorly on the driving outcome.

Also, there has been much efforts directed towards real time video processing systems to detect distracted driving situations. Authors in the paper [3] used dashboard-mounted closed-circuit camera for detecting mouth yawning movements. we recorded movements tend to be in exact proportions of the amount of fatigue drivers undergoing while driving. In another paper [4], authors make the use of two mobile based cameras to monitor and detect driver distraction and fatigue. we rear-facing camera of a cellphone is used for route mapping while the front-facing camera is used for monitoring driver's movements. In another work [5], the authors study different state of the art visual systems that are available to detect driver distraction. However, camera-based systems have serious privacy issues; which is a huge role block to be adopted by the end consumers. Also using cameras, detection of signs of distraction can be captured only when they became apparent through video recording; which might be rather late to avoid negative impacts. Therefore, we believe alternative research expeditions must be encouraged with major focus on the pre-requisites of driver distraction. We envision that there is a causality relation between physiological changes in the driver state to his/her behavior and eventually the driving outcome. Those physiological changes can be captured in forms of biometrics.

Electrocardiogram (ECG) on the other hand is a reliable physiological modality to measure driver distraction because of being an easy to wear and record technology, highly reliable with high signal to noise ratio and minimal intrusive implementation. Also, Entropy analysis of ECG data has yielded promising results with respect to modeling distraction. In a previous effort [6], the authors employed sample entropy (spectral feature) to analyze pre and post distraction behavior of ECG signals. It is observed that the sample entropy associated with post-distraction ECG data has a higher value than pre-distraction ECG data. Similarly, authors of the paper [7] compare the sensitivity of the extracted features (called indicators of sleepiness) in time and frequency domain analysis of ECG signal to detect sleepiness while driving. Also, in our previous works in [8, 9], we made the use of multiresolution spectral wavelet packet analysis as a comprehensive spectral filter bank on the ECG data to identify driver distraction in real-time. We observed identification accuracies characterizing distraction at different sub-bands of the spectrum with the average accuracy of ~90%. Considering the above-mentioned advantages of ECG signals and the promising multiresolution sub-band analysis in our previous works [8, 9], our goal in this paper is to extract well-established ECG specific temporal measures such as average heart rate and heart rate variability by localizing R-R intervals of recorded ECG during the naturalistic driving scenarios. We employ those indicators aiming to provide guidelines for ECG-based characterization of driver distraction. Our experimental results demonstrated ~83% average predictive accuracy of driver distraction identification in near real-time. Moreover, we observed consistent subject independent impacts on medically meaningful temporal indicators extracted from drivers' recorded ECG after being exposed to distracting elements. For instance, we observed an increase in average heart rate and decrease in hear rate variability while the driver is distracted by phone call or conversation with the passenger compared to the normal driving scenario.

2 Experimental Setup and Data Acquisition

2.1 Platform

At the Wearable Sensing and Signal Processing Lab (WSSP Lab); University of Michigan—Dearborn, we engineered a unique data acquisition platform, enabling synchronous data collection through different channels of heterogeneous physiological signals [10, 11]. We record integrated input from different sensors; recording signals such as ECG, EEG, Galvanic Skin Response (GSR), motion (acceleration data), and CAN BUS data synchronized in real-time.

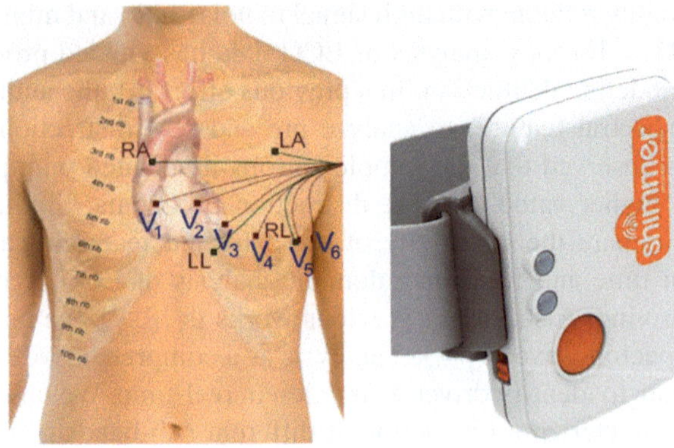

Fig. 1 Shimmer ECG electrode placement

2.2 *Signal Acquisition*

Real-time ECG signal is recorded at 250 Hz of sampling frequency. We used 3-lead configuration for ECG signal analysis, which will be referred to as LE1, LE2, and LE3, respectively. Location of the connected electrodes spans near the heart over chest area. LE2 is the electrode position closest to right arm (RA), LE3 positioned close to left arm (LA) and left leg. While LE1 refers to the voltage between RA and LA. We electrode positions are shown in Fig. 1. Along with the Shimmer ECG device used while conducting naturalistic driver data collection experiment [12]. All of the procedures and subsequent actions taken while conducting this experiment are approved by University of Michigan Institutional Review Board.

2.3 *Procedure*

The ECG readings were recorded within the time frame of 9 min: first 3 min were engaged in normal naturalistic driving (without distraction), remaining 6 min were engaged in performing secondary tasks with driving (phone and question explained in Table 1). We driving route was picked as a straight and almost empty road in order to minimize other external factors on driver state. Distracted driving experiment is illustrated in the following Fig. 2. There has been a ~5 min slot of driving only between each distracted driving scenario to minimize any gradual accumulation of impact from previous distracting elements.

Table 1 Driving tasks

Type	Duration (min)	Significance
Normal	3	When the driver is solely focusing on the task of driving
Phone	3	When the driver engages in a phone
Question	3	When the passenger asks a series of questions conversation while driving communication while driving in order to engage driver in an active

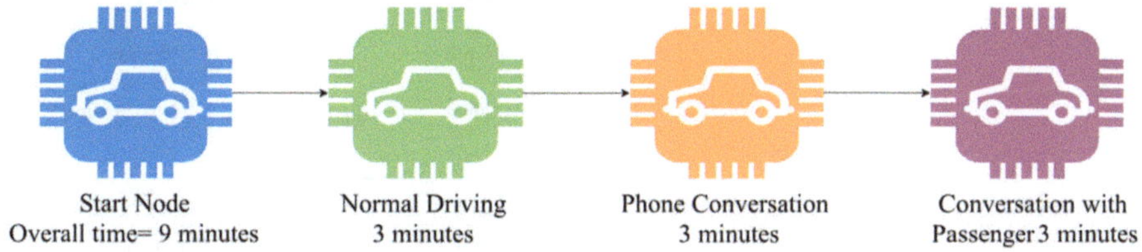

Fig. 2 Overview of distracted driving experiement

2.4 Participants

Six subjects in the age range of 22 ± 35, were actively participated in the experiment. From each driving session, ECG data was recorded for 9 min as discussed earlier from a during a naturalistic driving in a Ford Escape 2015 vehicle. In order to make sure that there is no external influence on driving outcome. We kept driving conditions identical for each subject in addition to it we made sure that the driver's health must not hamper the recorded data. A proper care was taken to avoid drinks and medication before driving which may potentially impact the performance of experiment.

3 Methodology

Our intention is to define concurrent relationship between the extracted well-established meaningful features and the driver's physiological state. In order to achieve this, we propose the following procedure for pre-processing the input data, segmentation, and classifier analysis. We aim to analyze all possible features, which can electively contribute in characterization as well as identification of the distraction scenarios. Figure 3 shows the detailed diagram of proposed methodology.

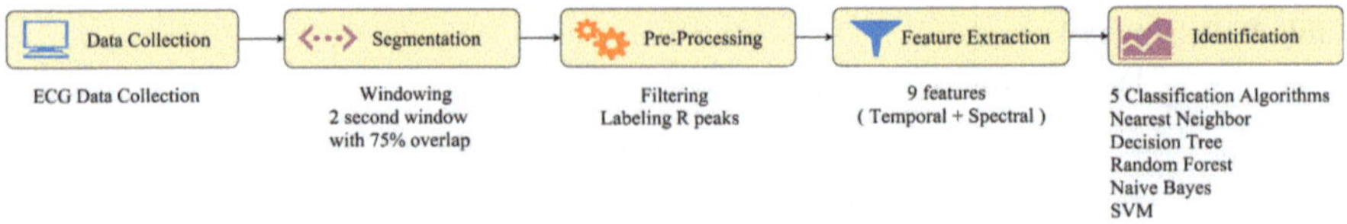

Fig. 3 Overall proposed methodology

Fig. 4 Segmentation and pre-processing

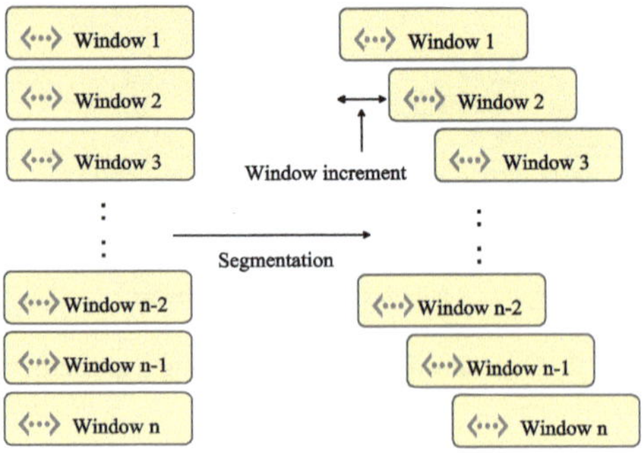

3.1 Segmentation and Pre-processing

Biological signal such as ECG and EEG, and they need to be preprocessed to be ready for the next analysis [12], and The ECG recorded signals were first segmented using 2 s windowing with 75% overlap (i.e. 25% increments). Each window was considered a unit of ECG data for the subsequent processing. A detailed illustration of the segmentation procedure is shown in Fig. 4. Due to the fact that driver was in a controlled experiment with the driver in his seat with limited movements, the raw ECG data has acceptable quality and consequently there is no need to perform advanced artifact removal methods. Pre-processing of the ECG data in this work consists of filtering, in which the recorded ECG data were pre-processed using sharp low-pass and high-pass filters having cut-off frequencies of 0.5–4 Hz. This process helps in removal of power line noise and high frequency noise. Then, we performed local maxima identification and labeling R peaks using [13] all of which was implemented in MATLAB.

3.2 Feature Extraction

After segmentation of the signal, various time domain as well as frequency domain features are extracted from each window. Feature name, unit of measurement, and explanation regarding each feature is provided in Table 2.

Table 2 Overview of temporal and spectral features

	Feature	Unit	Definition
Time domain analysis	AvgHR	bpm	Average heart rate
	MeanRR	ms	Mean of selected R-R series
	NN50	count	Number of consecutive R-R intervals that differs more than 50 ms
	pNN50	%	Percentage value of count of consecutive R-R intervals that differs more than 50 ms
	SD_HR	1/min	Standard deviation of heart rate
	SD_RR	1/min	Standard deviation of R-R interval
	RMSSD	ms	Root mean square of the differences of selected R-R interval series
	SE	–	Sample entropy
Frequency domain analysis	PSE	–	Power spectral entropy

3.2.1 Time-Domain Analysis

Our extracted time domain features are based on R to R peak intervals from each segment of the pre-processed ECG data. We used ECG preprocessing package authored by Chernenko [14] for localizing R peaks from the clean ECG signal. R peaks are integral part of our time domain feature analysis as most of the features are derived from it. For each of the following computations, let n_r = number of R peaks, and $d_{r+1} - d_r$ = successive distance between R-R peaks.

3.2.2 Frequency Domain Analysis

At the spectral part of ECG analysis, we focused mainly on entropy analysis. As entropy is the degree of dilution of specific component (here distraction) to the driving process. It will be easy for the predictive model to characterize effectively between the diluted component (distraction) from the driving process.

3.3 Identification of Driver Distraction

For the effective classification of the features, five classification algorithms were used including Decision Tree, Random Forest, Nearest Neighbor, SVM (Support Vector Machine) and Naïve Bayes.

Decision tree is the most popular and widely used algorithm as a predictive model [15]. Decision tree algorithm is a tree structured hierarchy where each non-leaf node denotes a test on an attribute, branches represent output of the test, and each terminal node has a class label. There are many algorithms to construct decision trees with different quality of splits depending on metrics applied to a measure.

Random forests algorithm follows ensemble learning method for classification [16]. Multiple decision trees are constructed during training phase and then provide decision in the classification phase using a voting mechanism.

Nearest neighbor classifiers [17] use a non-parametric approach in pattern recognition. It makes use of the label of the closest training sample point in the feature space for prediction. K-nearest neighbor classifier assigns a point to particular classes based on majority of votes among the k nearest training points.

Support Vector Machine (SVM) classifier is a linear discriminative classifier that aims to optimize the generalization performance of the trained model based on the concept of margin and maximizing it [18]. An optimal hyperplane is said to be good if it does not pass through majority of points leaving behind larger margin between the hyperplane and training examples.

4 Results and Discussions

4.1 Individual Feature Analysis

We carried out a detailed feature analysis of ECG signals that resulted in remarkable understanding of driver distraction when engaged in a secondary task. There is a direct relationship between heart rate and driver's cognition. Subjects participated in our experiments exhibited higher Average Heart Rate (AvgHR) when driving with distraction. Figure 5 shows AvgHR of all subjects with three inputs viz., LE1 = input from LARA node, LE2 = input from LLRA node and LE3 = input from V × RL node. As shown in Fig. 5, the increase in AvgHR is consistent over different leads and subjects. AvgHR corresponding feature values are also included in Fig. 8 which indicate relevancy between physiological changes and the observed heart rate when driver is exposed to distraction.

Figure 6 illustrates a continuous comparison between normal driving, driving while on a phone call, and driving while talking to the passenger during a 2 min segment of ECG recording. Observing Fig. 6, we can conclude that there is a consistent trend of change for Heart Rate Variability (HRV) and AvgHR. There is an increase in Herat Rate and a subsequent decrease in HRV which is expected when more

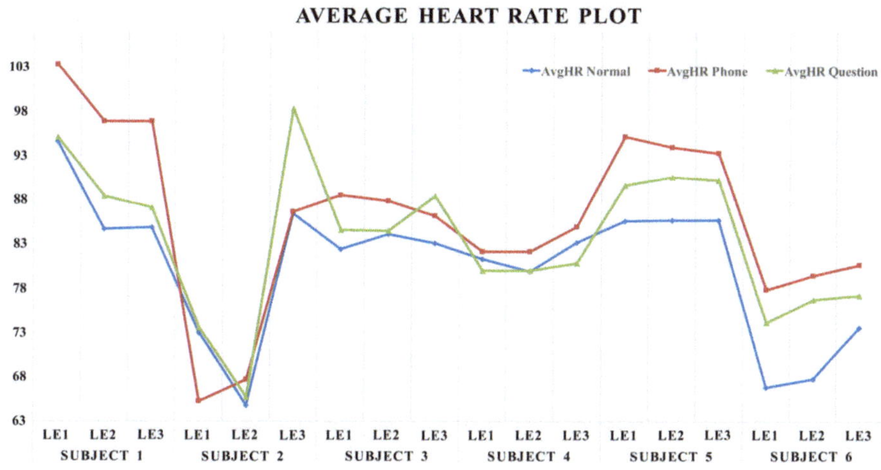

Fig. 5 Average heart rate plot

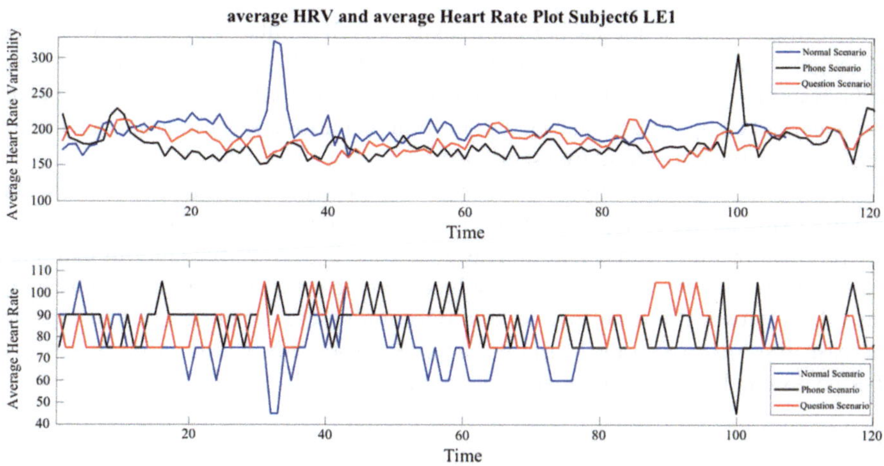

Fig. 6 Chnages in Average HRV with respect to average heart rate

mental effort is engaged while performing a more complex task. When subject drivers are engaged in a secondary task along with driving, their mental load is expected to increase compared to the load while driving without distraction. Our results are consistent with the previous literature [19, 20] where similar results were observed. Comparing phone vs. question scenarios, increase in AvgHR and decrease in HRV seems to be more distinguished that can indicate higher level of driver engagement.

In addition to the temporal distraction indicators, Complexity measures were also extracted including sample entropy (SE) and power spectral entropy (PSE). Particularly, sample entropy (SE) illustrated significant impact via the distracting elements; projecting higher values of SE when the driver is distracted. This observation

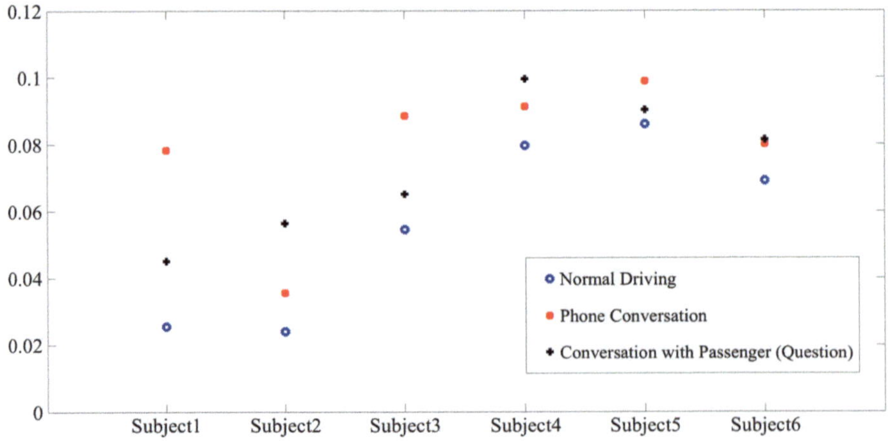

Fig. 7 Comparative analysis between sample entropies

demonstrates signs of increasing complexity in self-similar time series when the driver gets distracted. Figure 7 represents the values of SE containing normal, phone, and question scenarios with respect to each subject. According to Fig. 7, there is no specific trend comparing phone versus question scenarios and the amount of increase in SE is subjective for these two scenarios.

Figure 8 reports an analysis of each extracted feature where the rows are the features extracted from respective leads LE1, LE2 and LE3 and columns map driving environment. Similar consistent increase/decrease in various meaningful extracted features were observed from different subjects which is reported in Fig. 8. Observing the values between normal, phone, and question cases, it can be inferred that the increase in values of average heart rate, sample entropy, NN50, and variance between R-R peaks can be reliable indicators of characterizing distraction.

4.2 Multivariate Analysis and Identification Results

In this section, various extracted features are represented as dimensions of a multivariate space containing useful information to measure the changes from pre to post distraction cases. We used Weka to design and train the predictive models [21]. Main advantage behind using Weka is its extensive library of machine learning algorithms in order to evaluate the discriminative capability of all the extracted features. Figure 9 reports the identification accuracies of the windows of ECG data with different labels based on their corresponding scenarios including normal, phone, and question. Correct classification rate (CCR) is the percentage of correctly classified windows. From the prediction table, it can be inferred that Random Forest, Decision Tree, and 1-NN classifier generate reliable results consistently. Maximum classification

Features	Lead I/P	Subject 1			Subject 2			Subject 3			Subject 4			Subject 5			Subject 6		
		Normal	Phone	Question	Normal	Phone	Question	Normal	Phone	Question	Normal	Phone	Question	Normal	Phone	Question	Normal	Phone	Question
AvgHR	LE1	94.5631	103.275	95.02	73.02	65.25	73.58	82.38	88.47	84.57	81.26	82.12	79.97	85.57	95.08	89.64	66.84	77.86	74.11
	LE2	84.64	96.8	88.4	64.75	67.71	65.68	84.09	87.8	84.48	79.9	82.11	79.97	85.68	93.92	90.52	67.76	79.4	76.71
	LE3	84.85	96.82	87.09	86.4	86.62	98.2615	83.04	86.14	88.41	83.13	84.94	80.88	85.68	93.29	90.1732	73.51	80.69	77.24
MeanRR	LE1	153.135	140.012	151.96	197.53	221.24	196.13	175.34	163.37	170.78	177.78	175.62	180.47	169.33	151.76	160.94	217.11	185.54	194.96
	LE2	171.18	149.41	163.37	222.81	213.19	219.77	171.77	164.62	170.98	180.81	175.64	180.47	169.1	153.64	159.38	214.13	181.93	188.33
	LE3	170.76	149.37	165.83	166.98	166.58	146.79	173.95	167.82	163.35	173.77	169.78	178.43	169.1	154.68	160.01	197.27	179.02	187.04
RMSSD	LE1	157.665	144.08	156.4	215.03	239.5	213.32	178.63	164.95	172.63	179.68	176.61	181.03	173.32	154.83	163.83	222.93	189.75	196.02
	LE2	175.51	156.04	166.38	242.68	231.84	237.7	174.07	165.83	173.68	181.58	176.65	181.04	173.07	156.24	160.6	220.61	186.75	190.6
	LE3	175.34	156.35	168.91	185.01	185.75	158.41	176.26	168.3	170.31	176.07	172.27	179.45	172.96	158.46	162.13	207.23	185.95	189.95
NN50	LE1	162	221	337	384	392	386	298	312	313	293	598	429	158	445	505	118	278	281
	LE2	157	223	327	336	346	349	304	309	312	294	596	427	158	445	506	116	281	283
	LE3	157	223	324	357	369	452	303	316	321	292	596	430	157	445	506	115	279	279
pNN50	LE1	91.0112	91.7	94.13	67.01	63.63	65.91	95.51	97.33	94.5	96.69	98.19	98.84	98.75	97.58	99.01	92.91	92.3588	98.94
	LE2	98.74	98.67	98.19	66.14	66.5	66.08	95.29	97.31	97.19	98.65	97.86	98.38	98.75	98.66	98.25	89.92	91.53	96.25
	LE3	98.12	98.67	98.78	70.55	68.164	75.83	96.19	98.82	88.09	94.19	94.6	97.94	98.12	99.33	98.63	82.14	99.42	94.25
SD_RR	LE1	6.8652	20.98	17.03	209.46	36.75	25.86	8.3494	9.6274	12.78	53.78	0.376	82.47	13.33	0.0769	5.0529	18.88	7.5449	22.03
	LE2	13.18	10.58	4.627	160.18	38.8	51.77	85.77	8.3716	51.01	36.81	30.35	84.47	4.89	29.64	30.61	60.86	7.9381	25.66
	LE3	2.2375	13.62	43.16	14.93	110.43	12.8	79.95	5.1797	216.648	34.77	21.21	82.43	6.9	25.68	34.99	56.72	22.97	92.034
SD_HR	LE1	4.5631	13.83	9.81	37.64	9.4423	8.7187	3.8426	5.2402	6.5601	34.86	0.3098	66.96	6.7358	0.2756	2.8989	5.62	3.0318	7.7588
	LE2	6.4925	6.8046	2.6926	27.15	10.56	20.027	83.34	4.5679	19.61	20.09	12.21	70.02	2.92	22.2015	14.73	15.4	3.353	9.4291
	LE3	1.6177	8.484	18.19	8.3317	34.63	9.2012	70.15	2.9042	50.52	20.46	9.5555	69.11	3.8708	18.3292	16.3271	16.82	9.4046	74.33
SE	LE1	0.0255	0.0783	0.0451	0.0239	0.0223	0.0222	0.0545	0.0485	0.0651	0.0796	0.0813	0.0695	0.086	0.0988	0.0903	0.0691	0.08	0.0814
	LE2	0.112	0.1553	0.1344	0.01	0.0091	0.0108	0.0478	0.0724	0.694	0.0369	0.049	0.045	0.0733	0.0878	0.0736	0.07	0.0812	0.0763
	LE3	0.0752	0.0861	0.074	0.0085	0.0099	0.0075	0.0514	0.0536	0.0603	0.0517	0.0627	0.0518	0.0416	0.0784	0.0736	0.0536	0.0638	0.0548
PSE	LE1	4.2297	4.6289	5.125	5.9281	5.4706	5.6277	4.8246	4.0324	4.616	4.3683	4.789	4.0364	4.3716	5.2678	5.3246	3.9663	4.7606	3.7623
	LE2	4.2499	4.7135	4.9245	5.7591	5.5086	5.4882	4.5885	3.9068	4.8084	3.7025	4.9074	4.0631	4.2718	5.2158	4.7665	4.0252	4.8194	4.37
	LE3	4.2713	4.7475	4.8898	5.8134	5.839	5.9414	4.5479	3.3151	5.1407	4.4751	5.4088	4.4448	4.2452	5.3249	5.1727	4.2515	5.0408	4.4915

Fig. 8 Feature value analysis report

Subject	Lead	Decision Tree	Random Forest	1-NN	SVM	Naïve Bayes
1	LE1	90.13%	90.40%	94.13%	88.27%	73.87%
	LE2	79.55%	85.38%	84.09%	74.74%	68.44%
	LE3	88.14%	90.66%	94.19%	81.06%	78.28%
2	LE1	73.75%	79.85%	79.29%	61.88%	59.89%
	LE2	73.09%	81.83%	82.55%	78.90%	70.91%
	LE3	80.42%	76.23%	79.38%	73.37%	70.10%
3	LE1	67.32%	63.43%	83.27%	57.11%	52.42%
	LE2	77.54%	74.07%	77.54%	65.28%	57.75%
	LE3	67.21%	60.53%	58.75%	53.41%	49.25%
4	LE1	77.19%	82.75%	81.42%	68.19%	60.51%
	LE2	76.29%	81.59%	80.14%	68.25%	64.63%
	LE3	80.62%	84.48%	85.52%	66.15%	59.69%
5	LE1	90.77%	95.56%	94.73%	79.24%	75.62%
	LE2	91.87%	93.94%	94.49%	72.45%	71.70%
	LE3	90.22%	92.97%	92.59%	71.09%	65.43%
6	LE1	71.27%	71.11%	65.30%	59.65%	55.89%
	LE2	80.95%	84.63%	85.93%	73.16%	75.97%
	LE3	80.12%	85.47%	81.42%	71.69%	70.51%
Average		79.80%	81.94%	83.04%	70.22%	65.60%

Fig. 9 Multivariate analysis and identification accuracies (%) report

accuracy is observed in case of subject 5 LE1 ~96%, which is a high accuracy result for driver distraction identification. We observed that most of the errors induced was the confusion between phone and distraction labels that might be due to the fact that both distraction scenarios would put similar impact on various features.

5 Conclusion

The primary objective of this paper is to investigate the impact of driver distraction on their continuous recordings of ECG. We designed a methodology to pre-process and analyze short windows of recorded ECG toward real-time ECG-based identification of driver distraction. We characterized the impact of distraction using various well-established ECG-based features. We successfully localized the R-R peaks in ECG windows and extracted various features. Analysis of individual features showed a causality relationship between the driver's state of distraction and the physiological changes in ECG responses, as we envisioned. We also conducted multi-variate predictive analysis that showed the maximum subjective identification accuracy of ~96% while average accuracy of ~83% showed the promising performance of physiological measures (features) in order to characterize and identify the possible signs of distraction in driving data in near real time. However, there are still problems of subjective responses to distraction and motion artifacts which hampers the prediction accuracy. Our further research is spanning over bridging the gap our current identification rates toward a minimum system in real-time.

References

1. National Highway Traffic Safety Administration: Traffic safety facts 2011 data–pedestrians. Ann. Emerg. Med. **62**(6), 612 (2013)
2. Nakayama, O., Futami, T., Nakamura, T., Boer, E.R.: Development of a steering entropy method for evaluating driver workload (1999)
3. Rongben, W., Lie, G., Bingliang, T., Lisheng, J.: Monitoring mouth movement for driver fatigue or distraction with one camer. In: Proceedings the 7th International IEEE Conference on Intelligent Transportation Systems (IEEE Cat. No.04TH8749), pp. 314–319
4. You, C., et al.: CarSafe app: alerting drowsy and distracted drivers using dual cameras on smartphones categories and subject descriptors. In: Mobisys'13, pp. 1–14 (2012)
5. Fernández, A., Usamentiaga, R., Carús, J., Casado, R.: Driver distraction using visual-based sensors and algorithms. Sensors **16**(12), 1805 (2016)
6. Lin, C.-T., Chen, S.-A., Chiu, T.-T., Lin, H.-Z., Ko, L.-W.: Spatial and temporal EEG dynamics of dual-task driving performance. J. Neuroeng. Rehabil. **8**(1), 11 (2011)
7. Kim, J.Y., Jeong, C.H., Jung, M.J., Park, J.H., Jung, D.H.: Highly reliable driving workload analysis using driver electroencephalogram (EEG) activities during driving. Int. J. Automot. Technol. **14**(6), 965–970 (2013)
8. Yu, L., Sun, X., Zhang, K.: Driving distraction analysis by ECG signals: an entropy analysis. In: Lecture Notes in Computer Science (including Subser. Lect. Notes Artif. Intell. Lect. Notes Bioinformatics), vol. 6775 LNCS, pp. 258–264 (2011)
9. Mahachandra, M., Yassierli, Sutalaksana, I.Z., Suryadi, K.: Sensitivity of heart rate variability as indicator of driver sleepiness. In: 2012 Southeast Asian Network Ergonomics Societies Conference Ergonomics Innovations Leveraging User Experience Sustainable SEANES 2012, pp. 0–5 (2012)
10. Deshmukh, S.V., Dehzangi, O.: ECG-based driver distraction identification using wavelet packet transform and discriminative kernel-based features. In: 2017 IEEE International Conference on Smart Computing (SMARTCOMP), pp. 1–7 (2017)

11. Deshmukh, S., Dehzangi, O.: Identification of real-time driver distraction using optimal subband detection powered by Wavelet Packet Transform. In: 2017 IEEE 14th International Conference on Wearable and Implantable Body Sensor Networks (BSN), pp. 9–12 (2017)
12. Taherisadr, M., Dehzangi, O., Parsaei, H.: Single channel EEG artifact identification using two-dimensional multi-resolution analysis. Sensors **17**(12), 2895 (2017)
13. Alizadeh, V., Dehzangi, O.: The impact of secondary tasks on drivers during naturalistic driving: analysis of EEG dynamics. In: IEEE Conference on Intelligent Transportation Systems Proceedings, ITSC, pp. 2493–2499 (2016)
14. Chernenko, S., ECG processing— R-peaks detection— Librow— Software. Available from: http://www.librow.com/cases/case-2
15. Freund, Y., Mason, L.: The alternating decision tree learning algorithm. In: Proceeding Sixth International Conference on Machine Learning, p. 10 (1999)
16. Liaw, A., Wiener, M.: Classification and regression by randomForest. R news **2**(3), 18–22 (2002)
17. Keller, J.M., Gray, M.R.: A fuzzy K-nearest neighbor algorithm. IEEE Trans. Syst. Man Cybern. **SMC-15**(4), 580–585 (1985)
18. Cortes, C., Vapnik, V.: Support vector networks. Mach. Learn. **20**(3), 273–297 (1995)
19. da Silva, F.P.: Mental workload, task demand and driving performance: what relation? Procedia Soc. Behav. Sci. **162**, 310–319 (2014)
20. Hancock, P.A., Desmond, P.A.: Stress, Workload, and Fatigue (2001)
21. Hall, M., Frank, E., Holmes, G., Pfahringer, B., Reutemann, P., Witten, I.H.: The WEKA data mining software. ACM SIGKDD Explor. Newsl. **11**(1), 10 (2009)

Wearable Galvanic Skin Response for Characterization and Identification of Distraction During Naturalistic Driving

Omid Dehzangi and Vikas Rajendra

Abstract Fatalities due to road accidents are mainly caused by distracted driving. Driving demands continuous attention of the driver. Certain levels of distraction while driving can cause the driver lose his/her attention which might lead to a fatal accident. Thus, early detection of distraction will help reduce the number of accidents. Several researches have been conducted for automatic detection of driver distraction. Many previous approaches have employed camera based techniques. However, these methods might detect the distraction rather late to warn the drivers. Although neurophysiological signals using Electroencephalography (EEG) have shown to be another reliable indicator of distraction, EEG signals are very complex and the technology is intrusive to the drivers, which creates serious doubt for its implementation. In this study we investigate Galvanic Skin Responses (GSR) using a wrist band wearable and conduct an empirical characterization of driver GSR signals during a naturalistic driving experiment. We explored time and frequency domain to extract relevant features to capture the changes/patterns at the physiological level. Due to the fact that feature extraction is a manual process and to normalize the feature space toward the identification task, we then transform the feature space using linear discriminant dimensionality reduction to discover discriminative bases of the GSR multivariate feature space that identify distraction. That would eliminate both the computational complexity and the redundancies in the manually generated feature space. Due to multi-class nature of the identification task, there might be biases between the distracted and non-distracted categories that can bias the estimation of between- and within-class scatter matrices. Therefore, we incorporated a class dependent weight

O. Dehzangi (✉)
Rockefeller Neuroscience Institute, West Virginia University, Morgantown, WV 26505, USA
e-mail: omid.dehzangi@hsc.wvu.edu
URL: http://www.wssplab.net

V. Rajendra
University of Michigan Dearborn, Dearborn, MI 48126, USA
e-mail: vrajendr@umich.edu

G. Fortino and Z. Wang (eds.), *Advances in Body Area Networks I*, Internet of Things, https://doi.org/10.1007/978-3-030-02819-0_2

to calculate the within class scatter matrices. The proposed weight aims to increase the flexibility of the discriminative bases vectors to capture the factors that focus on eliminating the overlap between distracted versus non-distracted in the generalization phase. Our experimental results demonstrated high cross validation accuracies of distraction identification using GSR signals (i.e. 85.19%). Conducting dimensionality reduction using LDA resulted in slight improvement in accuracy (i.e. 85.94%) using only two discriminant bases. The generalization accuracy was further improved by applying our proposed weighting mechanism (i.e. 88.92%).

Keywords GSR driver distraction · Distraction detection with skin conductance Distraction detection

1 Introduction

As per a report in 2016 by National Highway Traffic Safety Administration (NHTSA) the fatalities on road has increased by 7.2% in 2015 losing 35,092 people compared to the year 2014. This is the largest percentage increase in past 50 years. Distracted driving constitutes major reasons for most of the fatal accidents on the roadway. Driving demands concentration and constant attention of the driver. Many researches have shown that the drivers performance and alertness is reduced during multitasking, including slower reaction time, decreased situational awareness, impairing judgments and narrowed visual scanning [1].

Distraction occurs when drivers divert their attention from driving task to a secondary activity instead such as having phone conversation, texting, using the infotainment system, using the GPS, talking to the passenger, eating or drinking while driving. Using phone for either calling or texting is one of the most common distracting tasks. In order to minimize the risk for accidents on road it is crucial to notify distraction at its early stages to the driver. To alert the driver at the early stages of distraction many researches have been conducted to develop reliable system.

Many of the previous works have been employed based on eye lid closure and movement tracking [2], lane tracking [3], and video cameras as image processing technique by periodically taking video image of the driver [4] to classify distracted state from non-distracted state of the drivers. Although good performance was achieved through above methods they had serious limitations such as privacy issues and delayed response. To overcome these limitation several physiological signals such as Electroencephalogram (EEG) were used. EEG based systems were very comprehensive, reliable and state-of-the-art. However, EEG has limitations of complexity of setup for collecting and analyzing the data [5, 6] which makes the system expensive and intrusive to implement. Galvanic Skin Responses (GSR) on the other hand is a non-obtrusive modality that has already been included in consumer electronic such as wrist watched and is easy to collect. GSR also known as skin conductance (SC) is one the most sensitive markers for emotional arousal. Unconscious response of our body to various stimuli through skin conductance is measured by GSR. Changes in skin conductance mainly in the hands and foot region triggers emotional stimulation [7, 8]. Higher arousal corresponds to higher skin conductance. The sympathetic activity which drives human behavior, cognitive, and emotional state on a subconscious level controls skin conductance autonomously. Recording of GSR is done by placing two electrodes on either fingers or palm or foot of the subject.

There have been several investigations on synchronously recorded GSR signals to analyze the correlation between the impacts of cognitive state change. The authors in [7] objectively evaluate users stress and arousal levels using GSR as an index of cognitive load. The features mean GSR (mGSR) and accumulated GSR was calculated (accGSR) where accGSR is the sum of all GSR values across all tasks of each broad load level (Low, Medium, High), and across all subjects. In paper [8] the authors evaluated cognitive load caused by arithmetic and reading tasks using GSR for detection. Temporal and spectral features of GSR against different task difficulty levels were examined. In the previous work [9], a novel method for analyzing skin conductance (SC) using Short Time Fourier Transform was employed to extract estimation of mental work load with high enough temporal bandwidth to be useful for augmented cognition application. Graphical data analysis of the STFT showed notable increase in the power spectrum across a range of frequencies directly following fault events.

Our aim in this paper is to conduct an empirical characterization of driver GSR signals during a naturalistic driving experiment via time and frequency domain analysis and feature extraction. In our experiments, we use a wrist band wearable GSR on a population of seven driver subjects that participated in this study during real driving experiments. Three scenarios were investigated in our experiments: (1) Normal driving focusing attention on the primary task of driving (2) Phone Distracted driving while having an engaging phone conversation and (3) Text distracted driving while writing and sending texts while driving. We aim to investigate phone distraction as cognitive distraction element in comparison to text distraction, which represent cognitive and visual distraction at the same time. We aim to evaluate GSR toward real-time identification of distraction using short term segments of GSR. We employed linear and kernel-based Support Vector Machine (SVM) and 10 fold cross validation (10-CV) to generate identification results. Upon evaluating the result, GSR showed promise as a reliable indicator of driver distraction by achieving an overall average accuracy of 85.19% to identify distraction elements under naturalistic driving condition.

Our experimental results demonstrated high average identification accuracy of 85.19% for all subjects. Due to the fact that feature extraction is a manual process, the redundancy and computational complexity of the feature space might limit the distraction identification performance. Therefore, we apply linear discriminant dimensionality reduction to discover discriminative bases of the GSR multivariate feature space to more efficiently identify distraction. We propose to incorporate a weight to the calculation of within class scatter matrices, aiming to alleviate limitations of LDA with multiple distracting categories and to eliminate the overlap between distracted versus non-distracted scenarios in the generalization phase. Our experimental results using LDA demonstrated slight improvement in accuracy of distraction identification (i.e. 85.94%) using only two discriminant bases. The generalization accuracy was further improved by applying our proposed weighting mechanism (i.e. 88.92%).

(a) A driver subject during our naturalistic driving experiment

(b) Wireless GSR Wearable Device on left hand wrist

Fig. 1 Experimental setup

2 Data Acquisition

We have designed a wearable data acquisition platform comprising a synchronized multi-modal solution to acquire the physiological signals using a comprehensive wearable sensor network. Our platform is capable of collecting large amount of heterogeneous drivers physiological including Galvanic Skin Response (GSR) during naturalistic driving, which was approved by Institutional Review Board of University of Michigan.

The subjects were provided with a naturalistic setting of driving a real car (Ford Escape 2015) and their GSR data were recorded at an experimental sampling frequency of 50 Hz. GSR signals were collected for three types of sessions namely normal, phone and text. Figure 1b shows our wireless GSR wearable device used to collect data.

Seven subjects of age ranging between 20 and 40 years, who were legally permitted to drive, participated in our experiment. Figure 1a shows one of the driver subjects during our naturalistic driving experiment. To eliminate discrepancies of results based on gender, only healthy male subjects were chosen for the experiment. These subjects were instructed to avoid any alcoholic beverages or medications that would compromise their alertness during the experiment.

Three types of driving tasks, each of two minutes duration were performed by the subjects and their corresponding GSR signals were recorded. The tasks performed includes (i) driving under normal condition, (ii) driving while engaging in a phone conversation and (iii) driving while using the phone for texting. Scenario (i) represents normal (non-distracted) while scenario (ii) and (iii) represents the distracted driving. Figure 2 shows data collection order during our naturalistic driving experiment.

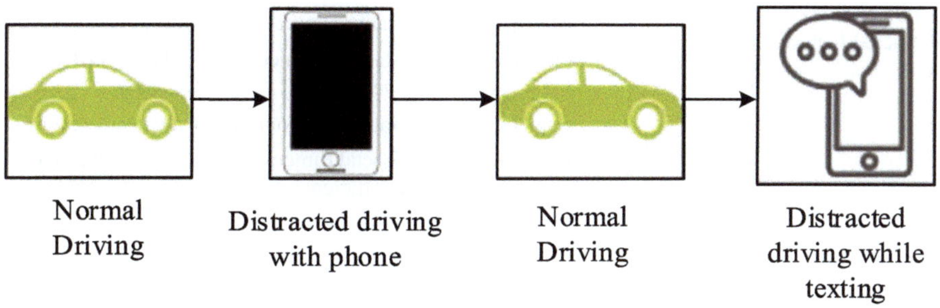

Fig. 2 Data collection order during our naturalistic driving experiment

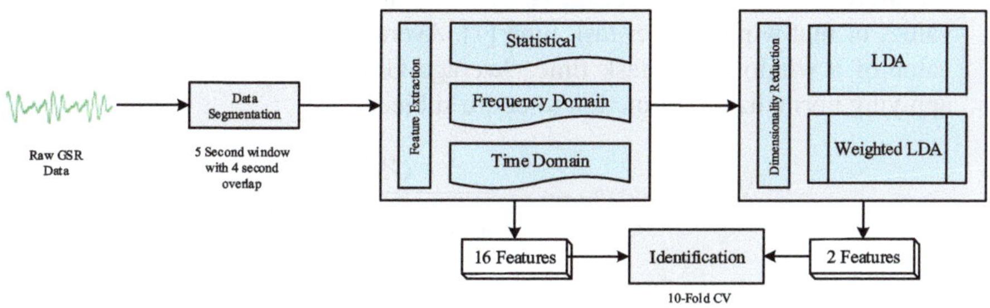

Fig. 3 Block diagram of our methodology

3 Methodology

3.1 Segmentation

Segmentation method of 5 s windows with 4 s overlap was employed to meet the requirement of short-response time for a driver monitoring or alerting system, bringing a real-time approach to our model. Figure 3 shows the block diagram of our methodology.

3.2 Feature Extraction

Various statistical, frequency and time domain features were extracted from every segmented window. The output of these features were labeled appropriately and considered as sample data point of our feature space. Extracted features are shown in Table 1 and explained below:

1. **Mean**: It is the average of all the data points present in the each window. Mean of each window was calculated using a simple mean function in MATLAB.

Table 1 List of all the extracted features

Feature domain	Feature names
Statistical	Mean, variance, accumulated GSR, average GSR, maximum value
Frequency	Short term fourier transforms (4 features)
Time	Fractal dimensions and auto regressive coefficients

2. **Variance**: It is the average of squared distance from the mean and was calculated using a MATLAB function.
3. **Accumulated GSR and Average GSR**: Accumulated GSR is summation of GSR values of that window over task time [7]. Average GSR is normalized GSR [8] value of a window over task time. Average GSR and accumulated GSR after applying normalization function to entire subjects

$$Normalise_{GSR(i,k,t)} = \frac{GSR\,(i,k,t)}{\frac{1}{m}\sum_{j=1}^{m}\sum_{t=1}^{Tij} GSR\,(i,j,t)} \quad (1)$$

GSR(i, k, t) is values of each data point at time t of task k of subject I and m is the number of tasks. They also calculated accumulated and average GSR for task k of subject I as below:

$$accGSR\,(i,k) = \sum_{t} Normalised_{GSR(i,k,t)} \quad (2)$$

$$avgGSR\,(i,k) = \frac{\sum_{t} Normalised_{GSR(i,k,t)}}{T} \quad (3)$$

4. **Maximum Value**: Maximum value of GSR in a window was the resulting value of this feature.
5. **Short Time Fourier Transform (STFT)**: Most frequently used technique to analyze physiological signal is STFT. STFT analyzes signals in frequency as well as time domain. It divides the long signal into equal smaller segments and applies Fourier Transform on those segments separately. Four features were extracted using STFT from each window of the signal. In [9] STFT analysis on skin conductance data showed effective results in detecting work load.
6. **Fractal Dimension (FD)**: Fractal dimension is used to determine the chaotic or fractal nature of a physiological signal. It is an impressive mathematical tool to model various complex physiological signals and an index to quantify the complexity of a fractal pattern. One of the common techniques to characterize the time-series data, Katz method was used to extract one fractal dimension feature.
7. **Auto-Regressive (AR)**: In an AR model of order p, the current output is estimated based on the linear combination of past p output values plus some noise term. The mean squared prediction error of an auto-regression model is minimized by

calculating the weights on the previous p outputs. The model with current output value $y(n)$ and zero mean white space noise input $x(n)$ is:

$$y(n) = \sum_{k=1}^{p} a(k)\, y(n-k) = x(n) \tag{4}$$

In this study the parameters of order p were selected as the features for each window. The order of p was set to 5 which generated five AR features.

3.3 *Discriminative Feature Space Analysis and Transformation*

An overall of 16 features (includes statistical, frequency and time domain features) were generated for each subject. To normalize the feature space towards the identification task, we transform the feature space using linear discriminant dimensionality reduction to discover discriminative bases of the GSR multivariate feature space purposed toward distraction identification. We propose to incorporate a class dependent weight to the calculation of scatter matrices, aiming to alleviate limitations of LDA with multiple distracting categories and to eliminate the biases between distracted versus non-distracted scenarios for unseen data at the generalization phase.

Linear Discriminant Analysis (LDA): LDA is a linear transformation technique that performs dimensionality reduction of the feature space while preserving the class discriminatory information as much as possible [10]. It assumes the distribution of each class is Gaussian and all the classes have same covariance matrix. LDA tends to find the directions also known as 'linear discriminants' that represents the axes along which classes are best separated [11]. A linear projection matrix $y = U^t x$ is computed by maximizing the ratio of between class scatter S_b and within class scatter S_w and expressed as follows: Within-class Scatter Matrix:

$$S_w = \sum_{i=1}^{c} \sum_{j=1}^{n_i} (Y_j - M_i)(Y_j - M_i)^T \tag{5}$$

Between-class scatter matrix:

$$S_b = \sum_{i=1}^{c} (M_i - M)(M_i - M)^T \tag{6}$$

A transformation space is computed by LDA that maximizes the between class scatter and minimizes the within class scatter, expressed as follows:

$$max \frac{\left|\tilde{S}_b\right|}{\left|\tilde{S}_w\right|} = max \frac{\left|U^T S_b U\right|}{\left|U^T S_w U\right|} \tag{7}$$

Weighted Linear Discriminant Analysis: In distraction identification task, there are multiple states to distinguish namely, phone and text. Phone conversation is expected to be less distracting since it only affects the drivers cognitively. On the other hand, texting has further effects on the driver due to the fact that the driver is visually distracted as well. Therefore, we hypothesize that the impact on our recorded signals would be different as well. We incorporate a weight in the calculation of within class scatter matrix for each class. The weight, w_i, represents the distance between the centroid of each class and the global mean.

$$S_w = \sum_{i=1}^{c} \frac{1}{w_i} \cdot \sum_{j=1}^{n_i} (Y_j - M_i)(Y_j - M_i)^T \tag{8}$$

3.4 *Identification Task*

We employed linear and kernel-base support vector machine with 10 fold cross validation to classify the transformed feature space. The Support Vector Machine is a state-of-the-art discriminative learning model. The SVM approach offer an effective classification strategy to separate input vector in a multiclass problem and is investigated in many different applications [12]. SVM projects input vector x into a scalar value $f(x)$ as the output score.

$$f(x) = \sum_{i=1}^{N} \alpha_i y_i K(x_i,\ x) + b \tag{9}$$

Where, the *vectors* $\{x_i | i = 1, \ldots, N\}$ are the support vectors, N is the number of support vectors, $\alpha_i > 0$ are adjustable weights, $y_i = \{-1,\ +1\}$, b is the bias term, and the function $K(x_i, x)$ is the kernel function. For the 2-class classification, the class decision is made based on the sign of $f(x)$. As it can be seen, the classifier is constructed from sums of the kernel function expressed as,

$$K(x_i, x) = \varphi(x_i)^t \varphi(x) \tag{10}$$

where φ(x) is a mapping from the input space to a possibly infinite dimensional space. Kernel SVM functions such as Radial Basis function and Polynomial function were also employed to model the non-linear feature space.

Table 2 Classification accuracies of all subject using 16 features

Subjects	Accuracy in %			
	Support vector machine			
	Linear	Poly d = 2	Poly d = 3	RBF
Subject 1	84.60	84.30	63.40	88.60
Subject 2	71.00	76.10	76.30	79.60
Subject 3	84.70	88.20	87.30	87.30
Subject 4	86.60	89.30	92.50	91.30
Subject 5	75.30	79.10	78.40	81.40
Subject 6	80.60	91.20	88.20	92.10
Subject 7	71.90	75.30	78.80	76.00
Average	79.24	83.36	80.70	85.19

4 Results

In order to characterize and identify driver distraction with short response time we first segmented the recorded GSR signal into small windows and then generated 16 different features which includes statistical, frequency and time domain features from each window. Linear and kernel-based support vector machine was employed on the data corresponding to each subject separately for training and then evaluation of the predictive model using 10 fold cross validation. Initially all the 16 features were used for predicting accuracy. Table 2 shows the classification models accuracy generated using 16 features. The scatter plot of subject 1 (fairly similar across all the subjects) generated using t-sne [13] shown in Fig. 4 depicts the classes normal (red), phone (green) and text (blue). The t-sne is a state of the art 2-D non-parametric representation of the manually generated 16-D feature space. Figure 4 shows a distinguished trend of change in the GSR-based feature space from normal driving toward phone distraction and further toward text distraction as hypothesized in Sect. 3.3: Weighted Linear Discriminant Analysis. This observation suggests that text distraction introduce higher amount of changes to the GSR-based features space, which might be due to the impact of visual distraction on top of cognitive distraction. The classification models resulted with an average accuracy of 85.19% averaged over all the subjects. A maximum accuracy of 92.5% was achieved with subject 4. The results reported in Table 2 demonstrated that GSR-based time and frequency feature extraction is useful to detect the state of distraction for each subject effectively.

Examining Fig. 4, there is still a great deal of overlap exists between the normal versus phone scenarios. Table II also reports dramatic improvement from linear SVM to the best kernel-based SVM 10-CV accuracy ($\approx$29% relative improvement) which is an evidence of high degrees of overlap in the feature space. Although

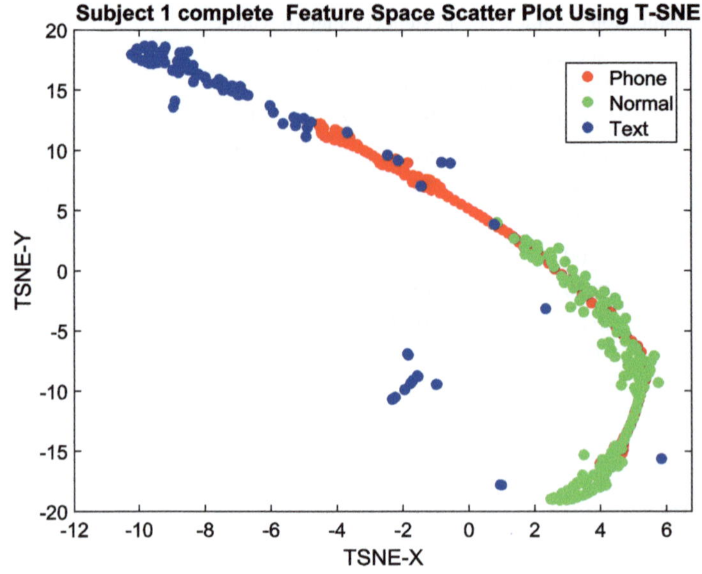

Fig. 4 Subject 1 scatter plot of complete feature space using t-sne representation

Table 3 10-CV identification accuracy (%) of subjects after applying LDA

Subjects	Accuracy in %			
	Support vector machine			
	Linear	Poly d = 2	Poly d = 3	RBF
Subject 1	86.40	89.40	88.90	87.50
Subject 2	75.50	76.30	69.30	77.40
Subject 3	85.60	89.40	89.40	89.50
Subject 4	92.30	95.30	95.50	95.00
Subject 5	77.10	80.20	80.30	78.90
Subject 6	88.00	91.20	91.20	92.30
Subject 7	75.80	78.30	74.20	81.10
Average	82.95	85.73	84.10	85.94

kernel-based SVM can alleviate the overlap in the feature space using kernel non-linear expansion. However, it can also cause a great deal of complexity in time and space of the detection procedure. LDA is an effective linear dimensionality reduction as it transforms the feature space into lower dimension by finding linear mapping that maximizes the class separability along with the direction in which most discriminative features of multiple classes are present. We conducted LDA in order to improve discrimination and reduce redundancy in the original 16-D space. Prediction accuracy after applying dimensionality reduction using LDA is shown in Table 3. LDA proved to be successful in transforming the high dimensional feature space into low dimension while preserving class discriminating features by demonstrating

Table 4 10-CV identification accuracy (%) of subjects after applying weighting

Subjects	Accuracy in %			
	Support vector machine			
	Linear	Poly d = 2	Poly d = 3	RBF
Subject 1	93.40	93.20	93.70	93.50
Subject 2	82.50	84.60	84.30	84.30
Subject 3	89.60	89.40	89.40	90.40
Subject 4	95.50	95.30	95.50	95.50
Subject 5	82.70	83.20	83.30	83.10
Subject 6	91.30	92.20	92.20	92.00
Subject 7	80.50	82.30	82.90	83.60
Average	87.93	88.60	88.75	88.91

an average accuracy of 85.94% averaged over all the subjects. A best case with maximum accuracy of 95.5% was demonstrated by Subject 4. This cost reduction in time, computational complexity and reduced feature space will able our current model to move towards real-time driver monitoring system. Table 3 also reports less improvement from linear SVM to the best kernel-based SVM 10-CV accuracy (≈17% relative improvement) which is an evidence of lower degrees of overlap in the feature space.

However, as shown in Fig. 4, phone class has more overlap with normal driving compared to texting. LDA can generate an optimistic estimate of between- and within-class scatter matrices if one or more of the classes are well-separated even if there are other highly overlapped classes. In order to take that into account, we employed our proposed weighting scheme to reduce the impact of within class scatter matrix for the text class in learning LDA transformation matrix as discussed in Sect. 3.3: Weighted Linear Discriminant Analysis. The aim of the presented weighting mechanism is to alleviate the biased estimation of the within class scatter matrices. and improve the prediction accuracy. We aim to focus on reducing the overlap between the normal vs phone classes while lowering the contribution of text class in the optimization, which might not introduce any negative impact. Table 4 also reports minimal improvement from linear SVM to the best kernel-based SVM 10-CV accuracy (≈8% relative improvement) which is an evidence of much lower degrees of overlap in the feature space. This observation is specifically significant since linear detection and identification systems are generally more preferable due to lower complexity and higher reliability in practice.

5 Conclusion

There has been much research conducted for automatic detection of driver distraction. Many previous approaches have been employed that either detect the distraction rather late to warn the drivers or while shown to be reliable indicator of distraction,

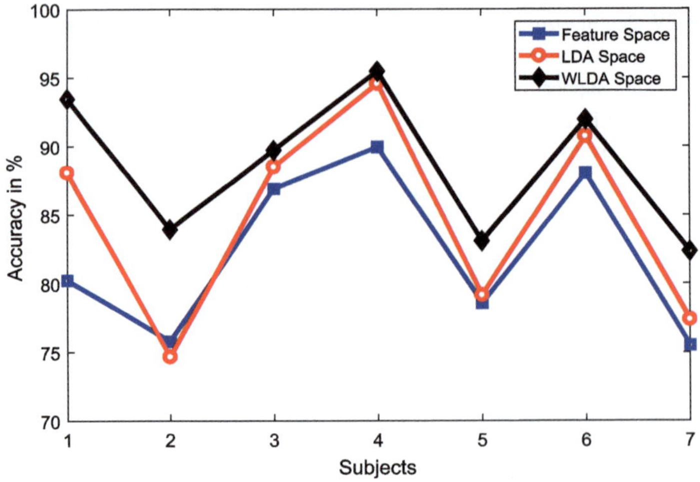

Fig. 5 Accuracy % chart of all subjects

the technology such as EEG is intrusive to the drivers. In this study, we investigated if GSR can be used for empirical characterization of driving task under naturalistic driving conditions using a wrist band wearable. Seven driver subjects participated in this study to perform three tasks i.e. driving normally, driving while having phone conversation and driving while using phone for texting. We evaluated GSR toward real-time identification of distraction using short term segmentation of 5 s windows of the GSR data with 4 s overlaps. We extracted 16 relevant features that include statistical, frequency domain and time domain features to capture changes/pattern at physiological level. As feature extraction is a manual process and also to normalize the feature space towards identification task we transform the original 16 dimensional feature space using linear discriminant dimensionality reduction to discover discriminative bases of the GSR multivariate feature space to more efficiently identify distraction. This also eliminates both the computational complexity and the redundancies in the manually generated feature space. We incorporated a class dependent weight to the calculation of scatter matrices, aiming to alleviate limitations of LDA with multiple distracting categories and to eliminate the overlap between distracted vs. non-distracted scenarios in the generalization phase. We employed linear and kernel-based Support Vector Machine (SVM) and 10 fold cross validation (10-CV) to generate identification results. The classification models of the original feature space of 16-D resulted with an average accuracy of 85.19% averaged over all the subjects. Our experimental results using LDA demonstrated slight improvement in accuracy of distraction identification (i.e. 85.94%) using only two discriminant bases. The generalization accuracy was further improved by applying our proposed weighting mechanism (i.e. 88.92%). Figure 5 shows the accuracy % comparison of the original 16-D feature space vs. reduced feature space with LDA and weighted LDA. This study provided evident results to use GSR as a reliable indicator of driver distraction.

References

1. Dawson, D., Searle, A.K., Paterson, J.L.: Look before you (s)leep: evaluating the use of fatigue detection technologies within a fatigue risk management system for the road transport industry. Sleep Med, Rev (2014)
2. Metz, B., Schmig, N., Krger, H.P.: Attention during visual secondary tasks in driving: adaptation to the demands of the driving task. Transp. Res. Part F Traffic Psychol. Behav. **14**(5), 369–380 (2011)
3. Young, K.L., Lenn, M.G., Williamson, A.R.: Sensitivity of the lane change test as a measure of in-vehicle system demand. Appl. Ergon. **42**(4), 611–618 (2011)
4. Wege, C., Will, S., Victor, T.: Eye movement and brake reactions to real world brake-capacity forward collision warnings–a naturalistic driving study. Accid. Anal. Prev. **58**, 259–270 (2013)
5. Wang, S., Zhang, Y., Wu, C., Darvas, F., Chaovalitwongse, W.A.: Online prediction of driver distraction based on brain activity patterns. IEEE Trans. Intell. Transp. Syst. **16**(1), 136–150 (2015)
6. Almahasneh, H., Chooi, W.T., Kamel, N., Malik, A.S.: Deep in thought while driving: an EEG study on drivers cognitive distraction. Transp. Res. Part F Traffic Psychol. Behav. *26*, no. PA, 218–226 (2014)
7. Nourbakhsh, N., Wang, Y., Chen, F.: GSR and blink features for cognitive load classification. In: Lecture Notes in Computer Science (including subseries Lecture Notes in Artificial Intelligence and Lecture Notes in Bioinformatics) LNCS, vol. 8117, no. PART 1, pp. 159–166 (2013)
8. Nourbakhsh, N., Wang, Y., Chen, F., Calvo, R.A.: Using galvanic skin response for cognitive load measurement in arithmetic and reading tasks. In: Proceedings of 24th Conference on Australian Computer Interaction OzCHI 12, pp. 420–423 (2012)
9. Lew, R., Dyre, B.P., Werner, S., Wotring, B.: Exploring the potential of short-time fourier transforms for analyzing skin conductance and pupillometry in real-time applications. In: Proceedings of the Human Factors and Ergonomics Society 52th Annual Meeting, vol. 52, no. 3, pp. 34–38 (2008)
10. Yan, S., Xu, D., Zhang, B., Zhang, H.J., Yang, Q., Lin, S.: Graph embedding and extensions: a general framework for dimensionality reduction. IEEE Trans. Pattern Anal. Mach. Intell. **29**(1), 40–51 (2007)
11. Farooq, M., Dehzangi, O.: High accuracy wearable SSVEP detection using feature profiling and dimensionality reduction. In: 2017 IEEE 14th International Conference on Wearable and Implantable Body Sensor Networks (BSN), pp. 161–164 (2017)
12. Vapnik, V.: Support vector machine. Mach. Learn. **20**(3), 273–297 (1995)
13. van der Maaten, L.J.P., Hinton, G.E.: Visualizing high-dimensional data using t-SNE. J. Mach. Learn. Res. **9**, 2579–2605 (2008)

References

[illegible]

A Wearable Multi-sensor IoT Network System for Environmental Monitoring

Fan Wu, Christoph Rüdiger, Jean-Michel Redouté and Mehmet Rasit Yuce

Abstract People spend more than 90% of their time indoor in Australia. Poor indoor air quality can cause severe health problems to individuals. It is necessary to develop a reliable and wearable systems for environmental monitoring. This chapter presents a low-power wearable sensor node for environmental Internet of Things (IoT) applications, forming wireless sensor network (WSN) based on XBee. Environmental data are monitored by the wearable sensor node and then transmitted to a remote cloud server via WSN. The data are displayed to authorized users through a web-based application located in cloud server. The experimental results indicate that the presented wearable sensor network system is able to monitor environmental conditions reliably.

1 Introduction

There are a number of air pollutants in both indoor and outdoor environments, such as carbon dioxide (CO_2) and particulate matter. CO_2 is a common indicator used for ventilation systems. High CO_2 levels can cause headaches, dizziness and a range of detrimental symptoms. Particulate matter (PM) is a mixture of small particles

F. Wu · J.-M. Redouté · M. Rasit Yuce (✉)
Department of Electrical and Computer Systems Engineering, Monash University, Melbourne, Australia
e-mail: mehmet.yuce@monash.edu

F. Wu
e-mail: fan.wu@monash.edu

J.-M. Redouté
e-mail: jean-michel.redoute@monash.edu

C. Rüdiger
Department of Civil Engineering, Monash University, Melbourne, VIC, Australia
e-mail: chris.rudiger@monash.edu

G. Fortino and Z. Wang (eds.), *Advances in Body Area Networks I*,
Internet of Things, https://doi.org/10.1007/978-3-030-02819-0_3

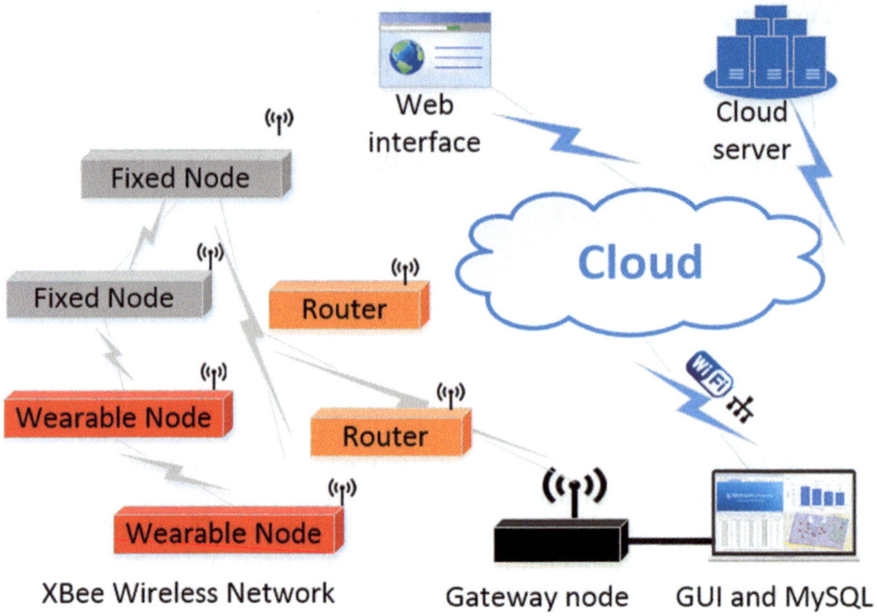

Fig. 1 Network diagram of the proposed wireless sensor network

found in the air [1]. Small particles with a diameter of 10 μA or less can be inhaled that causing serious health issues to lungs. Particles with diameter less than 2.5 μm ($PM_{2.5}$) can go deeper into lungs and can be more dangerous than bigger particles such as those between 2.5 and 10 μm (PM_{10}). The particle sources are ranging from forest fires, power plants, automobiles, organic compounds and also metals.

WSN is a promising solution for a number of monitoring applications including building structural health monitoring [2], urban environmental monitoring [3], indoor air quality monitoring [4], and fire monitoring [5]. Using WSN in indoor air quality (IAQ) monitoring is flexible in deployment and can cut costs in power wiring and related infrastructures [6]. WSNs usually have stringent power requirements to maintain continuous monitoring. Therefore, an efficient power management and energy source should be selected. XBee is a wireless technology widely used to create low-rate wireless personal area networks (LR-WPANs) for Internet of Things (IoT) applications.

Wearable sensor nodes are essential elements in wireless body area network in monitoring the human body [7]. In additional to medical signals, they can be used to collect environmental conditions around the human body as well, such as in safety applications [8]. Figure 1 shows the proposed WSN with wearable sensor nodes.

Different environmental parameters are measured in each system. For instance, [6] presents a WSN system that monitors volatile organic compounds (VOCs), and carbon monoxide (CO) in building. Work [9] measures methane and temperature in working environment. Reference [10] measures VOC, humidity, temperature and wind in industrial site. Finally, [8] measures CO_2 concentrations.

In this chapter, we developed a low-power WSN towards an IoT platform for environmental monitoring. The system is successfully tested at Monash University Clayton campus, Australia. Each sensor node includes multiple sensors, including CO_2, light, temperature, humidity and pressure sensor, one microcontroller unit (MCU), and one XBee module. The system has been validated in both field and laboratory environmental conditions. The wearable nodes are tested using an existing WSN [3] based on XBee. The complete network structure is shown in Fig. 1. The proposed network provides a reliable network solution for real-time environmental monitoring.

The remainder of this chapter is organized as follows: Sect. 2 outlines the system implementation; Sect. 3 discusses the experimental results, and a brief conclusion is given in Sect. 4.

2 System Architecture and Implementation

It is a challenging task to design and implement an environmental monitoring system. For example, power management, sensors' selection and networking type need to be considered. In this work, the wearable sensor nodes are integrated with an existing wireless sensor network infrastructure. Existing WSN is used to route the data from wearable nodes to the cloud server. Figure 2 shows the wearable node picture with components defined. Figure 3 provides the schematic diagram of the wearable node. Table 1 provides a summary of the components used in the wearable sensor node.

Fig. 2 Wearable node picture

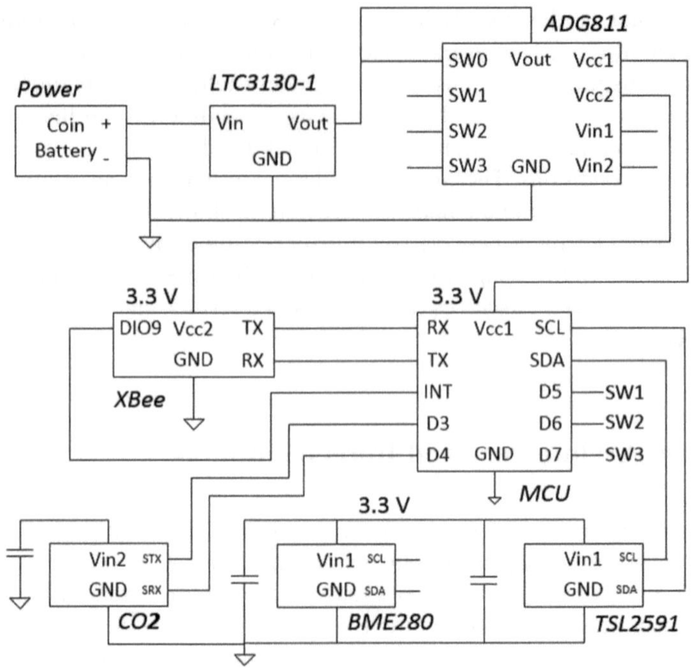

Fig. 3 Architecture of the wearable node

Table 1 Main components used in wearable nodes

Components	Model	Current consumption
Boost converter	LTC3130-1	1.6 μA
Switch	ADG811	0.003 μA
RF module	XBee-Pro 900HP	Sleep current: 2.5 μA
MCU	ATmega328p	Sleep current: 27.6 μA
CO_2	COZIR-GC0012	Peak current: 33 mA
Temperature	BME280	3.6 μA
Humidity	BME280	–
Pressure	BME280	–
Light	TSL2591	275 μA

2.1 *Hardware Implementation*

Instead of simply listing headings of different levels we recommend to let every heading be followed by at least a short passage of text. Further on please use the LaTeX automatism for all your cross-references and citations as has already been described in Sect. 2.

2.1.1 Power Management

The power management system includes a coin battery, a buck-boost converter (LTC3130-1), and a quad SPST switches (ADG811).

LTC3130-1 is a high efficiency (up to 95%) buck-boost converter which regulates the input voltage (1–25 V) to a constant 3.3 V for the entire circuit. ADG811 is a low voltage CMOS device containing four single-pole, single-throw (SPST) switches. The on resistance is typically 0.5 Ω. Switches are used to control sensors on or off according to different power consumptions and applications. In this chapter, only two switches are used to control the sensors on and off. XBee and MCU are configured to be on and off simultaneously.

2.1.2 Sensor Node

ATmega328p is used as MCU in this work. It collects the data from each sensors and then send the data to base station through RF link. It also controls the power consumption. It has low power consumption, low cost, and high performance features. It also has six 10-bit ADC and 14 digital pins which can support the various functions required by the sensor board. The nominal voltage for the MCU is 3.3 V when the clock speed is 8 MHz.

XBee-Pro 900HP is used as the RF module in this work. The module utilizes DigiMesh Networking Protocol. DigiMesh network enable every node to sleep or wake up at a same time. Every node within the network is peer-to-peer relation and does not require additional routers to be in the network that cannot be in sleep mode. It operates within 902–928 MHz ISM band which complies with the international standards. The XBee can transfer RF data at up to 200 kbps at a range of 305 m indoor and 6.5 km outdoor with line of sight (LOS). The XBee device is a low-power device. During sleep time, it only consumes 2.5 μA current.

Temperature, humidity and pressure data are achieved by using BME280 sensor. BME280 is a low-power sensor consuming only 3.6 μA when operating and 0.1 μA in sleep mode. It has an accuracy of ±0.5 °C for temperature, ±3% for relative humidity and ±1 hPa for barametric pressure.

The CO_2 sensor chosen is the COZIR CO_2 sensor. It is an ultra-low power (3.5 mW) and high performance (±50 ppm) sensor. The warm-up time for the sensor is less than 10 s, which is ideal for a remote sensor node. The measurement range is from 0 to 10,000 ppm, which can be used in both indoor and outdoor measurements.

Light Sensor chosen is TSL2591. It can measure LUX level at low power consumption (275 μA at 3.3 V supply voltage).

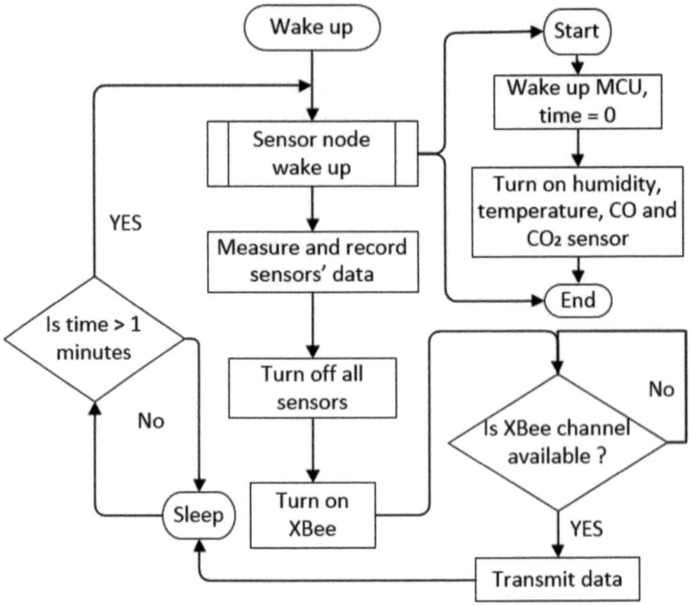

Fig. 4 Wearable node's software algorithm

2.2 *Software Implementation*

For fixed location sensor node, every node sleeps for 4 min and 50 s and then wakes up for 10 s for measuring. During wake up time, the MCU measures temperature, humidity, pressure, light and CO_2 value and then packetize the data. XBee will check if the RF channel is available for transmission. If yes, the data will be sent to the destination address. After successfully transmit the data, the entire sensor node will go back to sleep mode for another 4 min and 50 s.

Wearable sensor nodes are similar to fixed nodes in terms of functionality, except that they need to wake up more frequently. Wearable nodes are programmed to wake up a minute to keep track of the changing environmental data. It can also be programmed to run in continuous monitoring mode which will update the environmental data according to users' requirements. The detail software algorithm for wearable sensor node is presented in Fig. 4.

The base station receives the data from both fixed sensor nodes and wearable sensor nodes. It will display the data in a local Graphical User Interface (GUI) and store the data into a local MySQL database. The data will ultimately be transmitted to a cloud server via Ethernet.

The networking topology is a mesh + cluster type network. Fixed location sensor nodes are mesh type connections. They will wake up periodically to sense and send the data. There are also some fixed router nodes for supporting wearable sensor nodes' communication as shown in Fig. 1.

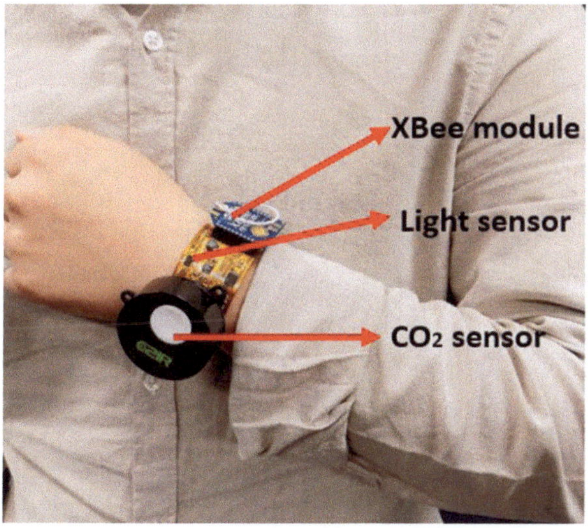

Fig. 5 Wearable sensor node attached to wrist

Table 2 Power comsumptions in different stages

Stage	Comments	Current (mA)	Power (mW)
1	MCU + Sensors	4.2	13.86
2	MCU + RF	65	214.5
3	MCU	4	13.2
4	Sleep	0.0276	0.0918

3 Experimental Results

3.1 Experimental Setup

The goal of this work is to implement a low power and reliable indoor air quality monitoring WSN. To test the reliability of the system, the WSN is deployed in Engineering Building, Monash University Clayton campus. The power consumptions of each node and sensors' performance are measured. The wearable sensor node is shown in Fig. 5, which is attached to wrist of the user.

3.2 Power Consumption

By using the low power components, low power management techniques, and proper software algorithm, the maximum current consumption while awake is measured at 64 mA, and the minimum current consumption while in sleep is measured at 27.6 μA. The power consumption of the wearable node is tabulated in Table 2. The lowest current consumption is considerably lower compared to our previous work in [3], which was 204.1 μA in sleep mode.

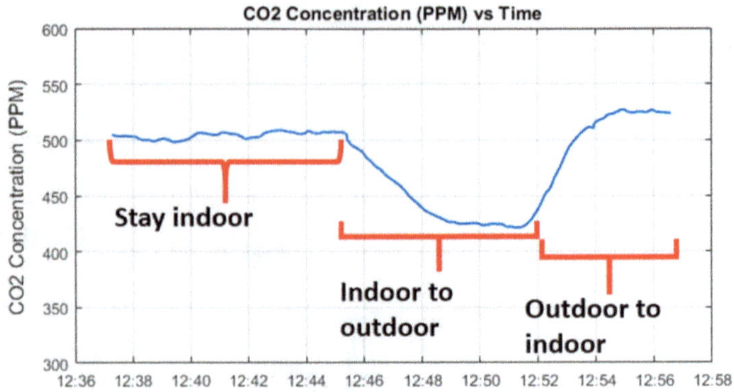

Fig. 6 Fluctuation of CO_2 concentration

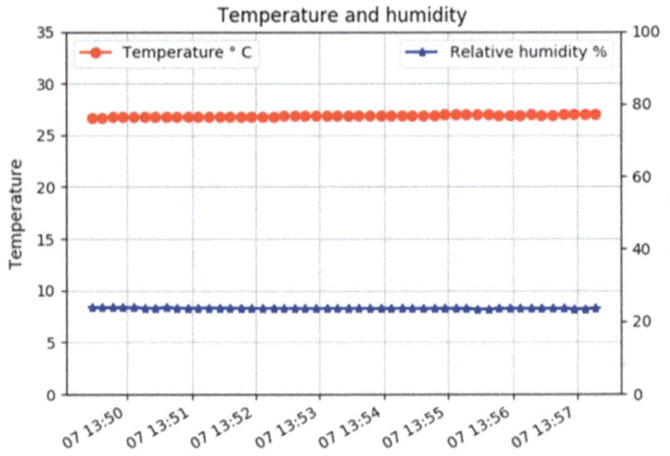

Fig. 7 Temperature and humidity (fixed sensor node)

3.3 Sensors' Preliminary Results

A selection of sensors' graph shown in the graphical user interface (GUI) in 10 s continuous monitoring mode is shown in Figs. 6, 7 and 8. The GUI is developed using Python to show the real-time graph for incoming data.

Figure 6 shows the change of CO_2 concentration as the user is moving from indoors to outdoors, and then entering indoors again. The CO_2 concentration remains constant at approximately 510 ppm in the indoor environment. As the user is moving outdoors, the CO_2 concentration decreases from 510 to around 420 ppm. After the user enters another building, the CO_2 concentration can be clearly seen to increase to around 525 ppm.

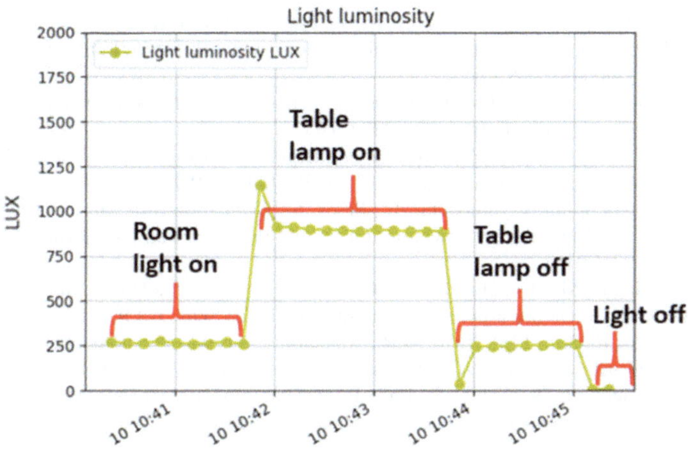

Fig. 8 Light luminosity (wearable sensor node)

Figure 7 shows the temperature (°C) and relative humidity (%) data. From Fig. 7, it can be seen that the data is quite consistent because the sensor node is fixed indoor during the time. This is because the air conditioning and ventilation systems are constantly on during the time of data acquisition.

Figure 8 shows the light luminosity (LUX) changing while the user's surrounding light environment changes. The figure shows that our wearable sensor node is able to capture the environmental conditions' change.

4 Conclusion

This chapter presents a wearable IoT wireless sensor network for environment monitoring. Using wireless sensor network for wearable environmental monitoring is discussed. Low power techniques, software, and sensing environmental parameters are described. The proposed work provides an effective and reliable solution for long term monitoring that presents many opportunities in safety related monitoring applications.

Acknowledgements M. R. Yuce's work is supported by *Australian Research Council Future Fellowships Grant FT130100430.*

References

1. Texas Instruments: PM 2.5 /PM 10 particle sensor analog front-end for air quality monitoring design. Technical Report, May (2016)

2. Torfs, T., Sterken, T., Brebels, S., Santana, J., van den Hoven, R., Spiering, V., Bertsch, N., Trapani, D., Zonta, D.: Low power wireless sensor network for building monitoring. IEEE Sens. J. **13**(3), 909–915 (2012)
3. Wu, F., Rüdiger, C., Yuce, M.R.: Real-time performance of a self-powered environmental IoT sensor network system. Sensors **17**(2), 282 (2017)
4. Kim, J.Y., Chu, C.H., Shin, S.M.: ISSAQ: an integrated sensing systems for real-time indoor air quality monitoring. IEEE Sens. J. **14**(12), 4230–4244 (2014)
5. Silvani, X., Morandini, F., Innocenti, E., Peres, S.: Evaluation of a wireless sensor network with low cost and low energy consumption for fire detection and monitoring. Fire Technol. **51**(4), 971–993 (2015)
6. Jelicic, V., Magno, M., Brunelli, D., Paci, G., Benini, L.: Context-adaptive multimodal wireless sensor network for energy-efficient gas monitoring. IEEE Sens. J. **13**(1), 328–338 (2013)
7. Wu, T., Wu, F., Redout, J.M., Yuce, M.R.: An autonomous wireless body area network implementation towards iot connected healthcare applications. IEEE Access **5**, 11413–11422 (2017)
8. Antoln, D., Medrano, N., Calvo, B., Prez, F.: A wearable wireless sensor network for indoor smart environment monitoring in safety applications. Sensors **17**(2) (2017)
9. Spiryakin, D., Baranov, A.: Uniform inbuilt wireless sensor node for working conditions monitoring. In: Proceedings of the Federated Conference on Computer Science and Information Systems, vol. 8, no. 1, pp. 1097–1102 (2016)
10. Manes, G., Collodi, G., Fusco, R., Gelpi, L., Manes, A.: A wireless sensor network for precise volatile organic compound monitoring. Int. J. Distrib. Sens. Netw. (2012)

Fabric Sensor Array Monitoring Pressure Distribution

Jiyong Hu, Hele Zhang, Yuanyuan Gu, Yinda Zhu, Xuyuan Guo and Xudong Yang

Abstract To improve the sensing performance of fabric sensor array monitoring pressure distribution, the structure of fabric sensor array was designed and optimized. The fabric sensor array was fabricated by seamlessly laminating multi-layers clothing fabric, and the optimized sensor array can sense both the size and position of the distributed dynamic forces. The performance of the designed fabric sensor array was evaluated. When the force is 5–25 N, the fabric sensor array monitoring pressure distribution has highly sensitive to the applied force. In terms of the material and construction of fabric sensor array, the mesh size of isolating layer determines the sensitivity and response range.

Keywords Sensor · Fabric · Array · Pressure · Distribution · Resistance

1 Introducation

Human body has curved and deformable body, and the traditional rigid sensors is limited to conform human body. In recent years, due to the flexibility and comfort, fabric pressure sensor is gotten more attention. The fabric-based pressure sensor can be applied in more and more fields, such as medical health, home life and sports.

Previous studies have developed some flexible sensors monitoring pressure distribution in wearable filed. However, most of these sensors are made of laminated films and they are impermeable [1]. To improve the comfort and air permeability, several fabric-based sensors have been designed.

Commonly, the fabric-based pressure sensor array is resistive and capacitive [2]. They have multi-layers structure or single-layer structure [3]. In the fabric-based resistive pressure sensor, the multi-layers structure is mostly three-layer array structure [4–8]. Philip et al. [8] proposed a typical technique fabricating fabric-based pressure sensor array. The commercial conductive fabric is used to create an array

J. Hu (✉) · H. Zhang · Y. Gu · Y. Zhu · X. Guo · X. Yang
Donghua University, Shanghai, China
e-mail: hujy@dhu.edu.cn

G. Fortino and Z. Wang (eds.), *Advances in Body Area Networks I*,
Internet of Things, https://doi.org/10.1007/978-3-030-02819-0_4

with 3-layer variable resistor that is approximately 10 cm in both length and width. The top/bottom layer consists of alternating thick conductive and thin nonconductive columns/rows of cloth fabric; the middle layer is a single sheet of piezoresistive fabric as the sensitive material. Each sensing element in the resistive array is defined by the intersection of a column with a row of conductive fabric. Büscher et al. [9] designed a smart tactile sensor with four-layer structure, which adds a layer of insulating cloth on the top of the piezoresistive layer. The sensor is in an open circuit when it is under zero load, which facilitates to save energy and to prolong the life-time of the sensor. These techniques fabricating pressure sensor pads and sheets typically cost less. However, it is lack of knowledge about the dependence of their performance on the material in each layer.

The other typical fabric sensor array is on the basis of capacitive principle. The capacitive pressure sensor array is generally three-layer structure which is a fabric-laminate structure or covering yarns [10–12]. Both upper and lower layers are the plate, the middle layer is the dielectric layer. The fabric- or yarn-based capacitive pressure sensor array commonly has large hysteresis. And also the moisture regain of textile materials greatly affects the sensing performance of capacitive sensors.

By comparing the structure and performance of fabric sensors in the existing literature, the existing fabric sensors have poor flexibility and comfort. Or some of them are not economic in power consumption, etc. In this paper, a kind of fabric sensor array monitoring pressure distribution is proposed to improve the sensitivity and decrease the power consumption. It also fully retains the comfort characteristics of fabric.

In the next section, the design and principle of the fabric sensor array monitoring pressure distribution would be introduced, and the measurement methods are showed. Section 3 gives the experimental results. Section 4 summarizes the conclusions.

2 Expermental Section

2.1 Structure Design

In order to monitor the pressure distribution, the fabric sensor is designed as array. It is a multi-layer fabric structure composed of upper electrode layer, conductive layer, isolation layer and lower electrode layer. Besides, the conductive yarn is used as electrode lead (Fig. 1).

The upper layer and the lower layer are the electrode layers, which are woven with conductive yarns and non-conductive yarns. Conductive yarns form the striped electrode and non-conductive yarns form a striped electrode separation. Two kinds of stripe could be alternatively arranged. This alternative structure can be flexibly designed with different density so that different sensitivity and resolution of the fabric sensors monitoring pressure distribution can be achieved. The second layer is conductive layer, which is conductive fabric coated with intrinsic conductive polymer.

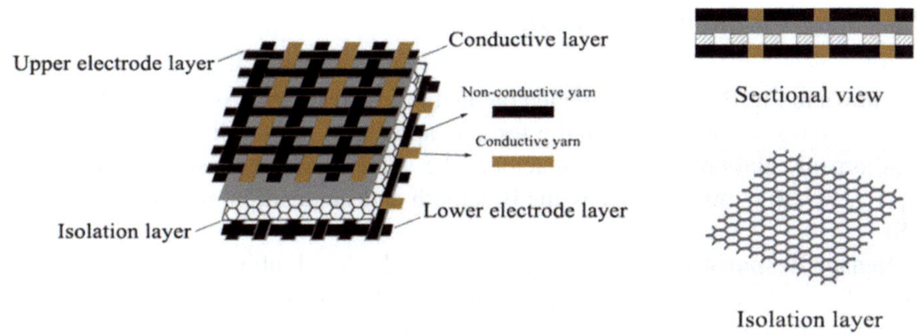

Fig. 1 Structure of fabric sensor array monitoring pressure distribution

The third layer is the isolation layer, and the nonconductive mesh fabric is selected. Due to the additive effect of the conductive layer and the isolation layer, the sensor array can perceive the size and position of the pressure. Finally, the electrode lead is a part of the conductive stripe, which simplifies the production of sensor and increases the sensor's flexibility.

2.2 Sensing Principle

The fabric sensor array monitoring pressure distribution is the piezoresistive sensor, and its equivalent circuit in principle is illustrated in Fig. 2. In Fig. 2, R'_1, R_1, R'_2 and R_2 is the resistor with constant resistance. They are used to indicate the voltage changes in the circuit. The variable resistors, R_a, R_b, R_c, R_d are the equivalent resistance of each of sensitive elements. When the external force is applied, the voltage of the constant resistor can be recorded. And then, an algorithm is applied to obtain the voltage changes because of the increment/decrement of variable resistors. So that the magnitude and position of pressure can be acquired. For example, in Fig. 2, in terms of the variable resistor, R_d, the formula for its voltage is followed as:

$$V(R_d) = U - V(R_2) - V(R'_2) \tag{1}$$

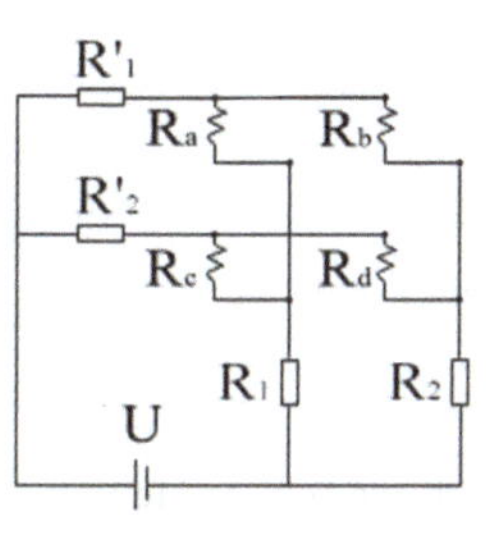

Fig. 2 The equivalent circuit diagram of fabric sensor array monitoring pressure distribution with 2 * 2 array

2.3 *Performance Evaluations*

As shown in Fig. 3, the force applied on the fabric sensor array is measured with the customized platform. The periodic force is applied to the fabric sensor array by a rigid organic glass plate with contactors. The voltages at the constant resistors (i.e. R_1, R_2, R'_1, R'_2) are recorded by the Bluetooth data acquisition module (BTH-1208 LS).

Sensor Responses. As shown in Fig. 2, the cylindrical rubber block as contactors in single-point and multi-point testing is fixed on the organic glass plate respectively.

In the preliminary experiment, the mesh fabric with a pore size of 2 mm is used as the isolation layer. The width of strip electrode is 6 mm. The inter-strips is 6 mm, too. The peak values of applied force change from 5 to 40 N (with an interval of 5 N). The frequency of pressure is 1 Hz. The circuit voltage is 5 V, and the resistances of constant resistors are $R_1 = R_2 = 200\ \Omega$, $R'_1 = R'_2 = 100\ \Omega$, respectively.

In order to explore the relationship between force and voltage change of fabric sensor array, single-point test and multi-points test are carried out. At the same time,

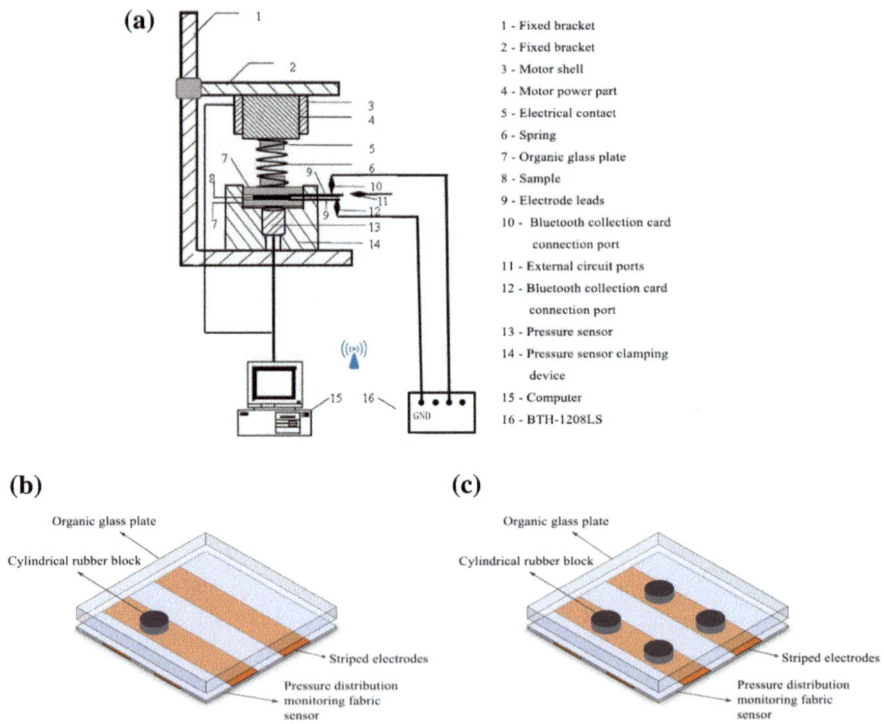

Fig. 3 Test schematic: **a** diagram of the customized testing platform, **b** single-point test, **c** multi-point test

in order to explore the repeatability of fabric sensors, single-point test is tested every two days. Three repeated experiments are performed in total.

Size Effect of Mesh Fabric. To explore influence of the aperture size of mesh fabric (as shown in Fig. 1) on the sensing responses of fabric sensor, the aperture size of 1 mm, 2 mm and 4 mm mesh fabrics, respectively, are used as the isolation layer. The single-point test is applied for it.

Frequency Effect of Activation Force. To explore the sensing performance under the same pressure at different frequencies, this study chooses 1 mm aperture size of mesh fabric as isolation layer. Single-point test at each of three activation frequency of 1, 2 and 5 Hz are performed.

3 Results and Discussion

3.1 Response of Sensing Element

Figure 4 shows the pressure and voltage variation at R_d by the single-point test. When the pressure is less than 25 N, the voltage changes rapidly. Once the force is more than 25 N, the voltage changes slowly. It also compares the difference of sensor responses by three repeated experiments, and no significant differences are observed.

Figure 5 shows the voltage variation of the fabric sensor by a periodic force of 1 Hz and 5 N. In Fig. 5a, the voltage between the resistance R_1 and R'_1 is always zero, and the voltage of the resistance R_2 is twice as high as the resistance R'_2. When the sensor element at the point d is pressed, Fig. 5b shows great repeatability. The voltage variation of every 10 s during the whole test process is 2.863 ± 0.084 V.

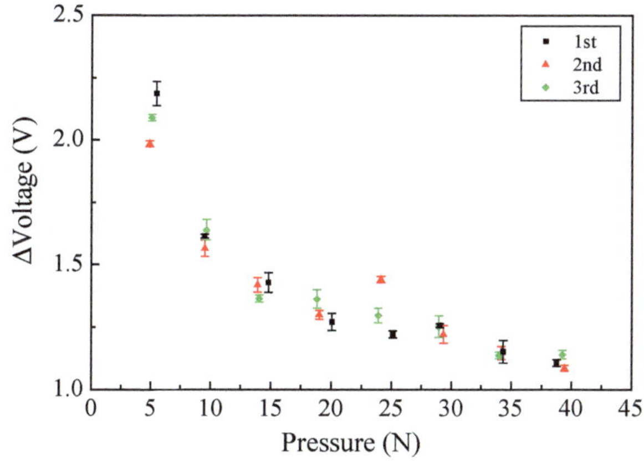

Fig. 4 Pressure and voltage variation of single-point test

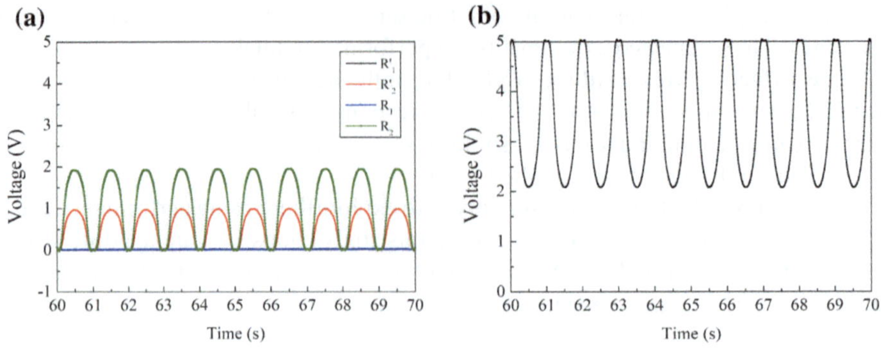

Fig. 5 Voltage change by a periodic force of 1 Hz and 5 N: **a** for each of resistors, **b** at location of d point

3.2 Response of Sensing Array

Figure 6 shows that the responses for each of four variable resistors is nearly same. When the pressure range increases from 5 to 25 N, they have fast and sensitive responses to the applied periodic forces. When the pressure was over 25 N, the responses become slowly. At the same time, as the pressure increases, the crosstalk effect between four sensitive elements increases gradually, and the voltage variation of each point varies. The reasons for this change may be the placement of cylindrical rubber blocks at the location of a, b, c and d. In Fig. 6, when the pressure is zero, the isolation layer has a great isolating effect to decrease the power consumption.

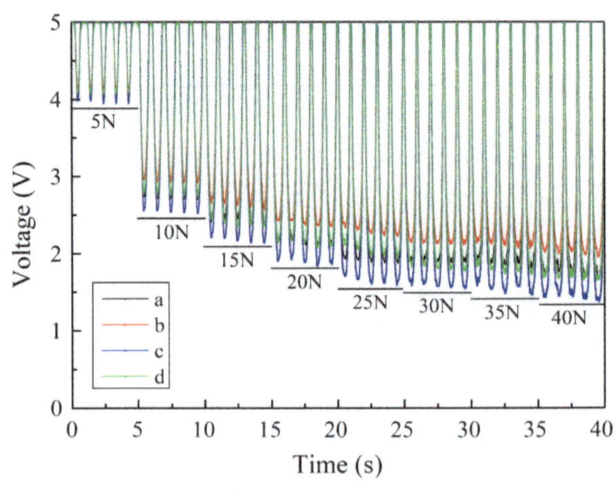

Fig. 6 Pressure and voltage variation diagram of multi-point test

Table 1 Mesh fabrics

	Fabric 1	Fabric 2	Fabric 3
Mesh size (mm)	1	2	4
Average porosity (%)	53.86 ± 0.036	58.59 ± 0.024	62.23 ± 0.022

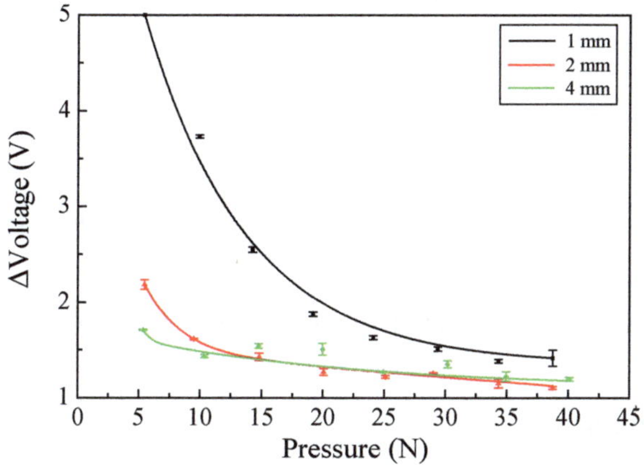

Fig. 7 The pressure and voltage variations of the mesh fabric test points with different aperture sizes at 1 Hz frequency

3.3 Influence of Mesh Size

Table 1 and Fig. 7 shows that voltage change of 1 mm mesh fabric is more than that of 2 mm and 4 mm mesh fabrics. Thus, the sensing sensitivity of sensitive element by 1 mm mesh fabric as isolating layer is better.

3.4 Dynamic Response to Periodic Force

As shown in Fig. 8, the dynamic responses of the fabric sensor monitoring pressure distribution change partially by the activation frequency, although there is no significant change. This change maybe from the mechanical properties of material made of the fabric-based sensors, or the assembled structure. The potential reasons will be investigated in the future.

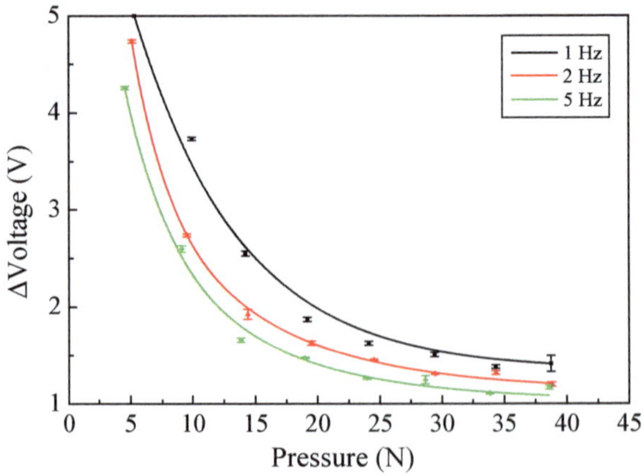

Fig. 8 Responses of sensors with 1 mm grid under different activation frequencies

4 Conclusions

The fabric sensor array monitoring pressure distribution is made of clothing fabric, and the optimized laminating structure improves the sensing performance. By a series of experiments, the fabric sensor array is demonstrated that it is not only sensitive to pressure size, frequency and location, but also can decrease energy consumption. The preliminary analysis shows that the effective response range is about 5–25 N. It is more suitable for 1 mm mesh fabric as the insulating layer. In the future, the effect of mechanical properties of the isolating and the electrode layer will be further evaluated on the response sensitivity and the hysteresis.

Refercences

1. Grau, A.M., et al.: Mechanical force redistribution: enabling seamless, large-format, high-accuracy surface interaction. In: Proceedings of the SIGCHI Conference on Human Factors in Computing Systems, pp. 4137–4146. ACM, Toronto, Ontario, Canada (2014)
2. Zhou, B., et al.: Textile Pressure Force Mapping. Springer International Publishing (2017)
3. Li, J.F., et al.: Novel highly sensitive and wearable pressure sensors from conductive three-dimensional fabric structures. Smart Mater. Struct. **24**(12), 125022 (2015)
4. Behroo, S.M., et al.: Fabric-based pressure sensor arrays and methods for data analysis. Patent, USA (2013)
5. Xu, W., et al.: eCushion: a textile pressure sensor array design and calibration for sitting posture analysis. IEEE Sens. J. **13**(10), 3926–3934 (2013)
6. Parzer, P., et al.: FlexTiles: a flexible, stretchable, formable, pressure-sensitive, tactile input sensor. In: Proceedings of the 2016 CHI Conference Extended Abstracts on Human Factors in Computing Systems, pp. 3754–3757. ACM, Santa Clara, California, USA (2016)
7. Danilovic, A.: SmartCast—novel textile sensors for embedded pressure sensing of orthopedic casts. University of California (2013)

8. Philip, C., et al.: Fabric-based pressure sensor array for decubitus ulcer monitoring. In: Annual International Conference of the IEEE Engineering in Medicine and Biology Society, pp. 6506–6509. IEEE, Osaka, Japan (2013)
9. Büscher, G.H., et al.: Flexible and stretchable fabric-based tactile sensor. Robot. Autonom. Syst. **63**, 244–252 (2015)
10. Saenz-Cogollo, J., et al.: Pressure mapping mat for tele-home care applications. Sensors **16**(3). https://doi.org/10.3390/s16030365 (2016)
11. Takamatsu, S., et al.: Meter-scale large-area capacitive pressure sensors with fabric with stripe electrodes of conductive polymer-coated fibers. Microsyst. Technol. **22**(3), 451–457 (2016)
12. Samad, Y.A., et al.: From sewing thread to sensor: Nylon (R) fiber strain and pressure sensors. Sens. Actuators B-Chem. **240**, 1083–1090 (2017)

Automatic EEG Blink Detection Using Dynamic Time Warping Score Clustering

Omid Dehzangi, Alexander Melville and Mojtaba Taherisadr

Abstract The electroencephalograph (EEG) is a powerful tool, involving multiple electrodes placed on the scalp, with the intention of measuring brain activity through the scalp. One significant application for EEG is to analyze the mental state of a subject. One of the challenges involved in using the EEG for identifying mental state in practical settings is ocular artifacts e.g. eye blinking. Eye blinks cause high amplitude noise in electroencephalograms (EEGs), the noise from these blinks can cause interference in several very important frequency bands and confuse predictive modeling e.g. introduce false positives. Prior works have employed independent component analysis (ICA) to decompose the noisy EEG signals into constituting sources and identify the eye blink sources. However, ICA requires off-line signal processing and is not suitable for online applications. More recently, time domain autoregressive features were used to model eye blink related segments in the recorded EEG data. While the autoregressive method showed high identification accuracy in isolated short trials, the goal of this work is to create a more advanced system capable of identifying and filtering blink noise in continuous trials during long and complex tasks. The proposed method detailed in this paper conducts automatic detection of eye blink noise using dynamic time warping (DTW) score clustering during wearable EEG-based cognitive workload assessment tests. The proposed eye blink detection system only uses EEG data for training and identification and does not require electrooculogram (EOG) data, which is particularly important for wearable systems. Our experimental results demonstrated the effectiveness of the proposed blink detection methodology by achieving 96.42% average accuracy of blink detection in the recorded EEG dynamics during a continuous workload assessment task.

O. Dehzangi
Rockefeller Neuroscience Institute, West Virginia University,
Morgantown, WV 26505, USA
e-mail: omid.dehzangi@hsc.wvu.edu
URL: http://www.wssplab.net

A. Melville · M. Taherisadr (✉)
University of Michigan, Dearborn, MI 48128, USA
e-mail: mojtabat@umich.edu

G. Fortino and Z. Wang (eds.), *Advances in Body Area Networks I*,
Internet of Things, https://doi.org/10.1007/978-3-030-02819-0_5

Keywords EEG-based monitoring · Ocular artifacts · Dynamic time warping Clustering · Eye blink detection

1 Introduction

Human Computer Interaction (HCI) in collaborative environments is becoming a crucial component in cyber-physical systems for decision making and operation. Human operator mental state e.g. workload level is found to be closely related to overall performance, attentiveness, and situational awareness e.g. events induced by automation failures. Therefore, there has been much recent interest to estimate human state using EEG dynamics [1–4]. One of the challenges involved in using the EEGs for identifying mental state in practical continuous settings is ocular artifacts e.g. eye blinking [5]. Eye blinks cause high amplitude noise in electroencephalograms (EEGs), the noise from these blinks can cause interference in multiple task related frequency bands and confuse predictive modeling of changes in the EEGs in those frequency bands. Trials that consider the subject's workload change over time involve sensitive calculations done continuously over long-term recordings, and blinking resides in frequency bands that are known to be useful in determining the subject's current workload. A blink in one of those bands could give a false positive result for workload, or it could confuse the predictive algorithm during training and degrade its generalization ability. Therefore, eye blink detection in continuous mental state assessment tasks during strenuous trials is necessary.

The Electroencephalogram (EEG) is a recording of the electrical potentials recorded from multiple electrodes placed on the scalp. While the skull is a very good insulator, some brain activity makes it through the bones and up to the scalp to be measured [6]. One of the challenges of using an EEG to analyze the mental state of a subject is that blinking and eye movement cause high amplitude noise in the frontal EEG channels [7]. This noise occurs in similar frequency bands to the task-related neurological data. Therefore, it is not possible to use a band pass filter to remove it. More sophisticated filters are typically implemented in order to filter out eye blinks without severely affecting useful parts of the EEG signal. There has been studies to detect isolated blinking vs. not blinking segmented data, in which subjects are following certain instructions with time marks [8]. However, our aim is to achieve blink detection during a continuous and complex task of workload assessment in a collaborative human-machine environment.

EOG regression is a common and lightweight solution for removing ocular noise as described in [9] and [10], this method has been used as a basis for comparison against newer methods [11]. While electrooculogram (EOG) regression is simple to implement and quick to process, this method has been known to remove neural data as well as ocular noise because some neural data propagates forward into the EOG channels. In addition to this, regressive methods have limitations on non-stationary signals such as the data collected during EEG trials.

Independent component analysis (ICA) has been shown in many papers to be a competent method for decomposing independent components from EEG signals. Authors in [12] demonstrated that it was possible to estimate EOG recordings using ICA and EEG data. In the early work in ICA and EEG research [13], the authors showed examples of independent components separating out various neural and noise activity. In much more recent work in [14], the authors proposed a method that employed both ICA and autoregressive methods for ocular noise removal, in order to build a model for noise correction that did not remove as much neural activity as ICA by itself. Authors in [15] designed a system for noise removal using ICA, including features such as kurtosis, modified multiscale sample entropy, and wavelet analysis. In another interesting study [16], an ICA method was combined with statistics and feature extraction for IC characterization. A window of independent component data was assessed based on its wavelet and kurtosis features, as well as its correlation to two chosen reference signals, the amplitude of the EOG, and mutual information between an independent component and the reference signals. Similar investigation via characterization of ICs was conducted in [17]. Despite the fact that ICA-based blink detection methods achieve high accuracies, the ICA algorithm involves computationally expensive and hard to predict processing times that would be difficult to bring into real time applications. Moreover, in order to have robust estimates of the independent noise sources, ICA requires higher number of EEG channels and longer recordings as input observations, which make it suitable for off-line applications. Therefore, a high accuracy blink detection system without the need for ICA would be far more useful in real time systems during continuous long-term trials.

In this study, we aim to design a real-time eye blink detection during a wearable EEG-based workload assessment experiment involving a low number of EEG channels (i.e. 7 channel and no EOG recording) via the proposed methodology of segmentation, scoring, and clustering. We propose a feasible system for annotation and segmentation of EEG data and then detection of blink activities in segments of recorded EEG data. The proposed learning system for blink detection presented in this paper has three major components including automatic eye blink data scoring and annotation, multi-score space clustering, and discriminative Eye blink Detection.

We first annotate the EEG recordings (blink or non-blink) by running dynamic time warping (DTW) on the EEG data from different subjects and trials. In order to make the detection model as accurate as possible, the model must foresee variation of eye blink patterns existing in the input data. Therefore in order to score segments of EEG data, we apply DTW scoring with multiple categories of eye blink templates. A theoretical ideal model for blink detection would capture all variants of an individual's eye blink that are needed to detect blinks with no error. However, the variants of the eye blinks that would create this ideal model are not known, and therefore it is necessary to extract and test different categories of templates from the input EEG data, so that the DTW score space contains as relevant of information as possible to improve the generalization ability of the predictor. On the other hand, in order to avoid the complexity of multi-dimensional score space, we conduct K-means clustering of DTW multi-score space into two clusters of blink and non-blinks. The final learned clusters can provide important clues regarding what DTW score combinations have more discriminative information to distinguish eye blink patterns.

2 Experiment and Data Acquisition

2.1 Experimental Setup

In the trials that make up the data for this study, the subject was instructed to track either one or two objects on a screen. After each experiment, the subject was asked to give their opinion on how strenuous the task was. These trials varied in length between 4 and 6 min. Since the trial details are associated with a workload assessment task, we skip those details and focus on eye blink detection during those continuous and unsupervised trials.

2.2 Data Acquisition

We recorded from 5 subjects and 4 trials of EEG recording per subject. The raw EEG data collected in this study is recorded during the trial. The EEG data was collected at 500 Hz on a *BioRadio* wet EEG (*Great Lakes Neuro Technologies*, *Valley View*, *OH*). The workload assessment task data collection consisted of seven EEG channels. Figure 1 shows the EEG channel locations on the scalp along with the electrodes names.

3 The Proposed Methodology

Our proposed methodology for blink detection presented in this paper includes two major modules including automatic eye blink data scoring and annotation, and multi-score space clustering. In this section, we will cover the details of the proposed system.

3.1 Annotation of Continuous Unsupervised EEG Data

This process creates aligned labels out of the unlabeled EEG data. Figure 2 illustrates the training data annotation process on the segmented data using a set of known blink patterns, selected as templates for use in the Dynamic Time Warping (DTW) algorithm. The generated labels, provide a set of initial labels for the subsequent blink characterization and detection method.

Scoring and Annotation using DTW DTW is an established algorithm that measures the similarity between two sequences of sample values. A major advantage of DTW is that these sequences can vary in speed. This means that a signal that is "stretched" or "shrunk" in the time domain can still be matched based on its similarity

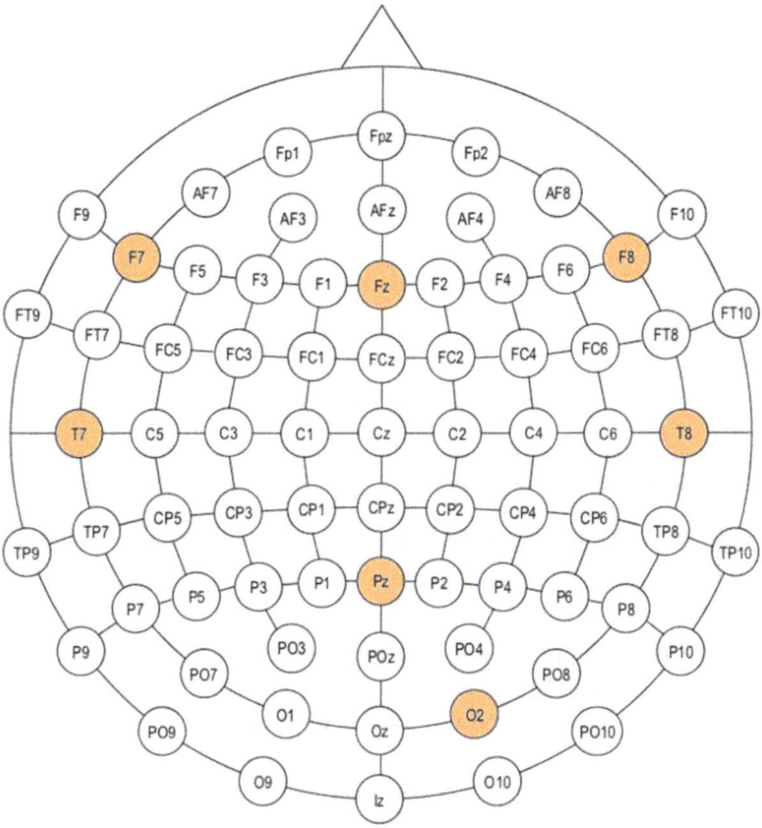

Fig. 1 Workload assessment task channel locations

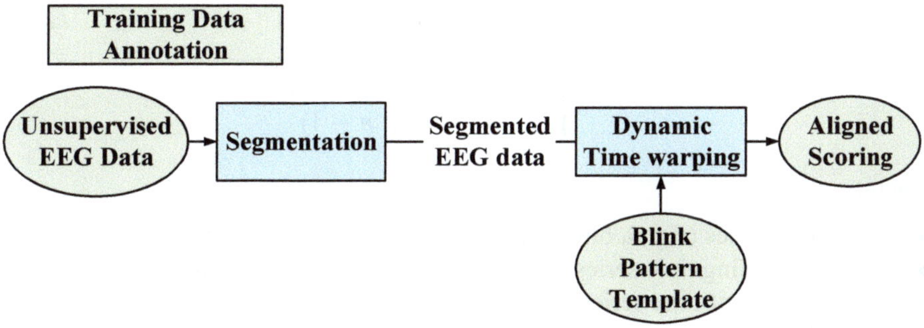

Fig. 2 Training data annotation process

to the unstretched signal. Figure 3 provides a graphical representation of DTW-based pattern alignment and matching.

Figure 3 illustrates how DTW would be able to relate the signal in green to the signal in blue. Even though the signals progress at three different rates (A: Compressed,

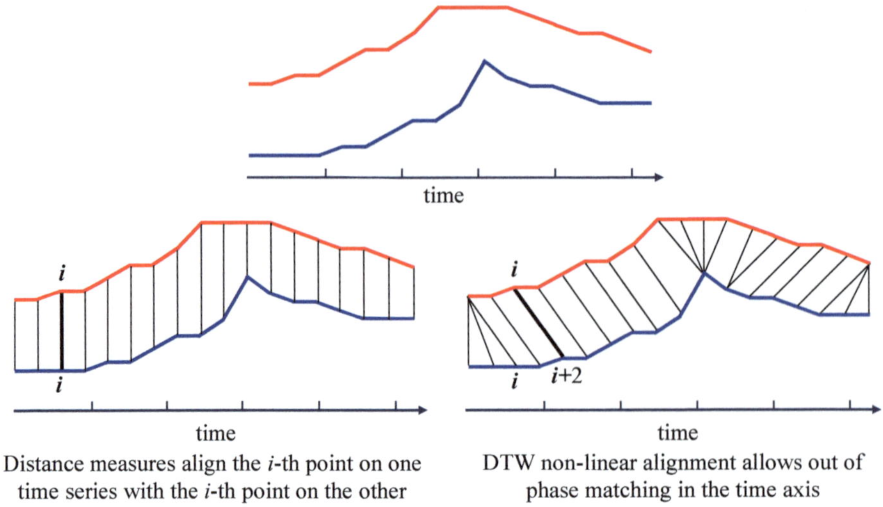

Fig. 3 A Graphical representation of DTW pattern matching by finding the shortest distances between graphs

B: Less compressed, and C: Stretched), dynamic time warping can still calculate how similar these signals are given the phase difference. The DTW algorithm has two parts to it, it is O(mn) in computational complexity:

$$C(x, y) = \sum_{k=1}^{K} c(M_{m_k}, N_{n_k}), \tag{1}$$

$$P_{k-1} = \begin{cases} C(1,\ n-1)\ if\ m=1 \\ C(m-1,\ 1)\ if\ n=1 \\ \quad otherwise \\ \min\begin{cases} C(m-1,\ n-1) \\ C(m,\ n-1) \end{cases} \end{cases} \tag{2}$$

where:

- M is a time series (e.g., a blink template).
- N is the matching time series (e.g., one sample window of data).
- m is the length of time series M.
- n is the length of time series n.
- m_k is a point in time series M, path index k.
- n_k is a point in time series N, path index k.
- K is equal to $n * m$.
- C is a cost matrix, each entry in $C(x, y)$ is the Euclidean distance between two points m_k and n_k.
- P_k is the smallest distance required to connect point m_k to n_k.

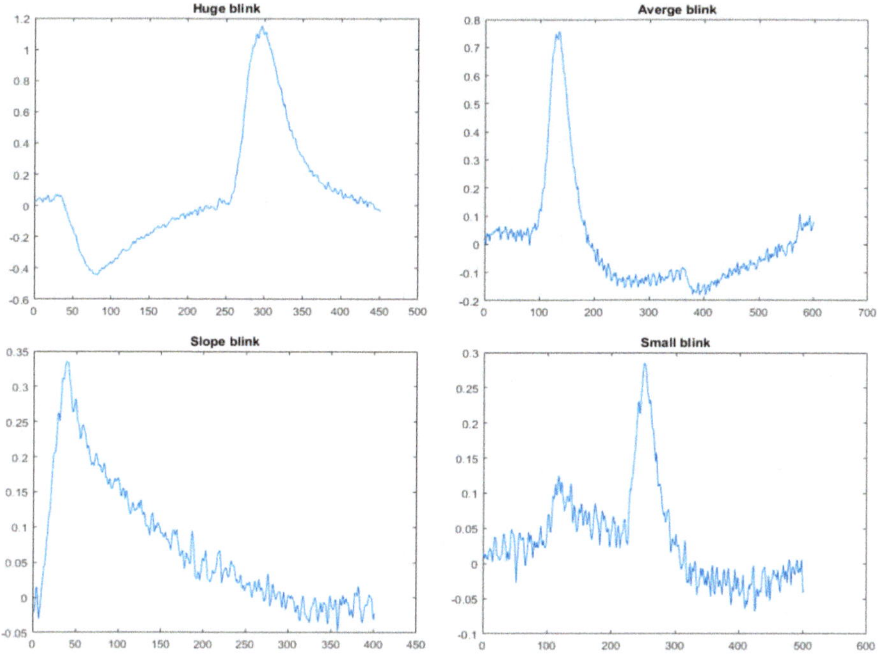

Fig. 4 Eye blink selected template patterns

This algorithm traverses the cost matrix C to find the minimum set of distances P between time series M and N.

Template Selection To label windows of EEG recordings as blink or non-blink, it is necessary to provide the DTW algorithm with representative templates of known blink activity. We conducted this task by defining four categories of blinks as follows: (1) Huge blinks, (2) Average blinks, (3) Slope blinks, and (4) small blinks. The four blink samples used as templates of the DTW algorithm for subject 1 are shown in Fig. 4.

3.2 *Eye Blink Clustering and Detection*

K-Means Clustering K-means clustering is an unsupervised learning method, which is used when input data is unlabeled (i.e., data without defined categories or groups). The goal of this algorithm is to find groups in the data, with the number of groups represented by the variable K. The algorithm works iteratively to assign each data point to one of K groups based on the input feature space similarities. The results of the K-means clustering algorithm are:

1. The centroids of the K clusters, which can be used to label new data.
2. Labels for the training data (each data point is assigned to a single cluster).

Each centroid of a cluster is a collection of feature values which define the resulting groups. Examining the centroid feature weights can be used to qualitatively interpret what kind of group each cluster represents. The K-means clustering algorithm uses iterative refinement to produce a final result. The algorithm inputs are the number of clusters and the data set. The data set is a collection of features for each data point. The algorithms starts with initial estimates for the centroids, which can either be randomly generated or randomly selected from the data set. The algorithm then iterates between two steps:

(1) *Data assignment step*: Each centroid defines one of the clusters. In this step, each data point is assigned to its nearest centroid, based on the squared Euclidean distance. If c_i is the collection of centroids in set C, then each data point x is assigned to a cluster based on:

$$arg_{c_i \in c}^{min}\ dist(c_i, x)^2 \tag{3}$$

where $dist(.)$ is the standard ($L2$) Euclidean distance. Let the set of data point assignments for each ith cluster centroid be S_i.

(2) *Centroid update step*: In this step, the centroids are recomputed. This is done by taking the mean of all data points assigned to that centroid's cluster.

$$c_i = \frac{1}{|S_i|} \sum_{x_i \in S_i} x_i \tag{4}$$

The algorithm iterates between steps one and two until a stopping criteria is met (i.e., no data points change clusters, the sum of the distances is minimized, or some maximum number of iterations is reached). This algorithm is guaranteed to converge to a result. The result may be a local optimum (i.e. not necessarily the best possible outcome), meaning that assessing more than one run of the algorithm with randomized starting centroids may give a better outcome.

The algorithm described above finds the clusters and data set labels for a particular pre-chosen K. To find the number of clusters in the data, the user needs to run the K-means clustering algorithm for a range of K values and compare the results. Our focus, in this work, is on clustering the EEG segments to blink and non-blink (i.e. detection task) clusters based on their DTW multi-score space positioning. Therefore, we set the parameter K = 2.

Discriminative Eye blink Detection At the detection phase, the set of distances between the data segments and the templates, used to determine the clusters (i.e. blink/non-blink), were labelled as blink or non-blink to derive the detection hyperplane. Training the detection hyperplane was conducted using the Support Vector Machine (SVM), which is a state of the art discriminative binary classifier [19]. We use soft-margin objective function for SVM optimization. SVM is always able to generate a hyperplane even if the problem is not linearly separable. It is also resilient to outliers, because they are taken into account in the objective function as shown below:

$$\begin{aligned} &\min\left(\tfrac{1}{2}\right)||w||^2 + C\sum_i \beta_i,\\ &y_i\,(w * x_i + b) \geq (1 - \beta_i)\,,\ \forall x_i,\\ &\beta_i \geq 0 \end{aligned} \tag{5}$$

Soft margin SVM seeks to minimize the sum of distances of misclassified patterns from the hyperplane, β_i, while maximizing the margin between classes, $\frac{2}{||w||}$. The constant C can be set to adjust the tradeoff factor between margin width and misclassifications. A grid-based parameter search is performed to find the best values for C and β_i.

4 The Experiment and Results

The sampling frequency of the EEG data recording was 500 Hz. The EEG data (from the F_z channel) is segmented into 250 sample size windows (0.5 s) with 90% overlap before it is passed into the DTW algorithm. The reason to set the size of the segments as 0.5 s is to make sure there would be no more than one eye blink in a single window at all times.

4.1 DTW Multi-score Clustering and Detection

The DTW algorithm processes each window and generated a distance metric for how different a given EEG segment was from each of the eye blink templates. The distance between the EEG segments with all four selected blink templates constitutes a data sample in a 4-D score space. The samples were fed to the subsequent K-mean clustering module with K=2 (Blink vs. non-blink). The EEG segments were labeled based on the learned clusters and employed to train the SVM detection hyperplane.

Performance on partial blinks Figure 5 shows a portion of EEG data from channel Fz that demonstrates the performance of DTW on windows containing only part of a blink. The distance measures from two templates and the blink/not-blink clusters separating threshold (in purple) calculated. It can be observed that the "*huge blink*" template distance (in orange) is below the SVM hyperplane (in purple) for windows 22 through 28. Figure 5 illustrates how DTW distance only based on one template has issues finding partial blinks whereas the cluster threshold based on all 4 templates is highly robust.

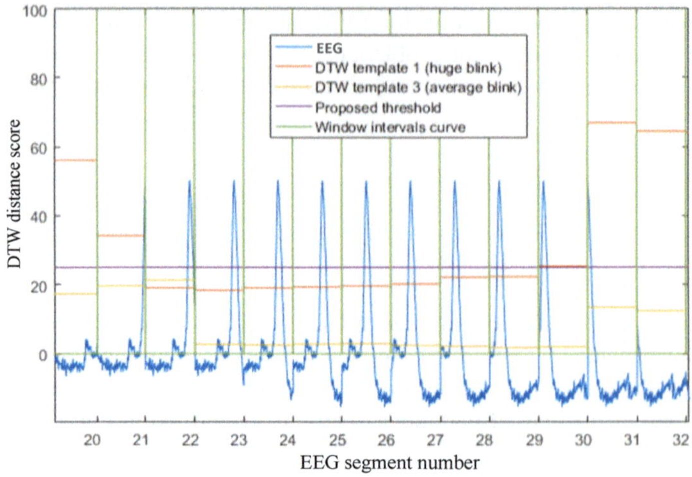

Fig. 5 DTW distances of all four templates on a sample blink

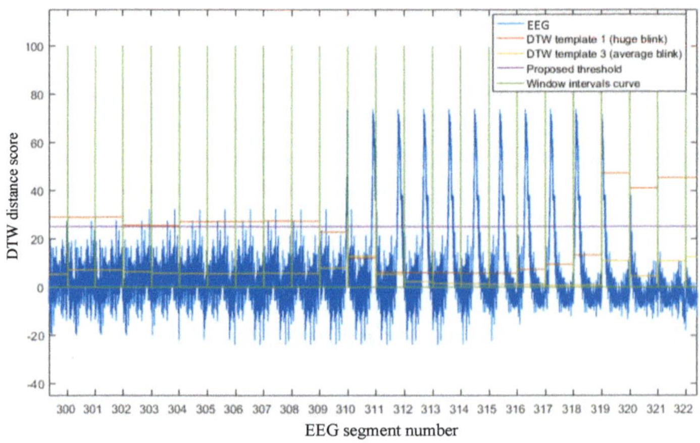

Fig. 6 Performance of DTW under noisy conditions

Performance during line noise There are several portions of data in this study that contain severe line noise. The segment shown in Fig. 6 illustrates how well DTW-based clustering performs under noisy conditions. The distance is much lower than normal for non-blink windows 300–309, nearing the threshold line. The distance goes down to expected levels during blink windows 310–320. The DTW-based clustering threshold also works for partial blink windows (windows 319 and 320 are labeled as blinks).

Blink detection Cross validation results We used expert decision to label the eye blinks as the ground truth. The experts went through the EEG segments and manually labeled the EEG data as the ground truth in order to evaluate the detection model generalization ability. We employed 10 fold cross validation (10-CV) to estimate the

Table 1 The blink detection 10-CV accuracy by channel

Channel#	Highest acc.	Lowest acc.	Average acc.
1: $F7$	95.10	87.89	92.62
2: $F8$	97.11	92.28	94.40
3: Fz	98.29	94.15	96.78
4: $O2$	81.56	75.45	79.43
5: Pz	82.52	76.64	79.92
6: $T7$	84.93	79.17	81.05

Table 2 DTW multi-score detection performance

	Temp#1	Temp#2	Temp#3	Temp#4	Temp 1–4
Subject1	92.75	81.28	70.65	64.19	96.78
Subject2	91.83	85.7	65.85	72.62	97.75
Subject3	90.35	83.52	60.64	70.27	95.62
Subject4	92.5	78.27	63.26	67.91	97.13
Subject5	89.52	82.29	67.15	69.36	94.79
Average	91.39	82.212	65.51	68.87	96.42

generalization capabilities of the trained model. Table 1 reports the blink detection 10-CV accuracy (The highest, lowest, and average 10-CV) using each of the EEG channel recordings. As shown in Table 1, frontal channels $F7$, $F8$, and Fz captures more detailed eye blinking patterns due to their proximity to the eyes. Among them, the Fz channel provides the highest accuracy.

The rest of the DTW multi-score detection analyses were conducted based on the EEG data from the Fz channel. Table 2 reports the proposed DTW multi-score space clustering and detection performance (10-CV average accuracy) in comparison with DTW score for each template individually. The results in Table 2 demonstrate that the proposed method considerably outperform individual templates (58% percent relative improvement compared to the best individual template). The results in Table 2 shows that the template#1, "huge blink", leads to the highest 10-CV average accuracy among the selected individual templates.

5 Conclusion

One of the challenges in using the EEG for identifying mental state in practical ubiquitous settings is ocular artifacts e.g. eye blinking that is present in important frequency sub-bands and can significantly degrade predictive state estimation accuracy. In this paper, we proposed automatic detection of eye blink noise during a wearable continuous EEG-based cognitive workload assessment tests using dynamic time warping (DTW) score clustering and categorization. Our proposed for blink detection presented in this paper has three major components: (1) Automatic eye blink data scoring and annotation, (2) Multi-score space clustering, (3) Discriminative Eye blink Detection. Our experimental results on a population of subjects demonstrated high performances of the proposed blink detection methodology via achieving 96.42% average accuracy of blink detection.

References

1. Yin, Z., Zhang, J.: Cross-session classification of mental workload levels using EEG and an adaptive deep learning model. Biomed. Signal Process. Control **33**, 30–47 (2017)
2. Ke, Y., et al.: Towards an effective cross-task mental workload recognition model using electroencephalography based on feature selection and support vector machine regression. Int. J. Psychophysiol. **98**(2), 157–166 (2015)
3. Yin, Z., Zhang, J.: Operator functional state classification using least-square support vector machine based recursive feature elimination technique. Comput. Methods Prog. Biomed. **113**(1), 101–115 (2014)
4. Wang, Z., et al.: Cross-subject workload classification with a hierarchical Bayes model. NeuroImage **59**(1), 64–69 (2012)
5. Urigen, J.A., Garcia-Zapirain, B.: EEG artifact removalstate-of-the-art and guidelines. J. Neural Eng. **12**(3), 031001 (2015)
6. Rana, A.Q., Ghouse, A.T., Govindarajan, R.: Basics of electroencephalography (EEG). In: Neurophysiology in Clinical Practice, pp. 3–9. Springer International Publishing (2017)
7. Taherisadr, M., Dehzangi, O., Parsaei, H.: Single channel EEG artifact identification using two-dimensional multi-resolution analysis. Sensors **17**(12), 2895 (2017)
8. Winkler, I., Haufe, S., Tangermann, M.: Automatic classification of artifactual ICA-components for artifact removal in EEG signals. Behav. Brain Funct. **7**(1), 30 (2011)
9. Croft, R.J., Barry, R.J.: EOG correction: a new perspective. Electroencephalogr. Clin. Neurophys. **107**(6), 387–394 (1998)
10. He, P., et al.: Removal of ocular artifacts from EEG: a comparison of adaptive filtering method and regression method using simulated data. In: 27th Annual International Conference of the Engineering in Medicine and Biology Society, 2005. IEEE-EMBS 2005. IEEE (2006)
11. Schlgl, A., et al.: A fully automated correction method of EOG artifacts in EEG recordings. Clin. Neurophysiol. **118**(1), 98–104 (2007)
12. Cheok, N.S., Raveendran, P.: Removal of EOG artifacts using ICA regression method. In: 4th Kuala Lumpur International Conference on Biomedical Engineering 2008. Springer, Berlin, Heidelberg (2008)
13. Makeig, S., et al.: Independent component analysis of electroencephalographic data. In: Advances in Neural Information Processing Systems (1996)
14. Wang, Z., et al.: Robust removal of ocular artifacts by combining independent component analysis and system identification. Biomed. Signal Process. Control **10**, 250–259 (2014)
15. Mahajan, R., Morshed, B.I.: Unsupervised eye blink artifact denoising of EEG data with modified multiscale sample entropy, kurtosis, and Wavelet-ICA. IEEE J. Biomed. Health Inform. **19**(1), 158–165 (2015)
16. Ghandeharion, H., Erfanian, A.: A fully automatic ocular artifact suppression from EEG data using higher order statistics: improved performance by wavelet analysis. Med. Eng. Phys. **32**(7), 720–729 (2010)
17. Kong, W., et al.: Automatic and direct identification of blink components from scalp EEG. Sensors **13**(8), 10783–10801 (2013)
18. Kanungo, T., et al.: An efficient k-means clustering algorithm: analysis and implementation. IEEE Trans. Pattern Anal. Mach. Intell. **24**(7), 881–892 (2002)
19. Cortes, C., Vapnik, V.: Support-vector networks. Mach. Learn. **20**(3), 273–297 (1995)

Continuous Blood Pressure Estimation Using PPG and ECG Signal

Bo Wang, Zhipei Huang, Jiankang Wu, Zhongdi Liu, Yuanyuan Liu and Pengjie Zhang

Abstract Continuous blood pressure monitor can detect the potential risk of cardiovascular disease and provide a gold standard for clinical diagnosis. The features extracted from photoplethysmography (PPG) and electrocardiogram (ECG) signals can reflect the dynamics of cardiovascular system. In this paper, 39 features are extracted from PPG and ECG signals and 10 features are chosen by analyzing their correlations with blood pressure. Several machine learning algorithms are used to predict the continuous and cuff-less estimation of the diastolic blood pressure and systolic blood pressure. The results shows that compared with linear regression and support vector regression methods, the artificial neural network optimized by genetic algorithm gives a better accuracy for 1 h prediction under Advancement of Medical Instrumentation and the British Hypertension Society standard.

Keywords Photoplethysmography · Electrocardiogram
Continuous blood pressure · Artificial neural network · Genetic algorithm

B. Wang · Z. Huang (✉) · J. Wu · Z. Liu · Y. Liu · P. Zhang
University of Chinese Academy of Sciences, Beijing, China
e-mail: zhphuang@ucas.ac.cn

B. Wang
e-mail: wangbo515@mails.ucas.ac.cn

J. Wu
e-mail: jkwu@ucas.ac.cn

Z. Liu
e-mail: liuzhongdi17@mails.ucas.ac.cn

Y. Liu
e-mail: liuyuanyuan115@mails.ucas.ac.cn

P. Zhang
e-mail: zhangpengjie15@mails.ucas.ac.cn

G. Fortino and Z. Wang (eds.), *Advances in Body Area Networks I*,
Internet of Things, https://doi.org/10.1007/978-3-030-02819-0_6

1 Introduction

According to the World Health Organization (WHO) report, the Cardiovascular diseases (CVD) has claimed 17.5 million lives in 2016 [1]. A noninvasive and continuous system for monitoring cardiovascular parameters has become particularly important. Blood pressure (BP) is a significant indicator of the status of the cardiovascular. Long time beat-to-beat BP monitoring is very useful in many areas, especially in medical domain, such as health assessment, medical diagnosis and surgical operation.

A number of approaches have been proposed for continuous and noninvasive BP estimation. Compared with Tonometry and Volume Clamp methods, the way of BP estimation by photoplethysmography signal is more promising [2].

The pulse waves of blood are generated by the beats of heart which push the blood flowing along the arterial tree. With the intermittent contraction and relaxation of heart, the propagation of blood volume produce a pulse wave which can be detected by PPG using infrared light. As BP being the pressure exerted by the blood against the vessel walls, a pulse wave velocity (PWV) method was proposed to measure the beat-to-beat BP signal through the velocity of pulse waves.

The PWV measurement method needs to know precisely the distance between sites where signals are collected, which requires the presence of a skilled operator to manually localize the carotid and femoral arteries [3]. The PWV's inverse, pulse wave transit time (PWTT), is commonly used to replace PWV.

The PWTT method to estimate blood pressure is based on Moens-Korteweg (M-K) equation, derived by Moens and Korteweg [4]:

$$PWV = \frac{D}{PWTT} = \sqrt{\frac{tE}{\rho d}} \tag{1}$$

where D is the length of the vessel, PWTT is the pulse wave transmit time, t is thickness of the vessel wall, E is Young's modulus, ρ is blood density and d is the vessel's diameter.

In recent years, there have been a lot of researches about the relationship between pulse wave transit time and blood pressure. A linear relationship proposed by Zhang et al. [5] has been widely recognized as below:

$$BP = a * PWTT + b \tag{2}$$

Practically, the PWTT is derived from the time delay between the R-wave of ECG and the peak of PPG in the same cardiac cycle. The R-wave occurs at the beginning of the heart systole when the blood ejected from heart to aorta. Therefore, the R-wave is a good reference for the beginning of the cardiac cycle.

A frequent re-calibration procedure must be applied to obtain the parameters in linear equation because of the physiological dynamics of vessel, which makes the PWTT method become inconvenient in our daily life.

The main cause of this disadvantage of traditional PWTT technic is that the single PWTT feature couldn't reflect the state of human body as a complex system, we need to seek more parameters which can represent patient's cardiovascular index at that moment. Experimental research shows that the features, extracted from PPG and ECG signals, can be used to predict many vital health concerned parameters such as blood pressure and heart rate [6]. In view of the complexity of the human's cardiovascular system, the relationship, linear or non-linear, between features and BP is uncertain. Machine learning provides a method that can accurately and effectively achieve the relationship between the features of physiological signals and BP.

In recent representative studies [7–11], researchers applied various machine learning algorithms for BP estimation from PPG and ECG signals. This study considered too much parameters and did not draw the correction between features and BP.

In this paper, the relationship between BP and the features extracted from PPG and ECG signals is investigated. Besides, various machine learning and signal processing algorithms are exploited in estimating 1-h continuous blood pressure. In summary, after denoising the PPG and ECG signals, their informative features are extracted, and finally, these features serve as an input to a regression model, which estimates the BP value. The accuracy of each algorithms was evaluated by percent of sets which reach the Association for the Advancement of the Medical Instrumentation (AAMI) and British Hypertension Society (BHS) standards among all available datasets to come to a conclusion.

The rest of this paper is organized as follows: Sect. 2 explains the whole experimental procedure and all the algorithms using this work. Section 3 presents the results and compares them with different machine learning algorithms and finally Sect. 4 concludes the paper.

2 Experimental and Computational Details

2.1 Data Acquisition

In this experiment, the Finometer [12] was used for noninvasive beat-to-beat blood pressure monitoring, which pressed the sensors on the skin surface of arteries. Meanwhile, the PPG and ECG signal were collected by a physiological signal synchronous collecting device, shown in Fig. 1. The sampling rate of ECG and PPG signal is 1000 Hz with 16 bit resolution. All 30 sets of data with 75 min long were collected in different person.

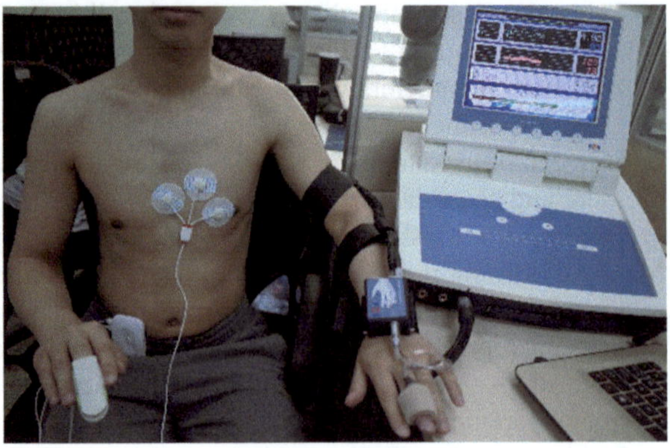

Fig. 1 Data collection during the experiment. By synchronizing the time of the two device's software platforms, we can obtain ambulatory blood pressure (ABP), PPG and ECG signal in the same cardiac cycle

2.2 Data Pre-processing

There are three kinds of noise which may be contained in the PPG and ECG signal [7]: (A) power-line interference–ambient electromagnetic signals picking up by the device; (B) baseline wandering caused by poor contact to the wearable sensor; (C) high frequency noise, such as environment interference and muscular activities. Through analyzed the spectrum of the raw signal, there is no power-line interference, thus, we should focus on eliminating the baseline wandering and high frequency noise in data pre-processing.

By analyzing and comparing, we selected the wavelet denoising method in pre-process because it provides various merits such as a better phase response and low computational complexity.

Figure 2 shows the preprocessing diagram of PPG and ECG signal. According to Singh BN [13] study, the raw signal is decomposed to 10 levels by Daubechies 8 (DB8) mother wavelet. Afterwards, the components corresponding to the very low frequency ranging from 0 to 0.25 Hz and the high frequencies between 250 and 500 Hz are eliminated with their coefficients zeroing. The conventional wavelet denoising is performed on the remaining decomposition coefficients with soft thresholding strategy [7]. Figures 3 and 4 represent the performance of wavelet denoising algorithm applied on the raw data.

2.3 Feature Extraction

According to Eq. (3), the value of blood pressure is determined by various cardiovascular parameters such as the thickness of the vessel wall, the blood density and the vessel's diameter. Besides, systemic vascular resistance and the action of the heart also affect blood pressure.

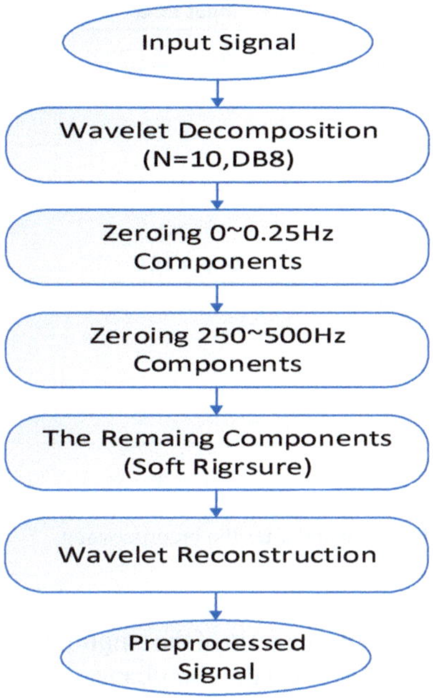

Fig. 2 Through the db8 wavelet, the baseline and high frequency noise are eliminated in raw PPG and ECG signal

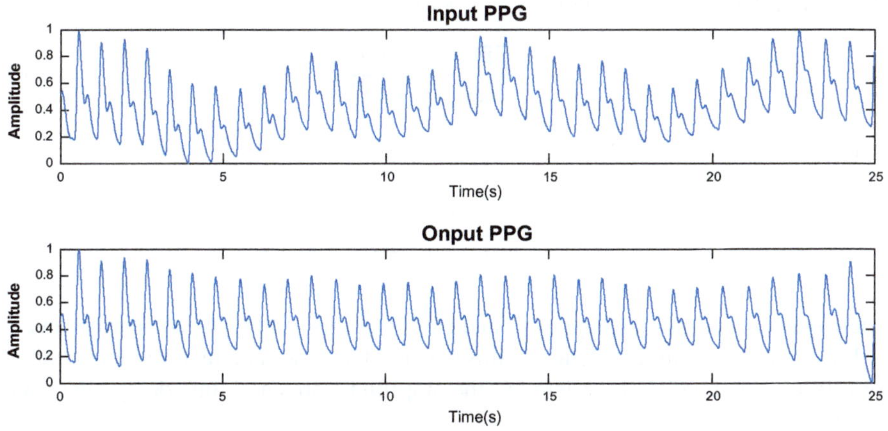

Fig. 3 Example PPG signals before and after the preprocessing

Dorlas and Nijboer [14] found that the PPG's peak amplitude is directly proportional to local vascular distensibility and the amplitude is potentially a more suitable measure than pulse arrival time for estimating continuous blood pressure [15]. Wang et al. [16] have divided the PPG area into two areas at the dicrotic notch and the ratio of the two areas can be used as an indicator of total peripheral resistance.

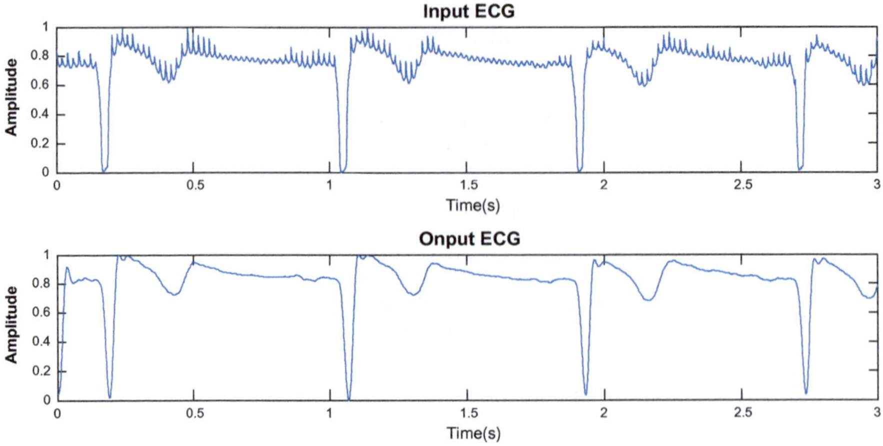

Fig. 4 Example ECG signals before and after the preprocessing

Takazawa et al. [17] believed that the ratio of the amplitude between peak and dicrotic notches in PPG can be served to reflect the compliance of the elastic arteries. The time delay between the peak and dicrotic notches in PPG is an index of arterial stiffness, proved by Millasseau et al. [18].

A K value derived from PPG also have significant correlation with the stiffness of arterial [19]. K value is calculated as below:

$$\mathrm{K} = \frac{\overline{ppg_{oo'}} - ppg(t_0)}{ppg(t_o) - ppg(t_{o'})} \tag{3}$$

In which

$$\overline{ppg_{oo'}} = \frac{1}{t_{o'} - t_o} \sum_{i=t_o}^{t_{o'}} ppg(i) \tag{4}$$

where o and o' are the starting points and ending points of PPG in each cardiac cycle.

Based on the above researches, in this paper, 28 time domain features are extracted form PPG signal which are related to the amplitude, area and time delay characters in PPG signal.

There are 4 frequency domain features extracted from PPG according to Enric's [9] study. E_n is the logarithm of energy in each cardiac cycle:

$$\log E_n = \log\left(\sum_{i=t_o}^{t_{o'}} ppg^2(i)\right) \tag{5}$$

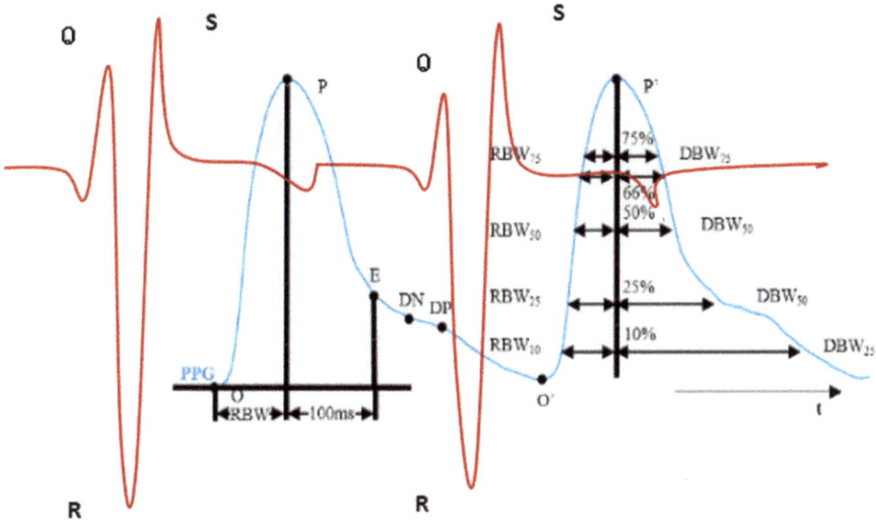

Fig. 5 Illustration of features extraction from data. The Q, R, S peaks in the ECG signal and the starting points, peaks and dicrotic notches in the PPG signal

Kaiser-Teager function is a transform applied on PPG signal:

$$\mathrm{KTE(i)} = ppg^2(i) - \mathrm{ppg}(i-1)ppg(i+1) \tag{6}$$

The statistical coefficients of KTE such as mean value and variance in each cardiac cycle are chosen. Besides, the energy of Fourier transform on PPG in per cycle is also selected.

A series of peak detection algorithm [20, 21] were used to detect Q, R, S peaks in the ECG signal and the starting points, Peaks and Dicrotic notches in the PPG signal. Altogether, there are 39 features extracted from PPG and ECG signals, which can be divided into three types, showed in Fig. 5 and listed in Table 1.

2.4 *Features Selection*

In order to reduce the feature's dimension, we analyzed the correction between each features and BP. Pearson Correction is a standard of relevance analysis:

$$\rho_{xy} = \frac{E((X-E(X))(Y-E(Y)))}{\sqrt{D(X)}\sqrt{D(Y)}} \tag{7}$$

where E is mathematical expectation and D is variance.

Table 1 Frequency of special characters

Source	Features abb.	Short description
PPG	PH/DNH/DPH	Height of peak point P/DN/DP
	DNHr/DPHr	The ratio of DNH/DPH and PH
	RBW/RBW10-75	Rising branch width and from 10%/25%/33%/50%/75% point of rising of branch to peak
	DBW10-75	From peak to 10, 25, 33, 50, 75% point of descending points
	PWA	Area between pulse wave curve O-O′ and line OO′
	RBAr/DBAr/DiaAr	Rising branch/Descending branch/Diastolic branch relative area
	SLP1/SLP2/SLP3	Slope of line OP/PO′/DP-O′
	K/AmBE	K and The mean value of P-E
	E/KTEMIU/KTEVAR/ENTROP	Frequency Domain Character
ECG	WQR	The width of Q-R
	WRS	The width of R-S
	HQR	The Height of Q-R
	HRS	The Height of R-S
PWTT	PWTTRO	The width of ECG-R to the onset off PPG
	PWTTRH	The width of ECG-R to the 50% point of rising of PPG
	PWTTRP	The width of ECG-R to the peak off PPG

Although the relevance of features to ABP could be manifestation by the Pearson correction, the correlation between different features still remained unknown. To avoid feature redundancy, referred to the research of He et al. [22], the correlation analysis was carried out between all the highest ranking features of systolic blood pressure (SBP) and diastolic blood pressure (DBP), which were significantly correlated ($|r| > 0.85$). For each group of significantly correlated features, only the one which has the highest score were selected.

Large number of features will lead to over-fitting in small train sets, so the features we chosen is the balance between accuracy and over fitting. Finally, 10 features are chosen to estimate DBP and SBP, respectively.

2.5 Regression Models

A series of machine learning algorithms can be used to represent such a relation in BP estimation. In this work, we selected 15 min data as the training set, which contains about 1000 pairs of feature vectors, while the rest 60 min were chosen as test set.

2.5.1 Linear Regression (LR)

LR is used to describe the linear relationship between the feature vector and the target variable. A gradient descent method is applied to search the minimum of cost function between the prediction and true values. Compared with other learning algorithms, LR is simple and easy to train. However, it is not applicable in the situation where features and target have a strong non-linear relationship [7].

2.5.2 Support Vector Regression (SVR)

SVR algorithm is a powerful technique based on statistical learning theory. The method is extremely robust and provides excellent generalization performance. There is only one category of the data in SVR and the optimal hyperplane is to make the sum of deviation in all samples can be minimized [23].

2.5.3 Artificial Neural Networks (ANN)

ANN is a powerful tool to realize a special non-linear transformation, from the input space to the output space. BP network is a kind of multilayer feed forward neural network. The error back propagation in weight updated process is the main character of BP.

Although the BP network has mature theory and wide application, it also suffers from specific problems, such as the node function, the number of hidden layers, the number of nodes per layer and the assignment of initial weight [24]. It is necessary to solve this problems and achieve optimal performance in BP network algorithm.

Genetic algorithm (GA) is an adaptive searching technique based on a selection and reproduction mechanism found in the natural evolution process. The purpose of using GA to optimize BP neural network is to get optimal weight and bias, which can avoid BP network trapping into local minima. Every individual in GA represents the initial weight and bias of BP and the fitness of the individual is the error computed by the network initialized by weights. Though Genetic operator like selection, cross-over and mutation, we can find the most optimal weights of BP in train set [25].

For each models, a fivefold separation of dataset samples to train and test sets is used. Unlike the traditional method, we separate the data into five parts, one is used to train and the rest four parts are used to test.

2.6 Models Evaluation

In order to compare the performance of different models in blood estimation, the root mean square error (RMSE) and the absolute error are used. The percent of set which reaches AAMI and BHS, the two main accuracy standards for blood pressure measuring devices, are also recorded for each model.

3 Results and Discussion

The relationship between features and BP are analyzed by computing Person correlation. The features lies on the top ten in the rank of p value are shown in Fig. 6. The correction between those features are also analyzed, with all the value lower than 0.85, implying all ten features in the figure can be used to BP estimation. From the figure, it is evident that the features we used to estimate DBP and SBP are different, so we must build models for DBP and SBP respectively.

Table 2 compares the performance of various learning algorithms using the root mean square error. It is evident that the BP estimation using non-linear learning algorithm is much better than linear regression. Therefore, it can be inferred that

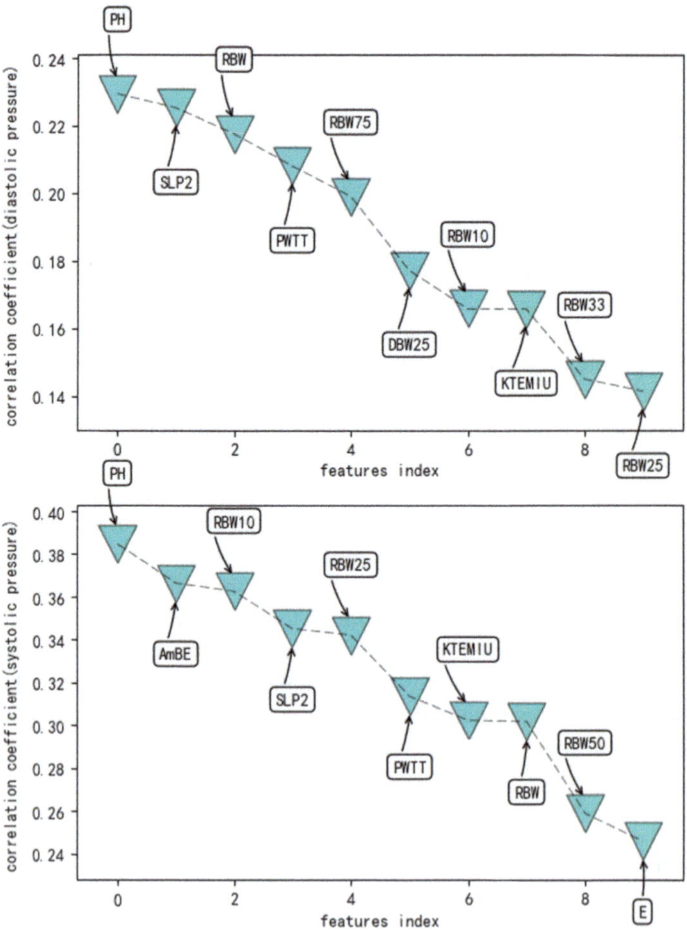

Fig. 6 Illustration of features extraction from data

Table 2 RMES (mean ± std.) in different learning models

	MODELS		
	ANN	SVR	LR
SBP	6.53 ± 2.21	7.09 ± 1.80	14.67 ± 16.94
DBP	3.17 ± 1.18	3.98 ± 1.21	9.38 ± 17.33

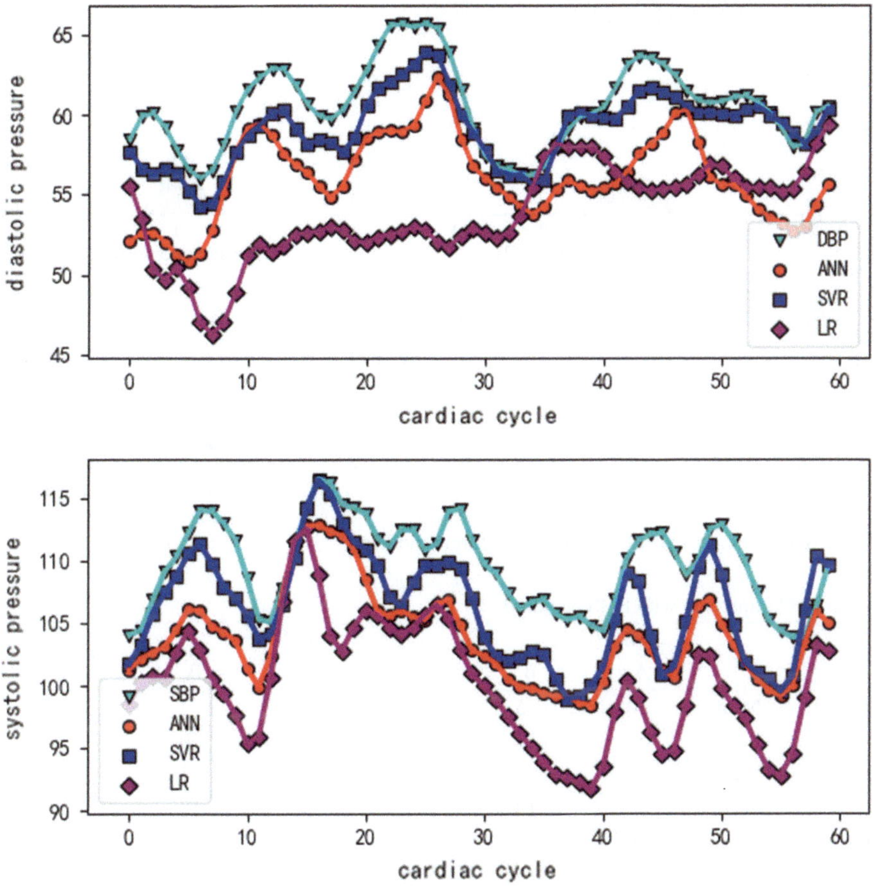

Fig. 7 One exemplary continuous blood pressure estimation with ANN (blue square), SVR (red sphere) and LR (purple diamond) compared with real pressure in 60 cardiac cycle

there exists a considerable non linearity between features and BP. The result of ANN method optimized by Genetic algorithm is a little better than SVR method.

Figure 7 shows a comparison between different models. With the finger pressure as the benchmark, BP estimation with the ANN method is better than SVR and LR. Figure 8 presents histogram of estimation absolute error of different models in one of the dataset. It can be seen that error values are normally distributed around zero.

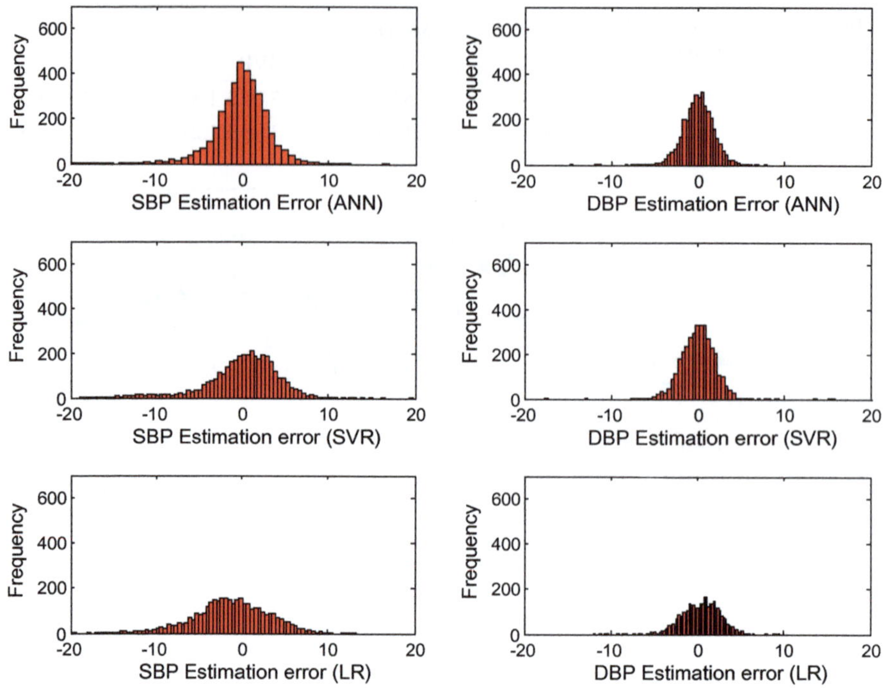

Fig. 8 Histograms of absolute errors in different models

Table 3 Comparison with BHS standard

		SBP			DBP		
		A	B	C	A	B	C
ANN	Num.	20	5	4	27	3	0
	Per. (%)	67	17	13	90	10	0
SVR	Num.	19	4	4	25	3	1
	Per. (%)	63	13	13	83	10	3
LR	Num.	16	5	3	23	4	1
	Per. (%)	53	17	10	77	13	3

Table 3 demonstrates a comparison between different models in blood pressure estimation using British Hypertension Society (BHS) standard [26], based on the percentage of absolute error under three different thresholds. All the methods were also evaluated by Advancement of Medical Instrumentation (AAMI) standard [27], listed in Table 4. It should be noted that all the models are tested in only 30 groups of dataset, so the numbers and the percentage of the total, satisfied with the standard, are both presented. As seen in the table, the estimation of BP with ANN optimized by Genetic algorithm have a better performance than other algorithms.

Table 4 Comparison with AAMI standard

	SBP		DBP	
	Num.	Per. (%)	Num.	Per. (%)
ANN	23	76	28	93
SVR	20	67	26	86
LR	18	60	26	86

4 Conclusions

In this paper, we explore the potential methods for cuffless continuous blood pressure estimation based on features extracted from electrocardiogram and photoplethysmogram using machine learning algorithms. All the data are collected from continuous blood pressure measuring instrument and physiological signal synchronous collecting device. 39 features are extracted form PPG and ECG signal; by correction analysis, 10 features are chosen to estimate DBP and SBP, respectively. The learning algorithms of Artificial Neural Networks, Support Vector Regression and Linear Regression are used to establish the real relationship between effective features and blood pressure. The results shows that the technique using ANN optimized by Genetic Algorithm performs better than the others, which could be potentially applied for continuous blood pressure estimation in the future.

Acknowledgements This work was supported by Special Fund for Scientific Research Cooperation of University Chinese Academy of Sciences.

References

1. World Health Statistics 2016 (2016)
2. Peter, L., Noury, N., Cerny, M.: A review of methods for non-invasive and continuous blood pressure monitoring: pulse transit time method is promising? IRBM **35**, 271–282 (2014)
3. Geddes, L.A., Voelz, M., James, S., Reiner, D.: Pulse arrival time as a method of obtaining systolic and diastolic blood pressure indirectly. Med. Biol. Eng. Comput. **19**, 671–672 (1981)
4. Isebree Moens, A.: Die Pulscurve
5. Yan, Y.S., Zhang, Y.T.: A model-based calibration method for noninvasive and cuffless measurement of arterial blood pressure. In: Biomedical Circuits and Systems Conference, 2006, BioCAS, pp. 234–236
6. Allen, J.: Photoplethysmography and its application in clinical physiological measurement. Physiol. Meas. **28**, R1–39 (2007)
7. Kachuee, M., Kiani, M.M., Mohammadzade, H., Shabany, M.: Cuff-Less Blood Pressure Estimation Algorithms for Continuous Health-Care Monitoring, pp. 1–1 (2016)
8. Kurylyak, Y., Lamonaca, F., Grimaldi, D.: A neural network-based method for continuous blood pressure estimation from a PPG signal. In: IEEE International Instrumentation and Measurement Technology Conference, pp. 280–283
9. Monte-Moreno, E.: Non-invasive estimate of blood glucose and blood pressure from a photoplethysmograph by means of machine learning techniques. Artif. Intell. Med. **53**, 127–138 (2011)

10. Ghosh, S., Banerjee, A., Ray, N., Wood, P.W., Boulanger, P., Padwal, R.: Continuous blood pressure prediction from pulse transit time using ECG and PPG signals. In: IEEE Healthcare Innovation Point-of-Care Technologies Conference, pp. 188–191
11. Su, P., Ding, X., Zhang, Y., Miao, F., Zhao, N.: Learning to Predict Blood Pressure with Deep Bidirectional LSTM Network (2017)
12. Imholz, B.P., Wieling, W., van Montfrans, G.A., Wesseling, K.H.: Fifteen years experience with finger arterial pressure monitoring: assessment of the technology. Cardiovasc. Res. **38**, 605–616 (1998)
13. Singh, B.N., Tiwari, A.K.: Optimal selection of wavelet basis function applied to ECG signal denoising. Digit. Signal Process. **16**, 275–287 (2006)
14. Dorlas, J.C., Nijboer, J.A.: Photo-electric plethysmography as a monitoring device in anaesthesia. Application and interpretation. Br. J. Anaesth. **57**, 524–530 (1985)
15. Chua, E.C., Redmond, S.J., Mcdarby, G., Heneghan, C.: Towards using photo-plethysmogram amplitude to measure blood pressure during sleep. Ann. Biomed. Eng. **38**, 945–954 (2010)
16. Wang, L., Pickwell-Macpherson, E., Liang, Y.P., Zhang, Y.T.: Noninvasive cardiac output estimation using a novel photoplethysmogram index. In: International Conference of the IEEE Engineering in Medicine & Biology Society, pp. 1746–1749
17. Takazawa, K., Tanaka, N., Fujita, M., Matsuoka, O., Saiki, T., Aikawa, M., Tamura, S., Ibukiyama, C.: Assesment of vasoactive agents and vascular aging by the second derivative of photoplethysmogram waveform. Hypertension **32**, 365–370 (1998)
18. Millasseau, S.C., Kelly, R.P., Ritter, J.M., Chowienczyk, P.J.: Determination of age-related increases in large artery stiffness by digital pulse contour analysis. Clin. Sci. **103**, 371 (2002)
19. Yang, H., Zhou, Q., Xiao, J.: Relationship between vascular elasticity and human pulse waveform based on FFT analysis of pulse waveform with different age. In: International Conference on Bioinformatics and Biomedical Engineering, pp. 1–4
20. Mahmoodabadi, S.Z., Ahmadian, A., Abolhasani, M.D., Eslami, M.: ECG feature extraction based on multiresolution wavelet transform. In: IEEE Engineering in Medicine & Biology Conference, pp. 3902–3905
21. Antonelli, L., Ohley, W., Khamlach, R.: Dicrotic notch detection using wavelet transform analysis. In: Engineering in Medicine and Biology Society, 1994. Engineering Advances: New Opportunities for Biomedical Engineers. Proceedings of the International Conference of the IEEE, vol. 1212, pp. 1216–1217
22. He, R., Huang, Z.P., Ji, L.Y., Wu, J.K.: Beat-to-beat ambulatory blood pressure estimation based on random forest. In: IEEE International Conference on Wearable and Implantable Body Sensor Networks, pp. 194–198
23. Chang, C.C., Lin, C.J.: LIBSVM: A library for Support Vector Machines. ACM (2011)
24. Peng, L.I., Liu, M., Zhang, X., Xiaohui, H.U., Pang, B., Yao, Z., Nhen, H.: Novel wavelet neural network algorithm for continuous and noninvasive dynamic estimation of blood pressure from photoplethysmography. Sci. China Inf. Sci. **59**, 1–10 (2016)
25. Goldberg, D.E.: Genetic Algorithm in Search, Optimization, and Machine Learning, vol. xiii, pp. 2104–2116 (1989)
26. O'brien, E., Petrie, J., Littler, W., Padfield, P.L., O'Malley, K., Jamieson, M., Altman, D., Bland, M., Atkins, N.: The British Hypertension Society protocol for the evaluation of automated and semi-automated blood pressure measuring devices with special reference to ambulatory systems. J. Hypertens. **8**, 607–619 (1990)
27. White, W.B., Berson, A.S., Robbins, C., Jamieson, M.J., Prisant, L.M., Roccella, E., Sheps, S.G.: National standard for measurement of resting and ambulatory blood pressures with automated sphygmomanometers. Hypertension **21**, 504 (1993)

The Prospect and Analysis of Nanogenerator for Wearable Devices

Jing-Yan Yu and Li Liu

Abstract With the emergence of wearable devices such as Google glasses, I Watch, Active trackers and etc., people are becoming increasingly interested in wearable device and an increasing number of electronics and functionalized components are being applied to wearable devices. Even though wearable devices do not consume too much energy, the requirement for sustainability still brings a big challenge for power supplying. In fact, the battery life of most wearable devices cannot satisfy the users' demand. On the other hand, because of the devices' characteristic of being "wearable", there is much higher demand for safety, life span, pollution and comfort in power supply components. Therefore, how to provide continuous and environmentally-friendly power supply for wearable devices becomes a popular research topic. In 2012, when the Triboelectric Nano generator (TENG) was first introduced, it drew everyone's attention because of its advantages of lightweight, safety, cleanness and sustainability. The possibility of using Nano generator technology as a self-driven power supply for wearable devices will be discussed in this paper.

Keywords Wearable device · Nanogenerator · Energy conversion · Self-energized

1 Nanogenerator

In today's fast developing informationalized society, the business models and technology featured with movement, timeliness and intelligence are subtly driving. And then new challenges are presented on the global energy system, the mobile energy characterized with mobile, all-weather, highly efficient supply model will define a

J.-Y. Yu (✉)
Jiangxi Normal University, Nanchang 330022, China
e-mail: 13161159355@163.com

L. Liu (✉)
Beijing Institute of Fashion Technology, Beijing 100029, China
e-mail: fzyll@bift.edu.cn

G. Fortino and Z. Wang (eds.), *Advances in Body Area Networks I*,
Internet of Things, https://doi.org/10.1007/978-3-030-02819-0_7

new chapter to energy demand [1]. Many types of mobile and distributed generation technologies are rising in the modern world, and one of them is the nanogenerator (NG) based on piezoelectricity, pyroelectricity, triboelectricity and electrostatic induction [2].

The nanogenerator is a technology device which can convert mechanical energy and thermal energy into electrical energy via small physical changes. There are three typical technical paths for nanogenerator: Piezoelectric, triboelectric, and pyroelectric, in which the piezoelectric nanogenerator and triboelectric nanogenerator (TENG) can convert mechanical energy into electrical energy, and pyroelectric nanogenerator (PENG) is the converting center in which the collected thermal energy during temperature fluctuations over time would be converted into electrical energy. Piezoelectric technology and triboelectric technology are commonly used in wearable equipment technological [3].

The operating principle of the nanogenerator is to use a unique coupling voltage and Zinc Oxide nanostructure, so that the array of Zinc Oxide nanowires would be vertically assembled on the top substrate to collect current. In the "Z-shaped" electrode made of silicon material, thousands of nanowires are arranged, which extend freely and generate electrical currents that drive external mechanical vibrations to continuously generate current.

2 Wearable Device

Wearable device means a kind of portable intelligent equipment which could be worn on the body directly or easy to be worn on the apparel and accessories. In general, Wearable Device is electronic communication gears which can be embedded into clothing. Some of wearable gears even are designed to be accessories or other types of wearable items directly [4], as shown in Figs. 1, 2, 3 and 4. They play a critical role in helping users to extend their perception, monitoring users' physical state and improving users' work efficiency, at the same time wearable gears could also make users feel comfortable when people are wearing them.

In recent years, the wearable device has been one of the focuses in the Information Technology area. However, it is still unclear that what makes people accept new electric devices like wearable devices today.

2.1 The Development Prospects of Wearable Device

Analysis shows, macro environment is friendly to the boom of wearable devices in which users have strong interest. The oscillation of the share price of some typical companies shows us a competitive environment. And we can find it in a big start with no ensured technology, standard product and high entry barriers. All these show us a chance to be an early bird in this industry [5]. After mapping the industry chain, we

Fig. 1 I watch

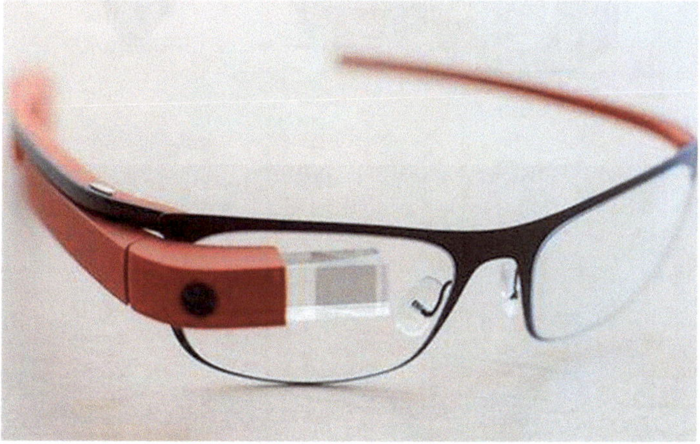

Fig. 2 Google glass

can find the companies bounded by components and service suppliers, and there's also a piece of good news that the new players in this industry coming from traditional industries may make a fast move. In one word, the longer chain, the more survivors [6].

So we can find it possible for wearable device industry to be a leading industry, and the independent device will play a role in the future. As long as the wearable device industry can promote each other with the internet of things?, the wearable device may finally become everything-wearable device and the data coming from wearable device will be most valuable at last [7].

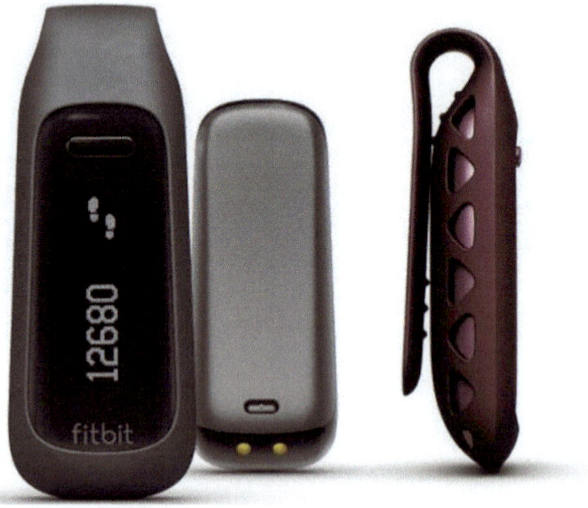

Fig. 3 Fitbit

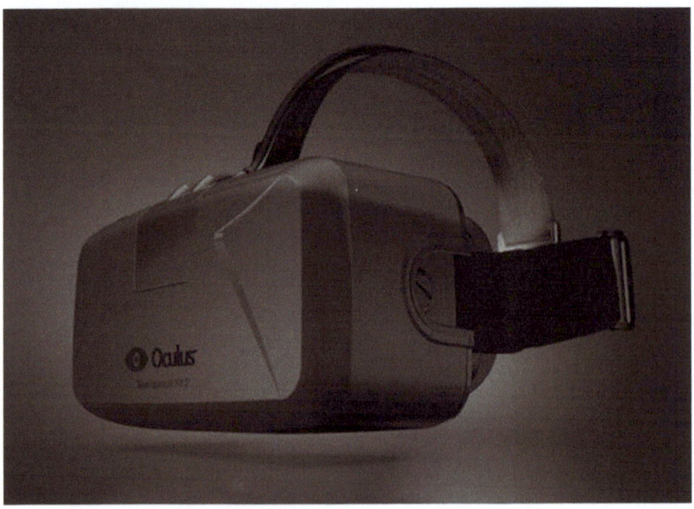

Fig. 4 Oculus rift

2.2 Development Status and Classification

In recent years, the development of wearable intelligent devices have been shifting to commercial application from academic research driven by multiple factors such as industry trends, technology and user demand, as shown in Figs. 5 and 6.

From technical view, multiple wearable gear technologies have become mature gradually. Components, operating system and the corresponding development tool

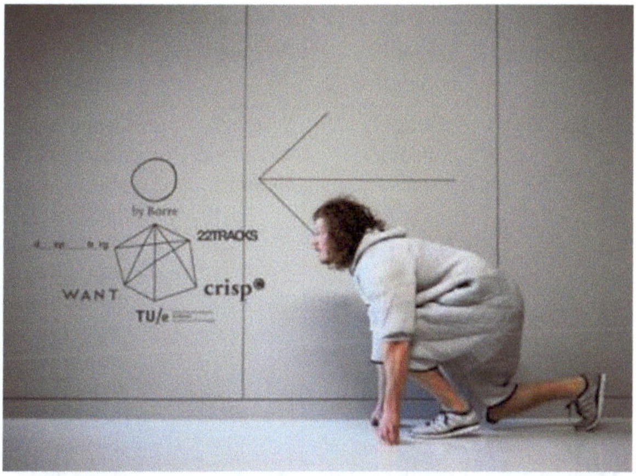

Fig. 5 BB Suit Smart Piece of clothing

Fig. 6 Gasport's Sports data vest

platform have been rapidly developing. From the market point of view, some of trans-boundary internet giants get into wearable devices field and have become the main enablers for market development. Nowadays, most wearable terminals are auxiliary mobile equipment [8]. The intelligent hand ring, smart watches and smart glasses are the most common wear terminals about sports and health, attracting huge concern, as shown in Tables 1 and 2. These phenomena show that consumers are no longer satisfied with the function of telephone and video for intelligent terminals but already increase their recognition and acceptance for other extended function of wearable devices.

Table 1 Wearable device type (by function)

Category	Target population	Product form	Interactive mode	Solve the problem	On behalf of the product
Fitness class	Young consumer groups	Watches, wristbands, shoes	Graphical interface, multi channel intelligent human-computer interaction, collect information and data through sensors	Collect motion information to achieve better exercise results	Nike+ training; Fitbit force
Medical safety category	For the public consumption, infants and young people as an important target population	Watch, wristband	Graphical interface, multi channel intelligent human-computer interaction, collect information and data through sensors	The collection of various health indicators, comparative analysis; dangerous alarm	360 children's hand ring; Garmin Vivo Fit; Samsung Gear Fit; Basis Carbon Steel
Information consultation	For mass consumption	Glasses, watches, wristbands, clothing	It is based on natural speech interaction and somatosensory interaction	Convenient, timely access to information; entertainment	Mote 360; Google Glass; MYO

3 Existing Problems

HIS Technology points out that the scales of battery markets around the world are expected to increase more than 10 times in a short span of four years driven by the development of wearable devices in its market research report. By 2018, the battery revenue of wearable devices is expected to reach $77 million [9].

Most wearable devices are using lithium polymer batteries now. The main reason is that this kind of battery is light and flexible for size and shape. According to survey, the income of this kind of battery has accounted for 73% of the whole earning for wearable gear battery.

However, battery is still a big bottleneck for the development of wearable devices field from the current market feedback. The main issue users complain about is the limited battery life, since the general situation of battery is that it needs recharging

Table 2 Wearable equipment type (by product type)

Wearable terminal		Features	On behalf of the product
Smart watch	Independent terminal class	Independent cloud interaction, independent call, precise positioning, data services	Omate True smart, GEAk Watch
	Mobile assistant class	Sports health, information viewing, event reminders, telephone answering	Samsung Galaxy Gear2, Pebble, LG G Watch
Smart bracelet		Fitness step, sleep monitoring, somatosensory control, vibration wake up, heart rate measurement, call reminder	MYO Wristbands, Jawbone UP, Fitbit Flex, Fuel band
Smart glasses		GPS navigation, information consultation, real-time camera, virtual reality, infrared medicine, synchronous dual-screen	Google Glass, Oculus Rift, DoCoMo Smart glasses
Smart belt		Sitting position monitoring, vibration warning	Lumoback
Smart clothing		Mobile Wi-Fi hotspots, music player	BB suit smart piece of clothing
Smart shoes and socks		Motion monitoring, map navigation, action guidance	No place like home, smart socks

every several hours when devices are used continuously, which is extremely inconvenient for users and exerts negative impact on user experience. In addition, the battery will bring serious pollution to the environment and huge consumption of material resources, which does not conform to the concept of sustainable development.

Based on the problems mentioned above, the suggestion is that integrating wearable energy-harvesting and energy-storing devices into a self-sustainable or autarky client power system [10].

Google Glass costs $1,500, and the pricing of Jawbone Can ring and Samsung Galaxy gear also reached the four-digit RMB. Such price could not do customers a favor but make ordinary public consumers discouraged. If you want to expand on a large scale exhibition market, to build civilian applications, we must reduce costs and achieve mass production. We can consider the creation of multi-level products to achieve differentiated pricing, thus not only for people with high incomes, high-profile and high-priced products for electronic products lovers, but also specifically designed products for the civilian population to create a more friendly product price, thus fully digging out consumption potential of different user groups to meet the needs of different groups of consumers [11].

4 Application of Nano-generator in Wearable Equipment

The basic principle for wearable equipment of nanogenerators is the conversion of bio mechanical energy (Such as human movement, muscle stretching, blood pressure) into electricity, which is concentrated in the flexible energy storage originals to supply the electronic devices worn on the human body, as shown in Fig. 7.

And based on Piezoelectricity, Pyroelectricity, Triboelectricity and electrostatic induction can collect the energy from the environment, and then drive the wireless mobile electronic device by the energy conversion to achieve energy supply. This mechanical energy is characterized by strong persistence and wide range of use, as shown in Fig. 8, including walking, running, typing, touching the phone screen and even blink, breathing and other activities can become a source of energy. The ideal energy source for wearing a product is the body movement energy. Starner [12] reported the energy consumed by human activities in daily life, as shown in Table 3.

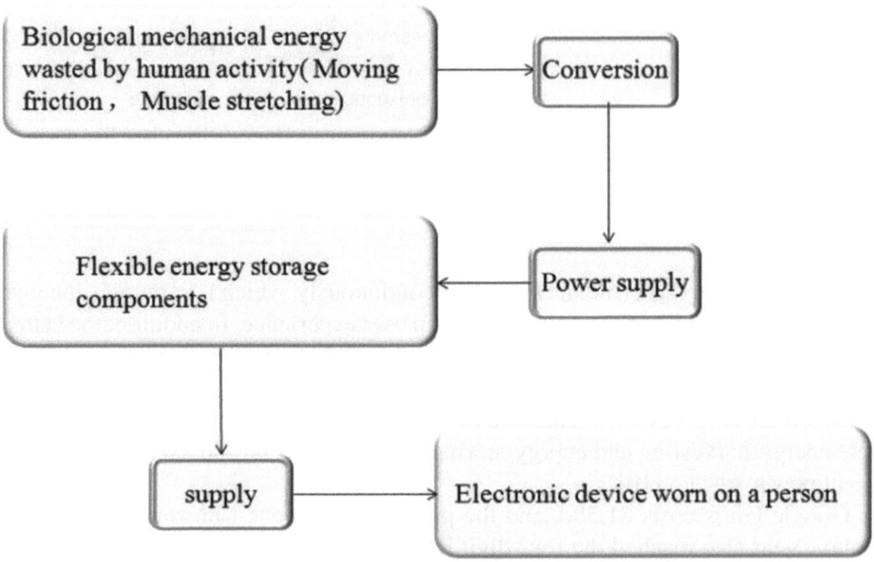

Fig. 7 The structure of nano-generator used in wearable equipment

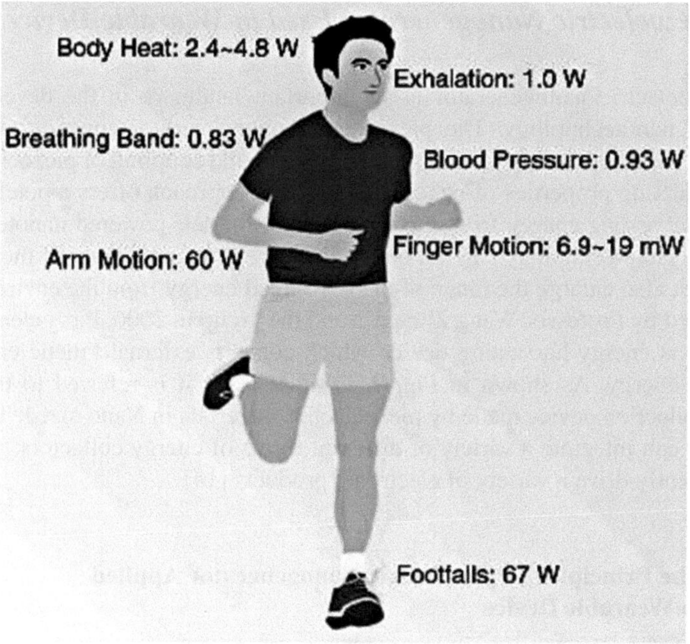

Fig. 8 The energy released by the daily

Table 3 Some of the energy consumed by human activities

Behavior	Power (W)
Sit	116
Stand up	128
Take a walk	163
Walk	407
Jogging	1048
Run away	1630
Climbing	698
Swim	582
Drive	163

The physical pathways for the conversion from small-scale mechanical energy into electrical energy supply and wearable devices mainly include piezoelectric effects and triboelectric effects.

4.1 Piezoelectric Nanogenerator Used in Wearable Devices

The piezoelectric nanogenerator is an important landmark in the developmental period of nanotechnology. The piezoelectric nanogenerator converts mechanical energy to electrical energy in nanoscale, relying on the coupling of piezoelectric and semiconducting properties of nanomaterials. This approach offers a potential solution for harvesting energy from the environment for self-powered nanotechnology [13]. This nanotechnology can not only reduce the size and weight of the electrical source, but also enlarge the range of the scavenged energy from the environment.

Invented by Professor Wang Zhonglin and the group in 2006, Piezoelectric nanogenerator is energy harvesting device which converts external kinetic energy into electrical energy. As shown in Fig. 9, in most cases it is referred to the kinetic energy collection device made by piezoelectric materials in Nano-sized. The Nanogenerator can integrate a variety of different forms of energy collectors, which can independently drive a variety of electronic products [14].

4.1.1 The Principle of Piezoelectric Nanogenerator Applied to Wearable Device

Piezoelectric Nano-generators are mainly used to device integration or fabric weaving with piezoelectric fibers. Piezoelectric fibers are used to convert the kinetic energy of the human body into electrical energy, thus being applied to the electronic device for driving the wearable device. Just like as a health management system with an intelligent wearable system, Piezoelectric Nano-generator is also easily integrated into shoes, collecting energy from people walking which would be converted into

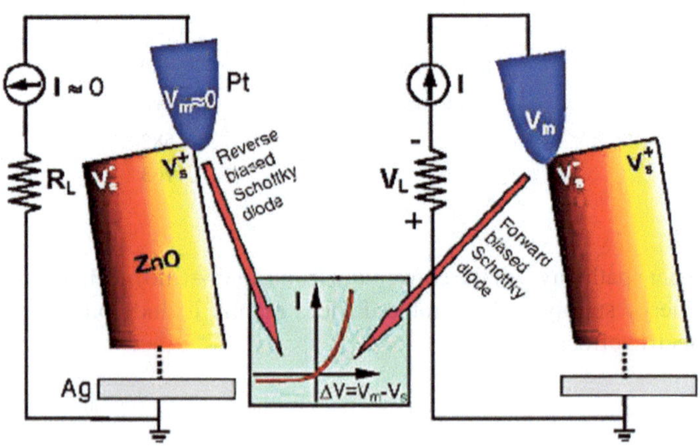

Fig. 9 The working principle of piezoelectric nano-generator when its external force is perpendicular to its axis

electricity to be used. But there are a lot of problems in the energy collection system of piezoelectric Nano-generator.

4.1.2 The Existing Problems of the Piezoelectric Nanogenerator in Wearable Applications

(1) Although the use of piezoelectric principles of mechanical energy into electrical energy research has been more in-depth, the power of piezoelectric nanogenerator is generally at the level of μW, nW, so the efferent energy is not enough to drive wearable equipment [15].
(2) The type of material with piezoelectric properties is relatively limited, resulting in its narrow range of applications; it cannot be used on a large scale in wearable equipment. The Triboelectric Nano-generator can solve this problem well.

4.2 Triboelectric Nanogenerator Used in Wearable Devices

The triboelectric nanogenerator (TENG) is a newly developed energy-harvesting technology that converts mechanical energy into electric power with a coupled effect of contact-electrification and electrostatic induction. This new type of nanogenerator was invented by Professor Dr. Zhong Lin Wang of Georgia Institute of Technology in the United States for the first time; it was proved to be a simple, low-cost, stable structure, and efficient way to collect mechanical energy. As shown in Fig. 10, the triboelectric nanogenerator (TENG), based on the coupling between tribe electrification and electrostatic induction, can convert mechanical energy into electrical energy which will provide a new way for the harvest and conversion of human activity. With liquid and air flow existing in the natural environment, it has a huge potential for application value in the field of new energy technology and self-powered active sensors [16].

It is critical and essentially necessary to develop the TENG extensively because TENG technology is a promising energy harvesting approach which can be used not only in large-scale energy scavenging, but also in tiny-scale energy harvest. It is possible to address worldwide energy shortage and to promote personal electronics and self-powered systems by TENG [17].

The human body will produce a lot of mechanical energy in people's daily life. TENG can convert mechanical energy into electric power with the friction electrification and electrostatic induction principle. With this feature, People can turn electrical energy by walking, typing, breathing, blinking and heartbeat into mechanical energy, continuously supplying stable power for wearable devices.

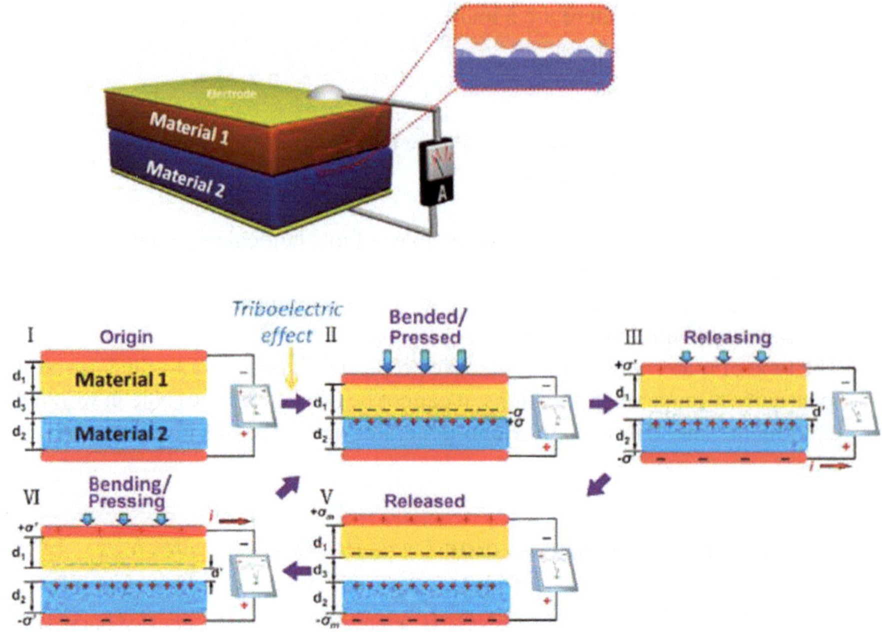

Fig. 10 The working principle of triboelectric nano-generator

4.2.1 The Advantages of Triboelectric Nanogenerator

(1) In a variety of energy harvesting devices, the ability of TENG to harvest human motion with high efficiency and large output power density has been proven. It has the features of high output voltage, environmental protection and safety [18]. (2) Compared with photoelectric and thermoelectric energy harvesting, the mechanical energy is almost independent of the weather and the working environment. (3) Because of its simple structure and large range of material selection, the entire device manufacturing process does not require expensive raw materials and advanced manufacturing equipment, which is conducive to large-scale industrial production. (4) The device is using flexible polymer film thin polymer as basic material. The device can be used for a long time since it is easy to process and easy to combine with other processing techniques [19]. So it is facile to be designed into textile cloths and to convert common cloth fabric into energy-storing devices will be better, thus reinforcing the combination with wearable products.

4.2.2 The Status Application of Triboelectric Nanogenerator to Wearable Device

(1) Recently, Professor Dr. Zhong Lin Wang led the research team to develop the nano-working mode, using vertical contact-separation mode composed of the PDMS film and aluminum foil to constitute a friction layer for triboelectric nanogenerator, which can be fixed in the clothes. As shown in Fig. 12, the triboelectric nanogenerator can be converted into a sustainable power supply by collecting the vibrational energy and the kinetic energy of the human bod, which provides the basis for the self-powered power supply mode of wearable equipment [20].

(2) Chinese Academy of Sciences Institute of Nano-Research in the recent research has reported paper or textile batteries with the utilization of graphene, carbon nanotubes (CNT), or carbon Nano fiber cloth as the flexible substrate [21]. As shown in Fig. 11, TENG cloth can be worn in many different regions of the body (for example, at the foot, under the arm and at the joint), and a self-charging power unit for smart electronics is developed to verify the viability of the whole wearable and self-charging power unit for future wearable smart electronics.

(3) A triboelectric nanogenerator (TENG) based on the contact-separation mode between a patterned polydimethylsiloxane (PDMS) film and an Al foil was fabricated between clothes for harvesting body motion energy. Under the general walking, the maximum output of voltage and current density are up to 17 V and 0.02 $\mu A/cm^2$ respectively. As shown in Fig. 12, the TENG with a single layer size of 2 cm × 7 cm × 0.08 cm sticking on the clothes was demon stated as a sustainable power source that not only can directly light up 30 light-emitting diodes(LEDs), but also can charge a lithium ion battery by persistently clapping clothes [22]. The detection of bioactive chemicals in our body using the energy harvested from body motion is demonstrated.

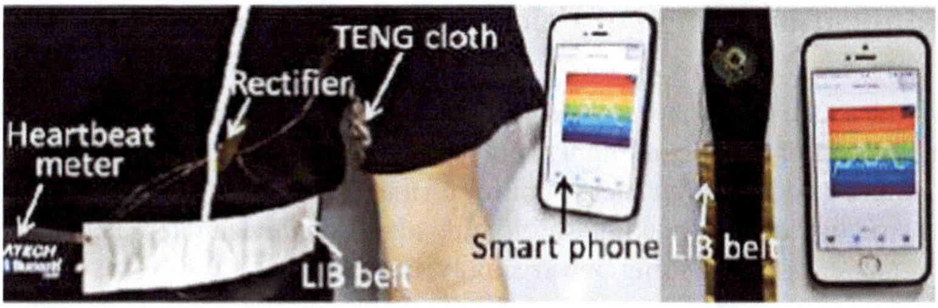

Fig. 11 Optical image of integrated self-charging power system

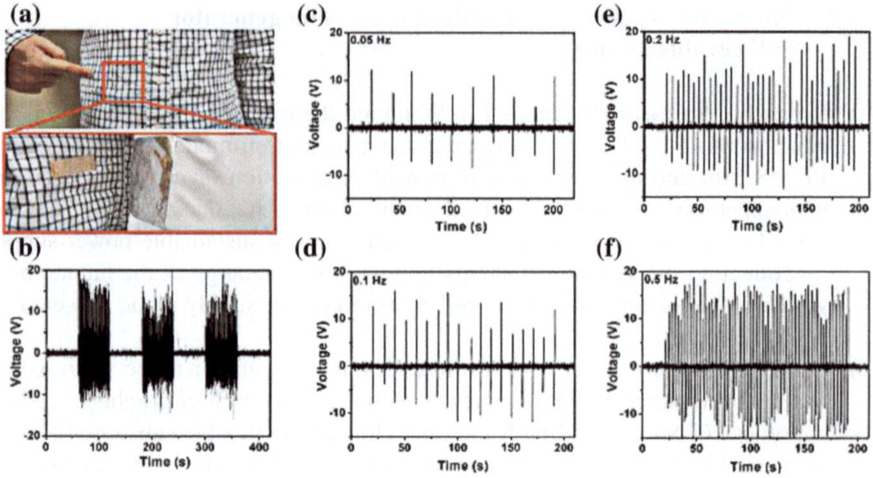

Fig. 12 **a** Photograph of a working TENG (2 cm × 7 cm × 0.08 cm in size) stuck on to the clothes with the shirt buttoned. **b** The output voltage of the TENG when walking. **c–f** The output voltage signals at different stride frequency, 0.05 Hz (**c**), 0.1 Hz (**d**), 0.2 Hz (**e**) and 0.5 Hz (**f**)

4.2.3 Current Problems of Triboelectric Nanogenerator in Wearable Applications

(1) There are many problems such as low energy conversion efficiency, high output power, and unstable pulse electric signal in TENG. (2) Considering the requirements of wear ability, the lightweight, high flexibility, stretchable requirements and washable power supply for power generation devices are key challenges.

5 Application Design of Friction Nanogenerator

According to the application principle of nanofriction generator, I put forward a design concept of intelligent night running clothes. The basic design principle is to make a human body motion energy collection and the storage into a one-dimensional fibrous structure device, making the clothing and nano device integrated into one, in order to achieve self-charging power supply system.

5.1 Design Scheme

The design method of the night running clothes is to use the friction plate to convert the body movement energy into power, collecting them in the flexible lithium-ion

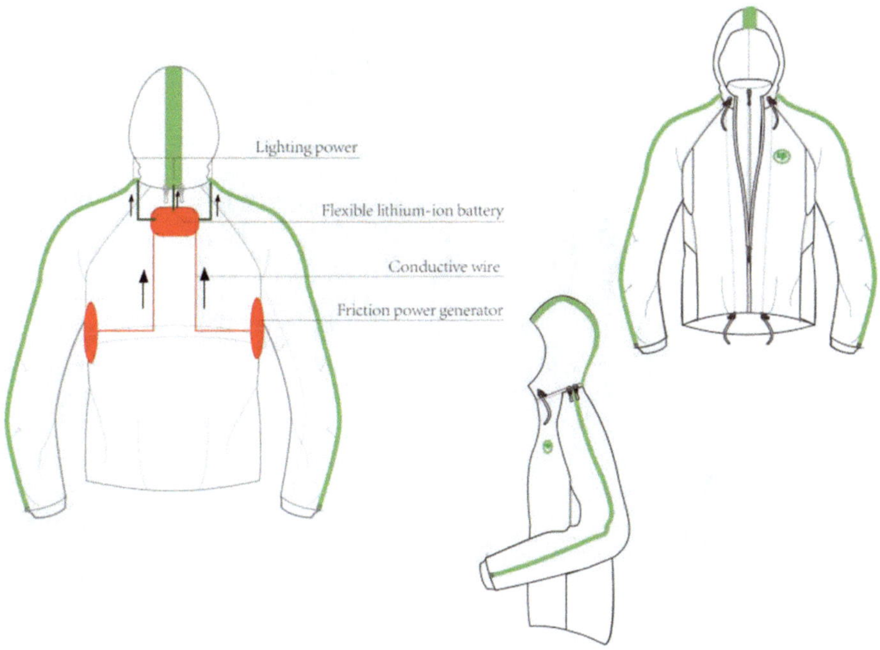

Fig. 13 Circuit design

battery, and transmitting power through the integrated conductive wire throughout the garment.

Because there is more friction between the upper arm and the armpit when human body is running, the friction plates are placed at two sides of the garment. The lights are placed in the middle of the hat and the outside of the sleeves, the flexible lithium-ion battery is at the back, then distribute the integrated conductive wire in the clothing lining, connecting the friction plates, flexible batteries and light power into a fabric-based wearable self-charging supply system. As shown in Fig. 13.

5.2 Design Renderings

The night running clothes is designed into double-placket look with drawstring on the hem to ensure the windproof effect when running. Using raglan sleeves so we can ensure that the upper arm has sufficient space for activity. The back of the garment is made of windproof fabric, separated from the lining which is made of mesh fabric in order to ensure the breathability of the back when exercising. The optical fiber ribbons located on the sleeve and the hat are assembled by zippers, which can be disassembled easily. The clothing effect diagram is shown in Fig. 14.

Fig. 14 Effect drawing

6 Conclusion

In order to effectively solve the problem of power supply for wearable devices, we can get various forms of waste energy from the surrounding environment. After years of development, researchers have more choices of materials, shape, structural design and the use of mixed devices into the study, therefore the output power and output stability of TENG are greatly improved. In the near future there is hope that it will be applied to the sustainable energy supply of wearable devices to achieve industrialization, greatly facilitating our daily life.

In this paper, by the comparison of piezoelectric nanogenerator and triboelectric nanogenerator, it is found that the fabric can be directly combined with the triboelectric nanogenerator for energy collection, and flexible power management circuit will be designed to realize a whole-flexible wearable self-charging power textile. Self-powered energy based on triboelectric nanogenerator will be a new direction for solving the problem of wearable equipment.

Acknowledgements This work was partially supported by the Beijing Municipal Education Commission Project KM201810012001.

References

1. Guo, Y., Zhang, Q., Li Y., Wang, H.: Progress of the research on wearable triboelectric nanogenerator. Prog. Chin. Mater. (02), 91–100+127 (2016)
2. He, X.: The Research of Flexible Triboelectricity Nanogenerator Based on PDMS Composite Film. Chongqing University (2015)
3. Wang, Z.L., Wu, W.: Nano-enabled energy harvesting for self-powered micro/nano-systems. Angew. Chem. Int. Ed. **51**, 11700–11721 (2012)
4. Huangwei Research on the Prospect and Investment of Wearable Device. Lanzhou University (2015)
5. Dong, F.L.: Research on the Industry Chain and Trend in Development of Industry of Wearable Device. Guangdong Academy of Social Sciences (2015)
6. Shuntian, F.: The present situation and trend of wearable equipment. Inf. Commun. Technol. **03**, 52–57 (2014)
7. Yi, G., Hui, A., Yang, H., Hua, J.: The present situation and prospect of wearable equipment. Electron. Sci. Technol. (02), 238–245 (2014)
8. Wikipedia [EB/OL] [2014-04-02]. http://zh.wikipedia.org/wiki/Interactiveparadigm
9. Feng, S.T.: The present situation and trend of wearable equipment. Inf. Commun. Technol. **03**, 52–57 (2014)
10. Business Insurance: Wearable Devices' Risks Emerge **48**(19) (2014)
11. Song, B.L.: Overview of the current situation and future development of wearable equipment. Silicon Valley **08**, 9–10 (2014)
12. Starner, T.: Human-powered wearable computing. IBM Syst. J. **35** (3/4), 618–629 (1996)
13. (a) Wang, Z.L.: Adv. Mater. **24**, 4632 (2012). (b) Wang, Z.L.: ACS Nano **7**, 9533 (2013)
14. Pu, X., Li, L., Song, H., Du, C., Zhao, Z., Jiang, C., Cao, G., Hu, W., Wang, Z.L.: A self-charging power unit by integration of a textile triboelectric nanogenerator and a flexible lithium-ion battery for wearable electronics. Adv. Mater. **27**(15), 2472–2478 (2015)
15. Wang, Z.L.: The principle and potential application of piezoelectric nano-generator. Physical (11), 897–903 (2006)
16. (a) Wang, Z.L.: The principle and potential application of piezoelectric nano-generator. (b) Guo, Y.B., Zhang, Q.H., Li, Y.G., Wang, H.Z.: Progress of the research on wearable triboelectric nanogenerator. Mater. China (02), 91–100+127 (2016)
17. Zheng, Z., Shao: Study on One-Dimensional ZnO Nanostructures Based Piezoelectric Nanogenerator. National University of Defense Science and Technology (2010)
18. Zang, Y., Zhang, F., Di, C., et al.: Mater. Horiz. **2**(2), 140–156 (2015)
19. Yang, Y., Wang, L.Y.: Composite electromagnetic—triboelectric nanogenerator. Chin. Sci. Bull. **12**, 1268–1277 (2016)
20. Xu, S., Qin, Y., Xu, C., Wei, Y.G., Yang, R.S., Wang, Z.L.: Self-powered nanowire devices. Nat. Nanotech. **5**, 366–373 (2010)
21. Jing, S.: Fabrication and performance of triboelectric generator based on biocompatible materials. Tianjin University of Technology (2015)
22. Wu, J., Xiao, D., Zhu, J.: Chem. Rev. **115**(7), 2559–2595 (2015)

References

[illegible]

EEG-Based Driver Distraction Detection via Game-Theoretic-Based Channel Selection

Mojtaba Taherisadr and Omid Dehzangi

Abstract In recent years, there has been much effort to estimate drivers' state with the goal of improving their driving behavior and preventing vehicle crashes in the first place. Physiological based detection has shown to be the most direct method of measuring driver state among which, electroencephalogram (EEG) is the most comprehensive method. However, EEG-based driver state detection faces the challenge of computational complexity of data mining algorithms given high density and resolution of EEG signals recorded from multiple channels. On the other hand, in order to early detection and prevention of driver critical states real-time responsiveness of the monitoring system is necessary. This challenges can be tackled by localizing the regional impact by selecting a small subset of coherent channels and reducing the processing load on all channels. In this paper, we present and investigate a Game-Theoretic-Based approach for EEG channel selection, in order to localize the most efficient sub-set of channels in addition to maximizing the driver distraction detection accuracy. In this way, we apply game theory based channel selection algorithm based on the utility measure, Shapley value, in exact to estimate overall usefulness of each EEG channel. We then consider the combination of channels and evaluate their performance. Empirical comparison of best combination of channels, best ordered channel based on Shapley value with another existing feature selection method shows that the sub-set of channels leads to the best detection performance in terms of accuracy (90.12% accuracy).

M. Taherisadr
University of Michigan, Dearborn, MI 48128, USA
e-mail: mojtabat@umich.edu

O. Dehzangi (✉)
Rockefeller Neuroscience Institute, West Virginia University,
Morgantown, WV 26505, USA
e-mail: omid.dehzangi@hsc.wvu.edu
URL: http://www.wssplab.net

G. Fortino and Z. Wang (eds.), *Advances in Body Area Networks I*,
Internet of Things, https://doi.org/10.1007/978-3-030-02819-0_8

Keywords EEG · Driver distraction · Game Theory · Channel selection Shapley value

1 Introduction

Wearable systems for sensing human physiological signals are becoming more available and reliable. This has improved our ability to record different physiological signals [1]. For instance, wearable *electroencephalogram (EEG)* is a noninvasive method for measuring brainwaves that has been broadly adopted for applications ranging from the diagnosis of epilepsy and sleep disorder to the monitoring of brain computer interfaces (BCIs) and human physiological state. In particular, there has been considerable growth in the use of wearable EEG to design and develop predictive solutions for applications involving human state monitoring [2, 3].

One instance, which is the basis of this paper, is *monitoring the state of human drivers*, or *driver-state monitoring* for short. Advances in this particular area have significant broad real-world implications. According to the *National Highway Traffic Safety Administration (NHTSA)*, there were at least 5.6 million vehicle crashes in the U.S.A. during the year 2015 alone [4]. According the National Safety Council (NCL), there were 6% increase in motor vehicle traffic deaths in 2016 reaching an estimated 40,200 deaths compared to 37,757 deaths the previous year. 94% of the crashes were estimated to be related to human choices/errors [5]. Therefore, there has been much effort recently to estimate driver state with the goal of improving human behavior while driving and preventing vehicle crashes in the first place.

Studies involving driver-state monitoring employ visual and non-visual characteristics of driver behavior, as well as vehicle-based features related to driving performance. Visual-based systems consider features such as eye-related measurements, yawning detection, facial expression, head pose, and gaze direction. Vehicle-based features include steering-wheel movement, standard deviation of lateral position, and car dynamics. Non-visual features or physiological signals, such as brain activities, have been found to be more useful in predicting driver states because they lead to fewer false positives compared to visual and vehicle-based features [6]. Determination of some states from visual or vehicle-based features is only after the driver is already well on the way to distraction. In other words, the prediction of driver state based on physiological signals makes it possible to warn a distracted driver in a more timely manner. Drivers' privacy is another related issue with visual features because drivers may be uncomfortable with a video monitoring of their activities.

Physiological signals may eventually provide a way around the problems and issues with visual and vehicle-based just described, and they can monitor state of the driver directly [7]. Hence, developing an effective wearable-channel or seat embedded system for detecting different driver states efficiently based on physiological signals in real-life driving environments is an important research direction. A large number of studies have been conducted on using physiological signals as a means of monitoring driver state [8–11]. EEG is one mechanism by which we can collect such type of physiological signals [12–15]. Despite the cumbersome nature of existing EEG technology, designing and studying EEG-based detection systems is viewed as an important step toward achieving the goal of understanding and detection of drivers' cognitive and physiological state [8].

However, EEG-based state detection faces the challenge of computational complexity and real-time responsiveness of data mining algorithms given high density and resolution of EEG signals recorded from multiple channels. The predictive models is also prone to overfitting caused by utilization of unnecessary channels. These challenges could be tackled by localizing the regional impact and finding small subset of coherent channels. As the number of channels increase, the number of possible combinations exponentially grows in order to find the global best sub-set of channels.

Different algorithms have been employed to find sets of coherent channels. In most channel selection algorithms, methods such as time-series analysis and prediction, power spectral estimation and common spatial pattern have been employed for effective EEG feature extraction, and hence indirectly for channel selection [16]. Due to heavy computational nature of such algorithms, selecting subset of relevant channels can dramatically improve the information transfer rate. In the literature, channel selection methods are mainly characterized as *wrapper* or *filter* approaches [17].

In wrapper approaches, channel selection is coupled with a particular *machine-learning (ML)* framework, model, and method; e.g., a *support-vector-machine (SVM)* classifiers [17]. The performance of wrapper approaches depends on the accuracy of the applied predictive model and the feasibility of examining the performance of more channel combinations. Although some methods have been proposed to avoid retraining classifiers [18], wrapper approaches generally involve intensive computations.

In filter approaches, channel selection is independent of the ML model or method. One example is to perform channel selection based on certain criteria between channels and class labels; e.g., *mutual-information (MI)* [19]. Filter approaches are much faster. However, they suffer more from local minima [18].

As our main contribution in this paper, we propose the application and empirically illustrate the effectiveness of a coalition-based method from cooperative game theory to specifically address the aforementioned problem of subset EEG-channel selection. We design a channel selection method based on game-theoretic approach as a way to account for the relevance between all potentially effective combinations of the channels. Our aim is to improve the performance of the detection system based on the selected subset of channels, toward reliable detection of driver distraction. Our proposed approach enables the channel selection algorithm to recognize the channels that, despite their *individually* weak contribution to the discriminative power of a final learned classifier, *globally* have a considerable impact on discrimination power when *grouped* with other channels.

In particular, we use a well-known solution concept from cooperative game theory called the *Shapley value* [20], as the basis for ranking the importance of each channel for the task at hand. By its very nature, the Shapley value takes directly into account interactions between channels.

1.1 Coalition Game Theory

In coalition games, groups of channels cooperate with each other by making different sub-sets of channels named coalitions. These coalitions are defined based on different situations that channels could form a sub-set and how the overall shared payoff is distributed between the channels. Let $K = 16$ be the number of channels and $S = \{S_1, \ldots, S_K\}$ be the set of players. The approach of player j, S_j is defined by Z_j. H indicates a coalition set, $H \subseteq W$. Generally, for a K-channel coalition game, 2^K combinations of coalitions with any size exists. For 16 channels, there is 65,536 possible coalitions. The coalition that indicates empty is presented by ϕ, and the coalition of all existed channels named grand coalition. $F(H)$ as a characteristic function provides the overall payoff that each member can gained through the coalition H. Characteristic of ϕ is zero, $F(\phi) = 0$. For two disjoint coalitions H_j and H_i, $(H_j, H_i \subseteq W)$ super-additivity would be a property for their characteristic function of their union, which, in turn, indicates $F(H_j \cup H_i) \geq F(H_j) + F(H_i)$.

A solution for a coalition game can be achieved by determining how channels can form coalition and how the overall payoff can be distributed among all 16 channels. Let vector $y = (y_i, \ldots, y_N)$, $y \in R^{16}$ indicates the payoff amount that each channel achieved in a coalition. Distributing payoff among the channels should be efficient enough such that $\sum_{j=1}^{16} y_j = F(W)$ and also this distribution should satisfies the condition of individual rationality, which means that each channel is being paid better in a coalition in comparison with being paid alone, $\forall_j \in W, y_j \geq F(W_j)$. Game theory also needs to indicate the amount of contribution for each channel to the game by establishing a value function. The value function allocates a value to each channel. The assigned value corresponds to the amount of contribution of the channel in acquiring a high payoff. Shapley value [20] is a solution to calculate the contribution of each channel in the coalition. Shapley value fairly and efficiently distributes the payoff among the channels based on their contributions. The Shapley value can be formulated as follows. Let the marginal importance of channel j to a coalition

$$\Gamma_j(H) = F(H \cup \{j\}) - F(H) \tag{1}$$

Therefore, Shapley value can be defined by the payoff equation as follows

$$\Delta_j(F) = \frac{1}{16!} \sum_{\mu \in \Psi} \Gamma_j(H_j(\mu)) \tag{2}$$

where Ψ is set which contains all 16! permutations over W, and $H_j(\mu)$ is a set of channels which appears before the jth channel in permutation μ. The Shapley value of a channel is an averaged weighted mean of its marginal value over all possible subsets of channels. Transforming these above-mentioned game theory concepts into the area of channel selection, in which we try to estimate the channels contribution in generating a predictive model, the channels E are mapped to the all channels and the payoff is represented by a characteristic function $F(H)$ based on *lineardiscriminantclassifier* that measures the performance of a predictive model generated by the sub-set of channels H.

Therefore, cooperative game theory concept can be used in channel selection algorithm. A coalition game can be defined by the set of channels and the characteristics function for every set $H \subseteq W$ as $(W;\ F(H))$. Using mentioned framework, we model the channels as game players, and the channels can be classified in different coalitions, while we can observe their individual impact on the predictive model and also their interdependency. Various possible set of the channels are evaluated to recognize the best classification. Amount of payoff for each coalition H, $F(H)$, measures the contribution for a each coalition of selected channels to the overall performance of the predictive model. If channel j joins a coalition H, it might increase the performance of this coalition. This effect is denoted as marginal importance.

2 Wearable Channel Platform and Experimental Setup

2.1 Data Acquisition

We have designed a wearable data platform comprising a synchronized multi-modal solution to acquire physiological signals using a comprehensive wearable channel network. Our platform is capable of collecting a large amount of heterogeneous drivers biometric data such as EEG, *galvanic skin response (GSR)*, *electrocardiogram (ECG)* and motion during naturalistic driving session. Sixteen channels EEG with a sampling frequency of 256 Hz were recorded for each subjects.

The electrodes were referenced to both of the earlobes and their connection impedances were kept to less than 5 kΩ. The location of these electrodes was distributed over the frontal, central, and parietal areas of the head crown. The channels names and numbers are 1-Fpz, 2-AF3, 3-AF4, 4-F3, 5-Fz, 6-F4, 7-FC3, 8-FCz, 9-FC4, 10-C3, 11-Cz, 12-C4, 13-CPz, 14-P3, 15-Pz, 16-P4. We will use these channels names and numbers hereafter. Figure 1 illustrates the channels location, respectively. Table 1 provides a high level description of the EEG recording parameters.

The driving experiments were conducted in a naturalistic setting in which the subjects drove a real car (Ford Escape 2015). The data acquisition and experiment procedures in this study were approved by University of Michigan, Institutional Review Board. The EEG data were collected while introducing different potentially distracting elements. Our goal was to remove biases incurred during simulated driving experiments and generate more accurate insights and in-depth results to use in practical systems.

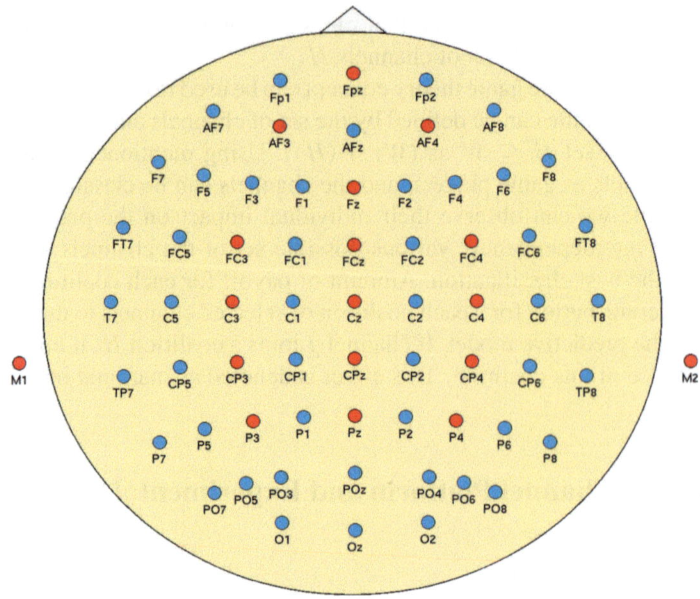

Fig. 1 EEG channels location and setup

Table 1 EEG recording parametrs

Amplifier	16 channel device (COGNIONICS, INC)
Sampling frequency	256 Hz
High-pass filter	0.5 Hz
Low-pass filter	50 Hz
Electrode placements	1020 systems (see Fig. 1)
Ground	Right-Left mastoid (M1, M2)
Electrode material	Ag/AgCl
Recording software	Our exclusive software

2.2 *Participants*

Ten subjects participated in our experiment with the conditions mentioned above. The subjects were legally allowed to drive. Their age range was 20–40. To make sure their performance was not affected by other factors unrelated to the experiment scenario, we asked them to avoid drinking alcoholic beverages and taking any medication that could make them feel drowsy or sleepy. All subjects were healthy with no mental or physical disabilities.

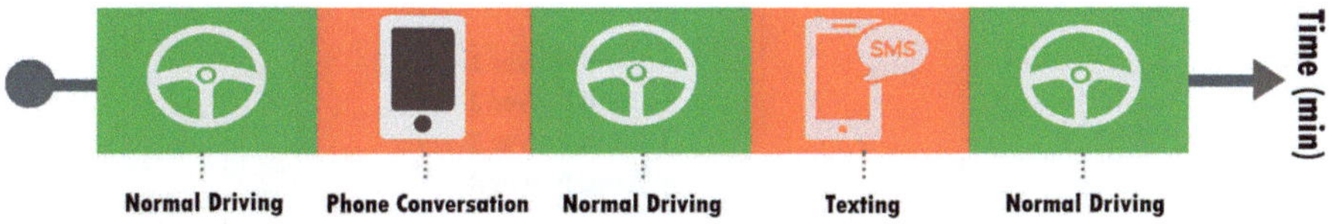

Fig. 2 Overview of the experiment. See main body for details

2.3 *Experimental Design/Conditions*

In this study, we aim to collect the EEG data from drivers intermittently involved with the following tasks:

(1) simultaneous talking on a hands-free cell phone.
(2) simultaneous texting on a cell phone.
(3) a baseline single-task condition (i.e., no simultaneous secondary task).

Each of the tasks has been designed to enables the driver to keep his or her eyes on the road and, with the exception of texting, both hands have been kept on the steering wheel, thus any wandering from driving must originate from cognitive sources associated with the diversion of attention from the task of operating the vehicle. All of the experiments were held on a straight route between 1 and 4 PM, which helps the subject experience lower traffic loads. First experimenter sitting on the passenger seat beside the subject and was responsible for handling the task conditions and also keeping the experiment safe. The other experimenter, sitting on the back right seat was responsible for recording and labeling the acquired data using software which we exclusively developed for this type of experiments.

Figure 2 provides an overview of this procedure. The order of the tasks is represented on the time axis. The experiment started from a baseline single-task condition (normal driving) as indicated by the green box.

2.4 *Data Pre-processing and Segmentation*

Due to the fact that EEG data always suffers from artifacts and noise [21], we first preprocessed acquired EEG data using high-pass and low-pass filters with cut-off frequencies of 0.5 Hz to 50 Hz to remove the line and high-frequency noise, respectively. Then, we segmented each resulting signals to generate a short-term time series and extract features. The application of driver monitoring systems depends crucially on a short response-time for practical driver feedback/alerting mechanisms. Therefore, the duration of the segments is critical. We considered segmentation size as 2 s window with 75% overlap based on our experience.

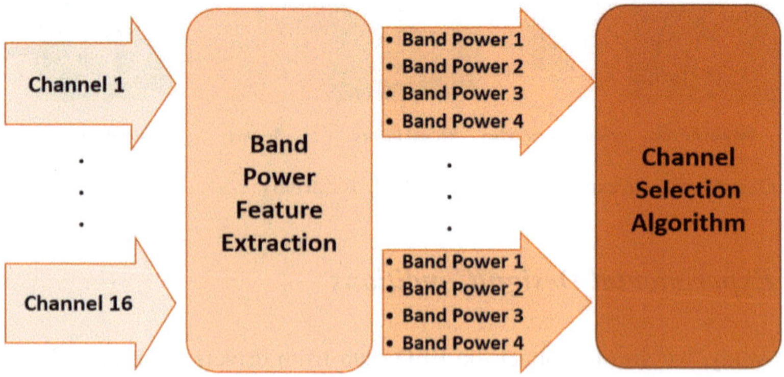

Fig. 3 Flowchart of band-power feature extraction. See main body for details

2.5 Band-Power Feature Extraction

Because of the high levels of noise in the time domain of the EEG signals, and the naturalistic experimental task of driving, we characterize each channel in the spectral domain using the power of the sub-bands of Alpha, Beta, Delta and Theta [22]. The power of a signal is the sum of the absolute squares of its time-domain samples divided by the signal length. *Band power* is the average power of the signal in the specific frequency range. We calculated band power for the four sub-bands of the EEG signal, $Delta$ [0.5–4 Hz], $Theta$ [4–8 Hz], $Alpha$ [8–16 Hz], and $Beta$ [16–31 Hz]. Then, we normalized them by the total power of the entire signal in each segment. We extracted the total of four band-power features from each segment of the signal. Then, instead of the raw data we considered each 4 sub-bands as a channel. Figure 3 shows the flowchart of the feature extraction process.

3 Results

In this section, we present the empirical analysis to evaluate the performance of the proposed game theory based channel selection method. The game-theoretic based channel selection method is performed on our empirically collected data set as described in Sect. 2. As stated previously, the impact of each channel on the classification performance is quantized by Shapely value through averaging over all coalitions up to size 16. In this study we used *linear discriminant classifier* as our hypothesis class in order to calculate Shapley values for 16 channels. Figure 4 shows the average Shapley values over the subject population for 16 channels for

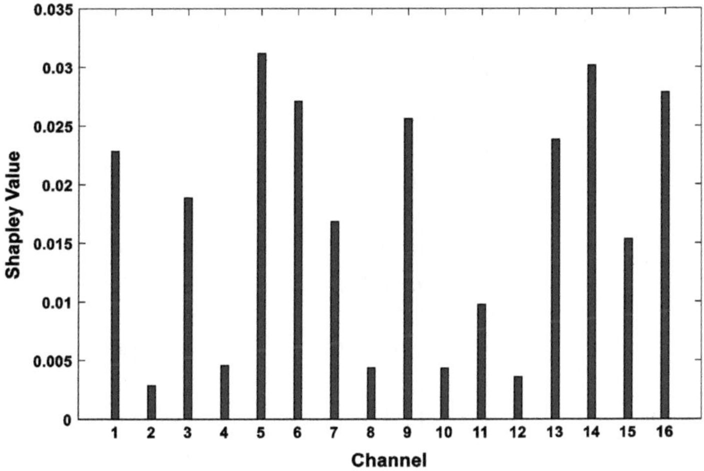

Fig. 4 Average Shapley values for 10 subjects

Table 2 Results for first and second best channels based on the Shapley values calculated in Fig. 4 for 16 channels

	1st best channel (Shapley value)	2nd best channel (Shapley value)
Subject 1	89.21 {14}	87.52 {9}
Subject 2	84.01 {9}	80.02 {15}
Subject 3	83.57 {13}	81.13 {9}
Subject 4	80.23 {6}	78.31 {5}
Subject 5	87.41 {5}	81.21 {13}
Subject 6	86.01 {14}	85.12 {5}
Subject 7	81.17 {16}	77.45 {7}
Subject 8	90.15 {5}	82.81 {14}
Subject 9	86.28 {13}	81.44 {6}
Subject 10	91.05 {15}	83.52 {13}
Average	85.92	81.48

the task of driver distraction. As Fig. 4 presents, channels 5, 14, and 16 are the most informative channels in capturing driver distraction over the subject population. This observation suggests that the frontal and parietal regions of the scalp are the most significant regions to capture the impact of distracting elements while driving.

Based on the calculated Shapley values, which is consistent with the observation with overall subject values that frontal and parietal regions demonstrate higher contribution (i.e. values) to capture distraction on a subject by subject basis.

Then Table 2 presents the results based on the first and second ranked channels based on their Shapley values.

Table 3 Results for first and second best combination of channels based on coalition game theory

	1st best combination (Cooperation)	2nd best combination (Cooperation)
Subject 1	94.61 {2, 8, 9, 11}	91.01 {10, 15}
Subject 2	89.23 {8, 12}	88.23 {1, 5, 11, 16}
Subject 3	95.34 {10, 13}	93.03 {5, 12, 13}
Subject 4	85.61 {9, 11}	84.23 {8, 10, 11}
Subject 5	96.73 {10 , 13}	92.23 {9, 7, 15}
Subject 6	93.71 {9, 11, 12}	91.65 {6, 8, 9}
Subject 7	87.18 {5, 7, 8, 12}	84.93 {2, 8, 13}
Subject 8	97.92 {5, 9, 12, 15}	94.23 {4, 6, 8, 9}
Subject 9	96.31 {7, 8, 9}	87.39 {6, 9, 11, 12}
Subject 10	93.53 {12, 13}	91.93 { 6, 7, 9}
Average	92.57	89.76

Table 3 shows the performance of the combination of the channels for all 10 subjects. As Table 3 presents, regardless of individually poor performance of some channels, they improve the overall performance of the system when they are combined with other channels. For instance, based on the Fig. 4, channels 2 has the least performance in comparison to other channels when they are ranked based on their Shapley values. But when channel 2 combines with channels 8, 9, and 11, it outperforms other individual channels or other combination of the channels.

3.1 Comparative Investigation

We also conducted a comparison of the performance of the proposed method with the other commonly used channel selection method namely, *information gain (IG)*.

Information Gain (IG) In general terms, the expected IG is the change in information entropy, H, from a prior state to a state that takes some information as given:

$$IG(T, a) = H(T) - H(T|a) \tag{3}$$

Let T denote a set of training examples, each of the form $(x, y) = (x_1, x_2, x_3, x_k, y)$, where x_a in $vals(a)$ is the value of the ath attribute of example x and y is the corresponding class label [23]. The IG for an attribute a is defined in terms of entropy $H()$ as follows:

$$IG(T, a) = H(T) - \sum_{v \in vals(a)} \frac{|x \in T|x_a = v|}{|T|} \cdot H(x \in T|x_a = v) \tag{4}$$

Table 4 Accuracy for 10 subjects and all methods

Method	First best combination (%)	IG (%)	1st best channel (%)
Subject 1	91.21	79.52	81.61
Subject 2	96.01	76.12	79.02
Subject 3	92.57	82.13	83.34
Subject 4	93.23	85.31	81.12
Subject 5	89.61	86.19	89.41
Subject 6	95.01	75.12	83.41
Subject 7	94.17	83.19	84.53
Subject 8	95,15	85.27	81.34
Subject 9	89.34	78.28	81.44
Subject 10	94.15	84.32	81.52
Average	92.57	81.42	85.92

For each channel, we calculated IG and ranked them based on the obtained IG. Higher IG for a channel means higher discriminative information it carries. Then, we selected 10 first subsets of the channels based on their IG ($\{1\}$, $\{1, 2\}$,$\{1, 2, 3\}$, …, $\{1, 2, …, 10\}$).

Then, we compare IG and our proposed method in terms of accuracy in Table 4. It illustrates accuracy for IG, first ranked channel based on the Shapley value, and first best combination of channels. As Table 4 shows, first best combination of channels outperforms other approaches. It shows how combining channels regardless of their individual performance is capable of increasing the system accuracy.

4 Conclusion

In this paper, we investigated a game theoretic channel-selection technique for evaluation of the most effective channels in driver distraction. For overall usefulness of each EEG channel, we designed a Cooperative-Game-Theoretic algorithm based on the Shapley contribution values. We evaluated the performance of the channels individually based on their Shapley value and also when they are combined with the other channels. Empirical comparison of the individual channels' performance and combination of the channels concluded that combination of the channels regardless of their individual low performance, in the task on hand, increased the overall performance of the system. According to our empirical results on a population of 10 subjects, the proposed system can decrease computational costs and reduce the number of channels. A comparison of the results was obtained between the proposed method in this study and the other common channel selection algorithms demonstrated that proposed algorithm outperformed it in terms of accuracy.

References

1. Kumari, P., Mathew, L., Syal, P.: Increasing trend of wearables and multimodal interface for human activity monitoring: a review. Biochannels Bioelectron. **90**, 298–307 (2017)
2. Sutter, R., Kaplan, P.W.: The history of continuous EEG monitoring. In: Continuous EEG Monitoring, pp. 3–12. Springer International Publishing (2017)
3. Wang, D., et al.: Monitoring workers' attention and vigilance in construction activities through a wireless and wearable electroencephalography system. Autom. Constr. (2017)
4. Singh, S.: Critical reasons for crashes investigated in the national motor vehicle crash causation survey. No. DOT HS **812**, 115 (2015)
5. Barnhizer, D.: Lawyers and Self-Driving Cars: An Example of Job Loss from Ai/Robotics
6. Kang, H-B.: Various approaches for driver and driving behavior monitoring: A review. In: Proceedings of the IEEE International Conference on Computer Vision Workshops (2013)
7. Sun, H., et al.: EEG-based safety driving performance estimation and alertness using support vector machine. Int. J. Secur. Appl. **9**, 125–134 (2015)
8. Alizadeh, V., Dehzangi, O.: The impact of secondary tasks on drivers during naturalistic driving: analysis of EEG dynamics. In: 2016 IEEE 19th International Conference on Intelligent Transportation Systems (ITSC). IEEE (2016)
9. Collins, RT., Liu, Y., Leordeanu, M.: Online selection of discriminative tracking features. IEEE Trans. Pattern Anal. Mach. Intell. **27**(10), 1631–1643 (2005)
10. Rajendra, V., Dehzangi, O.: Detection of distraction under naturalistic driving using Galvanic Skin responses. In: 2017 IEEE 14th International Conference on Wearable and Implantable Body Channel Networks (BSN). IEEE (2017)
11. Hwang, T., et al.: Driver drowsiness detection using the in-ear EEG. In: 2016 IEEE 38th Annual International Conference of the Engineering in Medicine and Biology Society (EMBC). IEEE (2016)
12. Fan, J., et al.: EEG-based affect and workload recognition in a virtual driving environment for ASD intervention. IEEE Trans. Biomed. Eng. **65**(1), 43–51 (2018)
13. Chai, R., et al.: Improving eeg-based driver fatigue classification using sparse-deep belief networks. Front. Neurosci. **11** (2017)
14. Barua, S., Ahmed, M.U., Begum, S.: Classifying drivers' cognitive load using EEG signals. Stud. Health Technol. Inf. **237**, 99–106 (2017)
15. Tu, W., et al.: A survey on mobile sensing based mood-fatigue detection for drivers. Smart City 360. Springer International Publishing (2016)
16. Alotaiby, T., et al. A review of channel selection algorithms for EEG signal processing. EURASIP J. Adv. Signal Process. **2015**(1), 66 (2015)
17. Lal, T.N., et al.: Support vector channel selection in BCI. IEEE Trans. Biomed. Eng. **51**(6), 1003–1010 (2004)
18. Guyon, I., Elisseeff, A.: An introduction to variable and feature selection. J. Mach. Learn. Res. **3**, 1157–1182 (2003)
19. Lan, T., et al.: Salient EEG channel selection in brain computer interfaces by mutual information maximization. In: 27th Annual International Conference of the IEEE Engineering in Medicine and Biology Society, 2005. IEEE-EMBS 2005. IEEE (2006)

20. Shapley, L.S.: A value for n-person games. Contrib. Theory Game. **2**(28), 307–317 (1953)
21. Taherisadr, M., Dehzangi, O., Parsaei, H.: Single channel EEG artifact identification using two-dimensional multi-resolution analysis. Channels **17**(12), 2895 (2017)
22. Klimesch, W.: EEG alpha and theta oscillations reflect cognitive and memory performance: a review and analysis. Brain Res. Rev. **29**(2), 169–195 (1999)
23. Russell, S., Norvig, P.: Artificial Intelligence: A Modern Approach, vol. 25, p. 27. Prentice-Hall, Egnlewood Cliffs (1995)

EEG Based Driver Inattention Identification via Feature Profiling and Dimensionality Reduction

Omid Dehzangi and Mojtaba Taherisadr

Abstract More than 90% of the persistently increasing traffic fatalities is related to human choice/error. Monitoring driver attention has a direct effect on decreasing injury/fatality rates. In recent years, there has been much effort to estimate drivers' state with the goal of improving their driving behavior and preventing vehicle crashes in the first place. Physiological based detection has shown to be the most direct method of measuring driver state among which, electroencephalogram (EEG) is the most comprehensive method. EEGs are recorded from multiple channels that are processed separately. However, contribution of a fairly large number of the channels might be minimal to the target application. The computational load and the redundancy induced by those channels can hurt the identification performance. In this study, we propose an EEG-based systematic methodology for the assessment of driver state of inattention. Our proposed framework includes three major modules: (1) We first characterize each EEG channel rigorously via extraction of various categories of descriptors as features, (2) we then capture the contribution of each channel toward the identification task via channel specific feature dimensionality reduction, (3) we then conduct channel selection in order to find key brain regions of impact. Eight subjects participated in our naturalistic driving study. Our proposed method resulted in the accuracy of $98.99 \pm 1.2\%$ inattention identification accuracy. We also discovered that the first and second best channels are consistently selected from frontal and parietal regions for participating subjects.

Keywords EEG · Dimensionality reduction · Linear discriminant analysis
Neighborhood preserving embedding · Driver distraction · ReliefF

O. Dehzangi
Rockefeller Neuroscience Institute, West Virginia University,
Morgantown, WV 26505, USA
e-mail: omid.dehzangi@hsc.wvu.edu
URL: http://www.wssplab.net

M. Taherisadr (✉)
University of Michigan, Dearborn, MI 48128, USA
e-mail: mojtabat@umich.edu

G. Fortino and Z. Wang (eds.), *Advances in Body Area Networks I*,
Internet of Things, https://doi.org/10.1007/978-3-030-02819-0_9

1 Introduction

Distracted driving is the act of driving while engaged in other activities such as texting, talking on the phone or to a passenger, or reading and etc. Distractions while driving can be separated into three distinct groups: visual, physical, and cognitive. Visual distraction involves taking one's eyes off the road, while physical distraction involves taking one's hands off the wheel. Cognitive distraction (mental workload or thinking about something other than driving) occurs when an individual's focus is not directly on the act of driving and his/her mind wanders.

Distractions influenced by the advancement of technology, especially text messaging or talking on the phone can require a combination of visual, manual, and cognitive attention from the driver, thus making these types of distractions particularly dangerous. Texting and calling while driving are the most widely reported form of distracted driving [1].

Many researches and intelligent Systems are being conducted and developed to assist the drivers better in avoiding car accidents due to fatigue [4, 5], drowsiness [6, 7], and distraction [8]. Hence, early detection of driver's distraction is a leading study concerning finding correlations between multi-tasking, driving performance, and driver's state. To reduce the risk of distraction, early detection of it in order to notify the driver in early stages of distraction is crucial.

Previous studies have employed various systems based on eye movements [9], lane tracking [10], and video cameras [11] for effective classification of distracted and non-distracted driving behaviors. Despite the fact that these methods achieved good performances, they have limitations such as privacy concerns, illumination effects, and responsiveness.

In recent years, many studies employed different biological sensors such as Electroencephalogram (EEG) [12], Electrocardiogram (ECG) [13], and Galvanic Skin Response (GSR) [14] to monitor the drivers' state and detect precursors of distraction at the physiological level to alleviate some of the limitations as mentioned above.

In [15], EEG power spectrum was used to analyze spatial and temporal brain signal dynamics in the presence of dual tasks (math questions and lane deviation) in a virtual reality environment. The results show an increase of theta power in the frontal area. In [16], the authors tried to create distracting stimuli in a simulation environment using different tasks (short message, mental task, and music player). In [17], a driver distraction prediction system based on EEG data was proposed to predict a specific distraction event. The event was starting to look at the map after at least 5 s of continuous driving. In another study, EEG data was recorded using a simulated driving session with a distracting task of questioning [18]. Then, driving performance and judgment capability were investigated. The results showed that right frontal cortex is the most affected area during distraction induced sessions.

Most of the previous works have been conducted using simulators and in a controlled manner where designing experiments and collecting data were under simulated experimental setups. However, due to the nature of these type of experiments, the results might not be realistic since the subjects may operate the vehicle differently knowing that they are under study in a safe and controlled environment. Therefore, designing the experiments in naturalistic driving settings, in which the data would be recorded in vehicles on the road, would be more beneficial toward designing practical driver monitoring and intervention systems [19].

For this purpose, we conducted our experiments by exposing the subject drivers to different distracting secondary tasks during naturalistic driving sessions. The main objective of this research is to design a system capable of distinguishing distracted driving (i.e. situations caused by secondary tasks) from not-distracted driving (i.e. Solely focusing on the primary task of driving) using EEG signal. In order to prevent computational load and the redundancy induced by less involved channels, we proposed a systematic methodology for the assessment of driver state of inattention. Our framework includes three major modules: We first characterize each EEG channel rigorously via extraction of comprehensive set of features including statistical features, measures of distribution, basis function representation, entropy, nonlinear time-series analysis and Model fitting. Then, we capture the contribution of each channel toward the identification task in one dimension via channel specific feature dimensionality reduction using 2 linear dimensionality reduction methods including linear discriminant analysis (LDA) and neighborhood preserving embedding (NPE), as parametric and non-parametric transforms, respectively. Finally, we conduct ReliefF channel selection algorithm in order to find key brain regions of impact in driver inattention task.

This paper is composed of the following structure; Sect. 2 presents the wearable sensor platform including the procedure of collecting data and experiment setup. Section 3 introduces pre-processing and data analysis including segmentation, feature extraction, feature reduction, and channel selection. Section 4 presents final results and discussion. Finally, this work is concluded in Sect. 5.

2 Study Design

The procedure of data collection and data analysis, including each step, is shown in the Fig. 1. Each of the steps is described in the following sections.

2.1 Wearable Sensor Platform

We have designed a wearable data acquisition platform comprising a synchronized multi-modal solution to acquire the physiological signals using a comprehensive wearable sensor network. Our platform is capable of collecting a large amount of heterogeneous drivers' biometric data such as Electroencephalogram (EEG), Skin response (GSR), electrocardiogram (ECG), Motion, and CAN-BUS during naturalistic driving session. Sixteen channels EEG with a sampling frequency of 250 (Hz) recorded. The electrodes were referenced to both of the earlobes and their

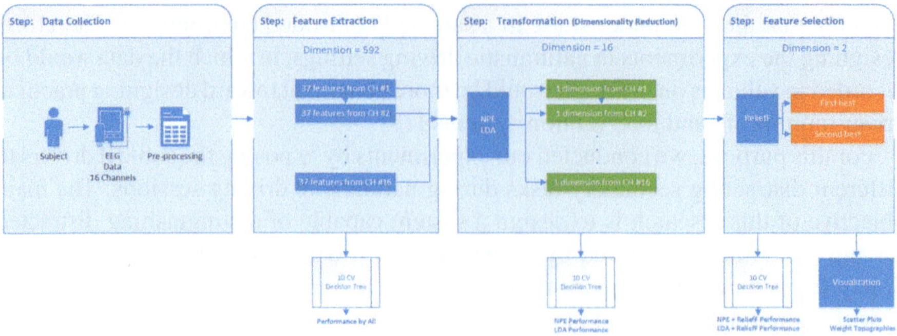

Fig. 1 Overview of the methodology. The investigation consists of 4 main steps: data collection, feature extraction, feature transformation, and feature selection

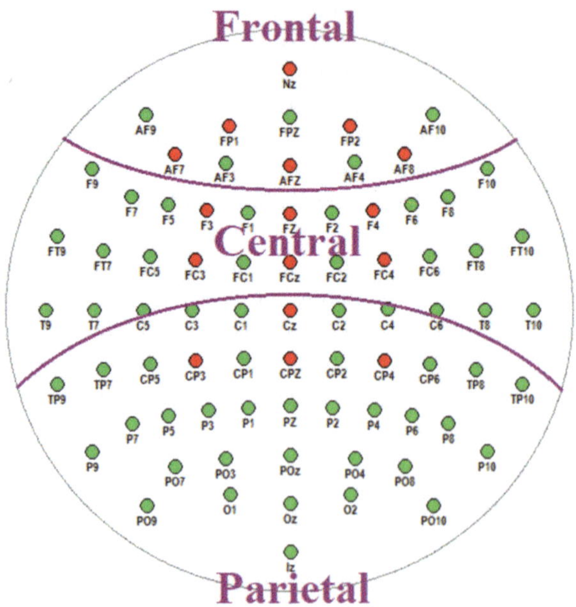

Fig. 2 EEG electrode locations on the scalp in frontal (F), central (C), and parietal (P) lobes

connection impedances were kept less than 5 (kΩ). The location of these electrodes was distributed over frontal (F), central (C), and parietal (P) areas of the head crown. The channels' name and number are as follow: (1-Fpz, 2-AF3, 3-AF4, 4-F3, 5-Fz, 6-F4, 7FC3, 8-FCz, 9-FC4, 10-C3, 11-Cz, 12-C4, 13-CPz, 14-P3, 15-Pz, 16-P4). We will use these channels' numbers and names hereafter. Figure 2 shows the channels' location and brain lobes.

Fig. 3 Data collection setup attached to a subject. Setup includes EEG, ECG, Motion, and GSR data acquisition consoles

2.2 Participants

8 subjects participated in our experiment with the following conditions. The subjects were legally allowed to drive, in the age range of 20–40. To make sure their performance is not impacted by other factors not related to the experiment scenario, we asked them to avoid drinking alcoholic beverages and taking any medication that makes them feel drowsy or sleepy. They were healthy with no mental or physical disabilities. Figure 3 demonstrates a data collection session and the setup with all devices attached to a subject. Our focus in this paper is solely on the analysis of EEG recorded signals.

2.3 Procedure

The driving experiments were conducted in a naturalistic setting in which the subjects drive a real car (Ford Escape 2015) and the signals were collected while introducing different potentially distracting elements. Our goal was to remove biases incur during simulated driving experiments and generate more accurate insights and in-depth results to use in practical systems. Procedure of collecting EEG data from driver while exposed to 5 different types of driving tasks is as follows:

- Normal Driving: 2 min of driving without any secondary task. The data from this session is used as the baseline or not-distracted driving behavior.

Fig. 4 Overview of experiment procedure

- Phone Conversation: Picking up the phone from the holder, dialing and talking for about 2 min.
- Texting: Picking up the phone from the holder, typing and sending a specific message.

The experiments were held between 3 and 6 PM during low traffic periods in a straight track. One person sitting on the passenger seat was responsible for preparing the task conditions and keeping the safety of the experiment. Another person, sitting on the back seat was recording and labeling the data using our exclusive software. The overview of this procedure is shown in Fig. 4 in which the order of the tasks is represented on the time axis. The experiment started from normal driving as indicated by the green box. We used normal driving as a gap session between distraction tasks to alleviate their effect on one another. The order of the distraction tasks as it is shown in Fig. 4 was phone conversation, texting, question, and spelling.

3 Methodology

3.1 Signal Enhancement and Pre-processing

Due to the always presented facial, muscle, and environmental artifacts in EEG signal [20], preprocessing is needed to EEG data was first preprocessed using high-pass and low-pass filters with cut-off frequencies of 0.5–50 Hz to remove the line and high-frequency noise, respectively. Then, we segmented each resulting signals to generate a short-term time series and extract features.

3.2 Segmentation

The EEG signal is nonstationary and, due to this, the features vary over time. Therefore, the most usual approach is to divide the whole period into different temporal windows and to consider statistically stationary each one [21]. The application of driver monitoring systems depends crucially on a short response-time for practical driver feedback mechanisms. Therefore, the duration of the segments is critical. We considered segmentation size as 2 s window with 75% overlap based on our experience.

Table 1 Number and brief description of the extracted features in this study

No.	Name	Dimension	Description
1	Skew	1	Pearson skewness measure
2	STFT	4	Short time Fourier transform
3	BP	4	Band power
4	Z-test	1	Statistical hypothesis test
5	Moment	3	Moments of distribution (3, 4, 5)
6	WMean	4	Wavelet, Mean
7	WPower	4	Wavelet, Power
8	WSTD	4	Wavelet, standard deviation
9	Shannon	1	Approximate Shannon entropy
10	Katz	1	Katz fractal dimension
11	AR	5	Auto-regressive parameters
12	Higuchi	1	Higuchi fractal dimension
13	EN	4	Entropy of sub-bands
	Total	**37**	

3.3 Feature Extraction

We will investigate a subject-dependent feature profiling and analysis to identify the channels and their corresponding features responsible for distinguishing different secondary tasks. We picked the features amongst the most common and wide variety of feature categories for analyzing the EEG signals used in the literature to constitute the feature profile as follows:

(a) Statistical features
(b) Measures of distribution
(c) Basis Function Representation
(d) Entropy
(e) Nonlinear time-series analysis
(f) Model fitting.

Table 1 shows the list of the name, number, and brief description of the extracted features in this study. We explain each feature briefly in the following:

Pearson Skewness: In probability theory and statistics, skewness is a measure of the asymmetry of the probability distribution of a real-valued random variable about its mean. The skewness value can be positive or negative, or even undefined.

Short Time Fourier Transform (STFT): STFT segments the signal into equal length parts and then applies the Fourier transform separately on each of these shorter signals. The EEG frequency spectrum is divided into four dominant sub-band frequency as Alpha (8–12 Hz), Beta (12–30 Hz), Delta (0–4 Hz), and Theta (4–8 Hz).

Then, we calculated the power spectral density (PSD) for the four frequency sub-bands, Delta, Theta, Alpha, and Beta.

Z-test: The p-value outputs of each segment of the signal calculated as a statistical feature. The z-test method is selected as the hypothesis function based on the following formula,

$$Z = \frac{\overline{x} - \mu}{\frac{\sigma}{\sqrt{n}}}, \tag{1}$$

where μ and σ are mean and standard deviation, respectively.

Moments: The moments, m of a distribution, for orders of $i = 3,\ 4,\ 5$ is calculated as

$$m_i = \frac{x_1^i + x_2^i + \cdots + x_n^i}{n}. \tag{2}$$

The 3rd and 4th moments called skewness and kurtosis, respectively. The 5th order is a measure of the relative importance of tails versus center (mode, shoulders) in causing skew. For example, a high 5th means there is a heavy tail with little mode movement and a low 5th means there is more change in the shoulders.

Band Power: The pre-processed EEG is then filtered into the delta, theta, alpha and beta wave via the equi-ripple band-pass filters [33]. Next, the EEG band waves are converted into the power spectral density (PSD) via the *Welch* method with Hamming window. In designing a Hamming window, the ideal frequency response is approximated by truncating the desired filter impulse response in time domain, $h_d[n]$ with the Hamming window function, $w[n]$. Hence the resultant filter impulse response, $h[n]$ can be represented by $H[n] = h_d[n] \cdot w[n]$.

Wavelet: The reason of choosing Discrete wavelet transform (DWT) over other methods (e.g. *BSS*, *EMD*, *AdaptiveFiltering*, etc.) is its ability to decompose EEG data into different frequency bands with high temporal resolution followed by easier denoising technique. This is done with reasonable computational complexity compared with *BSS* or *EMD* and without requiring any reference channel unlike adaptive filtering. DWT analyzes the signal at different resolutions through the decomposition of the signal into several successive frequency bands by utilizing a scaling and a wavelet function, associated with low-pass and high-pass filters.

After decomposing the signal in each segment, the following features were extracted from the sub-bands:

- Mean of the absolute values of the coefficients in each sub-band.
- Average power of the wavelet coefficients in each sub-band.
- Standard deviation of the coefficients in each sub-band [23].

Shannon Entropy: Entropy is the temporal distribution of signal energy in a given time window. The entropy of a random variable reflects the degree of disorder that the variable possesses.

Fractal Dimensions: Fractal dimension (FD) is a way to demonstrate the chaotic or fractal nature of a signal. The FD is a statistical measure that shows how a fractal fills the space in different scales. The FD easily gives us stable indices and time scale corresponding to the characteristic frequency even for a small number of data.

Auto-regressive: In an AR model of order p, the current output is a linear combination of the past p outputs plus a white noise input. The weights on the p past outputs are calculated in order to minimize the mean squared prediction error of the auto-regression.

We chose the parameters of AR model of order p as our features for each segment of the signal. In this study, we selected the order of 5 that generated five AR features. These features and their corresponding output dimensions are listed in Table 1.

3.4 Feature Dimensionality Reduction

Real-world data, such as EEG signals, usually has a high dimensionality. In order to handle such real-world data adequately, its dimensionality needs to be reduced. Dimensionality reduction is the transformation of high-dimensional data into a meaningful representation of reduced dimensionality. Ideally, the reduced representation should have a dimensionality that corresponds to the intrinsic dimensionality of the data. The intrinsic dimensionality of data is the minimum number of parameters needed to account for the observed properties of the data. It mitigates the curse of dimensionality and other undesired properties of high-dimensional spaces. Dimensionality reduction can be done either by Feature Selection (FS) or Feature Transformation (FT) to a low dimensional data space. In this study we focus on linear mapping techniques. Linear techniques are simpler to be implemented than non-linear transformations, and due to having less complexity, they are faster with higher generalization ability, which make them suitable. The different FT techniques can be categorized as either unsupervised or supervised depending on whether or not they use class-membership information, while computing the r-dimensional space. Unsupervised approaches may be not necessarily useful in classification problems since they do not take the target information into account.

Linear Dimensionality Reduction Problem: The generic problem of linear dimensionality reduction is the following. Given a set of points $(x_1, \ldots, x_m)$ in R^n, find a transformation matrix A that maps these m points to a set of points $(y_1, \ldots, y_m)$ in R^d $(d < n)$, such that y_i represents x_i, where $y_i = A^T x_i$. This method is of particular applicability in the special case where $(x_1, \ldots, x_m) \in M$, and M is a nonlinear manifold embedded in R^n.

Linear Discriminant Analysis (LDA): The most standard linear method of supervised dimensionality reduction is called linear discriminant analysis (LDA). It is designed to find low-dimensional projection that maximizes class separation. The objective of LDA is to perform dimensionality reduction while preserving as much of the class discriminatory information as possible. LDA can be interpreted as looking for projections having highest correlation with the dummy variables encoding group labels (in this sense LDA can be seen as a special case of canonical correlation analysis, CCA). If the classification error estimates establish that more features are needed, some other method must be employed to provide those additional features. LDA is a parametric method since it assumes unimodal Gaussian likelihoods. If the distributions are significantly non-Gaussian, the LDA projections will not be able to preserve any complex structure of the data that may be needed for classification. LDA will fail when the discriminatory information is not in the mean but rather in the variance of the data.

Neighborhood Preserving Embedding (NPE): We subdivide techniques for dimensionality reduction into convex and non-convex techniques. Convex techniques optimize an objective function that does not contain any local optima, whereas non-convex techniques optimize objective functions that do contain local optima. Convex techniques for dimensionality reduction optimize an objective function that does not contain any local optima, i.e., the solution space is convex. The objective function can be optimized by solving a generalized Eigen problem. One technique (Maximum Variance Unfolding) solves an additional semidefinite program using an interior point method. We subdivide convex dimensionality reduction techniques into techniques that perform an Eigen decomposition of a full matrix and those that perform an Eigen decomposition of a sparse matrix. The goal of sparse spectral techniques: retaining local structure of the data. NPE is a linear approximation to the Locally Linear Embedding (LLE) algorithm. Based on the categorizations, NPE is a convex and sparse spectral technique. LLE tries to preserve local structure and properties of the data by representing each point of high dimensional space as a linear combination of its nearest neighbor and attempting to keep this relation and weights between the points in the low dimensional space. Therefore, each data-point represented by its k nearest neighbors which means fitting a hyperplane. NPE algorithm has three straightforward steps.

(1) Constructing an adjacency graph
(2) Computing the weights (linear combination)
(3) Computing the Projections.

First step can be done by two common approaches. Considering a data-point x_i as a node in the adjacency graph G, an edge from node i to j is formed either x_j is among the K nearest neighbor of x_i, or if $\left|\left|x_i - x_j\right|\right| \leq \epsilon$. However, due to the difficulty of defining a proper constrain. The weight between two points is 0 if there is no edge linking them. The next step is to calculate the weight matrix, W. Suppose W_{ij} is the weight assigned to the edge between the nodes i and j which can be computed by minimizing

$$min \sum_i ||x_i - \sum_j W_{ij} x_j||^2 \tag{3}$$

with the following constraint

$$\sum_{j} W_{ij} = 1, \quad j = 1, 2, \ldots, m. \tag{4}$$

Linear projection to the low-dimensional space will be done in the last step by solving eigenvector equation

$$XMX^T a = \lambda XX^T a. \tag{5}$$

where,

$$X = (x_1, \ldots, x_m), andM = (I - W)^T (I - W). \tag{6}$$

If we order eigenvectors based on their eigenvalues and generate $A_{n \times d} = (a_1, \ldots, a_d)$, then the projection is as follows: $x_i \rightarrow y_i = A^T x_i$ Which results in a d dimensional space.

3.5 Channel Selection Using ReliefF

ReliefF algorithm has high efficiency and does not limit the characteristics of data types, the Relief enables it to deal with discrete or continuous data sets. When dealing with multi class problems, the ReliefF algorithm selects the nearest neighbor samples from each of the samples in different categories. At first, we randomly select a sample x from the training sample, then to find out k nearest neighbor samples from the kind of sample x, and randomly find out k non similar nearest neighbor samples from neighbors of different classes. To adjust a feature weighting vector to give more weight to channels by comparing within-class distance and between-class distance from neighbor samples. Repeat the above procedure on each channel dimension, finally get the weight value of each channel. The formula of ReliefF algorithm to update the weight value of channel is as following:

$$W_f^{i+1} = W_f^i + \sum_{c \neq class(x)} \frac{\frac{p(x)}{1-p(class(x))}}{\sum_{j=1}^{k} diff_f\,(x, M_j(x))} - \sum_{j=1}^{k} \frac{diff_f(x, H_j(x))}{n^* k} \tag{7}$$

where, $diff_f$ is the distance of two samples on the channel f, $H_j\,(x)$ is the neighbor samples from the kind of sample x, $M_j\,(x)$ is the neighbor samples from neighbors of different Classes, and $p\,(x)$ is the probability of class. In this study, the Euclidean distance is used to calculate the within-class distance and between-class distance of the samples.

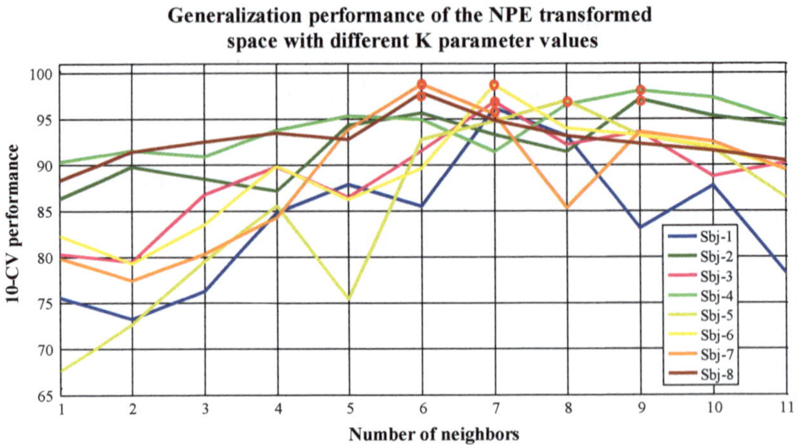

Fig. 5 Optimizing the *K* parameter of the NPE transformation on 8 subjects

4 Results and Analysis

In this section we assess the performance of the proposed method for brain region study of a distracted driver. Totally we extract 37 features from each channel, as explained in Sect. 3.3. Using extracted features, we generate feature sets of 592 dimensions for each subject. First, performance of driver distraction detection system is assessed using these feature sets via 10 cross validation (10-CV) decision tree classifier for all subjects. Then, feature reduction using FDE and LDA applied to each 37 dimension feature set to reduce it to lower dimensions. We optimize the *K* parameter of NPE transformation for each subject. Figure 5 depicts optimizing the *K* parameter of the NPE transformation for 8 subjects. As Fig. 5 shows, the number of optimal neighbors for all subjects are in the range between 6 and 9.

We reduce the number of features using LDA and NPE from 37 dimensions to 1 dimension (1 dimension per channel using FDE and 1 dimension per channel using LDA). Based on the reduced dimensions, we have 32 features per subject, 16 features for LDA and 16 features for FDE. Then, channel selection using ReliefF algorithm is applied to this feature sets to select the most efficient channels. Figure 6 presents the scatter plots for the best selected channels and topographies of the output weights from ReliefF method after applying NPE and LDA dimensionality reduction methods.

Table 2 presents the results of 10-CV accuracies using all feature spaces, reduced dimension with NPE, and reduced dimension with LDA.

In order to compare the NPE and LDA discriminatively transformed EEG spaces, we ran 10-CV evaluation 10 times and then conducted a statistical test with the null hypothesis that the performance accuracies of the LDA and NPE on different subjects comes from independent random samples from normal distributions with equal means, using the two-sample t-test. The alternative hypothesis is that

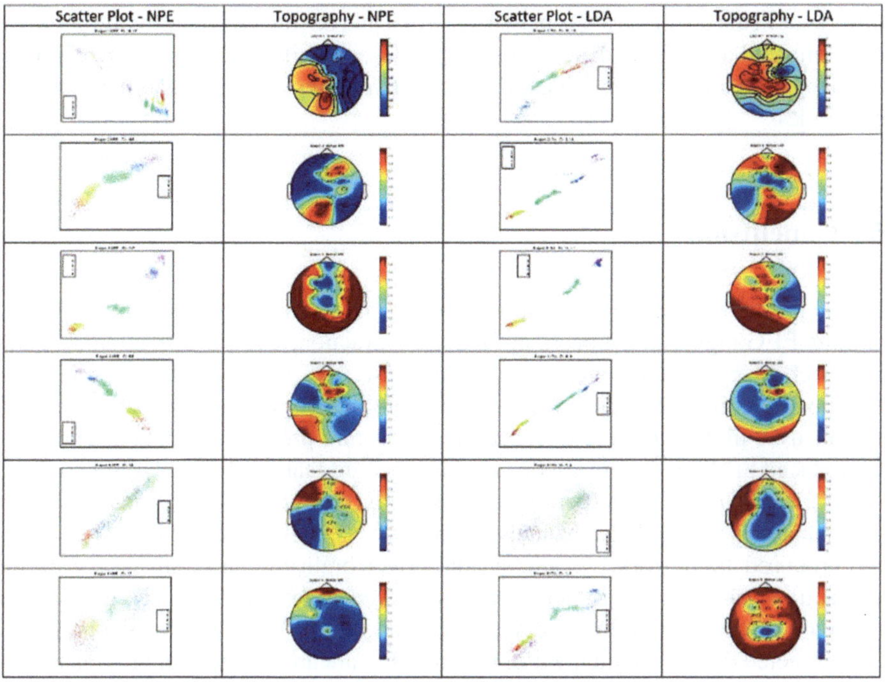

Fig. 6 Scatter plots for the best selected channels and topographies of the output weights from ReliefF method after applying NPE dimensionality reduction (left) and LDA dimensionality reductions (right)

Table 2 Results of 10-CV accuracies using all feature space, reduced dimension with NPE, and reduced dimension with LDA

Subject	All (%)	NPE (Best K) (%)	LDA (%)
1	97.75	96.11 (7)	94.01
2	98.43	97.14 (9)	97.11
3	97.72	96.78 (7)	95.85
4	98.53	98.02 (9)	97.78
5	97.45	96.95 (8)	95.48
6	99.73	98.63 (7)	96.77
7	99.46	98.76 (6)	96.13
8	98.69	97.81 (7)	96.45
Average	98.47	97.53	96.21

the LDA and NPE performance accuracies come from populations with unequal means. The test rejected the null hypothesis at the withe the p-value of 0.029 reports that the two selected channels are consistently from frontal and parietal regions for most of the subjects. It is consistent with the results reported in the previous studies [12, 26]. Figure 6 provides scalp topographical visualization of the scalp. The higher difference in the channel values between normal and distracted scenarios are shown in red compared to minimal differences in blue. Figure 6 illustrates that higher amount of change is captured in the linear discriminative bases of frontal and parietal EEG electrodes.

5 Conclusion

In this paper we proposed standard analysis framework to examine the impact of various EEG signal pre-processing, feature extraction, and classification methods to detect driver engagement in a secondary task. We picked the features amongst the most common and wide variety of feature categories for analyzing the EEG signals used in the literature to constitute the feature profile of each channel. Then, we reduced the high dimensionality of the extracted features by applying two linear transforms including LDA (parametric) and NPE (non-parametric) to one dimension per channel. We ranked the channels performance by ReliefF algorithm. Then, using the proposed framework, we studied the driver brain regions analysis during the distraction tasks. Our experimental results showed parietal and frontal regions of the brain are involved more in the distraction procedure. These areas of the brain that have affected during the moment of multitasking are areas of spatial processing and spatial attention. Even though driving becomes a primary cognitive function, when drivers are distracted (e.g. on their cell phones, talking to passengers, or texting), the regions of the brain that need to be activated to safely operate the vehicle are involved with the secondary tasks and will cause inattention.

References

1. Lenhart, A., et al.: Social Media & Mobile Internet Use among Teens and Young Adults. Millennials. Pew Internet & American Life Project (2010)
2. Green, L.: Texting and driving: a look at self-control. knowledge, and adherence to the law among young drivers. Diss. Soc. Learn. Theor. (2017)
3. Zhang, H., et al.: Effect of personality traits, age and sex on aggressive driving: psychometric adaptation of the driver aggression indicators scale in China. Accident Anal. Prevent. **103**, 29–36 (2017)
4. Lal, S.K.L., et al.: Development of an algorithm for an EEG-based driver fatigue countermeasure. J. Saf. Res. **34.3**, 321–328 (2003)
5. Yang, G., Lin, Y., Bhattacharya, P.: A driver fatigue recognition model based on information fusion and dynamic Bayesian network. Informat. Sci. **180**(10), 1942–1954 (2010)
6. Lin, C.T., et al.: EEG-based drowsiness estimation for safety driving using independent component analysis. IEEE Trans. Circuits Syst. I: Regul. Pap. **52.12**, 2726–2738 (2005)
7. Khushaba, R.N., et al.: Driver drowsiness classification using fuzzy wavelet-packet-based feature-extraction algorithm. IEEE Trans. Biomed. Eng. **58.1**, 121–131 (2011)
8. Dong, Y., et al.: Driver inattention monitoring system for intelligent vehicles: a review. IEEE Trans. Intell. Transp. Syst. **12.2** , 596–614 (2011)

9. Metz, B., Schmig, N., Hans-Peter, K.: Attention during visual secondary tasks in driving: adaptation to the demands of the driving task. Trans. Res. Part F Traffic Psychol. Behav. **14**(5), 369–380 (2011)
10. Young, K.L., Lenn, M.G., Williamson, A.R.: Sensitivity of the lane change test as a measure of in-vehicle system demand. Appl. Ergonom. **42**(4), 611–618 (2011)
11. Wege, C., Will, S., Victor, T.: Eye movement and brake reactions to real world brake-capacity forward collision warningsA naturalistic driving study. Accident Anal. Prevent. **58**, 259–270 (2013)
12. Alizadeh, V., Dehzangi, O.: The impact of secondary tasks on drivers during naturalistic driving: analysis of EEG dynamics. In: 2016 IEEE 19th International Conference on Intelligent Transportation Systems (ITSC) (2016)
13. Deshmukh, S., Dehzangi, O.: Identification of real-time driver distraction using optimal sub-Band detection powered by wavelet packet transform. In: 2017 IEEE 14th International Conference on Wearable and Implantable Body Sensor Networks (BSN). IEEE (2017)
14. Rajendra, V., Dehzangi, O.: Detection of distraction under naturalistic driving using galvanic skin responses. In: 2017 IEEE 14th International Conference on Wearable and Implantable Body Sensor Networks (BSN). IEEE (2017)
15. Lin, C.T., et al.: Spatial and temporal EEG dynamics of dual-task driving performance. J. Neuroeng. Rehabil. **8.1**(11) (2011)
16. Wali, M.K., Murugappan, M., Ahmad, B.: Subtractive fuzzy classifier based driver distraction levels classification using EEG. J. Phys. Ther. Sci. **25**(9), 1055–1058 (2013)
17. Wang, S., et al.: Online prediction of driver distraction based on brain activity patterns. IEEE Trans. Intell. Transp. Syst. **16.1**, 136–150 (2015)
18. Almahasneh, H., et al.: Deep in thought while driving: An EEG study on drivers cognitive distraction. Transp. Res. Part F: Traffic Psychol. Behav. **26**, 218–226 (2014)
19. Bach, K.M., et al.: Evaluating driver attention and driving behaviour: comparing controlled driving and simulated driving. In: Proceedings of the 22nd British HCI Group Annual Conference on People and Computers: Culture, Creativity, Interaction, vol. 1. British Computer Society (2008)
20. Taherisadr, M., Dehzangi, O., Parsaei, H.: Single channel EEG artifact identification using two-dimensional multi-resolution analysis. Sensors **17**(12), 2895 (2017)
21. Sugiyama, M.: Dimensionality reduction of multimodal labeled data by local fisher discriminant analysis. J. Mach. Learn. Res. 1027–1061 (2007)
22. Pampu, N.C.: Study of effects of the short time Fourier transform configuration on EEG spectral estimates. Acta Technica Napocensis **52.4**(26) (2011)
23. Safavian, S., Rasoul, S., Landgrebe, D.: A survey of decision tree classifier methodology. IEEE Trans. Syst. Man Cybernet. **21**(3), 660–674 (1991)
24. Loo, C.K., Samraj, A., Lee, G.C.: Evaluation of methods for estimating fractal dimension in motor imagery-based brain computer interface. Discret. Dyn. Nat. Soc. (2011)
25. Higuchi, T.: Approach to an irregular time series on the basis of the fractal theory. Phys. D: Nonlin. Phenom. **31**(2), 277–283 (1988)
26. Lei, S., Roetting, M.: Influence of task combination on EEG spectrum modulation for driver workload estimation. Human Fact. **53**(2), 168–179 (2011)

Context-Aware Sensor Solution for Remote Monitoring of Adolescent Idiopathic Scoliosis Brace Treatment

Omid Dehzangi, Bhavani Anantapur Bache, Omar Iftikhar, Jeffrey Wensman and Ying Li

Abstract The medical condition of Scoliosis occurs when an individual's spine develops curvature in adolescents. A Brace treatment is used to control the lateral curvature of the spine in scoliosis. However, brace treatment is a long and inconvenient process that demands strict compliance by the patients. In this work, we designed a wearable sensor solution to monitor the brace treatment compliance. The hardware is embedded into the patient's brace. The custom designed hardware consists of a sensor board, multiple sensors. The force sensor collects the force being exerted on the patient's back, while the motion sensor generates cues to determine the patient's activities and context. We aim to evaluate monitoring of the effectiveness of the brace treatment pervasively based on fusion of continuous force and motion recordings. The proposed method evaluates the duration of brace wear through the process of segmentation and calculates the level of tightness of brace by estimating the baseline force per segment in the presence of different activities including sitting, standing, climbing, walking, running and lying. We investigated an experimental scenario in which, the patient performs a series of pre-defined activities at home during day long segments of brace wear, during pervasive sensor data recordings. The experimental results demonstrated that we achieved an overall accuracy of a 96.1% for

O. Dehzangi
Rockefeller Neuroscience Institute, West Virginia University, Morgantown, WV 26505, USA
e-mail: omid.dehzangi@hsc.wvu.edu
URL: http://www.wssplab.net

B. A. Bache (✉) · O. Iftikhar
University of Michigan-Dearborn, Dearborn, MI 48128, USA
e-mail: bbache@umich.edu

O. Iftikhar
e-mail: oiftikha@umich.edu

J. Wensman
Orthotics and Prosthetics Center, University of Michigan, 2850 S.Industrial Hwy, Ann Arbor, MI 48104, USA
e-mail: jwensman@med.umich.edu

Y. Li
C.S. Mott Children's hospital, University of Michigan, Ann Arbor, MI 48109, USA
e-mail: yingyuli@med.umich.edu

G. Fortino and Z. Wang (eds.), *Advances in Body Area Networks I*,
Internet of Things, https://doi.org/10.1007/978-3-030-02819-0_10

unsupervised activity detection. Our trained model estimated a reduction in the level of tightness of brace by 30% during a period of 2 weeks while the compliance of brace treatment gradually increased.

Keywords Scoliosis · Brace treatment · Pervasive monitoring · Context aware sensing · Activity identification

1 Introduction

To ensure that the treatment is successful, it is important to monitor the quality and duration of brace-wear. The total number of hours of brace-wear correlates to the lack of curve progression [1, 2]. A braced curve that remains $\leq 45°–50°$ at skeletal maturity is considered a treatment success, as bracing is no longer effective once patients are skeletally mature. Curves that are $\leq 45°–50°$ at skeletal maturity are not likely to progress during adulthood [3]. Several studies provided convincing evidence about the effectiveness of this treatment given appropriate usage [4]. A Dose-response curve has been demonstrated by Katz et al. and Weinstein et al., in which duration of brace-wear is positively associated with the rate of treatment success [1, 5]. Weinstein et al. [5] reported that patients who wore the brace for 0–6 h daily had a success rate of 42%, whereas patients who wore the brace for at least 12.9 h had success rates of 90–93% [5]. The effectiveness of treatment depends on the duration of wear and tightness of the brace. However, monitoring brace compliance is a challenging task. Previous studies have demonstrated that when brace-wear is accurately monitored with a temperature sensor, adherence to brace-wear is frequently overestimated [1, 3, 5–9]. Morton et al. [10] found that patients actually wore the brace for only 47% of the prescribed time, even though physicians, orthotists, parents, and patients estimated that the brace was worn for 64%, 66%, 72%, and 75% of the prescribed time, respectively. Katz et al. reported that patients wore the brace for the same number of hours regardless of whether the prescribed time was 16 h or 23 h [10]. Recent advancements in the areas of wearable devices and data analysis techniques have greatly helped doctors in monitoring patient treatment. In this study, we aim to monitor and evaluate the duration and quality of brace wear during day to day activities of the patient by correlating the force, acceleration and angular velocity collected by a multi-modal sensor solution. There has been extensive research in the areas of activity detection from accelerometer and gyroscope readings. Ermes et al. developed a novel method to classify the unsupervised data obtained from a 3D accelerometer placed on the subject's wrist and hip. In addition, data from a GPS receiver was recorded and analyzed for activity identification [6]. The total classification accuracy achieved was 89%. Pereira et al. developed a MHARS or Mobile Human activity recognition system to monitor different activities performed by patients. They achieved an accuracy of 86.7% in classifying patient data into different activities. In this paper, we propose a novel method for pervasive monitoring, annotation and segmentation of patient data. We then use this information to design an activity-specific system and use force readings from inside the brace to determine the quality and duration of the brace wear, during first two weeks of treatment. Bennett et al. presented a MotionSynthesis toolset for data collection and algorithm development. They design a custom hardware which is used to collect Accelerometer and Gyroscope data [7]. The toolset is quite useful in generating varying data sets. Duration of brace wear is one of the major factors that impact the effectiveness of brace treatment. A study by Weinstein et al. showed that TLSO treatment significantly reduced the curve progression in patients [11] with AIS. They found that longer hours of brace wear proved to provide a greater benefit to the patients. Rahman et al. developed a methodology to determine the duration of brace wear by the patient by using a temperature sensor [2]. However, a temperature sensor can over-estimate the duration of brace wear. Moreover, it does not provide a reliable measure of brace fit quality. Due to this reason, we designed a custom-made force sensor to detect brace wear. In this work, we design a reliable heuristic method of segmentation to calculate the compliance. We develop two different methods of segmentation on a 4-s window: (1) Segmentation by mean of the force sensor values for each window (2) Segmentation by calculating the average power of the force sensor in the window. It was observed that segmentation by mean was more effective when calculating the duration for which the brace was worn.

The effectiveness of TLSO treatment in AIS patients also depends on whether the patient is wearing the brace at the prescribed tightness. Dehzangi et al. designed a two-stage algorithm in which the in-brace force data is classified into different activities including sitting, standing, walking, climbing stairs, lying and running [8]. The data corresponding to each activity is then classified into four different levels of tightness. In this work, we develop an adaptive methodology to calculate the baseline in-brace force each day. We then compare the baseline in-brace forces over consecutive days in a month period. In this way, we keep track of in-brace force estimates as an objective measure from the first day of the treatment on which, the physician has fitted and tightened the brace (i.e. the fitting session). Our pervasive experiment demonstrated an interesting relation between the number of brace wear with the trajectory of the in-brace force estimates. Our observations can be used to design a predictive model operated by the physicians to control the treatment management e.g. re-fitting scheduling.

The scope of this research paper intends to discover new and more effective treatment-monitoring methodologies by proposing a context-aware remote sensing solution. We conducted our experiments in a semi-supervised scenario, in which the patient volunteered to perform pre-defined activities including sitting, standing, walking, running, lying and climbing consecutively each for two minutes. They performed the activities for thirteen days and wrote notes on the duration of brace wear in their log book. In this work, we introduced a new priori decision tree to automatically annotate the semi-supervised data based on six different activities of sitting, standing, walking, running, lying and climbing. The effectiveness of the treatment in patients with scoliosis was evaluated using three main parameters: (1) Duration for which the brace was worn by the patient. This was measured by segmentation of the data from the force sensor to evaluate the total number of hours the brace was worn. (2) The level of tightness of the brace, by calculating the baseline force applied on the brace. (3) Quality of the brace fit. The patient volunteered to perform six pre-defined activities including Sitting (SIT), Standing (SND), Walking (WLK), Climbing (CLB), Lying (LIE) and Running (RUN). We developed the methodology of data segmentation for calculating the duration of brace-wear by mean value of the force exerted on the brace in a window. In order to determine the quality of how well the brace fits, the patient performed all six previously mentioned activities, for duration of two minutes daily for a period of 2 weeks. This data is used for creating a predictive model for activity detection for 13 days of semi-supervised data. Our proposed solution consists of two stages: (1) Remote sensing and data acquisition and (2) Data mining. For the remote sensing module, we designed and fabricated a sensor board to capture data from the patient's body. The data mining module handles processing and the analytical procedure performed on the captured data. To reach the quality measurement with minimum detection error, we employed a two-stage activity brace pressure monitoring design. In the first stage, we identified the current subject activity from 6 pre-defined classes of activities including lying down, sitting, standing, walking, running and climbing the stairs abbreviated as LIE, SIT, SND, WLK, RUN and STR in our context. In the second stage, we perform activity specific force estimation.

2 Proposed Methodology

In this section, we provide the details of our proposed method for pervasive and context aware monitoring of brace treatment. Figure 1 illustrates the end to end system including two main modules of: (1) Data Acquisition (2) Hardware architecture and (3) Signal processing and predictive modeling.

2.1 Data Acquisition

The patient is asked to perform the pre-defined activities: LIE, SIT, SND, WLK, RUN and STR for duration of 2 min each. The patient is required to wear the brace for either approximately 23 h a day, or the amount of time prescribed by the orthotist. During this time, data is being continuously recorded in a secure, digital SD card. The patient notes down the timings of specific activities they do each day, in a logbook provided. This helps us to make a connection of the data collected to the activities performed by the patient for evaluation purposes. The treatment continues for several months. The data is sampled at 40 Sample/s with 10-bit resolution. This data is passed through a Low Pass filter with cut off frequency of 10 Hz to filter high frequency noise. In this work, the analysis is carried out during 13 days of semi-supervised data.

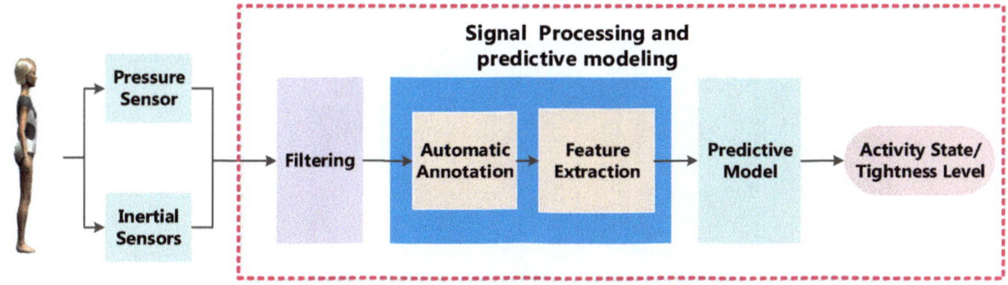

Fig. 1 End to end system with hardware architecture and signal processing modules

2.2 *Hardware Architecture*

The hardware design of the system consists of a multi-modal sensor data acquisition board, a force sensor and a motion sensor. The equipment is embedded in the brace, and data is continuously collected for analysis. The specifications of the sensor data acquisition board and the components associated with it are explained in great detail further in this article (Fig. 2).

The sensor board contains Atmega32u4, Atmel as a baseband data processor, MPU-9250 motion sensor from InvenSense, which is a 9 axis MEMS sensor using the I2C channel, a custom-designed Honeywell FSB1500NSB force sensor placed at the analog channel using 10-bit resolution A/D converter, and communication channels, which include a micro SD card as a data logger and nRF8001 from Nordic as a Bluetooth low energy module, BLE. Figure 4 displays the hardware schematic and Fig. 3 displays the sensor board that the proposed hardware architecture is imple-

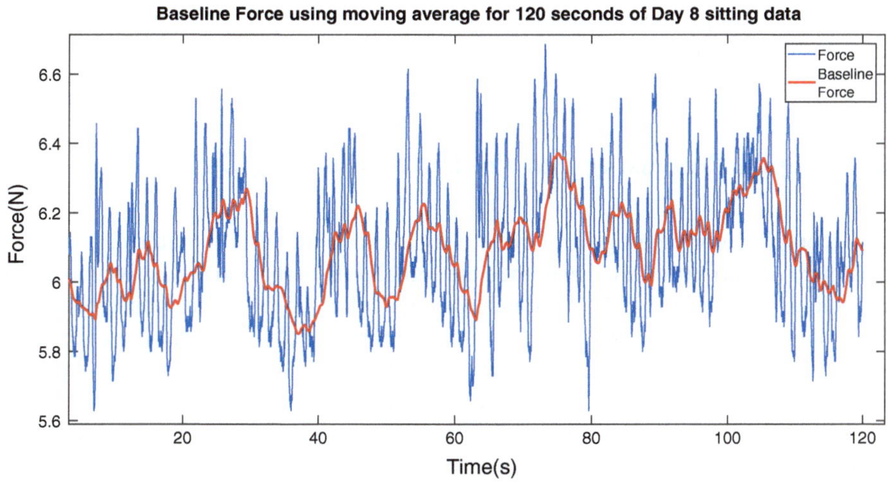

Fig. 2 Estimation of baseline pressure

Fig. 3 Picture of a PCB sensor board

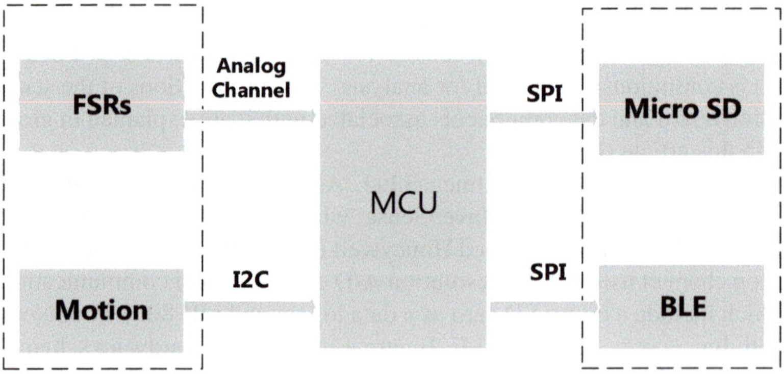

Fig. 4 Hardware design

mented in it. The proposed solution consists of our hardware design system embedded in a Boston-type TLSO.

2.3 Signal Processing and Predictive Modeling Design

To accomplish our target goal of effectively and continuously monitoring the brace treatment, we proposed and developed the signal processing and predictive modeling module as shown in Fig. 5. We input the data collected from the sensor board via Bluetooth or an SD card to the feature extraction module. The priori decision tree uses the x-axis acceleration, resultant acceleration, FFT, number of footsteps and FFT as features features to annotate the semi-supervised data automatically. Three of the features including x-axis acceleration, footsteps and pitch are selected by the feature selection module using brute force technique. The annotated data with the three features containing class labels is used to create the predictive model. The 13 day data is classified into different activities using the predictive model.

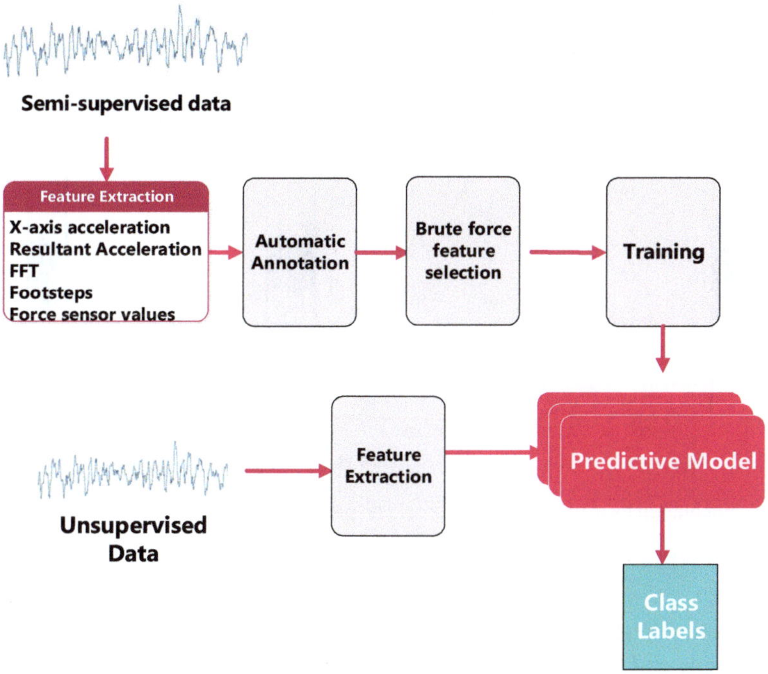

Fig. 5 Signal processing and predictive modeling design

Determination of Brace Treatment Compliance Through Segmentation
Duration of the brace wear is one of the major factors that impacts the treatment of patients with scoliosis. Compliance to the brace treatment is defined as the time for which the brace is worn relative to the prescribed time [12]. Compliance is important to physicians and patients as it provides important information about the patient's ability to wear the brace. If the patient has difficulties in wearing the brace, physicians can have a conversation and develop strategies to help increase the compliance. With higher compliance, curve progression of the spine and chances of requiring surgery are much lower [9]. In this paper, we design a methodology of segmentation to segment the force sensor data into windows where the brace was worn.

The initial step consists of filtering the force sensor data, which is sampled at 40 Hz, by the means of a 5 Hz Low-Pass Filter. This initial filtering process removes high frequency noise from the force sensor data. The force sensor data is divided into windows of 4 s with a window increment of 1 s. In the the process of segmentation, the average value of force inside the brace is calculated for each window. If the average value is high, the brace is worn. The markers are generated to clearly mark the windows for which the brace is worn, as shown in Fig. 6.

Calculation of Baseline Force
The force sensor measures the force exerted by the patient's back on the brace. In this work, we estimate tightness of the brace from the force sensor data. The force sensor data contains the patient's breathing patterns and high frequency noise. We employ

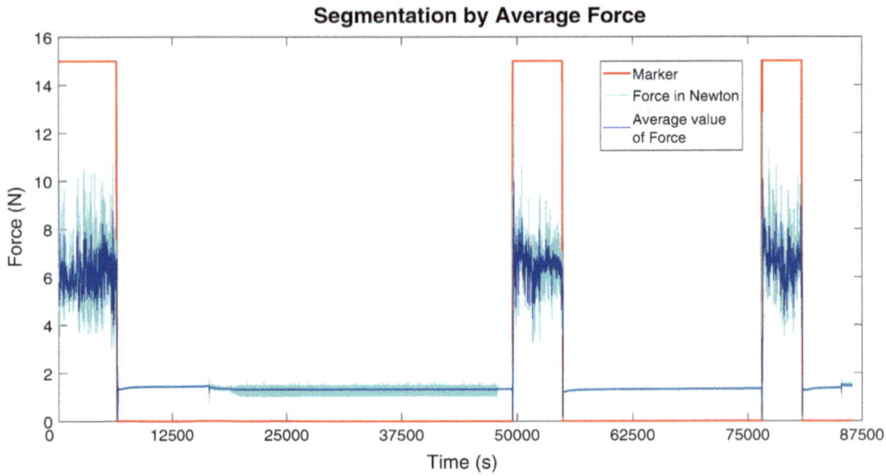

Fig. 6 Segmentation by average value of force. The part of the data when brace is worn is indicated by markers

moving average filter over each segment of data to calculate the baseline force. This filter is simple and effective in smoothing out the patient's breathing patterns or the peaks of waveforms during different activities. It computes the nth sample of the output sequence as the average of $M_1 + M_2 + 1$ samples of input sequence around the nth sample. In the equation form, it is written as [13]:

$$y[n] = \frac{1}{M_1 + M_2 + 1} \sum_{k=-M_1}^{M_2} x[n-k] \tag{1}$$

In this work, a moving average filter was applied on 10-s segments of data. Therefore, M is chosen as 400. Figure 2 shows the estimation of baseline force for 20 s of walking data.

Automatic Data Annotation

The patient volunteered to perform the activities: LIE, SIT, SND, WLK, RUN and CLB consecutively for 2 min each day. They performed the activities for 13 days, and wrote notes about the activities in their log book. We designed a novel priori decision tree to automatically annotate the training data. We specifically employ decision tree for this purpose in order to benefit from interpretability of the resulting model (i.e. rule-base) and use it for high level knowledge acquisition.

Feature Extraction

After segmentation, the in-brace data is passed to the feature extraction block. The direction of acceleration is along a specific axis for some of the day-to-day activities of the patient. For instance, when the patient climbs up or down the stairs, the direction of acceleration is mainly on the x-axis. When the patient walks, the acceleration along z-axis is dominant. These properties and values can help distinguish between the two activities. For stationary activities like sitting, standing or lying down, orientation of the brace or the angle of rotation of y-axis (pitch), can be used as a discriminative feature. For lying, the normalized pitch is almost 180°, whereas for sitting and standing, the normalized pitch is approximately 90°. The number of footsteps in a 10-s window is another discriminative feature that we employ in this work. To improve the identification accuracy of the predictive model, we use three features: (1) Resultant x-axis acceleration, (2) Pitch and (3) Number of footsteps in a 10-s window used to train our predictive model.

Detection of Footsteps
The number of footsteps is a discriminative feature used to differentiate stationary from non stationary activities. Sitting, standing and lying will have no footsteps as they are stationary activities. The number of footsteps is also used to differentiate walking from running. The number of footsteps in a 10-s window can be estimated using the x-axis, y-axis and z-axis vector from the accelerometer using Eq. 2.

$$a_r = \sqrt{(a_x)^2 + \left(a_y\right)^2 + (a_z)^2} \tag{2}$$

To remove the effect of acceleration due to gravity and the negative impact of sensor orientation on generalization of the predictive model, the mean of the resultant acceleration is calculated and subtracted from the actual acceleration.

$$a_{ng} = a_r - \sum_{i=1}^{M} a_i \tag{3}$$

where a_{ng} is the acceleration with no gravity of each sample, a_r is the resultant acceleration of the corresponding samples, and M is the total number of samples in the entire day, and a_i is the acceleration of the ith sample.

We assume that the data of resultant acceleration is normally distributed. The peaks which are above the mean and twice the standard deviation is considered as a footstep. Detected footsteps are shown in Eq. 4.

$$h_{min} = a_r - 2 \times \sum_{i=1}^{M} a_r + 2 \times std(a_r) \tag{4}$$

If the value of a_r is greater than the calculated value of h_{min}, this is considered as a footstep. This method proves to be highly accurate in calculating the number footsteps for noisy acceleration data, as noisy peaks are filtered out. Figure 7 shows the resultant acceleration and footsteps. The peaks marked in red represent each footstep.

Hoeger et al. have carried out a detailed study on the number of footsteps related to different activities like walking and running [14]. Walking has roughly 10–30 footsteps in a 10-s window, while running has 30–45 footsteps. Climbing is roughly 8–15 footsteps, thus, making it challenging to differentiate walking from climbing.

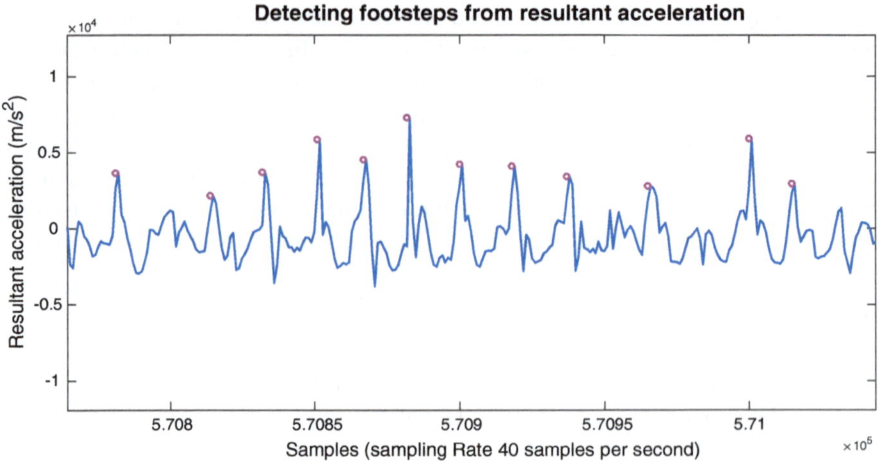

Fig. 7 Detection of footsteps from resultant acceleration. The markers show the footsteps detected

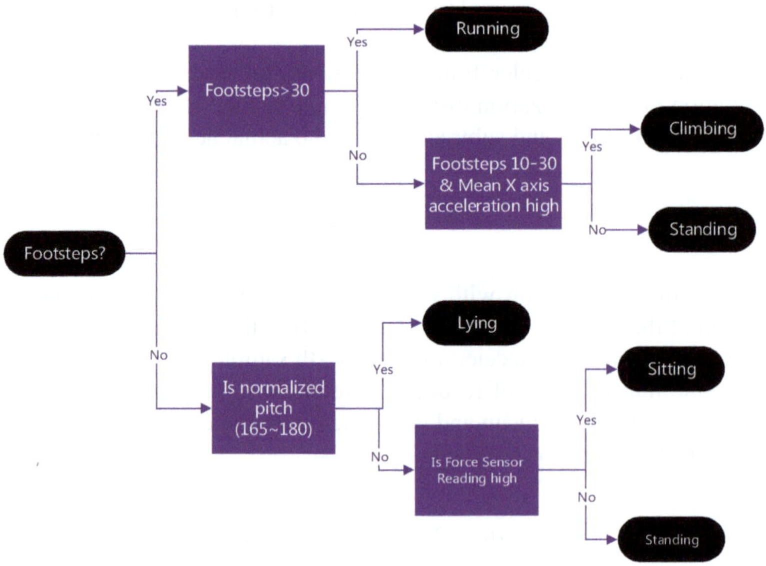

Fig. 8 The priori decision tree used in automatic data annotation

Automatic Annotation Using Priori Decision Tree
After the number of footsteps is detected, we can automatically annotate the semi-supervised data using priori decision tree. In the decision tree, we used three features: (1) Number of footsteps in a 10-s window, (2) X-axis acceleration and (3) Orientation of the brace which is measured by angle of rotation of y-axis (normalized pitch). The automatic semi-supervised data annotation is shown in Fig. 8.

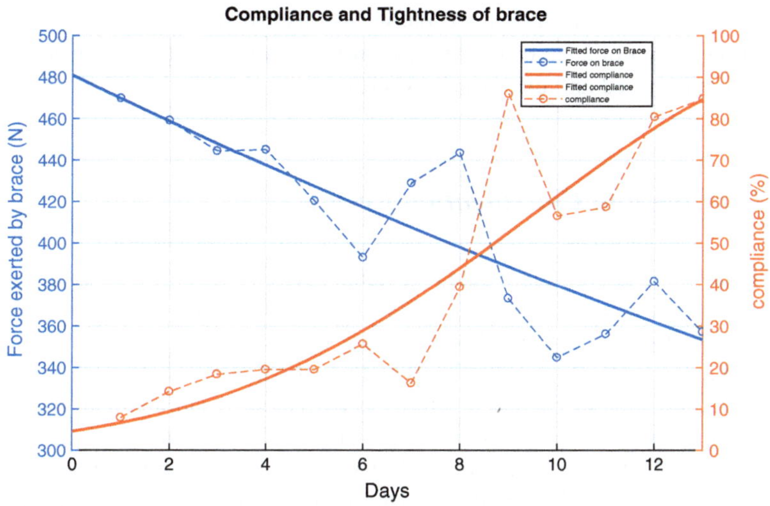

Fig. 9 Compliance and duration of the brace wear for 13 days

Table 1 Accuracy of activity detection: cross validation results

Accuracies	Gaussian SVM (%)	KNN with K=1 (%)	KNN with K=3 (%)	Complex decision tree (%)
Sitting	99.9	99.2	99.1	99.6
Standing	99.9	99.8	98.5	99.1
Walking	99.9	99.8	100	99.5
Climbing	99.9	99.9	99.2	99.9
Lying	100	99.5	99.7	99.4
Running	99.9	99.5	99.7	99.4

Classification
We use 10-fold cross-validation and employ a fine KNN classifier with different values of K. We also use a complex decision tree classifier, and a SVM classifier with a Gaussian Kernel for activity identification. The results of activity identification are shown in Table 1.

3 Experimental Results

In this section, we discuss the results of two different phases. In the first phase, we discuss the results related to automatic data annotation, training the model, cross validation results and the activity identification results in both, semi-supervised and unsupervised settings. In the second phase, we discuss the experimental results of the duration of compliance of the brace treatment and the tightness of the brace in both, semi-supervised and unsupervised settings.

3.1 Training and Classification Results

For each day's pervasive sensor data recordings, there is a window of approximately 15 min, when the activities: LIE, SIT, SND, WLK, RUN and STR were performed by the patient for 2 min each. The sequence of the activities and the duration is mentioned in the logbook provided to us by the patient. The accelerometer and force sensor data corresponding to each activity is extracted by automatically annotating the semi-supervised data. We observed that all activities follow a unique trend and pattern that can be further justified by observing the force readings associated with the respective activity. The breathing patterns of the patient can be observed for stationary activities like sitting, standing and lying. The peaks of the signal are at roughly 1.5 s apart. The frequency of the waveforms is much higher for walking and climbing. Running data shows highest frequency for the force sensor data.

The two-minute data extracted for each of activity for the first 13 days is used for training the predictive model. We used three features: (1) Number of footsteps, (2) X-axis acceleration and (3) Angle of rotation of y-axis (pitch) to train the predictive model. The cross-validation results are as shown in Table 1. Lying shows the highest classification accuracy among all the activities as the angle of rotation of y axis helps differentiate it from other activities. To get a better idea on which of the features are more suitable for creating a model, we generated a confusion matrix as shown in Fig. 10. The matrix shows the true positive rates for different activities. We observe that the two most challenging activities to distinguish between are: (1) Walking and climbing and (2) Sitting and standing.

After training the model, we use SVM with Gaussian kernel to classify semi-supervised and unsupervised data. The classification results for semi-supervised data are shown in Table 2. We observe that during the entire 13 days of data, the maximum duration of running is 25 min in an entire day. The patient was advised to take the brace off during running. We also observe that the duration of lying is about 8–10 h every day, which indicates that the patient sleeps at night with the brace on.

3.2 Experimental Results for Compliance of Brace Treatment and Tightness of the Brace

Table 3 shows force and the compliance percentages. The relation between compliance of brace treatment and tightness with which the brace is worn is shown in Fig. 9. We observe that initially, for the first 4 days the patient begins to become

TRUE CLASS \ PREDICTED CLASS	Sitting	Standing	Walking	Climbing	Lying	Running
Sitting	>99%	<1%				
Standing	<1%	>99%				
Walking			>99%	<1%		<1%
Climbing		<1%	<1%	>99%		<1%
Lying					100%	
Running			<1%			>99%

Fig. 10 Confusion matrix using three features including number of footsteps, x axis acceleration and pitch. The values shown in the table are true positives

Table 2 Classification results for 13 days data

Days	SIT	SND	WLK	CLB	LIE	RUN	Hours of brace wear
1	0.2	0.8	0.13	0.4	0.05	0.08	1.8
2	0.4	1.4	0.09	0.4	0.68	0.14	3.2
3	0.5	1.8	0.08	0.5	1.03	0.12	4.2
4	0.5	2.4	0.09	0.5	0.65	0.15	4.4
5	0.4	0.8	0.06	0.3	2.74	0.10	4.4
6	1.5	2.0	0.22	1.6	0.18	0.31	5.9
7	1.0	1.2	0.13	0.5	0.6	0.19	3.7
8	1.9	4.6	0.28	1.4	0.39	0.34	9
9	1.0	1.2	0.13	0.5	0.60	0.19	3.7
10	2.6	5.6	0.32	1.5	9.19	0.42	19.7
11	0.0	0.1	0.03	0.2	12.4	0.04	12.9
12	0.7	1.9	0.18	1.0	9.31	0.27	13.5
13	1.2	5.6	0	0.4	12.2	0	19.4

Table 3 Results of compliance and force on the brace for 13 days of data

Days	Force on brace	Compliance
1	469.8	7.91
2	459.1	14.1
3	444.5	18.3
4	445.1	19.5
5	420.4	19.5
6	393.1	25.6
7	428.9	16.2
8	443.4	39.4
9	373.2	86.5
10	344.8	56.2
11	356.1	58.6
12	381.5	80.4
13	357.1	84.7

Table 4 Comparison with the existing results

Activities	Ermes et al. [6] (%)	Pereira et al. [15] (%)	Kwapisz et al. [16] (%)	Our methodology (%)
Sitting	97	90	96	99
Standing	90	91	87	80
Walking	71	87	94	99
Climbing	–	80	61	99
Lying	98	89	–	100
Running	91	88	98	100
Total	89	83	92	96.1

accustomed to the brace. On day 1, they wear the brace for 2 h. We also observe a gradual reduction in the force exerted by the brace, which indicates that the brace is getting looser, which could be due to the patients wearing TLSO more loosely or due to patient's curve reducing while wearing the TLSO. This information is quite important for the physician as a loose brace may not be able to adequately control the spinal curvature. A comparison was made with the accuracy of activity identification in this study with the previous research work, as shown in Table 4. Our methodology and analysis is the most accurate to date (Fig. 10).

4 Conclusion

In this work, we propose new data-mining methodologies to evaluate the effectiveness of brace treatment pervasively based on force and inertial motion recordings. We evaluate three main aspects of brace treatment (1) Compliance of brace treatment by evaluating the duration of brace wear by the process of segmentation (2) Level of tightness of the brace by calculating the baseline force exerted by the patient on the brace (3) The quality of brace-fit by estimating the duration of activities: LIE, SIT, SND, WLK, RUN and STR performed daily. We achieved an overall accuracy of $\sim$100% activity recognition performance. Patients were instructed to wear the brace for 2 h a day initially, gradually increasing to 23 h a day. This was proven by our analysis as compliance was observed to increase from 7 to 90% after 2 weeks. Naturally, the brace is tightest when the patient begins treatment. As they continue wearing the brace daily, they start to break-in, making the brace more comfortable to wear. This can be calculated by measuring the baseline force exerted on the brace. Our analysis showed that the force decreased by 30% after 2 weeks.

References

1. Herring, A.J., Katz, D.E., Browne, R.H., Kelly, D.M., Birch, J.G.: Brace wear control of curve progression in adolescent idiopathic scoliosis. J. Bone Joint Surg. Am. **92**(15), 2616–2617 (2010)
2. Rahman, T., Richard Bowen, J., Takemitsu, M., Scott, C.: The association between brace compliance and outcome for patients with idiopathic scoliosis. J. Pediatr. Orthop. **25**(4), 420–422 (2005)
3. Asher, M.A., Burton, D.C.: Adolescent idiopathic scoliosis: natural history and long term treatment effects. Scoliosis **1**(1), 2 (2006)
4. Sapountzi-Krepia, D., Psychogiou, M., Peterson, D., Zafiri, V., Iordanopoulou, E., Michailidou, F., Christodoulou, A.: The experience of brace treatment in children/adolescents with scoliosis. Scoliosis **1**(1), 8 (2006)
5. Dolan, L.A., Wright, J.G., Weinstein S.L.: Effects of bracing in adolescents with idiopathic scoliosis. New Engl. J. Med. **370**(7), 681 (2014)
6. Ermes, M., Pärkkä, J., Mäntyjärvi, J., Korhonen, I.: Detection of daily activities and sports with wearable sensors in controlled and uncontrolled conditions. IEEE Trans. Inf. Technol. Biomed. **12**(1), 20–26 (2008)
7. Bennett, T.R., Savaglio, C., Lu, D., Massey, H., Wang, X., Wu, J., Jafari, R.: Motionsynthesis toolset (most): a toolset for human motion data synthesis and validation. In: Proceedings of the 4th ACM MobiHoc Workshop on Pervasive Wireless Healthcare, pp. 25–30. ACM (2014)
8. Dehzangi, O., Mohammadi, M., Li, Y.: Smart brace for monitoring patients with scoliosis using a multimodal sensor board solution. In: Healthcare Innovation Point-Of-Care Technologies Conference (HI-POCT), 2016 IEEE, pp. 66–69. IEEE (2016)
9. Aulisa, A.G., Giordano, M., Falciglia, F., Marzetti, F., Poscia, A., Guzzanti, V.: Correlation between compliance and brace treatment in juvenile and adolescent idiopathic scoliosis: Sosort 2014 award winner. Scoliosis **9**(1), 6 (2014)
10. Morton, A., Riddle, R., Buchanan, R., Katz, D., Birch, J.: Accuracy in the prediction and estimation of adherence to bracewear before and during treatment of adolescent idiopathic scoliosis. J. Pediatr. Orthop. **28**(3), 336–341 (2008)

11. Weinstein, S.L., Dolan, L.A., Wright, J.G., Dobbs, M.B.: Effects of bracing in adolescents with idiopathic scoliosis. New Engl. J. Med. **369**(16), 1512–1521 (2013)
12. Lou, E., Hill, D.L., Raso, J.V., Moreau, M.J., Mahood, J.K.: Smart orthosis for the treatment of adolescent idiopathic scoliosis. Med. Biol. Eng. Comput. **43**(6), 746–750 (2005)
13. Oppenheim, A.V., Schafer, R.W.: Discrete-time Signal Processing. Always learning, Pearson (2013)
14. Hoeger, W.W.K., Bond, L., Ransdell, L., Shimon, J.M., Merugu, S.: One-mile step count at walking and running speeds. ACSM's Health Fitness J. **12**(1), 14–19 (2008)
15. Pereira, J.D., da Silva e Silva, F.J., Coutinho, L.R., de Tácio Pereira Gomes, B., Endler, M.: A movement activity recognition pervasive system for patient monitoring in ambient assisted living. In: Proceedings of the 31st Annual ACM Symposium on Applied Computing, pp. 155–161. ACM (2016)
16. Kwapisz, J.R., Weiss, G.M., Moore, S.A.: Activity recognition using cell phone accelerometers. ACM SigKDD Expl. Newslett. **12**(2), 74–82 (2011)

Gait Analysis for Physical Rehabilitation via Body-Worn Sensors and Multi-information Fusion

Sen Qiu, Zhelong Wang, Hongyu Zhao, Long Liu, Jiaxin Wang and Jie Li

Abstract How to effectively use wearable sensors for medical rehabilitation is an interdisciplinary research hotspot of control subjects and biomedical engineering. This paper intends to integrate accelerometer, gyroscope and magnetometer to build a low-cost, intelligent and lightweight wearable human gait analysis platform. On account of complexity and polytopes of walking motion characteristics, the key is to solve the existing robustness and adaptability problems of current gait analysis algorithm. This project is starting from the sensor physical properties and human physiology structure, aiming to establish lower limb kinematics model constraint, and solving the applicability problem of the traditional zero velocity update algorithm. Digital filter and error correction of gait parameters could be done with multi-level data fusion algorithm. Preliminary clinical gait experiments results indicated the proposed method has great potential as an auxiliary for medical rehabilitation. The ultimate target is to realize auxiliary diagnosis and exercise rehabilitation plan formulation for patients with abnormal gait.

Keywords Human gait analysis · Biological information detection Multi-information fusion · Exercise rehabilitation · Micro-electro-mechanical sensor

S. Qiu (✉) · Z. Wang · H. Zhao · L. Liu · J. Wang · J. Li
School of Control Science and Engineering, Dalian University of Technology, Dalian 116024, China
e-mail: qiu@dlut.edu.cn

Z. Wang
e-mail: wangzl@dlut.edu.cn

H. Zhao
e-mail: zhaohy@dlut.edu.cn

G. Fortino and Z. Wang (eds.), *Advances in Body Area Networks I*, Internet of Things, https://doi.org/10.1007/978-3-030-02819-0_11

1 Introduction

Human walking contains important physiology, kinematic and dynamic information. There are many application prospect of human gait analysis in real life, such as monitoring the patient's recovery progress in clinical practice, the control strategy of bionic robot, etc. This paper established a wearable gait analysis platform based on MEMS sensor and body sensor network. The platform can be used to collect acceleration, angular velocity and the geomagnetic signals in the process of walking movement. Accurate gait parameters can be calculated through sensor data fusion algorithm and error correction process. This paper is composed on the basis of summarizing the existing research content at home and abroad, focusing on the following issues: how to reduce the orientation estimation error and increase the accuracy of gait phases partition; how to analyze the symmetry between dual feet and the gait stability; how to eliminate the sensor misalignment and binding position deviation among multiple sensors; Figures 1 and 2 show the mainstream quantitative gait analysis systems and the proposed body-worn sensor based system, respectively.

Pace or walking speed is the most commonly gait parameters used to predict the falls among community-dwelling elderly. Verghese et al. [1] calculated gait characteristics of 597 cases of elderly with an average age of 80.5 through base line measurement, and conducted follow-up observation of elderly fall events for 42 months, results show that the slow pace is associated with an increased risk of falls, i.e., fall risk increases by 7% when the pace slowdown 10 cm/s. Senden et al. [2] divided 100 cases of the elderly into three groups (high risk, medium risk and no risk) according to clinical commonly used Tinetti balance scale method, and then use the accelerometer to measure gait features in straight line walking trials. Results show that all the gait parameters are associated with Tinetti scores, with which the pace correlation intention reaches 0.73, and the fall risk threshold is 1.08 m/s. Mortaza et al. [3] proved the relationship between pace and fall through the literature review of 17 research articles.

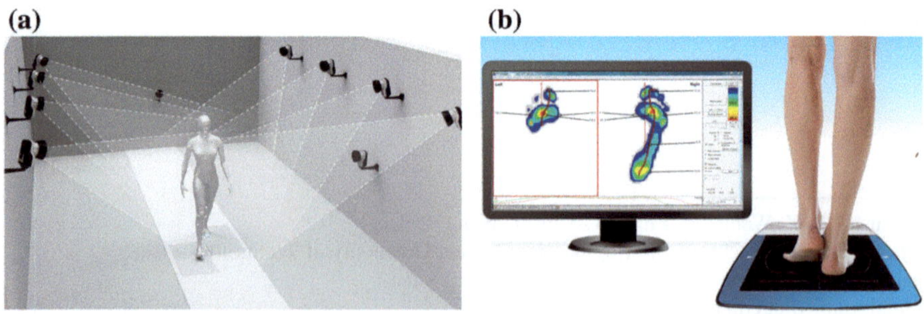

Fig. 1 The mainstream quantitative gait analysis systems: **a** video based gait analysis method, **b** volar stress distribution test system

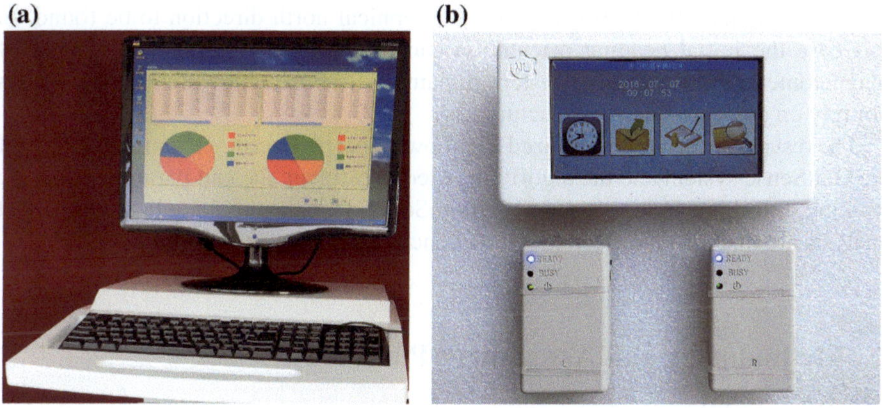

Fig. 2 The composition of gait analysis system: **a** system hardware, **b** gait data analysis and presentation

Quantitative gait analysis system mainly including camera system [4, 5], electromyography measuring system [6] and force platform [7–9]. Camera system consists of multiple high resolution cameras located in an indoor space, the orientation and position information of target subject can be calculated using attached highlighting reflective spots. Electromyography measuring system detects human lower limb muscle signals by surface electromyography in the waking process; Force platform reflects the change of plantar pressure during walking. However, the applications of above gait analysis system are limited in clinical practice, the main reasons lie in three aspects, firstly, the systems are expensive which might be a barrier to routine use, secondly, the usage of the systems are complex and need special operation, it usually takes hours to complete the whole gait measurement process. Finally, specific space is normally needed to perform gait analysis using the above systems, especially camera system may need more than 100 $100\,\mathrm{m}^2$ [10–12].

Due to the noticeable advantages of small size and low cost, wearable sensors can be mounted directly on the body segment with no need for specified test environment [13–16]. In our previous studies, adopted wearable inertial sensor in walking distance calculation and walking pattern classification [17–19]. Ambulatory measurement of the participant's trunk inclination using inertial measurement unit (IMU) was carried out by Farris et al. [20]. Bamberg et al. [21] developed a smart shoe for gait analysis using force sensitive resistors and IMU sensors. Luinge et al proposed the estimation of arm orientation by wearable inertial sensors. Dejnabadi et al. introduced an approach to accurate measurement of joint angles based on IMU. However, due to their inability to detect heading reference, inertial based systems generally fail to measure differential orientation, a prerequisite for computing the 3D knee joint angle recommended by the Internal Society of Biomechanics [22]. Roetenberg et al. [23] developed an ambulatory position and orientation tracking method fusing magnetic and inertial sensing. Since magnetometers measure the strength and direction of the local magnetic field, allowing the geographical north direction to be found. In this case the initial heading orientation can be obtained with the supplement of magnetometer. Moreover, the system remains self-contained, which means it does not rely on any external infrastructure [24].

The rest of this paper is organized as follows: Sect. 2 describes the methodology of the GaitSense system and the algorithms used to estimate the gait parameters during walking; Experimental results are given in Sect. 3; the potential applications of gait analysis are discussed in Sect. 4, which concludes the paper as well.

2 Methodology and Assessment Platform

2.1 Hardware Platform Based on MEMS Sensors

We have developed a gait analysis system based on the inertial/magnetic sensors, multi-sensor fusion algorithm is used to estimate gait parameters. The measurement unit are installed on the heels, and the installation is quite simple, which can be finished in several minutes. The sensor array includes triaxial accelerometers, gyroscope and magnetometer. The actual linear motion acceleration is used to calculate position. As shown in the following formula, Note that the gravity component ($\mathbf{g} = [0, 0, 9.81]^T$ m/s^2) is normally eliminated from the resultant acceleration signal before estimating position by twice integral operation of actual motion acceleration.

$$\begin{cases} \mathbf{a}_G(t) = q_S^G(t)[\mathbf{a}_S(t) - \mathbf{g}]q_S^{G-}(t) \\ \mathbf{v}_G(t) = \mathbf{v}_G(0) + \int_0^t \mathbf{a}_G(t)dt \\ \mathbf{p}_G(t) = \mathbf{p}_G(0) + \int_0^t \mathbf{v}_G(t)dt \end{cases} \tag{1}$$

where $\mathbf{v}_G(0)$ and $\mathbf{p}_G(0)$ represent the subjects' initial velocity and position, respectively.

2.2 Joint Angle Calculation

$$\theta_{knee} = \theta_{thigh} - \theta_{shank} + \theta^{ini} = \int (\omega_{thigh} - \omega_{shank})dt + \theta_{thigh}^0 - \theta_{shank}^0 \tag{2}$$

where θ_{thigh}^0 and θ_{shank}^0 are initial angles between lower limbs (thigh and shank, respectively) and the gravity direction when the subjects are standing still at ease, which can be calculated by the measurements of the accelerometer as proposed in the previous study [17, 18]. ω_{thigh} and ω_{shank} are angular velocity values of thigh fixed and shank fixed sensors.

2.3 Joint Position Calculation

$$\begin{cases} \frac{\partial x_k^G}{\partial \varphi} = \frac{2L_s c^2\varphi c^4\phi c^4\psi s\varphi}{(1+D^2)^2} - \frac{2L_s c^2\varphi c^4\phi c^4\psi E s\varphi}{(1+D^2)^2} + \frac{L_s c^2\varphi c^4\phi c^4\psi s\varphi[-c^4\psi+c^4\phi c^4\psi-2D^2]}{(1+D^2)E} \\ \frac{\partial x_k^G}{\partial \varphi} = \frac{2L_s c^4\varphi c^2\phi c^4\psi s\phi}{(1+D^2)^2} - \frac{2L_s c^4\varphi c^2\phi c^4\psi E s\phi}{(1+D^2)^2} + \frac{L_s c^4\varphi c^2\phi c^4\psi s\phi[1-c^4\psi+c^4\phi c^4\psi-D^2]}{(1+D^2)E} \\ \quad + \frac{L_s(-c^2\phi c^4\psi)s\phi+c^4\varphi c^2\phi c^4\psi s\phi}{E} \\ \frac{\partial x_k^G}{\partial \varphi} = \frac{2L_s c^4\varphi c^4\phi c^2\psi s\psi}{(1+D^2)^2} - \frac{2L_s c^4\varphi c^4\phi c^2\psi E s\psi}{(1+D^2)^2} + \frac{L_s c^4\varphi c^4\phi c^2\psi s\psi[1-c^4\psi+c^4\phi c^4\psi-D^2]}{(1+c^4\varphi c^4\phi c^2\psi)E} \\ \quad + \frac{L_s(c^2 s\psi-c^4\phi c^2\psi s\psi+c^4\varphi c^4\phi c^2\psi s\psi}{E} \end{cases} \tag{3}$$

Before the start of the walking motion, all sensor module are assumed in static state, the 3D location information of each sensor module is relative to the origin of G frame. Define $[x_h^G, y_h^G, z_h^G]^T$ as the calculated hip thigh angular velocity values of location information, $[x_k^G, y_k^G, z_k^G]^T$ as the calf sensor location information of the knee joint angular velocity observed value calculated, likewise, $[x_a^G, y_a^G, z_a^G]^T$ is through the inertia of the foot sensor signal observation combined with zero velocity update algorithm to calculate the location information in the ankle. So we have two constraints:

$$\int_0^t \int_0^t a_k^G dt = [x_k^G, y_k^G, z_k^G]^T + [x_h^G, y_h^G, z_h^G]^T \tag{4}$$

$$\int_0^t \int_0^t a_a^G dt = \int_0^t \int_0^t a_k^G dt + [x_a^G, y_a^G, z_a^G]^T \tag{5}$$

In this case, we can adopt a gradient projection based method [23]. Our goal is to determine the real 3D position of the knee joint from the sensor observations $\tilde{\boldsymbol{p}}$,

$$\tilde{\boldsymbol{p}} = \boldsymbol{p} + \boldsymbol{\varepsilon} \tag{6}$$

where ε is the unknown error in the process of lower limb movement, such as sensor bias or sensor sliding due to muscle strain, etc. This can be solved by $min F(\boldsymbol{p})$:

$$F(\boldsymbol{p}) = \int_\Omega \mid \nabla^\zeta \boldsymbol{p} \mid^\eta dx + \lambda \|\tilde{\boldsymbol{p}} - \boldsymbol{p}\|^2 \tag{7}$$

where λ is a nonnegative Lagrange multiplier and ζ is the regularization coefficient.

$$\mid \nabla^\zeta \boldsymbol{p} \mid^\eta = \sqrt{\boldsymbol{p}^2 + \zeta^2} \tag{8}$$

The Euler-Lagrange's equation can be used to solve the minimization problem as follows:

$$\boldsymbol{p}_{ideal} = \nabla\left(\frac{\nabla \boldsymbol{p}}{\mid \nabla^\zeta \boldsymbol{p} \mid^{2-\eta}}\right) + \varsigma(\tilde{\boldsymbol{p}} - \boldsymbol{p}) \tag{9}$$

where $\varsigma = 2\lambda/\eta$ is the direction of the gradient descent constraint factor, We can determine the differential of this minimum function and set it as zero to calculate λ. In the simplest case, $\zeta = 0$, $\eta = 1$. Then we will find the optimal solution of $\boldsymbol{p}$ through a series of iterative operation, i.e., $\boldsymbol{p}_{ideal}$. And this position estimation is the most close to the real value. As mentioned above, whenever we need to compute a joint location information, we can calculate the angular velocity integral of next level of connecting rod, so as to realize positioning calculation for the knee joint by means of the kinematics information of shank. Define φ, ϕ and ψ as Euler Angle around x, y and z axe, then we have (10) to calculate the knee joint:

$$\begin{cases} x_k^G = \frac{L_1 \pm L_1\sqrt{1-[1+(c\varphi c\phi c\psi)^2][1-c^2\psi+c^2\phi c^2\psi-(c\varphi c\phi c\psi)^2]}}{1+(c\varphi c\phi c\psi)^2} \\ y_k^G = \sqrt{L_1^2 - z_k^G s\psi} \\ z_k^G = \sqrt{L_1^2 - x_k^G s\varphi} \end{cases} \tag{10}$$

where c and s represent cosine and sine functions respectively.

Since the variable here is 3D direction angle, so we need to calculate separately the first order partial derivatives of knee joint position for three euler angles:
Set $D = c^2\varphi c^2\phi c^2\psi$,
$E = \sqrt{1-(1+D)(1-c^4\psi + c^4\phi c^4\psi - c^4\varphi c^4\phi c^2\psi)}$
and we have:

$$\begin{cases} \frac{\partial y_k^G}{\partial \varphi} = -\frac{z_k^G \frac{\partial z_k^G}{\partial \varphi} s\psi}{\sqrt{L_s^2-(z_k^G)^2}} \\ \frac{\partial y_k^G}{\partial \phi} = -\frac{z_k^G \frac{\partial z_k^G}{\partial \phi} s\psi}{\sqrt{L_s^2-(z_k^G)^2}} \\ \frac{\partial y_k^G}{\partial \psi} = -\frac{z_k^G \frac{\partial z_k^G}{\partial \psi} s\psi}{\sqrt{L_s^2-(x_k^G)^2}} + \sqrt{L_s^2 - (x_k^G)^2 c^2\psi} \end{cases} \tag{11}$$

$$\begin{cases} \frac{\partial z_k^G}{\partial \varphi} = -\frac{x_k^G \frac{\partial x_k^G}{\partial \varphi} s\varphi}{\sqrt{L_s^2-(x_k^G)^2}} + \sqrt{L_s^2 - (x_k^G)^2 c^2\varphi} \\ \frac{\partial z_k^G}{\partial \phi} = -\frac{x_k^G \frac{\phi x_k^G}{\partial \varphi} s\varphi}{\sqrt{L_s^2-(x_k^G)^2}} \\ \frac{\partial z_k^G}{\partial \psi} = -\frac{x_k^G \frac{\psi x_k^G}{\partial \varphi} s\varphi}{\sqrt{L_s^2-(x_k^G)^2}} \end{cases} \tag{12}$$

Based on these restricted condition, we can use the formulae (4) and (5) to correct the position estimation error of knee joint $[x_k^G, y_k^G, z_k^G]^T$ Similarly, one can use this method to calculate the accurate 3D hip position $[x_h^G, y_h^G, z_h^G]^T$.

Where $\left[x_k^G, y_k^G, z_k^G\right]^T$ is the knee joint displacement vector based on shank motion, while $\left[x_h^G, y_h^G, z_h^G\right]^T$ is the hip joint displacement vector based on thigh motion.

3 Experimental Results and Analysis

Gait analysis system have been adopted to carry out preliminary clinical trials at the Second Affiliated Hospital of Dalian Medical University, Fig. 3 shows two of the patients with diseases of the nervous system. The left one in Fig. 3 is a 35 years old male with 176 cm and 75 kg. This patient has a clinical symptoms of cerebral hemorrhage, during the recovery process, he has been suffered from hemiplegia on the right part of body. While the right one is a 67 years old male with 170 cm and 68 kg, who has a clinical symptoms of cerebral infarction and the left part of the body showed hemiplegia gait character.

Table 1 is the gait cycle partition result of the patient with cerebral hemorrhage. As shown in the table, there exists obvious differences between left foot and right foot. As for the left foot, the swing phase is shortened and the stance phase is prolonged, which means that the center of gravity become unstable when the patient performs swing motion. And the ankle control ability is worse due to the neurological damage (Table 2).

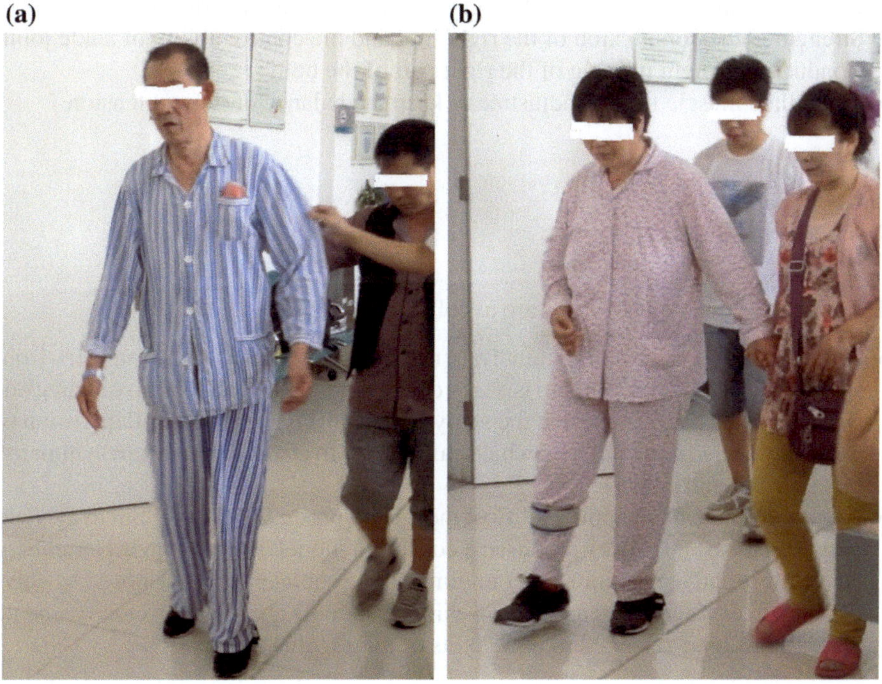

Fig. 3 Clinical gait analysis application. **a** A patient with cerebral hemorrhage. **b** A patient with cerebral infarction

Table 1 Joint angle calculation results of linear walking experiments

Percentage	Stance (%)	Heel strike (%)	Heel off (%)	Swing (%)
Left foot	47.61	15.53	15.82	27.82
Right foot	29.90	15.46	17.87	36.77
Normal value	40	15	20	40

Table 2 Joint angle calculation results of linear walking experiments

Joint angle (°)	Heel strike	Heel off	Swing
Hip joint	25.8 ± 6.32	-20.05 ± 5.29	5.47 ± 4.54
Knee joint	7.26 ± 3.61	39.45 ± 6.14	66.23 ± 5.42
Ankle joint	8.75 ± 4.49	-19.56 ± 6.81	-8.36 ± 4.24

On account of the above gait analysis, the therapist made the following rehabilitation training plan:

- Correct the strephenopodia on the right foot, adopting ankle orthoses if necessary;
- Strengthen the dorsiflexion of the right foot and the control ability of ankle joint. Conducting weight training of the right part of the body;
- Strengthen muscle tension adjustment, keep the balance of muscle tension.

4 Conclusions

The main research contributions are as follows:

- In literature, most of the gait analysis researches are based on parameters from single foot. Which means there is a lack of study about the collaborative strategy of both feet. This paper presented the study of bipedal coordination by data fusion of gait parameters from dual feet, which can be used to evaluate biped gait symmetry, gait stability and consistency.
- In order to solve the inapplicability of the widely used ZUPT algorithm in the field of gait analysis, this paper proposed a constraint model using Denavit-Hartenberg convention, and the position error minimization problem can be disposed by solving the Euler-Lagrange equation, in this way, the usage of zero velocity update algorithm can be extended to other parts of lower limbs.

Acknowledgements This work was supported by the National Natural Science Foundation of China under Grant 61473058, and in part by the Fundamental Research Funds for the Central Universities under Grant DUT18RC(4)034. This project was supported by China Postdoctoral Science Foundation under Grant 2017M621132 and 2017M621131. The authors would like to express their thanks to these funding bodies.

References

1. Verghese, J., Holtzer, R., Lipton, R.B., Wang, C.: Quantitative gait markers and incident fall risk in older adults. J. Gerontol. Ser. A: Biol. Sci. Med. Sci. **64**(8), 896–901 (2009)
2. Senden, R., Savelberg, H., Grimm, B., Heyligers, I., Meijer, K.: Accelerometry-based gait analysis, an additional objective approach to screen subjects at risk for falling. Gait Posture **36**(2), 296–300 (2012)
3. Mortaza, N., Abu Osman, N., Mehdikhani, N.: Are the spatio-temporal parameters of gait capable of distinguishing a faller from a non-faller elderly. Eur. J. Phys. Rehabil. Med. **50**(6), 677–691 (2014)
4. Zhou, H., Hu, H.: Reducing drifts in the inertial measurements of wrist and elbow positions. IEEE Trans. Instrum. Meas. **59**(3), 575–585 (2010)
5. Prakash, C., Gupta, K., Mittal, A., Kumar, R., Laxmi, V.: Passive marker based optical system for gait kinematics for lower extremity. Proc. Comput. Sci. **45**(3), 176–185 (2015)
6. Park, S.Y., Lee, S.Y., Kang, H.C., Kim, S.M.: EMG analysis of lower limb muscle activation pattern during pedaling: experiments and computer simulations. Int. J. Precis. Eng. Manuf. **13**(4), 601–608 (2012)
7. Wang, Z., Zhao, C., Qiu, S.: A system of human vital signs monitoring and activity recognition based on body sensor network. Sens. Rev. **34**(1), 42–50 (2014)
8. Qiu, S., Wang, Z., Zhao, H., Liu, L., Li, J., Jiang, Y., Fortino, G.: Body sensor network based robust gait analysis: toward clinical and at home use. IEEE Sens. J. 1–9 (2018)
9. Yu, L., Zheng, J., Wang, Y., Song, Z., Zhan, E.: Adaptive method for real-time gait phase detection based on ground contact forces. Gait Posture **41**(1), 269–275 (2015)
10. Qiu, S., Wang, Z., Zhao, H., Hu, H.: Using distributed wearable sensors to measure and evaluate human lower limb motions. IEEE Trans. Instrum. Meas. **65**(4), 939–950 (2016)
11. Chen, S., Lach, J., Member, S., Lo, B., Member, S.: Sensors: a systematic review. IEEE J. Biomed. Health Inf. **20**(6), 1521–1537 (2016)
12. Qiu, S., Wang, Z., Zhao, H., Liu, L., Jiang, Y.: Using body-worn sensors for preliminary rehabilitation assessment in stroke victims with gait impairment. IEEE Access **6**, 31249–31258 (2018)
13. Fortino, G., Giannantonio, R., Gravina, R., Kuryloski, P., Jafari, R.: Enabling effective programming and flexible management of efficient body sensor network applications. IEEE Trans. Hum. Mach. Syst. **43**(1), 115–133 (2013)
14. Qiu, S., Yang, Y., Hou, J., Ji, R., Hu, H., Wang, Z.: Ambulatory estimation of 3D walking trajectory and knee joint angle using MARG sensors. In: Fourth International Conference on Innovative Computing Technology (INTECH), pp. 191–196 (2014)
15. Wu, D., Wang, Z., Chen, Y., Zhao, H.: Mixed-kernel based weighted extreme learning machine for inertial sensor based human activity recognition with imbalanced dataset. Neurocomputing **190**, 35–49 (2016)
16. Gravina, R., Alinia, P., Ghasemzadeh, H., Fortino, G.: Multi-sensor fusion in body sensor networks: state-of-the-art and research challenges. Inf. Fusion **35**, 68–80 (2016)
17. Wang, Z., Qiu, S., Cao, Z., Jiang, M.: Quantitative assessment of dual gait analysis based on inertial sensors with body sensor network. Sens. Rev. **33**(1), 48–56 (2013)
18. Qiu, S., Wang, Z., Zhao, H.: Heterogeneous data fusion for three-dimensional gait analysis using wearable MARG sensors. Int. J. Comput. Sci. Eng. **14**(3), 222–233 (2017)
19. Qiu, S., Wang, Z., Zhao, H., Qin, K., Li, Z., Hu, H.: Inertial/magnetic sensors based pedestrian dead reckoning by means of multi-sensor fusion. Inf. Fusion **39**, 108–119 (2018)
20. Farris, R.J., Quintero, H.A., Murray, S.A., Member, S., Ha, K.H., Hartigan, C., Goldfarb, M.: A preliminary assessment of legged mobility provided by a lower limb exoskeleton for persons with paraplegia. IEEE Trans. Neural Syst. Rehabil. Eng. **22**(3), 482–490 (2014)
21. Bamberg, S.J.M., Benbasat, A.Y., Scarborough, D.M., Krebs, D.E., Paradiso, J.A.: Gait analysis using a shoe-integrated wireless sensor system. IEEE Trans. Inf. Technol. Biomed. **12**(4), 413–23 (2008)

22. Favre, J., Aissaoui, R., Jolles, B.M., de Guise, J.A., Aminian, K.: Functional calibration procedure for 3D knee joint angle description using inertial sensors. J. Biomech. **42**(14), 2330–2335 (2009)
23. Roetenberg, D., Luinge, H., Slycke, P.: Xsens MVN: Full 6DOF human motion tracking using miniature inertial sensors, pp. 1–9. XSENS TECHNOLOGIES (2013)
24. Wang, Z., Li, J., Wang, J., Zhao, H., Qiu, S., Yang, N., Shi, X.: Inertial sensor-based analysis of equestrian sports between beginner and professional riders under. IEEE Trans. Instrum. Meas. **14**(8), 1–13 (2018)

An Embedded Risk Prediction System for Wheelchair Safety Driving

Congcong Ma, Wenfeng Li, Qimeng Li, Raffaele Gravina, Yi Yang and Giancarlo Fortino

Abstract Development of intelligent wheelchairs can increase mobility and independence of impaired individuals. As there exist dangerous driving risks such as improper driving postures, moving too fast or on rough road, it is useful to monitor the wheelchair user's driving conditions and, particularly, predict potential driving risks. This paper proposes a risk prediction system for the wheelchair safety driving. A novel designed smart cushion is used to evaluate dangerous drivings risks. Our cushion is able to better combine the pressure sensors and accelerometer and it can detect sitting postures, wheelchair accelerations, and terrain conditions. In addition, we propose a prediction system to monitor the wheelchair driving status. Features are extracted and a fuzzy inference system is used to quantify the dangerous driving risks. Warnings or intervention control strategies will be trigged to increase wheelchair driving safety.

Keywords Smart cushion · Pressure sensor · Accelerometer
Wheelchair safety driving · Risk prediction · Fuzzy inference system

C. Ma (✉) · W. Li · Y. Yang
Wuhan University of Technology, Wuhan, China
e-mail: macc@whut.edu.cn

W. Li
e-mail: liwf@whut.edu.cn

Y. Yang
e-mail: yangyi_wl@whut.edu.cn

Q. Li · R. Gravina · G. Fortino
University of Calabria, Rende, Italy
e-mail: liqimeng@msn.com

R. Gravina
e-mail: rgravina@dimes.unical.it

G. Fortino
e-mail: g.fortino@unical.it

G. Fortino and Z. Wang (eds.), *Advances in Body Area Networks I*,
Internet of Things, https://doi.org/10.1007/978-3-030-02819-0_12

1 Introduction

With the development of Smart Cities and the Internet of Things (IoT), the quality of life for mobility impaired individuals can be improved by the increasingly advanced designed smart wheelchairs. Safe operation and driving risks prediction of wheelchair moving condition is very important for the users.

In our view, the development of smart wheelchairs is tightly coupled with the concept of smart cushion that can play an important role in the wheelchair system, thanks to its non-invasive characteristic. As a matter of fact, many studies are interested in improving the smart cushion as it can effectively provide several useful information of the seated user and the wheelchair itself. Kumar et al. [1] have designed the Care-Chair that can recognize different kinds of sedentary activities, and also detect user functional and emotion-based activities. Barba et al. [2] used a cushion with 16 pressure sensors to detect postural changes of users and to recognize three kinds of affective states in an e-learning environment. Furugori et al. [3] used seated posture information to detect mental fatigue on the basis of driver's posture change.

In our previous research, the smart cushion was used to detect postures [4], assess activity level [5], and fatigue condition [6]. These studies obtained good results and high recognition accuracy, but for more complex activities recognition or specific needs of users, there still exist some shortcomings.

In this paper, we propose a risk prediction system for wheelchair safety driving based on a novel smart cushion designed to better combine pressure sensors and inertial measurement sensor. The system can detect driving postures, measure wheelchair acceleration and assess terrain conditions.

The main contributions of the paper are twofold:

- A novel smart cushion proposed to can better combine the pressure sensors with inertial measurement sensor.
- A driving risks prediction system based on Fuzzy Inference System (FIS) suited with the smart cushion to support the wheelchair users.

The remainder of the paper is organized as follows. Section 2 discusses the related work on wheelchair safety driving. Section 3 introduces the development of the smart cushion and its implementation. Section 4 describes the prediction system using FIS. Section 5 is focused the experiments and preliminary results. Finally, Sect. 6 presents the conclusion and outlines future works.

2 Related Work

To date, most of the researchers focused on safety driving of vehicles, fewer focused on the safety driving of wheelchairs. In order to increase the safety of wheelchair users, researches on obstacle detection and road condition detection are examined.

Tyagi et al. [7] proposed a smart wheelchair system using fuzzy inference to decide the control method. If the obstacle is far, fingertip is used to control the movement, while if the obstacle is close, obstacle avoidance system is activated. Dryvendra et al. [8] proposed a low cost assistance kit for manual wheelchair users with enhanced obstacle detection. Ultrasonic sensor was used to detect the presence of obstacles behind the wheelchair and a LED emergency light system attached to the wheelchair can provide a bright and clear path for the user. Rockey et al. [9] proposed an obstacle avoidance method using Microsoft Kinect, laser scanner and sonar.

Candiotti et al. [10] designed and evaluated the seat orientation controller for uneven terrain driving of the electric-powered wheelchairs (EPWs); their system can provide sensing for a dynamic self-leveling seat application so to maintain the center of mass within the boundaries of the wheelchair with the aim to improve the EPW safety. Seki and Tanohata [11] proposed an assist driving control scheme for EPWs on various unsafe roads. Fuzzy rules were designed from human input torque, driving velocity and assisted torque reduction rate, so as to adjust the velocity also according to road conditions. Pineau et al. [12] collected the accelerometer data of EPW with 35 different activities (such as rolling on incline plane, rolling across multiple types obstacles) and used Support Vector Machine (SVM) binary classification to distinguish safe and unsafe events.

In this paper, we design a novel smart cushion that can detect postures, wheelchair driving acceleration, and terrain conditions. Using the cushion, we build a risk prediction system to guard the safety of the wheelchair users.

3 Novel Designed Smart Cushion

3.1 Hardware Design

Figure 1 shows the deployment of the sensors on the base board in the cushion. There are two types of sensors constituting our novel smart cushion:

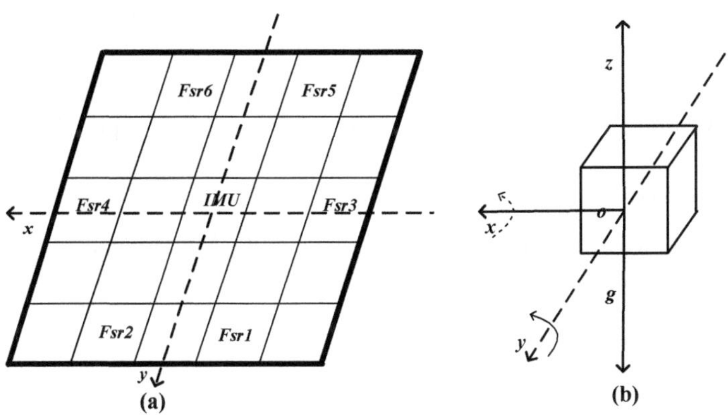

Fig. 1 Sensor deployment of the smart cushion

Table 1 Comparison of the state-of-the-art on smart cushions

	Pressure detection unit location	Accelerometer location	Integration unit
Yu et al. [14]	Seat and backrest	Seat	No
Zemp et al. [15]	Seat and backrest	Backrest	No
Ma et al. [13]	Seat and backrest	Waist	No
Proposed cushion	Seat	Seat	Yes

- a pressure sensors array, and
- an inertial measurement unit.

Compared with the former cushion designed using pressure sensors and accelerometer [13], our novel cushion has the advantage of one integration unit and easier deployment in the seat. Table 1 compares three representative smart cushions using the two kinds of sensors against our novel designed cushion.

3.1.1 Pressure Detection Unit

With respect to our former research [4] and [13], we designed the cushion with pressure sensors deployed as shown in Fig. 1a. Assuming a conventional size for the seat of 40 cm × 40 cm, we split it into 5 × 5 square zones (i.e. each zone is 8 cm × 8 cm). The six pressure sensors are separately placed as in Fig. 1a. Specifically, two pressure sensors ($Fsr3$ and $Fsr4$ in the figure) have been deployed to the left and right sides of the seat to better account for individual physiognomy differences.

3.1.2 Inertial Measurement Unit

We selected the MPU9250 (see Fig. 1b), a 9-DoF IMU [16], and we placed it in the center of the bottom side of the cushion. It consists of two chips:

- *MPU-6500*—contains a 3-axis gyroscope, a 3-axis accelerometer, and an on-board Digital Motion Processor (DMP) capable of processing complex Motion Fusion algorithms;
- *AK8963*—3-axis digital compass.

In this work, we use the inertial measurement unit (IMU) to measure acceleration on the three axes (a_x, a_y, a_z).

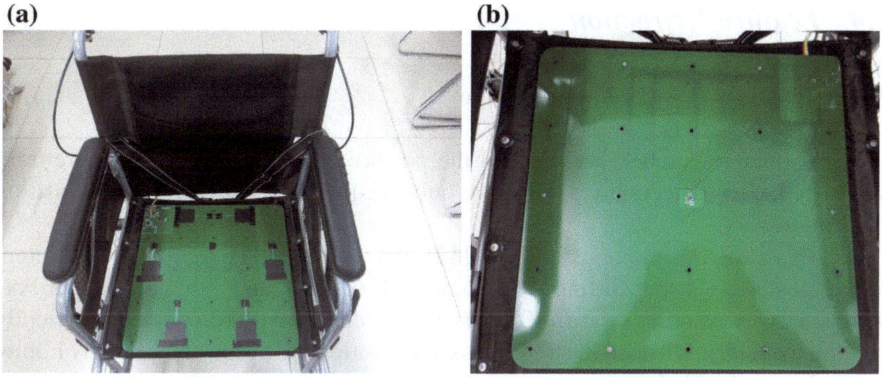

Fig. 2 Prototype of smart cushion on a real wheelchair

3.2 *Implementation*

The circuit board of the smart cushion is shown in Fig. 2, where Fig. 2a is top side of the board. Six pressure sensors were deployed as previously described.

On the left-top of the board is the control circuit unit, which is composed of:

- Micro Control Unit (MCU), to process the sensor data using Arduino pro mini;
- Bluetooth communication unit, to send the processed results to smart devices (such as computer, smartphone, or tablet);
- Vibration Motor, controlled by the MCU, is used to alarm the users;
- Power supply circuit, to provide power to the board.

Figure 2b is the opposite side of the circuit board; the IMU is placed in the center of the board. The circuit board will be eventually embedded in the foam filling of the cushion to make users comfortable.

3.3 *Data Collection*

Using the smart cushion, a set of time-series sample data can be obtained. Instance vectors are generated from the pressure sensors and IMU. In order to capture small movements of the user body, data was collected at a frequency of 10 Hz. At time t, a collection of raw data can be represented as $R_t = \{Fsr_{1,t}, Fsr_{2,t}, Fsr_{3,t}, Fsr_{4,t}, Fsr_{5,t}, Fsr_{6,t}, a_{x,t}, a_{y,t}, a_{z,t}\}$.

For risk prediction, it is very important to monitor data over a correct period in order to have actually the time to identify the dangerous situation and allow the user taking the right actions. Therefore, choosing the right time period is important. We set the time slide window as T s, with no overlap. The feature extraction method will be discussed in the following subsection.

3.4 Feature Extraction

3.4.1 Posture Transition Percentage

In our former researches, we realized the posture detection [17] and posture transition frequency detection [5]. Here we only focus on the sitting posture transition percentage.

Firstly, using our previous cushion [17], we are able to recognize five different sitting postures (Proper Sitting (PS), Lean Left (LL), Lean Right (LR), Lean Forward (LF) and Lean Backward (LB)). The aforementioned postures can be efficiently recognized with a J48 decision tree. This classification algorithm can be easily implemented on embedded devices such as Arduino platform. Using the instance vectors generated from the pressure sensors signals, postures can be recognized.

In a time slide window, composed of $10T$ postures, posture transition percentage (PTP) can be calculated as follows (see Eq. 1):

$$PTP = \frac{number\ of\ posture\ transitions}{10T} \tag{1}$$

From our previous research [5], we calculated the following results:

- *PTP range*—(0, 0.27), wheelchair user is in stationary or moderate activity and the user is driving in normal sitting postures;
- *PTP range*—(0.23, 0.55], wheelchair user is in hyperactivity and the user is driving in abnormal sitting postures.

3.4.2 Max Wheelchair Acceleration

Wheelchair overturn is an accident usually caused by rough road or unbalanced weight distribution due to improper posture. Therefore monitoring wheelchair acceleration (WHA) is very important. WHA is the Signal Magnitude Vector of IMU placed on the wheelchair; it can provide a measurement of the wheelchair movement intensity [18] and is calculated over the tri-axial acceleration values as shown in Eq. 2.

$$WHA_t = \sqrt{a_{x,t}2 + a_{y,t}2 + a_{z,t}2} \tag{2}$$

Then, Max Wheelchair Acceleration (MWA) value can be found among the instantaneous WHA_t values in each time slide window.

As in our previous research [13], when the wheelchair is in static state or moving slowly, the *WHA* value is in the range of [0.7g, 1.5g]. In the probability of wheelchair overturn, each axis will have high fluctuation and the *WHA* will reach a peak value exceeding 2.5g. So, in the range of (1.5g, 2.5g), the wheelchair has real probability of overturn. During actual wheelchair overturn, the *WHA* value might reach a value of 4g.

According to this, two states of wheelchair acceleration can be defined as follows:

- *MWA range*—$[0.7g, 2.5g)$, the wheelchair is in safe status;
- *MWA range*—$(1.5g, 4g)$, the wheelchair is in unsafe status.

3.4.3 Max Incline Angle

When the wheelchair moves on a flat floor, it can be considered in stable equilibrium. However, when wheelchair is moving uphill, downhill or on uneven roads, the wheelchair will not keep a continuously safe equilibrium; thus, there is probability to overturn. The wheelchair incline angle can be abstracted as shown in Fig. 3.

The incline angle of each axis can be calculated as follows (see Eqs. 3–5).

$$\alpha_t = \arctan \frac{a_{x,t}}{\sqrt{a_{y,t}2 + a_{z,t}2}} \tag{3}$$

$$\beta_t = \arctan \frac{a_{y,t}}{\sqrt{a_{x,t}2 + a_{z,t}2}} \tag{4}$$

$$\gamma_t = \arctan \frac{a_{z,t}}{\sqrt{a_{x,t}2 + a_{y,t}2}} \tag{5}$$

As we are focused on left/right and front/back overturn, we need to calculate α_t and β_t. Then, we choose the Max Incline Angle (MIA) of α_t and β_t in one time slide window. As discussed in [11], most typical roads have less than 10° inclination. Trials on uphill roads with incline angle of 4° and 6° were examined and their results showed that such inclinations constitute already a risk of wheelchair overturn.

By means of their empirical observation, we define the following MIA ranges:

- *MIA range*—[0°, 6°), safe driving terrain incline angle;
- *MIA range*—(4°, 10°), unsafe driving terrain incline angle.

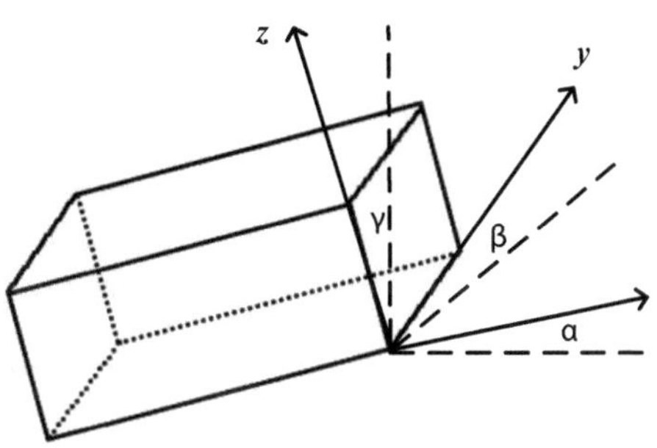

Fig. 3 The abstraction of wheelchair incline angle

4 FIS-Based Risk Prediction System

4.1 System Design

As described before, the proposed system can detect posture transition percentage, max wheelchair acceleration, and max incline angle. We can therefore develop a risk prediction method with alarm services using these information. The method design is shown in Fig. 4.

The risk prediction method consists of the following steps:

1. Detect the posture transition, wheelchair acceleration and terrain incline using the smart cushion.
2. Use the FIS designed in Sect. 4.2 to predict the driving risks.
3. If the wheelchair user is in safety driving condition, return to (2) and continue to monitor the driving condition. If the user is in warning condition, alarm the user to adjust the posture and to possibly take a safer road. If the user is in danger, the intervention control system will be activated automatically to slowdown the speed (only available for EPWs), or it will alarm the user to slowdown the speed (in case of manual wheelchairs); meanwhile, the system will send alarms to warn the user as in the warning condition.

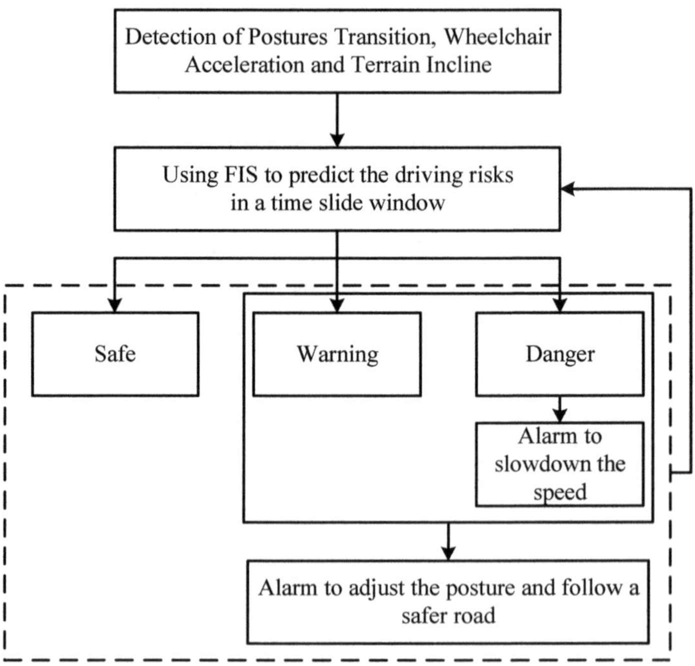

Fig. 4 Risk prediction method for wheelchair users using smart cushion

We performed wheelchair moving experiments with different time windows (T), chosen as in former research. Specifically, in [19], three kinds of wheelchair physical activities (PAs) were defined: (i) nearly-stationary PAs ($speed < 0.03\,\mathrm{m/s}$), (ii) PAs involving wheelchair movement in the range of ($0.03\,\mathrm{m/s} < speed < 0.2\,\mathrm{m/s}$), and (iii) PAs with consistent wheelchair movement ($speed > 0.2\,\mathrm{m/s}$). In brief, we considered time windows of 10, 30 and 60 s. Empirical results showed that 10 s is indeed sufficient to capture the dynamic movement for predicting the risks.

4.2 *Fuzzy Inference System*

In some cases the available information might not so precisely evaluated. This makes fuzzy logic the ideal methodology as it bring manageability, robustness, low cost solution, and a better relation with the reality [20].

Initially, using the selected features, we defined the fuzzy sets and include different types of the membership functions to fuzzificate the input features in the FIS. After fuzzification, *if-then*-based rules that control fuzzy logic between the input and output features can be defined. Consequently, output features can be inferred from the rules according to the defuzzification process. Finally, we can apply the FIS classification.

4.2.1 Input Features to Fuzzy Sets Conversion

Fuzzy membership function is usually determined by the perception of experimental data and simulation. Triangular, trapezoidal, S-shape and gauss fuzzy membership functions have been widely used in the construction of fuzzy membership functions.

This work proposed the division of the three parameters (i.e. the feature defined in Sect. 3.4) by two different membership functions: safe and unsafe. Table 2 presents the details of each fuzzy set configuration.

As trapezoidal fuzzy membership function is easy to construct. Also it's rational for two groups which are safe and unsafe, as shown in Fig. 5a–c. The driving safety condition can be predicted based on the inputs. As to the outputs, we defined three levels of driving conditions: safe, warning, and danger. The fuzzy membership function is shown in Fig. 5d.

Table 2 Fuzzy sets of each feature

Features	Range	Safe	Unsafe
PTP	(0, 0.55]	(0, 0.27)	(0.23, 0.55]
MWA	[0.7, 4)	[0.7, 2.5)	(1.5, 4)
MIA	[0, 10)	[0, 6)	(4, 10)

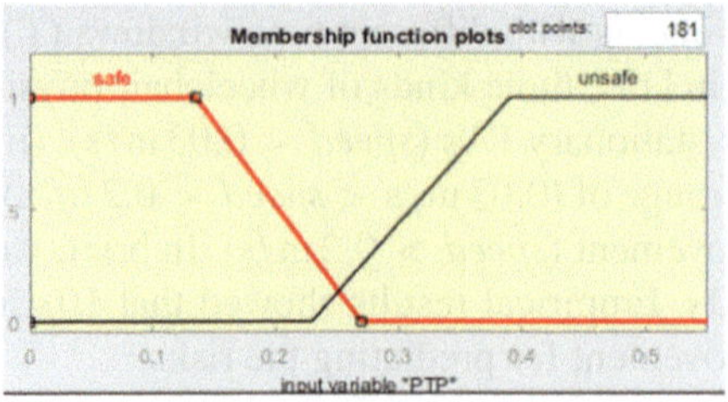

(a) Membership function of PTP

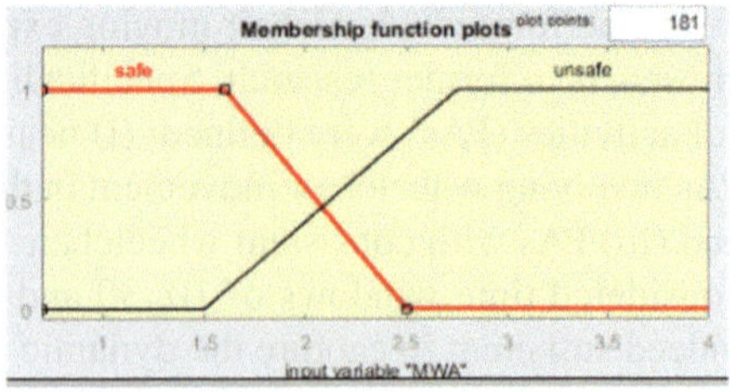

(b) Membership function of MWA

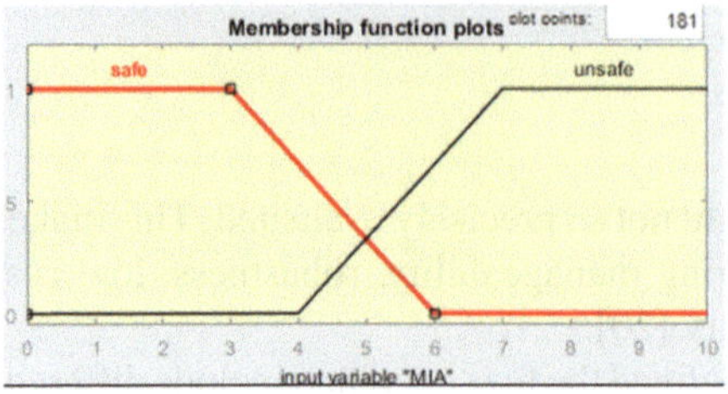

(c) Membership function of MIA

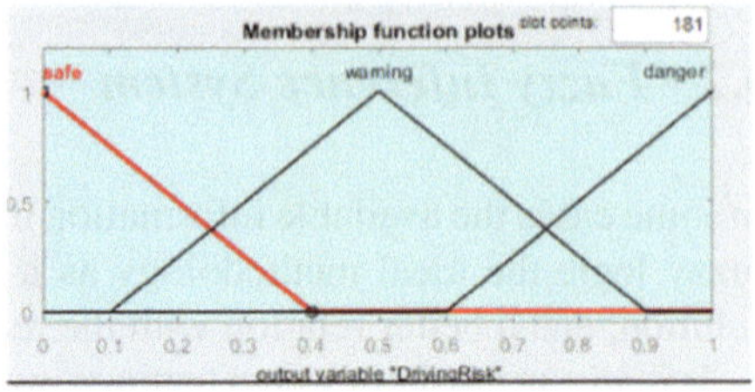

(d) Membership function of Driving Risk

Fig. 5 Membership function of each feature and output

Table 3 Fuzzy inference rules of the wheelchair system

Rules	Input factors			Driving risk
	PTP	MWA	MIA	
Rule 1	Safe	Safe	Safe	Safe
Rule 2	Unsafe	Safe	Safe	Warning
Rule 3	Safe	Unsafe	Safe	Warning
Rule 4	Safe	Safe	Unsafe	Warning
Rule 5	Unsafe	Unsafe	Unsafe	Danger

4.2.2 Fuzzy Inference

Mamdani's fuzzy system [21] has been widely used in expert systems due to their comprehensible rules; thus, in this paper we used such algorithm to describe the membership functions and output distributions. The rules of predicting the driving risk are shown in Table 3.

4.2.3 Defuzzification

Defuzzification is applied to convert the fuzzy result into a matching value; it is carried out by a centroid method. We defined five levels of risks: No Risk (NR), Low Risk (LR), Medium Risk (MR), Medium-High Risk (MH), and High Risk (HR), as shown in Table 4.

Table 4 Risk levels of defuzzification

No risk	Low	Medium	Medium-high	High
(0, 0.1]	(0.1, 0.4]	(0.4, 0.6]	(0.6, 0.9]	(0.9, 1.0]

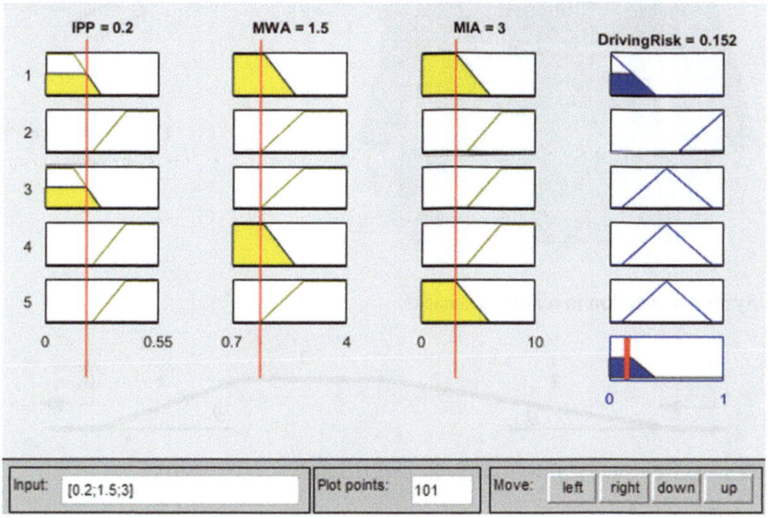

Fig. 6 Substitution of fuzzy parameters for quantification of driving risk

For example, as shown in Fig. 6, if the values of the feature set PTP, MWA, MIA from the measured data are set to {0.2, 1.5, 3} respectively, then the output value is 0.152, which is in the range of (0.1, 0.4] and it is therefore recognized as a low risk condition.

5 Experiments and Results

5.1 Implementation in Real Life-Scenario

We realized a prototype system (see Fig. 7) and carried out preliminary experiments. Using the smart cushion placed on the wheelchair seat, we can compute the three features introduced in Sect. 3.4, which are transmitted via Bluetooth to the computer, where the trained FIS timely detects the dangerous driving risks.

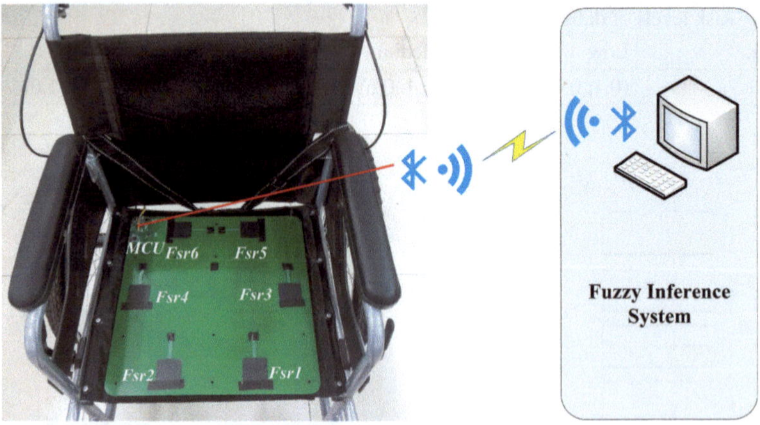

Fig. 7 System realization in real life scenario

Fig. 8 Descriptive schema of the road in the experiment

5.2 *Preliminary Results*

In order to simulate incline roads, we used wooden boards to construct the obstacles as shown in Fig. 8. For the first segment, the length and incline angle are $l_1 = 100$ cm and $\theta_1 = 3°$, respectively. For the second segment, the length and incline angle are $l_2 = 50$ cm and $\theta_2 = 0°$, respectively. For the last segment, the length and incline angle are $l_3 = 50$ cm and $\theta_3 = 6°$, respectively.

The experiment was carried out in a corridor to simulate different road conditions. We recruited 4 students that were asked to go across the obstacle at sustained speed, in order to simulate the dangerous risks. Each participant repeated the experiment 10 times. As in Sect. 4.1, the time slide window is set as 10 s. For a constant push of the wheelchair, we observe that each experiment could go across the obstacle in 10 s (a time slide window duration).

As the obstacle is fixed, terrain' Max Incline Angle is constant ($MIA = 6°$), so *PTP* and *MWA* are the two main variables. In the following, we give preliminary experiment results as shown in Fig. 9. The x-axis represents the number of samples. The y-axis are *PTP*, *MWA* and risk probability separately.

The *PTP* is in the range of [0.01, 0.38], as shown in Fig. 9a. The *MWA* of each experiment is shown in Fig. 9b. Finally, Fig. 9c shows the risk prediction probability determined by *PTP* and *MWA*. We can see that the risk probability is in the range of (0.50, 0.59) which can be classified as Medium Risk condition (see Table 4). As

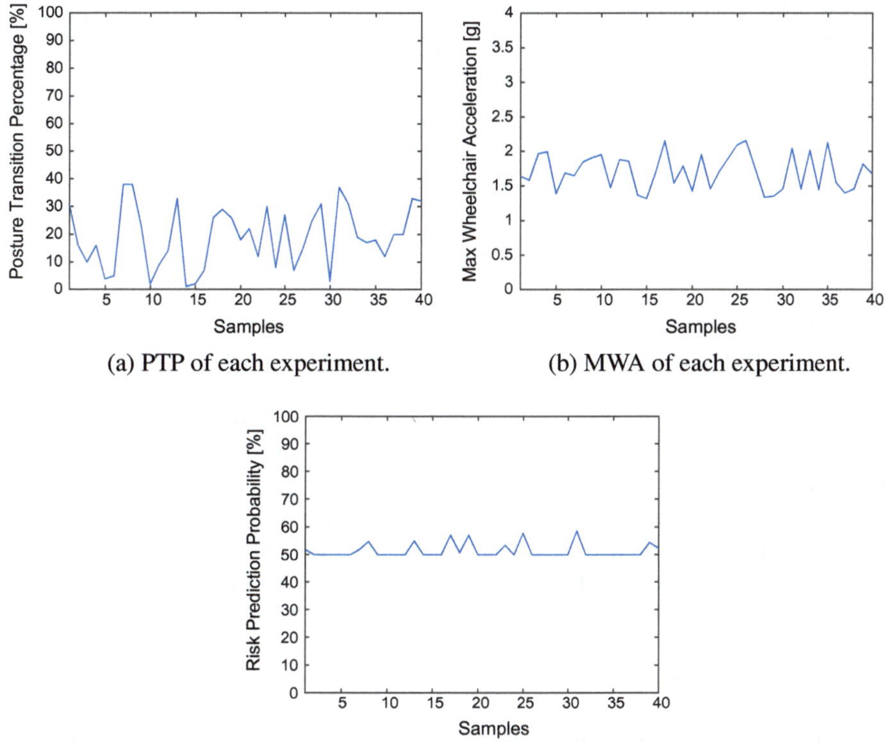

(a) PTP of each experiment.

(b) MWA of each experiment.

(c) Risk prediction probability results.

Fig. 9 Features value and risk prediction results of each experiment

Table 5 Wheelchair moving with similar *PTP*

Sample	*PTP*	*MWA*	Risk probability (%)
7	0.38	1.65	51.78
8	0.38	1.85	54.69
31	0.37	2.04	58.57

we can see in Fig. 9, most of the results fall into the value of 50%(risk prediction probability).

For a more comprehensive exemplification, we compared some sample results as shown in Tables 5 and 6. In Table 5, the value of *PTP* is a bitter large, the larger value of *MWA* can cause the higher probability of driving risk. Also in Table 6, with the value of *MWA* a bitter large, the larger value of *PTP* can cause the higher probability of driving risk. But, in the condition of the *PTP* value a bitter small, the driving risk can be reduced to a relatively lower probability.

Table 6 Wheelchair moving with similar *MWA*

Sample	*PTP*	*MWA*	Risk probability (%)
17	0.26	2.16	57.11
26	0.07	2.16	50
35	0.18	2.13	50

6 Conclusion

This paper presented a risk prediction system using a novel designed smart cushion. With such cushion, three features like posture transition percentage, max wheelchair acceleration, and max incline angle can be calculated. Fuzzy Inference System was adopted to better fuse the three kinds of features to quantify the risk probability. The proposed system has been prototyped and preliminary experimentation carried out in real-life scenario.

In the near future, more studies on the smart cushion will be carried out to reveal further useful information related to the user general health. In order to make the wheelchair driving prediction system portable, the FIS will be implemented on Raspberry Pi rather than on the PC. Finally, to determine empirically the membership function for each feature, further experiments will be carried out so to develop an adaptive control system.

Acknowledgements The research is financially supported by China-Italy S&T Cooperation project "Smart Personal Mobility Systems for Human Disabilities in Future Smart Cities" (China-side Project ID: 2015DFG12210, Italy-side Project ID: CN13MO7), and the National Natural Science Foundation of China (Grant Nos: 61571336 and 61502360). This work has been also carried out under the framework of INTER-IoT, Research and Innovation action—Horizon 2020 European Project, Grant Agreement #687283, financed by the EU.

References

1. Kumar, R., Bayliff, A., De, D., Evans, A., Das, S.K., Makos, M.: Care-chair: sedentary activities and behavior assessment with smart sensing on chair backrest. In: 2016 IEEE International Conference on Smart Computing, pp. 1–8. IEEE (2016)
2. Barba, R., Ángel, P., Boticario, J.G.: Development of an inexpensive sensor network for recognition of sitting posture. Int. J. Distrib. Sensor Netw. 1–13 (2015)
3. Furugori, S., Yoshizawa, N., Iname, C., Miura, Y.: Measurement of driver's fatigue based on driver's postural change. In: SICE Annual Conference, pp. 264–269 (2003)
4. Ma, C., Li, W., Gravina, R., Fortino, G.: Posture detection based on smart cushion for wheelchair users. Sensors **17**(4) (2017)
5. Ma, C., Gravina, R., Li, W., Zhang, Y., Li, Q., Fortino, G.: Activity level assessment of wheelchair users using smart cushion. In: IEEE International Conference BodyNets, pp. 104–110 (2016)
6. Ma, C., Li, W., Cao, J., Wang, S., Wu, L.: A fatigue detect system based on activity recognition. In: IEEE International Conference IDCS, pp. 303–311. Springer (2014)

7. Tyagi, V., Gupta, N.K., Tyagi, P.K.: Smart wheelchair using fuzzy inference system. In: Global Humanitarian Technology Conference: South Asia Satellite, pp. 175–180 (2013)
8. Dryvendra, R.M., Chinnavan, E., Puviarasi, R.: A better engineering design: low cost assistance kit for manual wheelchair users with enhanced obstacle detection. J. Eng. Technol. Sci. 389–405 (2015)
9. Rockey, C.A., Perko, E.M., Newman, W.S.: An evaluation of low-cost sensors for smart wheelchairs. In: IEEE International Conference on Automation Science and Engineering, pp. 249–254 (2013)
10. Candiotti, J., Wang, H., Chung, C.S., Kamaraj, D.C., Grindle, G.G., Shino, M., Cooper, R.A.: Design and evaluation of a seat orientation controller during uneven terrain driving. Med. Eng. Phys. **38**(3), 241–247 (2016)
11. Seki, H., Tanohata, N.: Fuzzy control for electric power-assisted wheelchair driving on disturbance roads. IEEE Trans. Syst. Man Cybern. Part C (Applications and Reviews) **42**(6), 1624–1632 (2012)
12. Pineau, J., Moghaddam, A.K., Yuen, H.K., Archambault, P.S., Routhier, F., Michaud, F., Boissy, P.: Automatic detection and classification of unsafe events during power wheelchair use. IEEE J. Trans. Eng. Health Med. **2**, 1–9 (2014)
13. Ma, C., Li, W., Cao, J., Gravina, R., Fortino, G.: Cloud-based wheelchair assist system for mobility impaired individuals. In: IEEE International Conference IDCS, pp. 107–118 (2016)
14. Yu, H., Stoelting, A., Wang, Y., Yi, Z., Sarrafzadeh, M.: Providing a cushion for wireless healthcare application development. IEEE Potent. **29**, 19–23 (2010)
15. Zemp, R., Tanadini, M., Plüss, S., Schnüriger, K., Singh, N.B., Taylor, W.R., Lorenzetti, S.: Application of machine learning approaches for classifying sitting posture based on force and acceleration sensors. Biomed. Res. Int. **2016**(1), 1–9 (2016)
16. MPU9250: https://www.invensense.com/products/motion-tracking/9-axis/mpu-9250/, April 2017
17. Ma, C., Li, W., Gravina, R., Fortino, G.: Activity recognition and monitoring for smart wheelchair users. In: IEEE International Conference CSCWD, pp. 664–669 (2016)
18. Ni, Q., Patterson, T., Cleland, I., Nugent, C.: Dynamic detection of window starting positions and its implementation within an activity recognition framework. J. Biomed. Inf. **62**, 171–180 (2016)
19. Hiremath, S.V., Intille, S.S., Kelleher, A., Cooper, R.A., Ding, D.: Detection of physical activities using a physical activity monitor system for wheelchair users. Med. Eng. Phys. **37**(1), 68–76 (2015)
20. Zadeh, L.A.: Fuzzy logic = computing with words. IEEE Trans. Fuzzy Syst. **4**(2), 3–23 (1999)
21. Guney, K., Sarikaya, N.: Comparison of mamdani and sugeno fuzzy inference system models for resonant frequency calculation of rectangular microstrip antennas. Prog. Electromag. Res. B **12**(12), 81–104 (2009)

[illegible]

A Wearable, Low-Power, Real-Time ECG Monitor for Smart T-shirt and IoT Healthcare Applications

Taiyang Wu, Jean-Michel Redouté and Mehmet Yuce

Abstract A wearable health monitoring system combined with Internet of Things (IoT) is going to be a promising alternative to the conventional healthcare systems. In this work, a small, flexible and wearable real-time electrocardiograph (ECG) monitoring system integrated on a T-shirt is proposed. It uses an off-the-shelf biopotential analog front end (AFE) chip, AD8232, to collect subjects' ECG data with satisfactory quality. The collected ECG data are transmitted through Bluetooth low energy (BLE) to an end device for real-time display. A PC graphical user interface (GUI) and a smartphone application are designed for indoor and outdoor real-time visualisation respectively. The power consumption of the proposed wearable ECG monitoring system can be as low as 5.2 mW. Powered by a 240 mAh rechargeable battery, it can operate for more than 110 h continuously. To prolong the lifetime of the battery, a flexible solar energy harvester is also adopted within this system.

1 Introduction

The population aging draws more and more attention into the research of wearable healthcare monitoring systems [1]. Among the vital signals of the human health, the electrocardiogram (ECG) signal is of great significance as the cardiovascular disease (CVD) is one of the principle causes of catastrophic mortality [2]. The early detection and treatment of CVDs will help a lot in increasing the patients' survival

T. Wu · J.-M. Redouté · M. Yuce (✉)
Electrical and Computer Systems Engineering, Monash University,
Wellington Road, Clayton, VIC, Australia
e-mail: mehmet.yuce@monash.edu

T. Wu
e-mail: taiyang.wu@monash.edu

J.-M. Redouté
e-mail: jean-michel.redoute@monash.edu

G. Fortino and Z. Wang (eds.), *Advances in Body Area Networks I*,
Internet of Things, https://doi.org/10.1007/978-3-030-02819-0_13

chances. Therefore, the real-time and long-term monitoring of ECG is important in the promptly intervention of CVDs [3]. With the emerging of Internet of Things (IoT) technology, the remote health monitoring is enabled by transmitting the real-time ECG data through the cloud to healthcare centres [4–6].

Implementation of a wearable ECG monitoringsystem should output a good quality ECG signal, be comfortable to wear and be low-power consumption for long-term and continuous use. Lots of research has been carried out to develop wearable ECG systems in the recent years. A low-power and miniaturized ECG data acquisition system is proposed in [7], which uses textile electrodes to collect ECG data and store them on a SD card. In [8], a flexible finger-based ECG acquisition system that can send data through Bluetooth low energy (BLE) is proposed and compared with commercial portable ECG systems. E. Spano et al. propose a remote ECG system dedicated to long-term monitoring in residential environment [4]. It aims to be integrated in a broader IoT infrastructure with Zigbee and Internet server. In [9], the effectiveness of using ubiquitous smartphones in the real-time ECG recording and viewing is studied. This will reduce the cost of personnel required in the daily ECG monitoring and notification. Y. Wang et al. evaluate the BLE transmission and smartphone application in the real-time wireless ECG monitoring in [10]. An IoT-cloud based wearable ECG monitoring system is also presented in [5]. It gathers the ECG data by a wearable sensor node and send them directly to the IoT cloud using Wi-Fi. As most of the power consumed in wearable ECG monitoring systems is the wireless transmission, some studies have focused on data compression and low-power wireless technologies and infrastructures [4, 10]. Another possible improvement is adopting energy harvesting technique as presented in [11], where a solar energy harvesting module for vital signal measurement is designed, manufactured and tested.

This chapter presents a wearable ECG monitoring system integrated on a T-shirt towards the IoT based healthcare applications. It collects the ECG data using an off-the-shelf biopotential IC chip with satisfactory accuracy. With the wireless transmission of BLE, the collected real-time ECG data can be displayed on a PC GUI and a smartphone application. The proposed wearable ECG monitoring system is designed with very low power components and the overall power consumption is 5.2 mW. It is able to operate for 110 h continuously when powered by a 240 mAh rechargeable battery. The battery lifetime is further extended by a wearable solar energy harvester. At the end of the chapter it gives a comparison of the recent wearable ECG monitoring systems.

The rest of the chapter is organized as follows. The next section presents the system architecture of the wearable ECG monitor. Section 3 shows the experimental results of the real-time ECG monitoring and low power consumption. The conclusion and future improvement of the work is discussed in the last section.

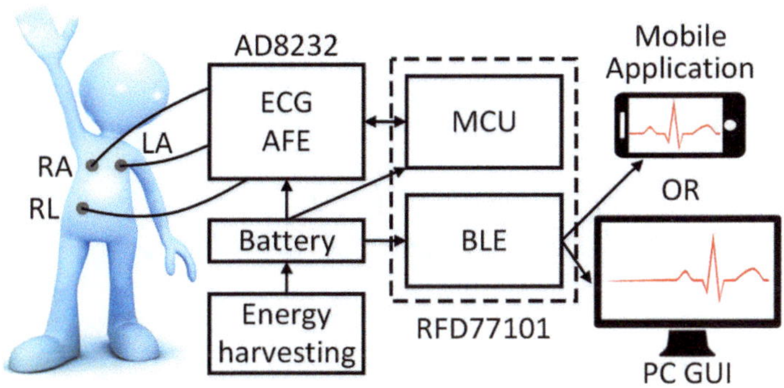

Fig. 1 The proposed wearable ECG monitoring system

2 System Architecture

The proposed wearable ECG monitoring system consists of three main parts: (1) ECG data acquisition circuit; (2) ECG data processing and transmission module and (3) power management with energy harvesting, as shown in Fig. 1. The ECG data of the subject is collected by the analog front end (AFE) circuit, which will be processed by the microcontroller (MCU) and transmitted through BLE to a smartphone or a PC for visualization and storage. The ECG AFE, MCU and BLE are powered by a 2.4 V 240 mAh rechargeable battery. To prolong the lifetime of the battery, a solar energy harvester is adopted to charge the battery when the subject is outdoor. The whole system is integrated on a T-shirt to demonstrate its wearability and usability.

2.1 ECG Data Acquisition Circuit

The ECG AFE circuit is the foundation of the wearable ECG monitoring system. In this work, an off-the-shelf chip, AD8232 from Analog Device®, is chosen for the ECG data acquisition. AD8232 is a fully integrated signal conditioning block that can extract, amplify and filter small biopotential signals for ECG and other biopotential measurements [12]. It can operate from 2 to 3.5 V with a low current of 170 μA. Its small size, low-power consumption and low cost makes it suitable for wearable healthcare applications. To minimize the distortion of the ECG waveform, a high-pass filter of 0.5 Hz and a low-pass filter of 40 Hz are used. The detected biopotentials from electrodes are further amplified with the built-in operational amplifier to better display the ECG waveform.

With the aid of AD8232 and some peripheral components, the proposed monitoring system can detect weak ECG signals with satisfactory quality. Figure 2 shows the raw measurement data of the proposed ECG AFE recorded from a healthy 26-year old man using DPO7104 oscilloscope from Tektronix®. The QRS complex, the P

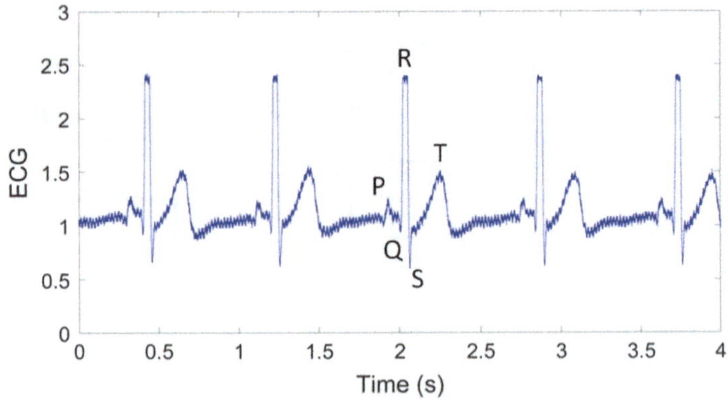

Fig. 2 The raw ECG measurement of the proposed system

and T waves and the ST segment can be clearly identified. The detected ECG data will be further processed and transmitted for real-time visualization.

2.2 Data Processing, Transmission and Visualization

The core of the wearable ECG monitoring system is based on RFD77101, also known as Simblee. It is a high performance, professional grade BLE transceiver with built-in ARM Cortex M0 microcontroller (MCU). The current of the Simblee transmission is about 8 mA and its power consumption can be further reduced by an ultra-low-power sleep mode with 600 nA. This module is highly suitable for wearable and IoT applications because of its small size, low-power consumption and integrated BLE transceiver with MCU. In this work, the collected ECG data from the AFE circuit will be processed by the built-in MCU of Simblee, and then transmitted to a PC connected with another Simblee module or a smartphone through BLE.

The real-time ECG data can be visualized and stored with a PC graphical user interface (GUI) developed with Python. A smartphone application is also developed for data display when the PC access is difficult to achieve for outdoor situations. The smartphone application is based on the Evothings® platform, where it enables wrapping the CSS, HTML5 and Javascript code according to the device platform. The real-time ECG data will be displayed on the smartphone, which will help the subject or medical personnel to know better of heart when daily ECG monitoring needed.

2.3 Energy Harvesting

To prolong the lifetime of the proposed wearable ECG monitoring system, a solar energy harvester is used to charge the battery when the subject is outdoor. The solar energy harvester is proposed and verified in our previous work [13]. It uses a flexible solar panel, the MP3-37 from Sundance Solar®, as the power source, which suits well with wearable applications. Figure 3a shows the solar panel, which has a size of 11.3 cm * 3.6 cm and is bendable to conform to human body contour. The proposed ECG monitoring system with energy harvesting is fabricated on a flexible PCB to make it wearable, as shown in Fig. 3. The PCB is small (diameter: 3.2 cm), flexible and can be easily sewed and integrated on a T-shirt as a wearable device. Figure 3b depicts the top layer of the PCB, which shows the ECG AFE and the Simblee module. The energy harvesting circuit is on the bottom layer in Fig. 3c.

(a) Solar panel

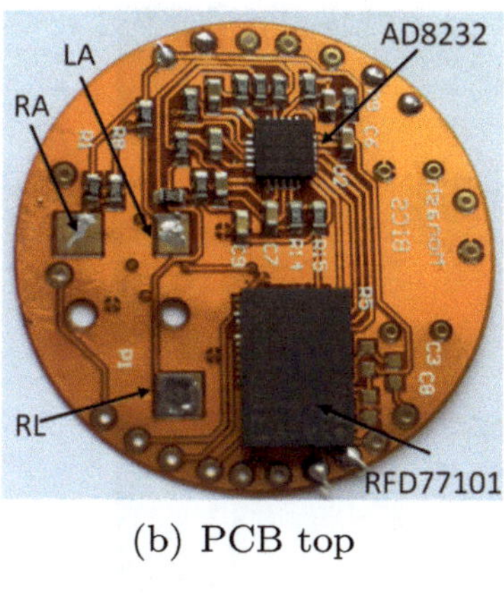

(b) PCB top

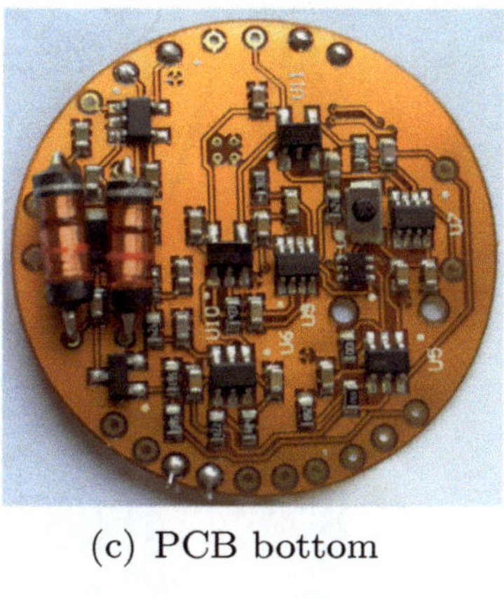

(c) PCB bottom

Fig. 3 The flexible solar panel and PCB design: **a** solar panel; **b** the ECG AFE and Simblee module on the top; **c** the energy harvesting circuit on the bottom

3 Experimental Results

The experimental setup is shown in Fig. 4, where the flexible ECG PCB and solar panel are sewed on a T-shirt. The flexible solar panel is used as the energy harvesting source to power the ECG monitor. During the experiment, the ECG data are directly obtained through the electrodes connected to the ECG AFE. The collected ECG data are to be transmitted wirelessly to an end device. When the subject has access to a PC, the data are transmitted to another Simblee module connected to the PC. The real-time ECG data will be displayed on a designed GUI based on Python, as shown in Fig. 5a. When the PC access is difficult to achieve, the ECG data are transmitted through BLE to a mobile device, like a smartphone. Figure 5b shows the real-time ECG signal of the subject displayed on the smartphone application. Experimental results show that both of the ECG data transmitted to PC and smartphone can display the QRS complex clearly, which means the proposed wearable device can be used in daily ECG monitoring requirements.

Low-power consumption is another critical requirement of wearable devices aiming for long-term operation. In this work, the wearable ECG monitoring system is powered by a 2.4 V rechargeable battery. The current of the proposed wearable ECG system including the BLE transmission is measured by the 34460A Digit multimeter from Keysight®. Experimental measurements show that the average current of the wearable ECG monitoring system 2.1 mA, as depicted in Fig. 6. Therefore the average power consumption is around 5.2 mW. As the capacity of the battery is 240 mAh, the wearable ECG monitoring system can operate for more than 110 h. To prolong the battery's lifetime, a flexible solar energy harvester is adopted in the wearable ECG monitoring system. It can output around 80 mW power on a sunny day with a flexible solar panel, as described and verified in our previous work [13]. Table 1

Fig. 4 The wearable ECG monitor and solar energy harvester integrated on a T-shirt

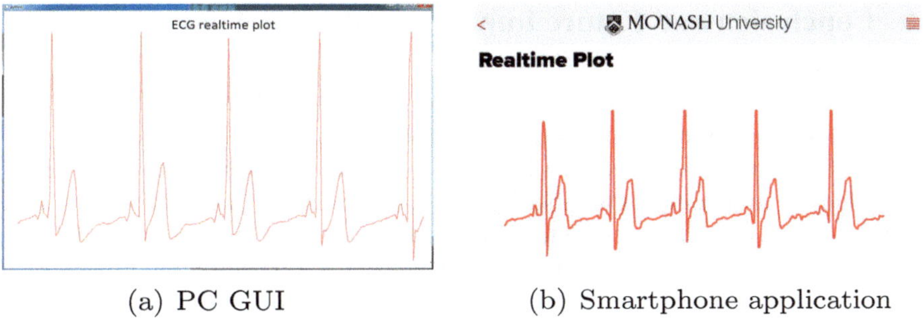

Fig. 5 The real-time ECG data visualisation: **a** PC GUI; **b** smartphone application

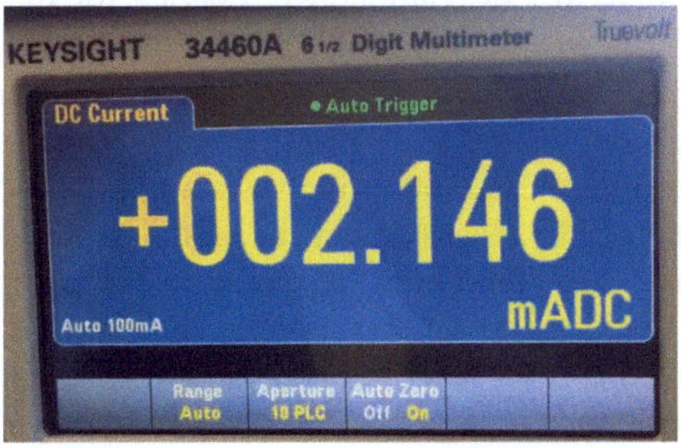

Fig. 6 The average current of the proposed wearable ECG monitoring system

Table 1 Comparisons of recent wearable ECG monitoring systems

	[4]	[10]	[7]	[11]	This work
Size (cm^2)	6.5 * 3.4	5.5 * 2.5	5.5 * 8.0	5.5 * 4.5	3.2 (diameter)
Voltage (V)	3	—	3	3.3	2.4
Power consumption (mW)	12	—	29.74	17	5.2
Communication protocol	Zigbee	BLE	BLE	MAX1472-433MHz	BLE
Mobile application	No	Yes	Yes	No	Yes
Battery (mAh)	650	600	310	—	240
Battery life (h)	160	67	31	—	110
Energy harvesting	No	No	No	Yes	Yes

presents a comparison of the recent wearable ECG monitoring systems in terms of size, power consumption, wireless protocol and battery lifetime.

4 Conclusion and Future Improvement

Wearable health monitoring devices play an important role in people's daily life with the increase of aging population worldwide. In this chapter, a wearable ECG monitoring system integrated on a T-shirt is proposed. All the electrical components are soldered on a compact flexible PCB to make it usable in daily life. The ECG signals of the subject are collected, filtered and amplified by a biopotential AFE chip. The ECG data are transmitted to an end device for visualisation through BLE. A PC GUI and a smartphone application are designed to display the real-time ECG waveform for indoor and outdoor situations respectively. The average current of the proposed wearable ECG monitoring system is 2.1 mA, and the overall power consumption is 5.2 mW. It can continuously operate for more than 110 h when powered by a 2.4 V 240 mAh rechargeable battery. To prolong the lifetime of the battery, a flexible solar energy harvester is also adopted to charge the battery when the subject is outdoor.

The proposed wearable ECG monitoring system will be further improved in the future work. For example, dry electrodes can be used to replace the conventional wet electrodes, which may cause discomfort on the skin. The compression of raw ECG data will be studied to reduce power consumption during the wireless transmission. Finally, a cloud database is to be developed to store the real-time ECG data for long-term healthcare applications.

References

1. Chen, M., Ma, Y., Song, J., Lai, C.F., Hu, B.: Smart clothing: connecting human with clouds and big data for sustainable health monitoring. Mob. Netw. Appl. **21**(5), 825–845 (2016)
2. Jeon, T., Kim, B., Jeon, M., Lee, B.G.: Implementation of a portable device for real-time ECG signal analysis. Biomed. Eng. Online **13**(1), 160 (2014)
3. Nemati, E., Deen, M.J., Mondal, T.: A wireless wearable ECG sensor for long-term applications. IEEE Commun. Mag. **50**(1), 36–43 (2012)
4. Spanò, E., Di Pascoli, S., Iannaccone, G.: Low-power wearable ECG monitoring system for multiple-patient remote monitoring. IEEE Sens. J. **16**(13), 5452–5462 (2016)
5. Yang, Z., Zhou, Q.H., Lei, L., Zheng, K., Xiang, W.: An IoT-cloud based wearable ECG monitoring system for smart healthcare. J. Med. Syst. **40**(286) (2016)
6. Wu, T., Wu, F., Redouté, J.M., Yuce, M.R.: An autonomous wireless body area network implementation towards IoT connected healthcare applications. IEEE Access **5**, 11413–11422 (2016)
7. Dai, M., Xiao, X., Chen, X., Lin, H., Wu, W., Chen, S.: A low-power and miniaturized electrocardiograph data collection system with smart textile electrodes for monitoring of cardiac function. Australas. Phys. Eng. Sci. Med. **39**(4), 1029–1040 (2016)
8. Zhang, J., Chan, S.C., Li, H., Wu, H.C., Wu, J., Wang, L.: A flexible and miniaturized wireless ECG recording system with metal-skin contacts input for wearable personalized healthcare. In: Proceedings of the 21st IEEE International Conference on Digital Signal Processing, pp. 318–321. IEEE Press (2016)
9. Haberman, Z.C., Jahn, R.T., Bose, R., Tun, H., Shinbane, J.S., Doshi, R.N., Chang, P.M., Saxon, L.A.: Wireless smartphone ECG enables large-scale screening in diverse populations. J. Cardiovasc. Electrophysiol. **26**(5), 520–526 (2015)

10. Wang, Y., Doleschel, S., Wunderlich, R., Heinen, S.: Evaluation of digital compressed sensing for real-time wireless ECG system with bluetooth low energy. J. Med. Syst. **40**(7), 170 (2016)
11. Dionisi, A., Marioli, D., Sardini, E., Serpelloni, M.: Autonomous wearable system for vital signs measurement with energy-harvesting module. IEEE Trans. Instrum. Meas. **65**(6), 1423–1434 (2016)
12. Analog Devices: Single-Lead, Heart Rate Monitor Front End AD8232 Datasheet. http://www.analog.com/media/en/technical-documentation/data-sheets/AD8232.pdf
13. Wu, T., Arefin, M.S., Redouté, J.M., Yuce, M.R.: A solar energy harvester with an improved MPPT circuit for wearable IoT applications. In: Proceedings of the 11th EAI International Conference on Body Area Networks, pp. 166–170 (2016)

JMMM: A Mobility Model for WBANs Based on Human Joint Movements

Chengjie Guan, Bin Liu, Zhiqiang Liu, Yufei Zhang and Xiaoyu Zhang

Abstract In wireless body area networks (WBAN), the sensors are usually attached on or implanted in the human body to monitor different vital signals. During routine activities, there is high mobility in WBANs, which results in frequent changes of the network topology. An accurate mobility model plays a vital role in protocol simulation and performance evaluation for WBANs. In this paper, we propose JMMM, a mobility model for WBANs according to the movement of human joints. Via mimicking the real motions of the human body, the proposed model can model the movement of any nodes in any place on body more accurately, and the changes in distance between different nodes are more realistic compared to previous models, which is of great importance in accurate simulation for WBANs. Moreover, the model is configurable, thus it is usable for a large variety of applications.

1 Introduction

Wireless Body Area Network (WBAN) consists of micro sensor devices deployed on or implanted in the human body to monitor different physiological signals, such as glucose level, blood oxygen level, heart rate, etc. Sensors form a network to communicate with each other and transmit the monitored data to the personal device (e.g. smartphone), which can further connect to the Internet. Since the whole system

C. Guan · B. Liu (✉) · Z. Liu · Y. Zhang · X. Zhang
CAS Key Laboratory of Electromagnetic Space Information, University of Science and Technology of China, Hefei, China
e-mail: flowice@ustc.edu.cn

C. Guan
e-mail: gcj@mail.ustc.edu.cn

Z. Liu
e-mail: lzhq28@mail.ustc.edu.cn

Y. Zhang
e-mail: zyf001@mail.ustc.edu.cn

X. Zhang
e-mail: zxyisland96@gmail.com

G. Fortino and Z. Wang (eds.), *Advances in Body Area Networks I*,
Internet of Things, https://doi.org/10.1007/978-3-030-02819-0_14

is flexible, effective and comfortable for human, the WBAN can be adopted for many applications, such as health care, smart home, military and sports.

Proper architectures and protocols for WBANs are still under active research, and performance evaluation is an important phase of conducting such research. Performance evaluation of the WBAN-related systems is typically conducted using hardware experiments or software simulations. Although the hardware experiments may provide more precise and reliable results, software simulations are still of significance because they are extensible, less expensive, and can be done in a shorter period. We can also identify potential flaws more easily via the observation of the simulation results.

The mobility model plays an important role on the accuracy of the simulations for the wireless networks with mobile nodes. Since the sensors are placed on the body in the WBAN, there exists high mobility as the body moves. In the WBAN, the channel quality between nodes strongly depends on the relative positions of the sensor nodes. The frequent change of the node positions will lead to the instability of the wireless links, thus resulting in the frequent change of the network topology. Therefore, an accurate mobility model is even more critical for WBANs than other wireless networks, and the simulations without an accurate mobility model are not credible at all.

A mobility model should consider all the characteristics of nodes' movements in the network. Meanwhile, sufficient flexibility should be provided in the model for researchers to adapt it to their specific scenarios.

In last decades, quite a number of mobility models have been proposed in research of mobile ad-hoc and wireless sensor networks. The proposed mobility models can be generally classified into two major classes, namely singular node mobility models and group mobility models. In the singular node mobility models, e.g., Random Walk Mobility Model (RWMM) [15], Random Waypoint Mobility Model (RWPM) [5], Random Direction Mobility Model (RDMM) [10], and Random Gauss-Markov Mobility (RGMM) [6], there is no correlation between the movements of different nodes. However, there are dependencies among the movement patterns of different nodes in the group mobility models. The Column Mobility Model, the Pursue Model, and the Nomadic Community Model, all presented in [11], are some early models to depict such patterns. The Small World In Motion (SWIM) [7] and The N-Body model [14] are established based on the analysis of real human traces in society. The Reference Point Group Mobility model (RPGM) presented in [4] is a general model, which can be adapted to many scenarios.

The traditional models try to statistically model the movement patterns of the nodes, but they are not suitable for WBANs. Since the sensor nodes in the WBAN are tightly attached to the body, their movements are subjected to the physical constraints of the human body. Moreover, the movement of the individual node strongly depends on the specific human action, which cannot be characterized well by all the models mentioned above. Several mobility models specially designed for WBANs have been presented in the literature during recent years. H. Ren and M. Meng introduced a model for WBANs in [9]. The authors utilized a specially configured RPGM model for WBANs but did not explain clearly why these configurations were the proper choices, and it was unadaptable for different applications and sometimes far from the real world mobility patterns. The MoBAN model presented in [8] is also based on RPGM but introduced postures with individual mobility parameters. It firstly selected a posture and then determined the random movement of each node according to the selected posture. The MMoW model [12, 13] used a similar posture selection strategy, but takes the relative distances between the normal nodes and the sink node into account in each posture instead of using a random model. Both MoBAN and MMoW are of completeness and configurable, however, they are established without enough consideration on the regularity of the body movements.

In this paper, we present JMMM, a mobility model for WBANs based on the movements of the human joints. We firstly simplify the human body into a combination of trunk, limbs and main joints and propose a method to calculate the positions of sensor nodes according to the positions of the main joints. Then we analyze different postures based on the real-world movement patterns and determine the corresponding human-joint-based mobility models. Since the model is established based on the real motion of the human body, it can determine the position of any node attached on the body more truly and precisely. To the best of our knowledge, we are the first to compare the mobility models to the movements in reality and the results demonstrate the superiority of our model in reliability and accuracy. Furthermore, abundant parameters could be adjusted, which makes the model configurable and could be applied for various applications.

The rest of this paper is organized as follows. Section 2 describes the mobility model in details. Section 3 gives the suggestions on parameter configurations. Section 4 implants the model in OMNeT++ [3] and evaluates its performance together with MoBAN and MMoW. Section 5 concludes the paper.

2 Mobility Modeling

An accurate mobility model for mobile networks should be able to simulate the positions of nodes at any instance of simulation time as precisely as possible. To achieve this goal, the unique characteristics of WBANs must be taken into consideration. In the WBAN, the network moves as a whole while the movement of the individual node follows its own patterns, which are both important for model establishment. Essentially, the limb movement determines the movement of the nodes, and the movement patterns differ from each other for different types of human motions. Therefore, via mimicking the rotation of main human joints, we establish the movement patterns in different postures thus generating a mobility model for WBAN nodes.

In this section, we first introduce a general method for determining the position of any node placed on the human body. Then we introduce the mobility model from two parts, i.e., the global movement and the individual movements. In the former part, the movement of the whole WBAN is determined. And in the latter part, the movements of individual nodes in each posture are determined based on the joints movement.

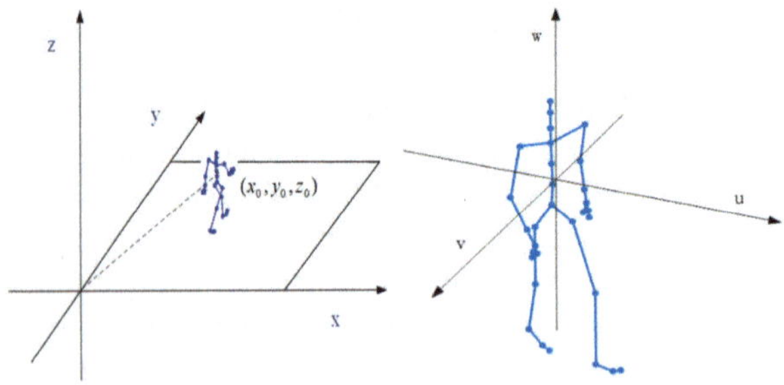

(a) Global Coordinate System (b) Relative Coordinate System

Fig. 1 The global coordinates system and the relative coordinates system

2.1 Position Determination

We discuss the mobility in a global coordinates system (GCS) as shown in Fig. 1a. Axis x and y form a plane on which the human moves, and axis z is perpendicular to that plane. How to determine the positions of the sensor nodes in the GCS is the most important phase in the mobility model. To further study the movement patterns of the human body, here we also adopt a relative coordinates system (RCS) with its origin at the mid chest as the local coordinates system. As shown in Fig. 1b, the positive direction of the axis u is to the front of human body, and that of the axis v and w is toward the right and upward, respectively. Assume that ψ is the angle between axis x of the GCS and axis u of the RCS, we can convert the coordinates of any point p_k in the relative coordinate system into the global coordinates system by:

$$\begin{cases} x_k = x_0 + \sqrt{u_k^2 + v_k^2}\cos(\psi + \phi) \\ y_k = y_0 + \sqrt{u_k^2 + v_k^2}\sin(\psi + \phi) \\ z_k = z_0 + w_k \\ \phi = \arctan\left(\frac{v_k}{u_k}\right) + \arctan\left(\frac{y_0}{x_0}\right) \end{cases} \quad (1)$$

where (x_k, y_k, z_k) and (u_k, v_k, w_k) are the coordinates of p_k in the GCS and the RCS, respectively, and (x_0, y_0, z_0) are the coordinates of the origin of the RCS in the global coordinate system. In the rest of this section, we focus on the position determination in the RCS.

To evaluate its effect on WBANs, the human body can be simplified into a combination of trunk, limbs and main joints. We first consider the position determination of the main joints. The main joints include shoulder joints, hip joints, elbow joints, knee joints, wrist joints and ankle joints, and we number them from JN_1 to JN_{12} in this order. The odd-numbered joints are at the left side of the body while the even-numbered joints are at the right side. In most applications, the trunk cannot be curved, thus the joints on it are relatively stationary and their positions can be easily determined based on the chest location. Meanwhile, except for the movement

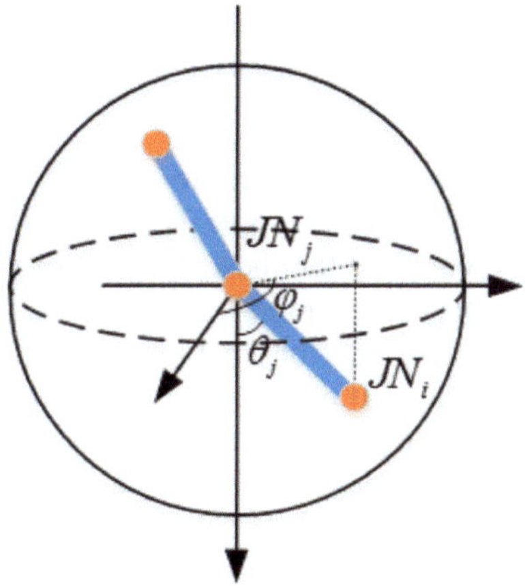

Fig. 2 Spherical coordinates system

of the whole body, the relative movements of limbs are the main movements during routine activities. Limb movements can be treated as the rotations around certain joints. Therefore, the positions of the joints in limbs can be determined using the spherical coordinates.

Here we assume that one joint is represented by a single point in the space. As different joints at the limbs have different rotation centers, e.g., the rotation center of the elbow joint is the shoulder joint at the same side, we use a four-tuple $(JN_j, L_{i,j}, \theta_j, \varphi_j)$ to represent the joint JN_i in the limbs, where JN_j is the rotation center of JN_i, $L_{i,j}$ is the distance between these two joints, and θ_j and φ_j are the polar and azimuth angles of the center joint, respectively. As shown in Fig. 2, the polar angle is the angle between the opposite direction of axis w and the radial from JN_j to JN_i, and the azimuth angle is the angle between the direction of axis v and the projection of that radial on the plane perpendicular to human body. Then, the coordinates of JN_i in the RCS can be calculated based on the coordinates of its rotation center JN_j:

$$\begin{cases} u_i = u_j + L_{i,j} sin\theta_j sin\varphi_j \\ v_i = v_j + L_{i,j} sin\theta_j cos\varphi_j \\ w_i = w_j - L_{i,j} cos\theta_j \end{cases} \tag{2}$$

We can determine the position of any sensor node after determining the positions of the main joints. The coordinates of node S_i deployed on the body can be denoted as a linear combination of the coordinates of all main joints. Since we regard the joints as the single points in space in the previous discussion, the position calculated in this way is inside the body, thus an offset ought to be added to represent the difference between the inside position and the actual deployed position. Therefore, the coordinates of S_i in the RCS can be calculated as:

$$P^R_{S_i} = \sum_{j=1}^{12} a_j \cdot P^R_{JN_j} + \mathit{offs}_{S_i} \tag{3}$$

where $P^R_{JN_j}$ is the coordinates of the joint JN_j in the RCS, a_j is the corresponding coefficient with value $0 \leq a_j \leq 1$, and offs_{S_i} is the offset. For example, if one node is deployed on the chest, it's position can be calculated from the coordinates of shoulder joints and hip joints as they form a plane where the node locates above, and the value of the offset from that plane to its actual location is determined by the thickness of the trunk. If node S_j is attached to the left upper arm and the distances from S_j to the shoulder joint JN_1 and the elbow joint JN_5 are c and d respectively, its position in the RCS can be calculated as:

$$P^R_{S_j} = \left(\frac{c}{c+d}\right) \cdot P^R_{JN_1} + \left(\frac{d}{c+d}\right) \cdot P^R_{JN_5} + \mathit{offs}_{S_j} \tag{4}$$

where the former part is the linear combination of two joints' coordinates and the latter part is the offset determined by the wearing position around the arm. Since the position of any sensor node can be calculated from the positions of the human joints, we only focus on the movement of main joints in the following sections.

2.2 *Model Structure and Global Movement*

WBAN nodes move as a whole while individual node has its own mobility pattern. Thus the proposed mobility model consists of two phases: global movement phase and individual movement phase. During simulations, global movement phase is responsible for controlling the mobility of all sensor nodes as a whole. Then individual movement phase further determines the position of each node. As the posture strongly influences the network topology and node connectivity, it is necessary to establish individual movement patterns for different postures. In fact, the posture not only determines the individual movement but also affects the global movement to a large extent because the body usually moves at different speeds in different postures.

Here we adopt five typical postures in our model, namely sitting, standing, walking, running and lying-down. Each posture is defined by a set of parameters which can be configured by users. These parameters can be classified into two parts: initial angles and mobility patterns. When the simulation starts with or switch to a particular posture, human joints are set to the initial angles configured by users to characterize the initial pose of current posture. Then the model determines the movements of individual joints and nodes under the defined mobility pattern. The postures can be determined based on the applications, i.e., a single posture or a sequence of postures can be adopted according to the specific application scenario.

We use the Random Waypoint Model [5] to model the global movement pattern in the GCS. The movement pattern determines the movement of the whole body and

Table 1 Configurable parameters for static postures

Notation	Description
θ_i^{ini}	Initial polar angle of joint JN_i
φ_i^{ini}	Initial azimuth angle of joint JN_i
ω_i	Angular velocity of joint JN_i
θ_i^{min}	Minimum polar angle of joint JN_i
θ_i^{max}	Maximum polar angle of joint JN_i
φ_i^{min}	Minimum azimuth angle of joint JN_i
φ_i^{max}	Maximum azimuth angle of joint JN_i

the new position of the origin of the RCS in the GCS is calculated as:

$$P_o(t_0 + t) = P_o(t_0) + \overrightarrow{V_g}(t) \cdot t \tag{5}$$

where $P_o(t_0)$ is the position of the origin at t_0, $\overrightarrow{V_g}(t)$ is the global velocity vector and t is the time after which the new position is calculated. The coordinates of the joints on the trunk can then be calculated based on their relative positions to the origin.

2.3 *Individual Movements in Different Postures*

Individual mobility pattern in different postures differs from each other. A posture can be dynamic or static based on the global velocity. In the static postures like standing, the trunk nearly keeps stationary while the limbs trembling without regularity. And in the dynamic postures, the limbs wave periodically under certain rules. Therefore, it is necessary to distinguish the mobility patterns in different postures and to pay enough attention to the rules they follow. As the coordinates of the main joints on the trunk are calculated with the determination of the global mobility, we focus on the joints in the limbs in this phase.

2.3.1 Standing, Sitting and Lying-Down

With static postures including standing, sitting and lying-down, the movements of the human body follow similar patterns. In the standing posture, there are few movements of trunk, but the arms and legs exhibit motion in three dimensions occasionally.

To simulate the joint movements, the initial angles of the main joints need to be firstly set. The initial angles can be configured according to specific applications to characterize the pose in which a man stands. Then we adopt a movement pattern similar to the RDMM model [10] to determine the random rotation of the joints in the static postures. Here we use the left-side elbow joint JN_5 as an example to explain how to calculate its position. The positions of the other joints could be determined in a similar way. Suppose at instance s, the state of the left-side shoulder joint JN_1 is determined by its angle (θ_1^s, ϕ_1^s) and the position of JN_5 is denoted as $(JN_1, L_{5,1}, \theta_1^s, \varphi_1^s)$. To determined the position of JN_5 at the next instance, the model will firstly select a direction randomly for joint JN_1 to rotate. The direction is measured in degrees and can be denoted as $(\Delta\theta_1^s, \Delta\varphi_1^s)$ where $\Delta\theta_1^s$ and $\Delta\varphi_1^s$ are the polar and the azimuth components of the direction respectively. Then, the upper arm swings to the selected direction with a configured angular velocity ω_1 during the time interval $[s, s+1]$, the new state $(\theta_1^{s+1}, \phi_1^{s+1})$ at the next instance can be thus determined and the new position of JN_5 can be denoted as $(JN_1, L_{5,1}, \theta_1^{s+1}, \varphi_1^{s+1})$. Once joint JN_1 rotates to the next state, the model would select a new direction for it again to determine the next position of JN_5. The ranges $[\theta_1^{min}, \theta_1^{max}]$ and $[\varphi_1^{min}, \varphi_1^{max}]$ are the boundaries of the rotation of JN_1, and we should make sure the next state is inside the ranges while selecting the directions. The ranges of the joint angles have to meet a lot of constraints due to the physical factors. For example, the range of polar angle for JN_1 can be set to $[-60°, 120°]$ since the upper arm can freely wave within this range. But for JN_5, the relative polar angle must be greater than zero as the lower arm cannot bend against the upper arm. The configurable parameters for standing posture are summarized in Table 1. The individual mobility in sitting and lying-down is similar and only the parameter values need to be changed to fit their situations.

2.3.2 Walking

The movements of human limbs follow certain principles in the dynamic postures. During walking, the arms and the legs show repetitive movement patterns. When the left arm moves forward, the right leg moves forward as the right arm moves backward, which is helpful to model the mobility of the human body efficiently. Furthermore, the swing of the limbs is not a constant process. The angular velocity varies greatly at different moments. We use the parameters as shown in Table 2 to characterize these patterns during the walking posture. Because of the periodicity of the human motion, we defined T as the action period in which the upper limb finishes an entire wave from behind to front. With the study in human movements in reality, the period T is divided into a sequence of stages $[t_1^i, t_2^i, \ldots, t_n^i]$ for each joint JN_i, which corresponds to the sequences of angular velocities $[\omega_{\theta_i}^{t_1}, \omega_{\theta_i}^{t_2}, \ldots, \omega_{\theta_i}^{t_n}]$ and $[\omega_{\varphi_i}^{t_1}, \omega_{\varphi_i}^{t_2}, \ldots, \omega_{\varphi_i}^{t_n}]$.

Initial angles of all joints ought to be set firstly at the start of the simulation for the walking posture. Since walking is an ongoing process, the pose at any instance can be selected as the initial pose. In the proposed model, we define the initial pose as the pose in which the arms stretch to the limit. A typical setting of initial angles is shown in Fig. 3. Then, the joints begin to rotate at the configured angular velocities. In the RCS, the new position of joint JN_i which has its rotation center in joint JN_j is calculated as:

Table 2 Configurable parameters for dynamic postures

Notation	Description
θ_i^{ini}	Initial polar angle of joint JN_i
φ_i^{ini}	Initial azimuth angle of joint JN_i
T	Action period
$[t_1^i, t_2^i, \ldots, t_n^i]$	Partition of period T for joint JN_i
$[\omega_{\theta_i}^{t_1}, \omega_{\theta_i}^{t_2}, \ldots, \omega_{\theta_i}^{t_n}]$	Polar angular velocities of joint JN_i
$[\omega_{\varphi_i}^{t_1}, \omega_{\varphi_i}^{t_2}, \ldots, \omega_{\varphi_i}^{t_n}]$	Azimuth angular velocities of joint JN_i

Fig. 3 Typical setting of initial polar angles in walking posture

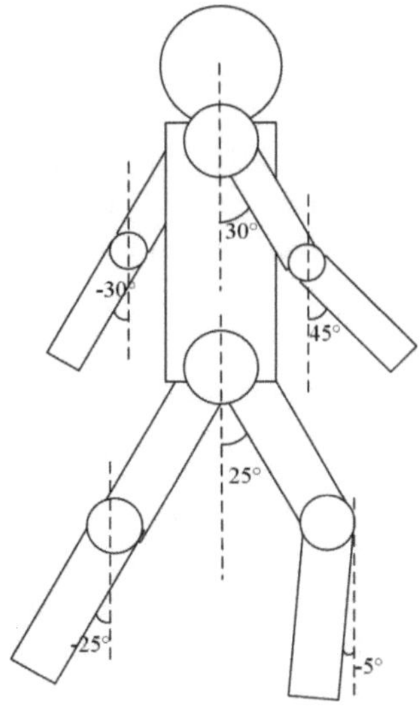

$$\begin{cases} u_i^{t_0+t} = u_j^{t_0+t} + L_{i,j} sin\theta_j^{t_0+t} sin\varphi_j^{t_0+t} \\ v_i^{t_0+t} = v_j^{t_0+t} + L_{i,j} sin\theta_j^{t_0+t} cos\varphi_j^{t_0+t} \\ w_i^{t_0+t} = w_j^{t_0+t} - L_{i,j} cos\theta_j^{t_0+t} \\ \theta_j^{t_0+t} = \theta_j^{t_0} + \omega_{\theta_j}^{t} \cdot t \\ \varphi_j^{t_0+t} = \varphi_j^{t_0} + \omega_{\varphi_j}^{t} \cdot t \end{cases} \quad (6)$$

where $(u_i^{t_0+t}, v_i^{t_0+t}, w_i^{t_0+t})$ and $(u_j^{t_0+t}, v_j^{t_0+t}, w_j^{t_0+t})$ are the new coordinates of JN_i and JN_j respectively, $\theta_j^{t_0+t}$ and $\varphi_j^{t_0+t}$ are the new polar and azimuth angles of center joint JN_j, and $\omega_{\theta_j}^{t}$ and $\omega_{\varphi_j}^{t}$ are the angular velocities of the center joint in its corresponding stage during walking. New coordinates of all joints in the limbs can be calculated in the order from JN_5 to JN_{12}. After that, the coordinates of sensor nodes in the RCS can be determined by Eq. (3) and then be converted into the GCS using Eqs. (1) and (5).

Fig. 4 The human body in walking process

The action period T is divided into different stages according to the actual rotation of each joints in the real life. During walking, for upper limbs, the polar angular velocity of rotation is nearly constant and a single stage is enough to fit the wave forward or backward within one period. However, the rotations of the leg joints are fairly different. In terms of the polar angles, four stages are used to characterize the mobility pattern as shown in Fig. 4. In the first stage the back shin moves back to lift the foot, then in the second stage the upper leg moves to front, after that the knee joint rotates swiftly to move the shin to front in the third stage and lastly, it move back to step on the ground. The period partitions and the corresponding angular velocities for azimuth angles can be determined by users but the values have to satisfy:

$$\sum_{s=1}^{n} \omega_{\varphi_i}^{t_k} \cdot t_i^k = 0 \tag{7}$$

2.3.3 Running

Compared to walking, running is a posture with much higher node mobility. The movement pattern in running is similar to walking in some extent, but the values of the parameters are quite different. First of all, the period T in running is much shorter than in walking, and the initial values of the polar angles are different as shown in Fig. 5 because the human body shows different poses during the whole running process.

The parameters defined in Table 2 can also been used to characterize the running posture. Figure 5 shows a typical setting of initial angles in running and the new coordinates of joint JN_i with its rotation center in JN_j can be calculated by Eq. (6) where $\omega_{\theta_j}^t$ and $\omega_{\varphi_j}^t$ are the angular velocities of JN_j in running. And the positions of sensor nodes in the GCS can be calculated in a similar way with walking. The angular velocities and the period partitions in running are far from those in walking because the rules of movements are not same. As human usually keep the elbow bent during running for a while, a new stage ought to be added for the shoulder joints to characterize the moments that upper limbs don't rotate. Meanwhile, four stages are still used to characterize the movement pattern of legs, but the duration of each stage and the corresponding angular velocity are quite different and the whole running process is shown in Fig. 6.

3 Suggested Configuration

The configuration of the parameters is of great importance to form a realistic mobility model. Even in a certain posture, the movements vary greatly with different parameters. For instance, if one sits on a chair in front of a desk, the legs are curved so the initial angle is close to 90° and the arms should be specially configured to set on the desk. And if one sits in a sofa, the initial angles of hip joints and shoulder joints should change to characterize a more relax pose. The ranges of the joint angles are also adjustable by researchers according to their application scenarios.

In dynamic postures, the parameters are more important in determining the movements than in static postures. Here we provide the empirical configuration on parameters, aiming to assist the researchers to build a more realistic model. The empirical parameters, as shown in Tables 3, 4, 5 and 6 respectively, are analyzed based on the real movement data in CMU motion capture database [1], which is a dataset of motions captured on various subjects. The period is normalized and the values of angular velocities are with the unit of degree/s.

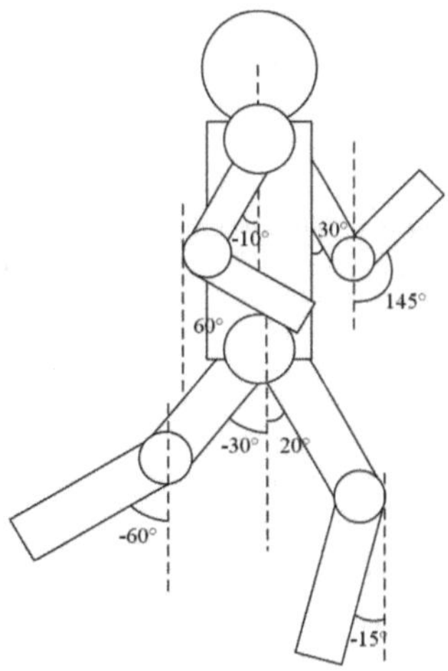

Fig. 5 Typical setting of initial polar angles in running posture

Fig. 6 The human body in running process

Table 3 Suggested configuration for shoulder joints

Items	Walking	Running
T	[0.50 s, 0.65 s]	[0.3 s, 0.4 s]
θ_1^{ini}	[−25°,−21°]	[−32°, −20°]
θ_2^{ini}	[24°, 27°]	[22°, 34°]
$[t_1^1, t_2^1, \ldots, t_n^1]$	1	[0.3, 0.7]
$[\omega_{\theta_1}^{t_1}, \omega_{\theta_1}^{t_2}, \ldots, \omega_{\theta_1}^{t_n}]$	85	[0, 180]
$[t_1^2, t_2^2, \ldots, t_n^2]$	1	[0.3, 0.7]
$[\omega_{\theta_2}^{t_1}, \omega_{\theta_2}^{t_2}, \ldots, \omega_{\theta_2}^{t_n}]$	−85	[0, −180]

Table 4 Suggested configuration for elbow joints

Items	Walking	Running
T	[0.50 s, 0.65 s]	[0.3 s, 0.4 s]
θ_5^{ini}	[−25°, −21°]	[42°, 63°]
θ_6^{ini}	[34°, 50°]	[120°, 128°]
$[t_1^5, t_2^5, \ldots, t_n^5]$	1	[0.2, 0.8]
$[\omega_{\theta_5}^{t_1}, \omega_{\theta_5}^{t_2}, \ldots, \omega_{\theta_5}^{t_n}]$	140	[0, 165]
$[t_1^6, t_2^6, \ldots, t_n^6]$	1	[0.2, 0.8]
$[\omega_{\theta_6}^{t_1}, \omega_{\theta_6}^{t_2}, \ldots, \omega_{\theta_6}^{t_n}]$	−140	[0, −165]

Table 5 Suggested configuration for hip joints

Items	Walking	Running
T	[0.50 s, 0.65 s]	[0.3 s, 0.4 s]
θ_4^{ini}	[−31°, −20°]	[−41°, −28°]
θ_3^{ini}	[19°, 24°]	[17°, 22°]
$[t_1^4, t_2^4, \ldots, t_n^4]$	[0.5, 0.2, 0.2, 0.1]	[0.7, 0.3]
$[\omega_{\theta_4}^{t_1}, \omega_{\theta_4}^{t_2}, \ldots, \omega_{\theta_4}^{t_n}]$	[180, 70, 0, −40]	[234, −155]
$[t_1^3, t_2^3, \ldots, t_n^3]$	1	[0.7, 0.3]
$[\omega_{\theta_3}^{t_1}, \omega_{\theta_3}^{t_2}, \ldots, \omega_{\theta_3}^{t_n}]$	−100	[−243, 160]

Table 6 Suggested configuration for knee joints

Items	Walking	Running
T	[0.50 s, 0.65 s]	[0.3 s, 0.4 s]
θ_8^{ini}	[−31°, −20°]	[−61°, −54°]
θ_7^{ini}	[−4°, 5°]	[−15°, −10°]
$[t_1^8, t_2^8, \ldots, t_n^8]$	[0.2, 0.3, 0.3, 0.2]	[0.2, 0.1, 0.6, 0.1]
$[\omega_{\theta_8}^{t_1}, \omega_{\theta_8}^{t_2}, \ldots, \omega_{\theta_8}^{t_n}]$	[−200, 160, 270, −180]	[−315, 0, 350, −255]
$[t_1^7, t_2^7, \ldots, t_n^7]$	[0.6, 0.2, 0.2]	[0.4, 0.3, 0.3]
$[\omega_{\theta_7}^{t_1}, \omega_{\theta_7}^{t_2}, \ldots, \omega_{\theta_7}^{t_n}]$	[−50, −15, −100]	[−180, 60, −255]

4 Simulation Results

As the node positions have a large impact on the communications in WBANs, it is important to evaluate how the mobility model matches the realistic movements. We implanted the proposed model under MiXiM framework [2] on top of the OMNeT++ discrete event simulator. MoBAN and MMoW are also implanted to present comparisons. In simulations, all sensor nodes are set at the place of main joints. We use the normalized distance between particular nodes as the metric to evaluate the performance of the three models. And the results calculated from the samples in CMU database are taken as the baseline. Main parameters of the three models are determined based on the anthropometric data in the database.

We focus on the results of dynamic postures. Figure 7 shows the change of the straight-line distance value in GCS between different nodes in walking postures. We can see that the proposed model matches the real data very well in all four cases. Performance of MMoW, in which the periodicity of the motion is taken into consideration, is better than MoBAN but worse than the proposed model. MoBAN

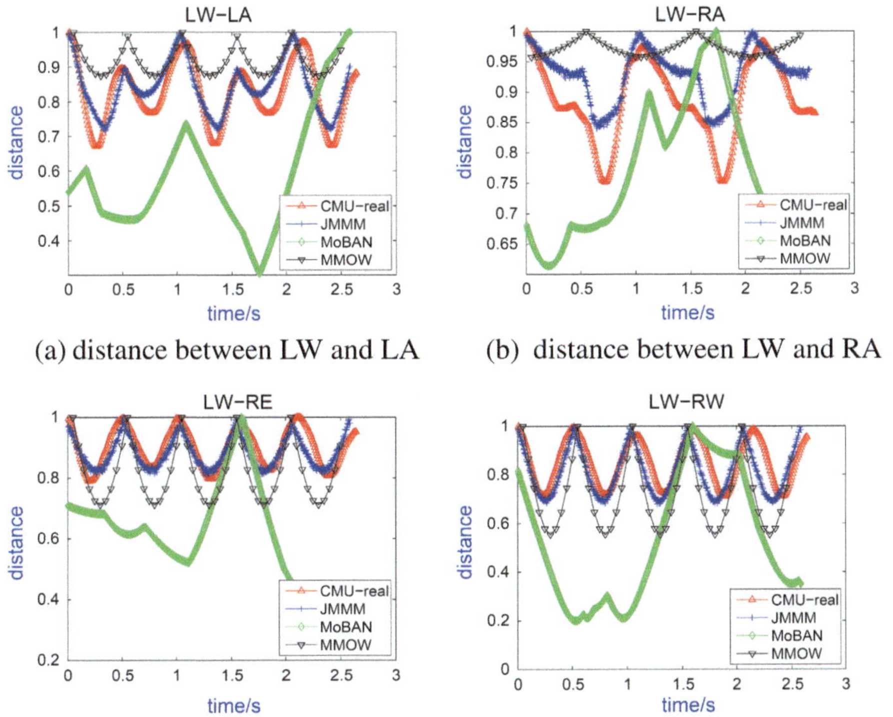

(a) distance between LW and LA
(b) distance between LW and RA
(c) distance between LW and RE
(d) distance between LW and RW

Fig. 7 Simulation results during walking posture (LW: left wrist, LA: left ankle, RA: right ankle, RE: right elbow, RW: right wrist)

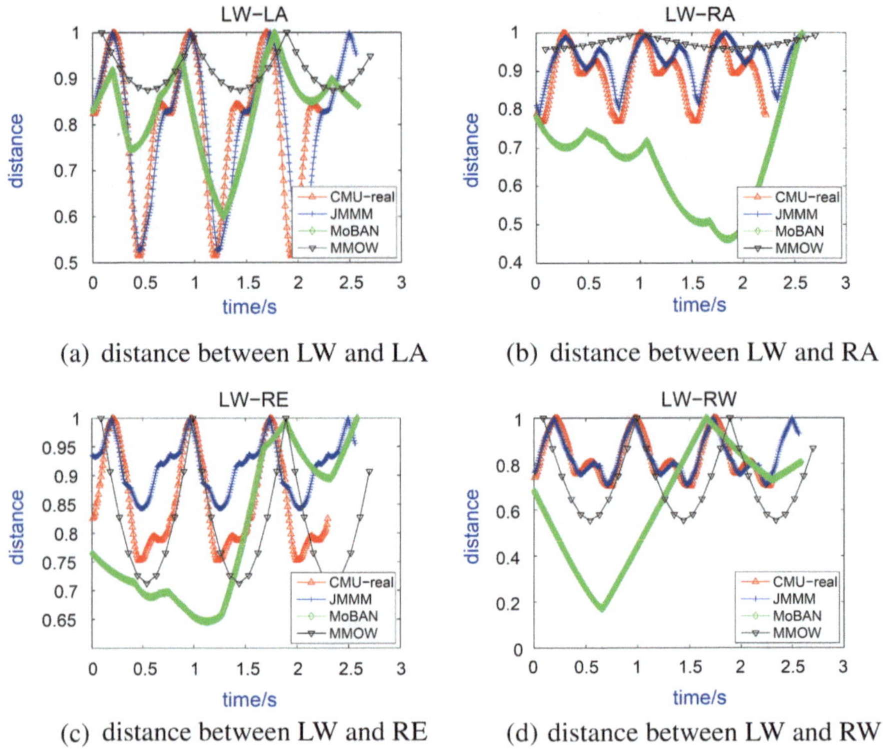

(a) distance between LW and LA (b) distance between LW and RA

(c) distance between LW and RE (d) distance between LW and RW

Fig. 8 Simulation results during running posture (LW: left wrist, LA: left ankle, RA: right ankle, RE: right elbow, RW: right wrist)

performs the worst in all cases as it is a random model and no regular pattern is adopted. The simulations for the running posture show similar results as shown in Fig. 8. The proposed model achieves good performance in case LA-LW, LW-RA and LW-RW, and matches the trend in case LW-RE. The results of the other two models are far away from the real data, which illustrates that they can't deal with the high mobility in running posture.

5 Conclusion

In this paper, a mobility model is proposed for wireless body area networks based on the actual movements of the human joints. The model mimics the real actions of the human in specific postures, thus determining the node positions in WBANs during the movements. Compared to previous models, it can characterize the patterns of node movement in WBANs more precisely and truly. Furthermore, the model can simulate the real-time positions of any node deployed on body, and configurable parameters ensure that the model can be adapted to many scenarios. Finally, simulation results show the practicability and superiority of the proposed model. The model can be used in simulating WBAN protocols for a large variety of applications.

References

1. "Cmu motion capture database," http://mocap.cs.cmu.edu
2. "Mixim project website," http://mixim.sourceforge.net
3. "Omnet++ network simulator website," http://www.omnetpp.org
4. Hong, X., Gerla, M., Pei, G., Chiang, C.-C.: A group mobility model for ad hoc wireless networks. In: Proceedings of the 2nd ACM International Workshop on Modeling, Analysis and Simulation of Wireless and Mobile Systems, pp. 53–60. ACM (1999)
5. Hyytiä, E., Virtamo, J.: Random waypoint mobility model in cellular networks. Wirel. Netw. **13**(2), 177–188 (2007)
6. Liang, B., Haas, Z.J.: Predictive distance-based mobility management for pcs networks. In: INFOCOM'99. Eighteenth Annual Joint Conference of the IEEE Computer and Communications Societies. Proceedings. IEEE, vol. 3, pp. 1377–1384. IEEE (1999)
7. Mei, A., Stefa, J.: Swim: a simple model to generate small mobile worlds. In: INFOCOM 2009, IEEE, pp. 2106–2113. IEEE (2009)
8. Nabi, M., Geilen, M., Basten, T.: Moban: a configurable mobility model for wireless body area networks. In: Proceedings of the 4th International ICST Conference on Simulation Tools and Techniques, pp. 168–177. ICST (Institute for Computer Sciences, Social-Informatics and Telecommunications Engineering) (2011)
9. Ren, H., Meng, M.Q.-H.: Understanding the mobility model of wireless body sensor networks. In: 2006 IEEE International Conference on Information Acquisition, pp. 306–310. IEEE (2006)
10. Royer, E.M., Melliar-Smith, P.M., Moser, L.E.: An analysis of the optimum node density for ad hoc mobile networks. In: IEEE International Conference on Communications, 2001, ICC 2001, vol. 3, pp. 857–861. IEEE (2001)
11. Sánchez, M., Manzoni, P.: Anejos: a java based simulator for ad hoc networks. Future Gener. Comput. Syst. **17**(5), 573–583 (2001)
12. Sandhu, M.M., Akbar, M., Behzad, M., Javaid, N., Khan, Z.A., Qasim, U.: Mobility model for wbans. In: 2014 Ninth International Conference on Broadband and Wireless Computing, Communication and Applications (BWCCA), pp. 155–160. IEEE (2014)
13. Sandhu, M.M., Javaid, N., Jamil, M., Khan, Z.A., Imran, M., Ilahi, M., Khan, M.: Modeling mobility and psychological stress based human postural changes in wireless body area networks. Comput. Human Behav. **51**, 1042–1053 (2015)
14. Zhao, C., Sichitiu, M.L.: N-body: social based mobility model for wireless ad hoc network research. In: 2010 7th Annual IEEE Communications Society Conference on Sensor Mesh and Ad Hoc Communications and Networks (SECON), pp. 1–9. IEEE (2010)
15. Zonoozi, M.M., Dassanayake, P.: User mobility modeling and characterization of mobility patterns. IEEE J. Select. Areas Commun. **15**(7), 1239–1252 (1997)

A Noninvasive Continuous Fetal Heart Rate Monitoring System for Mobile Healthcare Based on Fetal Phonocardiography

Pengjie Zhang, Shiwei Ye, Zhipei Huang, Dina Jiaerken, Shuxia Zhao, Lingyan Zhang and Jiankang Wu

Abstract Although the noninvasive continuous fetal heart rate (FHR) monitor is often recommended, the Doppler Ultrasonographic Cardiotocography (CTG) is improper for long-term monitor due to the less safety and the requirement of professional operation skill. In this paper, we design a noninvasive, continuous and real-time FHR monitoring system based on fetal phonocardiography by stationary wavelet denoising and cyclostationary process. Good agreement with CTG is obtained by Bland Altman analysis. Besides, quantitative results show that the FHR has an average accuracy of 97% compared with CTG on clinical data sets. The proposed system provides an alternative for CTG.

Keywords Fetal heart rate · Fetal phonocardiography (FPCG) Stationary wavelet process · Cyclostationary process · Bland Altman analysis

---

P. Zhang · S. Ye · Z. Huang (✉) · D. Jiaerken · J. Wu
University of Chinese Academy of Sciences, Beijing, China
e-mail: zhphuang@ucas.ac.cn

P. Zhang
e-mail: zhangpengjie15@mails.ucas.ac.cn

S. Ye
e-mail: shwye@ucas.ac.cn

D. Jiaerken
e-mail: dinajiaerken15@mails.ucas.ac.cn

J. Wu
e-mail: jkwu@ucas.ac.cn

S. Zhao · L. Zhang
Department of Gynecology and Obstetrics, Beijing Huairou Hospital of University of Chinese Academy of Sciences, Beijing, China
e-mail: zhaoshuxiajia@sina.com

L. Zhang
e-mail: zhanglingyanmang@126.com

G. Fortino and Z. Wang (eds.), *Advances in Body Area Networks I*,
Internet of Things, https://doi.org/10.1007/978-3-030-02819-0_15

1 Introduction

In many countries, pregnant women are required to take pregnancy tests by monitoring the variations in fetal heart rate periodically. This method provides up-to-date information to prevent intrauterine death or permanent damages to the fetus. The widespread clinical diagnostic tool for recording and analyzing the FHR is Doppler Ultrasonographic Cardiotocography [1].

Although CTG provides noninvasive and accurate way for the clinical diagnosis, it is inadequacy for long-term FHR monitoring for the risk of the exposure to ultrasound energy. Furthermore, ultrasound monitors are expensive and require a professional doctor to operate, thus making it unsuitable and inconvenient to use outside the hospital.

For all the reasons mentioned above, researchers have made many efforts to find a suitable alternative for CTG. To date, Electronic Fetal Monitoring (EFM) has been used in 90% of the labor diagnosis procedures in the United States [2], including fetal Phonocardiography, fetal Electrocardiography (FECG) [3], fetal Magnetocardiogram (FMCG) [4], Pulse Oximetry and Tocodynamometer. Among those methods, FPCG, FECG and FMCG can be considered as feasible plans for FHR monitoring. However, FMCG requires expensive devices and complex method for data acquisition, which is inapposite for daily monitoring. Since the Signal Noise Ratio (SNR) of FECG is very low because of the maternal abdominal attenuation, it is difficult to estimate FHR using FECG. FPCG, which is a recording of fetal heart sounds originated from interaction between fetal heart and blood flow therein, is a low-cost and relatively simple solution to monitor FHR. Besides, it is easier to record the acoustic signal from maternal abdominal surface. Given those, FPCG can be chosen as an alternative for CTG for long-term monitoring of the FHR.

FHS is a periodic signal which has a band range from 20 to 200 Hz [1]. Although FHS captured from maternal abdominal by a sensor has relatively high SNR compared to the FECG, there are still much noise in the FPCG in consideration of fetus in a particular environment and the method of data collection. The main noise sources include [1, 5]:

- maternal heart sounds (MHS)
- maternal respiratory sounds
- maternal digestive sounds
- acoustic noise produced by contraction of mother's uterus
- acoustic noise produced by fetal movements
- movements of maternal abdominal surface sensor during recording
- ambient noise

Several signal processing methods have been proposed for long-term monitoring. The FHS data analysis is classified into different categories based on the signal processing and classification techniques [6], such as direct analysis [7, 8], adaptive filtering [9, 10], linear decomposition including short time Fourier transform [11], Wigner Ville distribution [12], wavelet transforms [13–16] and blind source separation [17], nonlinear decomposition [18] and multimodal analysis [19]. However, these are not feasible for FHR detection of mobile healthcare in noisy environment due to lack of the high precision and real time characteristic.

Considering the above reasons, we develop a noninvasive and continuous phonocardiographic FHR monitoring system, which provides accurate and real-time FHR measurement and can be implemented as a low-cost device for both clinical and home use.

2 Experimental and Computational Details

2.1 System Architecture

The system architecture and all the modules of "UBabycare" are illustrated in Fig. 1. The UBabycare consists of a single channel recording device with wooden receiver and electret microphone, a smartphone and a server. The microphone is fixed in the narrow side of a wooden double-cone measurement. The other side acquires the acoustic signals on the maternal abdominal surface.

The acquisition equipment is connected to smartphone by standard headphone jack. The signal is digitized by a mobile phone (sampling rate f_s = 16,000 Hz, 16 bits resolution).

Figure 2 illustrates the signal processing component of UBabycare. The fetal heart sound is processed in the smartphone through three modules: preprocessing, stationary wavelet denoising and FHR detection. The device starts to collect data for

Fig. 1 System architecture of UBabycare

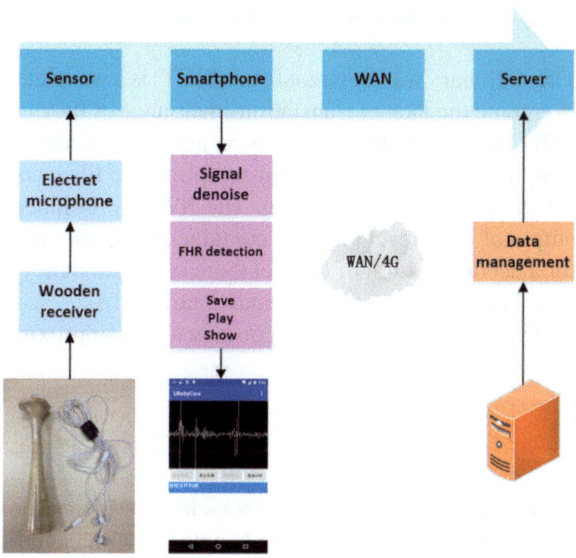

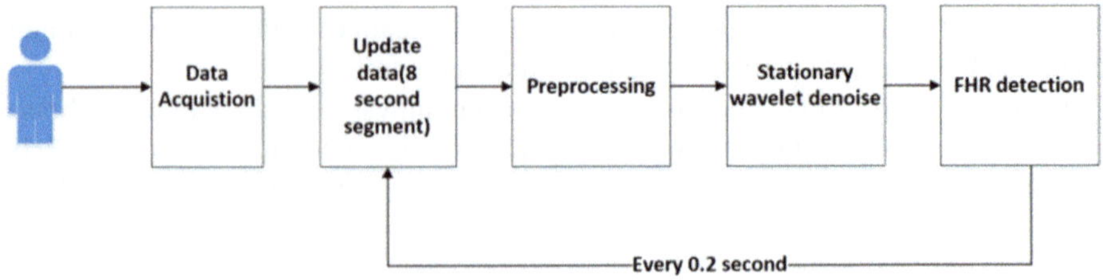

Fig. 2 Function flow diagram of the signal processing

8 s and then to update the data every 0.2 s, implementing real time FHR monitoring function.

2.2 *Framework of Algorithm*

2.2.1 Preprocess

From the signal processing perspective, the FPCG recording is an acoustic signal with its energy concentrated between 20 and 200 Hz. The interferences are confined below 20 Hz (including maternal heart sounds and digestive sounds) and over 200 Hz (ambient noise). We design a sixteen-order Butterworth digital band-pass filter with a 3 dB band equal to 20–200 Hz.

2.2.2 Stationary Wavelet Denoising

The stationary wavelet transform (SWT) is a wavelet transform algorithm designed to overcome the lack of translation-invariance of the discrete wavelet transform (DWT) [20]. Translation-invariance is achieved by removing the down-sampling and up-sampling in the DWT and up-sampling the filter coefficients by a factor of 2^{j-1} in the *j*th level of the algorithm. The SWT is an inherently redundant scheme as the output of each level of SWT contained the same number of samples as the input, so for a decomposition of N level there is a redundancy of N in the wavelet coefficients called approximations $A_j(n)$ and details $D_j(n)$.

The signal y(n) can be reconstructed by approximations and details:

$$\mathrm{y(n)} = \sum_{j=1}^{J} D_j(n) + A_J(n) \tag{1}$$

It's vital to choose a proper wavelet family to improve the efficiency of denoising. Different kinds of wavelets in Coieflet, Symlet and Daubechies families are recommended for signal denoising in fetal phonocardiogram [14]. In this paper, we select

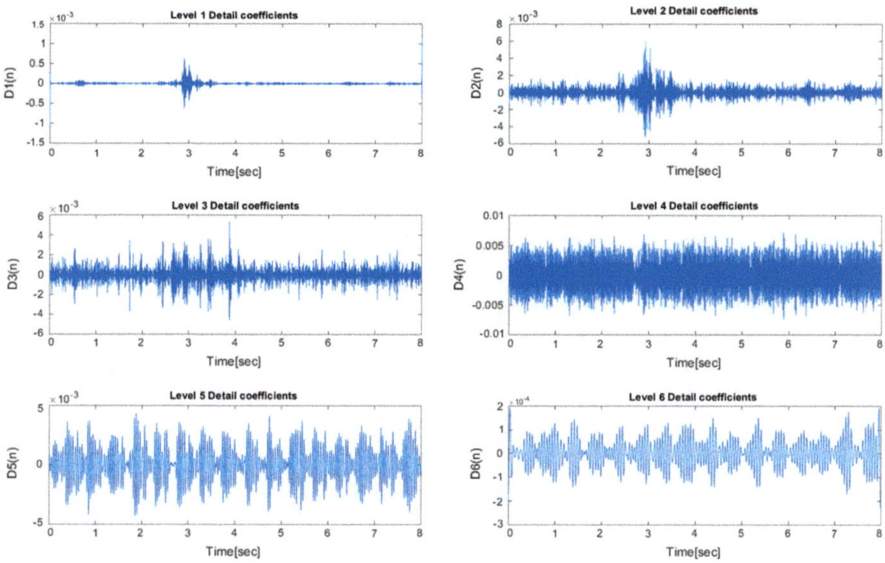

Fig. 3 6-level wavelet decomposition of the signal. 8 s interval of clinical record No. 10 is shown. Note that y-axis scale is not the same for different levels

Daubechies-11 wavelet and do a decomposition level 6 due to the feature of fetal heart sound. Figure 3 illustrates the detail coefficients of SWT and fetal heart signals have been identified in the details $D_5(n)$ and $D_6(n)$.

The other critical procedure of wavelet denoising is thresholding involved threshold value determination and the method for applying the threshold. A comparative test between different threshold value including Fixed threshold, Rigrsure threshold, Heursure threshold and Minimax threshold, the Rigrsure threshold with the method of the hard thresholding is selected [14]. Rigrsure threshold is based on Stein's Unbiased Estimated of Risk [21] and can be selected from the coefficients.

According to the result of decomposition and threshold filter, we choose the $D_{5sf}(n)$ coefficients for reconstruction. The reconstructed signal $y_r(n)$ combines the $D_{5sf}(n)$ coefficients with $A_6(n)$:

$$y_r(n) = D_{5sf}(n) + A_6(n) \tag{2}$$

Figure 4 shows the original signal, the preprocessing signal and the reconstructed signal. The fetal heartbeat can be identified from the reconstructed signal clearly compared to the original signal.

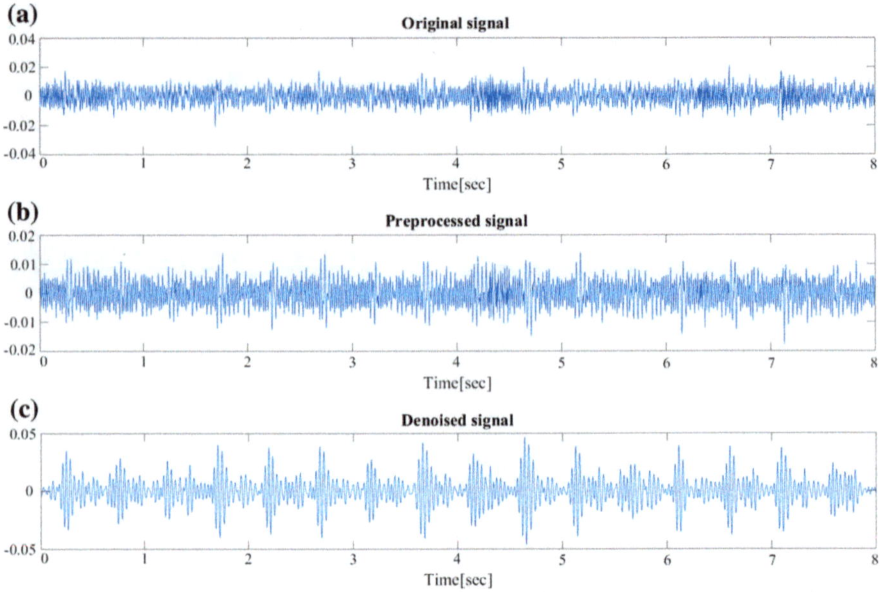

Fig. 4 An example of stationary wavelet analysis application. 8 s interval of clinical record No. 10 is shown. Note that y-axis scale is not the same for different levels. **a** The original record signal **b** the preprocessed signal after a 20–200 Hz bandpass filter **c** the denoised signal by stationary wavelet analysis

2.2.3 FHR Calculation

The filtered signals have filtered the most of the noise. However, it may still contain some false peaks leading to making s1 and s2 peak identification difficult. Therefore, a robust FHR detection algorithm has been developed to calculate FHR based on $y_r(n)$. In consideration of the interference of false S1 and S2 peak in time domain, we determine to adopt a method of cyclostationary process to detect in frequency domain [22].

The cyclostationary process is a signal having statistical properties that vary cyclically with time [23]. We assume that $x(t)$ is a FPCG signal and its expectation is $E[x(t)]$, so autocorrelation function can be written as

$$R_x(t, \tau) = E\big[x(t + \tau)x^*(t)\big] \tag{3}$$

where * indicates complex conjugation, τ is time delay. It is called wide-sense cyclostationary if $E[x(t)]$ and $R_x(t, \tau)$ are cyclic in t with period T_0.

$$E[x(t)] = E[x(t + T_0)] \quad for\, all\, t \tag{4}$$

$$R_x(t, \tau) = R_x(t + T_0; \tau) \quad for\, all\, t, \tau \tag{5}$$

The autocorrelation function is thus periodic in t and can be expanded in Fourier series by assuming that the Fourier series expansion of $R_x(t, \tau)$ is convergent to $R_x(t, \tau)$.

$$R_x(t, \tau) = \sum_{n=-\infty}^{+\infty} R_x^{\frac{n}{T_0}}(\tau) e^{j2\pi \frac{n}{T_0} t} \quad (6)$$

where $R_x^{\frac{n}{T_0}}(\tau)$ is called cyclic autocorrelation function and equal to:

$$R_x^{\frac{n}{T_0}}(\tau) = \frac{1}{T_0} \int_{-\frac{T_0}{2}}^{\frac{T_0}{2}} R_x(t, \tau) e^{j2\pi \frac{n}{T_0} t} dt \quad (7)$$

The frequencies $\alpha = \frac{n}{T_0}, n \in Z$, are called cyclic frequencies. Wide-sense stationary processes are a special case of cyclostationary process with only $R_x^0(\tau) \neq 0$.

The Fourier transform of the cyclic autocorrelation function at cyclic frequency α is called cyclic spectrum density function and is equal to:

$$S_x^\alpha(f) = \int_{-\infty}^{\infty} R_x^\alpha(\tau) e^{-j2\pi f t} d\tau \quad (8)$$

If $x(t)$ has any cyclic component at cycle frequency α, $S_x^\alpha(f)$ must have a value. For the FPCG signals, $S_x^\alpha(f)$ must have a sharp peak because the fetal heart sounds are dominant cyclic components in the filtered FPCG signals. The normal FHR generally ranges from 110 to 160 bpm. Thus, the range of cycle frequency α is between 1.83 and 2.67 Hz. The α of the algorithm is set in 1.63–3.38 Hz in case of some extreme situations. By integrating $S_x^\alpha(f)$, we get:

$$\emptyset_x(\alpha) = \int_{-\infty}^{\infty} \left|S_x^\alpha(f)\right| df \quad (9)$$

Thus, the FHR can be calculated by the dominant peak of $\emptyset_x(\alpha)$:

$$\text{FHR} = [\max(\emptyset_x(\alpha))] * 60 \quad (10)$$

where [] returns the value of abscissa when the function meets the condition in the bracket and FHR indicate beat per minute.

For example, an 8 s segment of the clinical record No. 10 is adopted to analyze. Figure 5 shows the cyclic spectral density. Figure 6 shows an integral for the cyclic spectral density to remove frequency based on the result of Fig. 5. The abscissa of the peak in the Fig. 6 is 2.06 Hz. Thus, the instantaneous FHR is 124 bpm.

In order to realize a real-time detection, the algorithm can continually calculate a FHR from an 8 s segment and update the segment every 0.2 s by instead of the oldest 0.1 s segment.

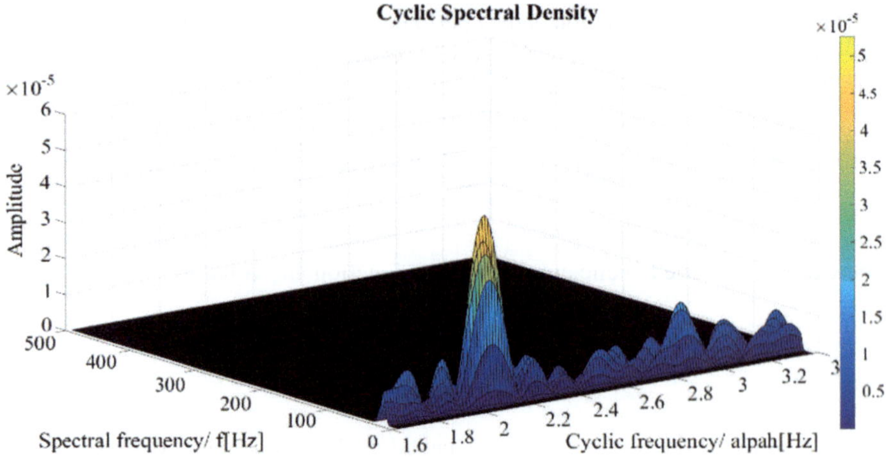

Fig. 5 The cyclic spectral density of 8 s interval signal

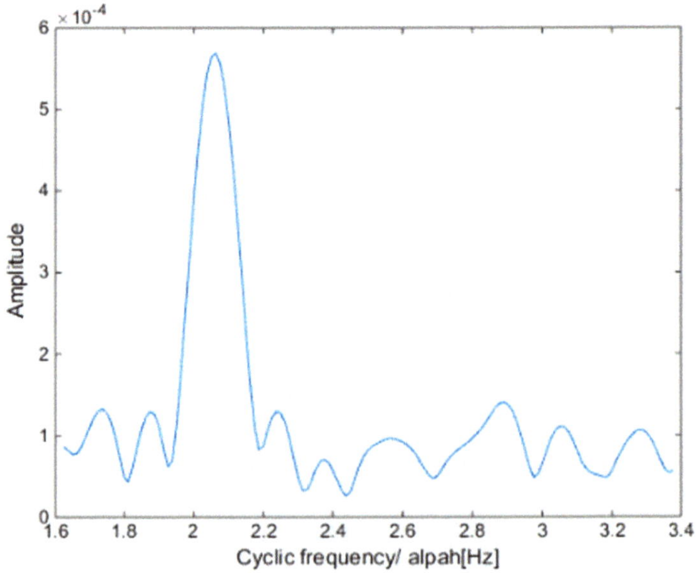

Fig. 6 An integral for the cyclic spectral density of 8 s interval to remove frequency

2.3 *Experiment*

2.3.1 The Simulated FPCG Signal

To evaluate the performance of the proposed algorithm, it is necessary to verify hypothesis by fetal PCG signals. In this paper, we adopt the simulated FPCG signals provided by Physiobank [24, 25], which has 1 kHz sampling frequency and lasts 478 s. The SNR varies from −4.4 to −26.7 dB. The signals simulate the fetal physiological, maternal physiological, transmission condition and ambient condition. The conclusion of consistency check shows expert obstetricians cannot distinguish simulated FPCG signals from the true FPCG recordings, ensuring the simulated signals on science and reliability.

2.3.2 The Clinical FPCG Signal

To verify the performance of the developed equipment and method, experimental clinical tests have been done. The equipment and Ultrasonic Doppler baby's heart instrument were synchronously placed on the abdominal surface and tightened by an elastic belt at Beijing Huairou Hospital of University of Chinese Academy of Sciences. The acquired signal was sampled at 16 kHz and recorded for further processing. 10 records from 36 to 39 weeks' gestational age pregnant women have been collected. Each record lasted 91.8 s.

3 Results and Discussion

3.1 Bland Altman Analysis

Bland Altman analysis is widely used in analytical chemistry and biostatistics, which describes agreement between two quantitative measurements [26]. 10 clinical data sets are chosen for Bland Altman analysis. The differences between the reference (the Doppler) and the predicted traces, ref − predic, are plotted against the average, (ref + predic)/2. The reproducibility is considered to be good if 95% of the results lie within a ±1.96SD (Standard Deviation) range.

A test for normal distribution called Kolmogorov-Smirnov test have been done. Figure 7 presents histogram for the reference and the predicted FHR. From the histogram of ref − predic, it shows that the values of ref − predic are normally distributed around zero. After ensuring that the values of ref − predic are normally distributed, we can use the SD to define the limits of agreement. Figure 8 illustrates the Bland Altman statistics for reference and predicted values of FHR. For the clinical data set (9190 samples), 96.18% of values lie within the ±1.96SD range for the FHR determination, indicating that our method is in good agreement with the Doppler method. A synchronous experiment with UBabycare and CTG have been done. Figure 9 shows a detail comparison of Record No. 6 between the reference and the estimated FHRs. It is evident that the variation between the reference and the estimated FHRs has the similar trend.

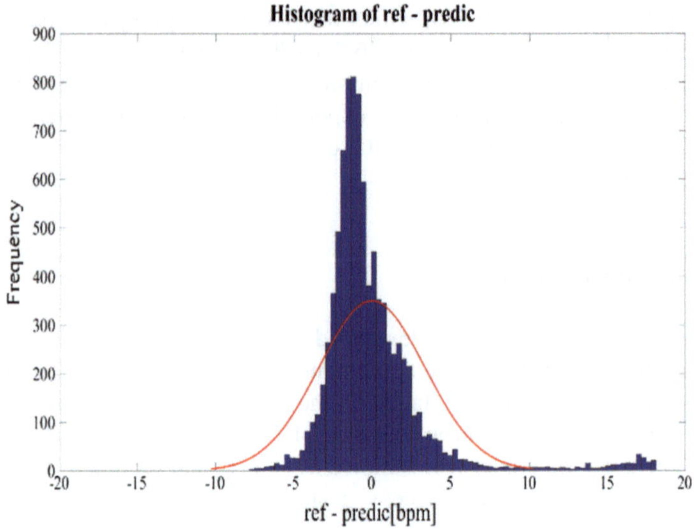

Fig. 7 Histogram of ref − predic. The red line represents Normal distribution. Kolmogorov-Smirnov test for normal distribution accepted normality (P = 0.8991)

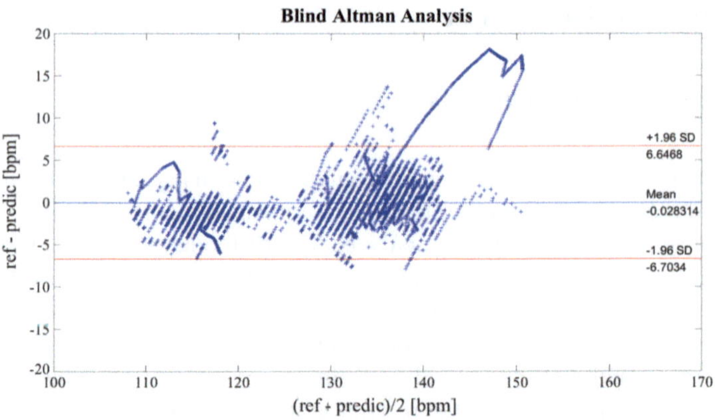

Fig. 8 Bland Altman analysis between the Doppler method and the proposed method

3.2 Performance of UBabycare Comparison with the Doppler

Among all the record signals, 20 simulated data sets and 10 clinical data sets are chosen for analysis. Due to the 16 kHz sampling rate of the clinical data and considering the fact that the valuable information of the FPCG is under 1 kHz, the signals are resampled to 1 kHz to reduce the processing load. If the difference between the detected value and the benchmark is ±5 bpm, the detected fetal heart rate is considered accurate. The accuracy rate is define as

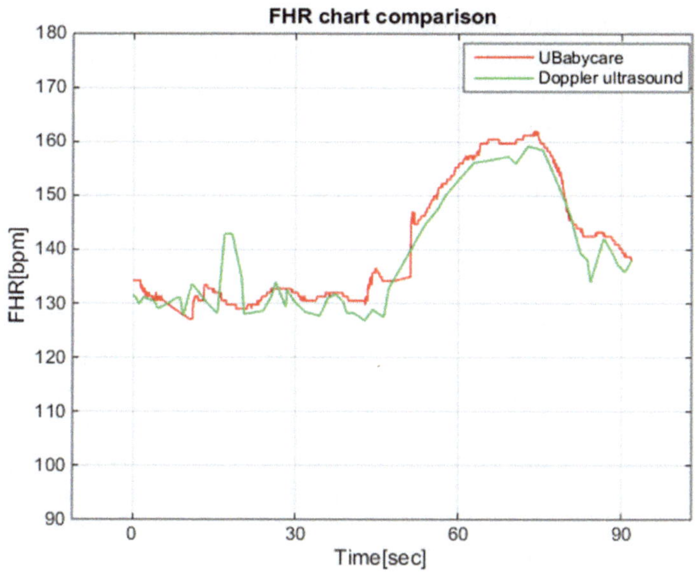

Fig. 9 FHR chart comparison. Record No. 6 of the clinical FPCG is shown. Doppler ultrasound chart is shown in red color. UBabycare chart is shown in green color

$$\text{Accuracy} = \frac{M\left(|FHR_D - FHR_S| \le 5bpm\right)}{N} \tag{11}$$

where FHR_D indicates the fetal heart rate calculated by the proposed method, FHR_S indicates the fetal heart rate provided by the Doppler, M indicates the number of the difference under ±5 bpm, and N indicates the total number of fetal heart.

Quantitative results show that the proposed method has 90.93–99.86% accuracy from the simulated data as the SNR ranges from −4.4 to −26.7 dB. A summary of all the simulated FPCG signal tests is presented in Table 1. With clinical data, FHR detection accuracy of 89.45–100% was obtained in different segments by comparing to the Doppler. On the average, the proposed method gets a 96.99% accuracy from clinical data sets. A summary of all the clinical FPCG signal tests is presented in Table 2.

The noninvasive continuous monitoring system results in good performance for simulated data sets and clinical data sets, which can identify FHR in a noisy condition accurately.

Table 1 Summary of simulated FPCG tests

Record number	SNR (dB)	FHR variations (bpm)	Accuracy rate (%)
1	−4.4	136–157	99.66
2	−6.6	136–157	99.86
3	−8.1	136–157	99.82
4	−10.2	136–157	99.82
5	−11.3	133–157	99.72
6	−14.8	133–158	97.90
7	−15.1	133–157	97.87
8	−15.4	133–158	97.64
9	−15.7	133–157	97.18
10	−16.4	133–158	97.07
11	−17.2	133–158	96.61
12	−19.1	134–158	95.95
13	−20	134–158	95.36
14	−21	135–159	94.88
15	−22	136–160	94.87
16	−23.1	134–161	94.70
17	−24.4	137–162	94.67
18	−25.9	139–166	94.84
19	−26.3	137–168	93.00
20	−26.7	140–169	90.93

Table 2 Summary of the clinical FPCG tests

Record number	Gestational age (weeks)	FHR variations (bpm)	Accuracy rate (%)
1	35	131–143	99.02
2	38	135–141	100
3	38	130–141	96.08
4	38	128–140	99.78
5	38	131–142	100
6	35	127–162	89.45
7	39	114–122	100
8	39	112–122	96.84
9	39	114–119	96.95
10	39	108–129	91.84
Average			97.00

4 Conclusions

In summary, a noninvasive continuous FHR monitoring system for mobile healthcare based on fetal phonocardiography has been presented. The system combined traditional wooden receiver and smartphone for mobile healthcare creatively. Application of the method is demonstrated by a comparative performance analysis with Doppler ultrasound monitor. The effect of method is verified by simulated FPCG recordings and clinical FPCG recordings. The system realizes the function of recording, saving, playing and analyzing simultaneously. Our noninvasive monitoring system will be suitable for both clinical and home use for continuous fetal heart rate monitoring.

Acknowledgements This work was supported by Special Fund for Scientific Research Cooperation of University Chinese Academy of Sciences.

References

1. Várady, P., Wildt, L., Benyó, Z., Hein, A.: An advanced method in fetal phonocardiography. Comput. Methods Programs Biomed. **71**, 283 (2003)
2. Martin, J.A., Hamilton, B.E., Osterman, M.J., Curtin, S.C., Matthews, T.J.: Births: final data for 2012. National vital statistics reports: from the Centers for Disease Control and Prevention, National Center for Health Statistics. Natl. Vital Stat. Syst. **59**, 3–71 (2010)
3. Mantini, D., Comani, S., Alleva, G., Romani, G.L.: Fetal cardiac time intervals: validation of an automatic tool for beat-to-beat detection on fetal magnetocardiograms. Int. Res. J. Pharm. **3** (2005)
4. Sameni, R., Clifford, G.D.: A review of fetal ECG signal processing; issues and promising directions. Open Pacing Electrophysiol. Therapy J. **3**, 4 (2010)
5. Mittra, A.K.: System simulation and comparative analysis of foetal heart sound de-noising techniques for advanced phonocardiography. Int. J. Biomed. Eng. Technol. **1**, 73–85 (2007)
6. Adithya, P.C., Sankar, R., Moreno, W.A., Hart, S.: Trends in fetal monitoring through phonocardiography: challenges and future directions. Biomed. Signal Process. Control **33**, 289–305 (2017)
7. Talbert, D.G., Davies, W.L., Johnson, F., Abraham, N., Colley, N., Southall, D.P.: Wide bandwidth fetal phonography using a sensor matched to the compliance of the mother's abdominal wall. IEEE Trans. Bio-med. Eng. **33**, 175 (1986)
8. Bassil, H.E., Dripps, J.H.: Real time processing and analysis of fetal phonocardiographic signals. Clinical physics and physiological measurement: an official journal of the Hospital Physicists' Association, Deutsche Gesellschaft für Medizinische Physik and the European Federation of Organisations for Medical Physics **10**(Suppl B), 67 (1989)
9. Soysa, W.N.M., Godaliyadda, R.I., Wijayakulasooriya, J.V., Ekanayake, M.P.B., Kandauda, I.C.: An eigenfilter based approach for extraction of fetal heart signals under noisy conditions using adaptive filters. In: Fourth International Conference on Computational Intelligence, Modelling and Simulation, pp. 254–259
10. Martinek, R., Nedoma, J., Fajkus, M., Kahankova, R., Konecny, J., Janku, P., Kepak, S., Bilik, P., Nazeran, H.: A phonocardiographic-based fiber-optic sensor and adaptive filtering system for noninvasive continuous fetal heart rate monitoring. Sensors **17**, 890 (2017)
11. Mittra, A.K., Choudhari, N.K.: Time-frequency analysis of foetal heart sound signal for the prediction of prenatal anomalies. J. Med. Eng. Technol. **33**, 296–302 (2009)
12. Balogh, Á.T., Kovács, F.: Application of phonocardiography on preterm infants with patent ductus arteriosus ☆. Biomed. Signal Process. Control **6**, 337–345 (2011)

13. Chourasia, V.S., Mittra, A.K.: A comparative analysis of de-noising algorithms for fetal phonocardiographic signals. IETE J. Res. **55**, 10–15 (2009)
14. Chourasia, V.S., Mittra, A.K.: Selection of mother wavelet and denoising algorithm for analysis of foetal phonocardiographic signals. J. Med. Eng. Technol. **33**, 442–448 (2009)
15. Vaisman, S., Yaniv Salem, S., Holcberg, G., Geva, A.B.: Passive fetal monitoring by adaptive wavelet denoising method. Comput. Biol. Med. **42**, 171–179 (2012)
16. Chourasia, V.S., Tiwari, A.K., Gangopadhyay, R.: A novel approach for phonocardiographic signals processing to make possible fetal heart rate evaluations. Digit. Signal Process. **30**, 165–183 (2014)
17. Samieinasab, M., Sameni, R.: Fetal phonocardiogram extraction using single channel blind source separation. In: Electrical Engineering, pp. 78–83
18. Warbhe, A.D., Dharaskar, R.V., Kalambhe, B.: A single channel phonocardiograph processing using EMD, SVD, and EFICA. In: International Conference on Emerging Trends in Engineering and Technology, pp. 578–581
19. Noorzadeh, S., Rivet, B., Guméry, P.Y.: A multi-modal approach using a non-parametric model to extract fetal ECG. In: IEEE International Conference on Acoustics, Speech and Signal Processing, pp. 832–836
20. Fowler, J.E.: The redundant discrete wavelet transform and additive noise. IEEE Signal Process. Lett. **12**, 629–632 (2005)
21. Donoho, D., Johnstone, I.: Adapting to unknown smoothness via wavelet shrinkage. Publ. Am. Stat. Assoc. **90**, 1200–1224 (1995)
22. Li, T., Li, T., Qiu, T., Park, Y.: Fetal heart rate monitoring from phonocardiograph signal using repetition frequency of heart sounds. J. Electr. Comput. Eng. **2016**, 3 (2016)
23. Gardner, W.A., Napolitano, A., Paura, L.: Cyclostationarity: half a century of research. Signal Process. **86**, 639–697 (2006)
24. Cesarelli, M., Ruffo, M., Romano, M., Bifulco, P.: Simulation of foetal phonocardiographic recordings for testing of FHR extraction algorithms. Comput. Methods Programs Biomed. **107**, 513–523 (2012)
25. Ruffo, M., Cesarelli, M., Romano, M., Bifulco, P., Fratini, A.: An algorithm for FHR estimation from foetal phonocardiographic signals. Biomed. Signal Process. Control **5**, 131–141 (2010)
26. Giavarina, D.: Understanding Bland Altman analysis. Biochemia Medica **25**, 141 (2015)

Medical Quality of Service Optimization over Joint Body Sensor Networks and Internet of Multimedia Things

Ali Hassan Sodhro, Aicha Sekhari, Yacine Ouzrout, Gul Hassan Sodhro, Noman Zahid, Sandeep Pirbhulal and M. Irfan Younas

Abstract This paper proposes novel Rate Control Video Transmission Algorithm (RCVTA) to optimize medical quality of service (m-QoS) in terms of network metrics such as, standard deviation (Std dev), throughput, peak-to-mean ratio (PMR), delay, average delay, jitter and average jitter during transmission of high-definition (HD) video stream named 'Tracking and Retargeting in GI endoscopy' over joint Internet of Multimedia Things (IoMT) and Body Sensor Networks (BSNs). Experimental results reveal that m-QoS is optimized with workahead transmission over joint BSN and IoMT networks for Tele-surgery.

Keywords Pre-recorded · High definition · Telemedicine · RCVTA · IoMT · BSN Telemedicine · Tele-surgery

---

A. H. Sodhro (✉) · N. Zahid · M. Irfan Younas
Sukkur IBA University, P.O. Box 65200, Sukkur, Sindh, Pakistan
e-mail: ali.hassan@iba-suk.edu.pk; aliso78@liu.se

N. Zahid
e-mail: noman.mece17@iba-suk.edu.pk

M. Irfan Younas
e-mail: irfan.younas@iba-suk.edu.pk

A. H. Sodhro · A. Sekhari · Y. Ouzrout
DISP LAB, University Lumiere Lyon 2, P.O. Box 69500, Lyon, France
e-mail: aicha.sekhari@univ-lyon2.fr

Y. Ouzrout
e-mail: yacine.ouzrout@univ-lyon2.fr

G. H. Sodhro
Shah Abdul Latif University, P.O. Box 87036, Khairpur, Sindh, Pakistan
e-mail: hassangul183@gmail.com

S. Pirbhulal
SIAT, CAS, P.O. Box 518000, Shenzhen, China
e-mail: sandeep@siat.ac.cn

G. Fortino and Z. Wang (eds.), *Advances in Body Area Networks I*,
Internet of Things, https://doi.org/10.1007/978-3-030-02819-0_16

1 Introduction

As a dynamic and state-of-the art technology, Internet of Things (IoT) conveys message about Telemedicine as a rising star in this digital and sensor world. Traditional medical devices are upgraded by adding sensors and networking features to them. Thus, devices are able to interact intelligently between patients and distant medical staff. As health industries interconnect wearable devices to medical networks where patient's health statistics are recorded through sensors and sent to distant medical locations. Hence, doctors can monitor and prescribe their suggestions to patients remotely. Alvi et al. [1], present a combined referred application with many important methods such as multimedia streaming and inter-operability among IoT Gateway and Telemedicine peripherals.

Wireless Sensor Networks (WSNs) are readily deployed at large as it requires less budget to employ sensor technology in small devices for recording data from different environments. Atzman et al. [2], design a method of sending recorded data at distant locations for various applications. IoT gives any device or thing ability to connect its sensors to the Internet. Shah [3], design the concept of green video transmission over IoT and also described the new features (online) to devices like CMOS camera, sensor devices, mobile devices, and many more other devices for home automations and smart grids.

Nowadays, the multimedia devices such as CMOS camera and microphone are easily available at very nominal cost. As a consequence, the Wireless Multimedia Sensor Networks (WMSNs) have achieved too much attention. In these networks multimedia content is extracted from physical environment through low cost sensor devices in formats such as audio video streams, images or scalar sensor data. These devices communicate to share this extracted data among the WMSN. This data is transmitted to a centralized application server or network manager by these very small devices, where data is stored and fused together. Gubbi et al. [4], develop a framework which provides the capacity to implement and utilize IoT at large extent, in order to allow computation, networking, storage, interaction of different clouds in shared environment. The IoMT is an improvement to the IoT, which focuses in providing video streams as a part of IoT implementation. Although IoMT devices are heterogeneous and have power limitations, but still they can be accessed through a unique public IP in the same way as the computers and other networking devices are accessed through the Internet. This accessibility benefits many applications in the fields of home automation, factory plants monitoring, smart city projects, transport industry, smart grids and efficient management of energy. The heterogeneity, limited power and capacity of devices hinder in the implementation of IoMT which gives Internet connectivity to WMSN. The issue of power consumptions in WMSNs is discussed in many researches. But, integration of IoM with BSN for Tele-surgery in WMSN is not discussed.

The inter-operability problem is among Personal Area Network (PAN) medical devices and IoT gateway, this can be employed mobile devices or Set Top Box (STB) and forwards the data to peripheral health devices and distant medical service providers. Streaming problem is among IoT gateway and distant back-end Wide Area Network (WAN) devices e.g. doctor's PC or Notebook used for multimedia communications. Hassan and Li [5], propose optimal smoothing algorithm for stored video transmission to optimize medical quality of service (m-QoS) and examine how the client buffer space can be used efficiently and effectively towards reducing the rate variability of the compressed variable bit rate (VBR) video.

Though it is hard to keep high quality real time video streaming, but few protocols such as, RTP and RTCP are considered for m-QoS control and optimization during video streaming over IoMT network. At patient's end, the video stream is first recorded and Data Source object encoding is applied, and stored into a server's buffer which is managed by Session Manager Object. After network availability, the buffered data is forwarded to receiver's buffer, where if receiver's buffer is filled or wait time is over then data is retrieved and Data Source object decoding is applied. Finally, it is played to the doctor's device. Buffer is simultaneously managed at Session Manager Object patient's side as well as doctor's end. This controls the QoS for streaming. For instance, if RTCP reports that jitter is very big for ongoing stream then Session Manager can widen the buffer size for uninterrupted data flow and may also decrease it to minimize streaming delay. Westerborn et al. [6], address algorithms which have linear computational complexity, and very limited memory requirements.

The remaining of this paper is organized as follows. Section 2 summarizes literature survey, Sect. 3 discusses convergence and inter-operability between IoT, BSNs and Telemedicine. Section 4, proposes algorithm for m-QoS Optimization over joint IoT and BSNs. Section 5, examines m-QoS mapping over joint Telemedicine and IoT network. Section 6, discusses experimental results and discussion. Paper is concluded in Sect. 7.

2 Literature Survey

With the increase in health requirements the Telemedicine is no more only a concept but a reality. But still its systems are employed using old vide conference applications which are not able to cope up with its requirements. So, this issue is addressed by revealing a Telemedicine system based on two tiers, one of old multimedia consultation systems and other of physiological data collection and transmission. Authors in [1], present in detail the vision and challenges during operation and implementation of internet of multimedia things. [7], surveyed about the IoT technologies and their convergence in future healthcare. Moreover, they discuss the converged transmission of Telemedicine and health IoT. Researchers in [2], reported about the green video streaming over the internet of things. In [3], detailed survey about IoT is condcuted, its implementation and future advancement with broad vision. Authors in [5, 6, 8], address this issue by employing personal area network standard, which supports higher data rates and can be feasible for WMSNs applications. Besides, they propose transmission power control-based energy efficient adaptive power control algorithm to save the energy in WBANs. Furthermore, a novel and efficient key management scheme for WSNs which manages keys with less storage, less communication and computation complexity. As Internet infrastructure provides "best effort delivery" but guaranteed packet transmission is not promised, although this framework fits well for non-real time data and inappropriate for time sensitive multimedia data and video. In [9], a coding solution is developed for minimization of resources and power consumption by considering path loss model in WBAN.

Researchers in [10–13], propose the battery-friendly packet transmission strategies for the wireless capsule endoscopy, they used the battery model and measure the energy and battery charge consumption of different algorithms. Besides, green and battery-friendly video streaming in WBSNs is discussed by developing separate algorithms for different applications to get energy-efficient and longer battery lifetime. Also, they present the routing issues in the IoT by examining the impact of window size on the routing protocols of the WSNs, they used different metrics for performance analysis of the routing protocols. Most standards for providing Internet access to Things are not considering the issues of providing multi-media objects over IoT. In [14, 15], the integration of Telemedicine system with IoT is discussed with novel framework, besides, they discuss the main issues in the inter-operability and implementation of both Telemedicine and IoT for remote healthcare. Researchers in [16], propose framework for the internet of multimedia nano- things. Also, they discuss in detail about the IoT from the data-centric aspect to manage and collaborate for different sensor devices.

Authors in [17–19], present the global internet mechanism for the IoT networks. Also, they reported that video streaming plays a major part in today's emerging heterogeneous technologies for the development of robust and efficient mechanism for IoT, its inter-operability and functionalities. In [20–22], a promising sensing cloud platform for the IoT and its paradigm versions, a cloud-based IoT framework for the control and monitor of the sensor networks, and the data-centric platform of IoT for the accurate performance analysis of the heterogeneous networks are proposed accordingly. Researchers in [23–27], propose online smoothening algorithm based on markov model, novel research direction for fusion based BSNs, and framework and energy management algorithms for the body sensor networks, respectively. Researchers in [28–31], present the various techniques of data fusion, IoT and software-based tools for the medical healthcare. Besides, they proved that their approaches for power aware and resource efficient for BSNs. Authors in [32–34], develop the innovative energy efficient, battery aware and consumer electronics-based methods with detailed review for healthcare especially, BSNs. In addition, they compare the several algorithms for energy and battery lifetime extension in healthcare applications. Furthermore, very remarkable contribution is made on IoT, Green, sustainable and reliable applications and protocols in medical healthcare environment [35–40]. They clearly described the emerging trends and practices achieved by their proposed, designed and developed strategies in the distinct applications of medical IoT, human biological signal transmission, medical media streaming, power-aware and battery-efficient methods.

After rigorous review of literature, it is found that many researchers addressed different opinions about the role of Telemedicine and IoT in healthcare applications.

They do not focus at the integration of Telemedicine with IoT and then to optimize the m-QoS at remote location. Our research is the first step towards achieving this target. Moreover, joint Telemedicine and IoT framework is proposed for HD video transmission and m-QoS optimization. Also we propose a novel algorithm for Tele-Surgery application.

3 Convergence Between Telemedicine and IoT

This paper discusses that Telemedicine should integrate multimedia and different medical peripherals using IoT gateway. In this regard, an integrated reference Telemedicine system is suggested to research issues of interoperable systems, streaming and security. Figure 1 depicts the IoT and BSNs based combined organizational framework for streaming and m-QoS improvements.

Mostly, ZigBee (IEEE 802.15.4 standard) is recommended for IoT due to its simple design fast execution of tasks. On the other hand, IoM applications need more data rate then ZigBee which is one of the bottlenecks for better user experience. In many studies, minimizing collision probabilities, delay, duty cycle, etc. are discussed but no one discussed about resource constrained WMSN devices on the basis of real time streaming. We present the convergence and inter-operability between Telemedicine and IoT for proper and robust operation for the entire network. It is the great and amazing convergence! That is, the convergence of the IoT and Telemedicine. "From smart business and smart industry to Telemedicine, smart hospitals, smart medical operation theaters, digital homes and medical health monitoring", the explosive

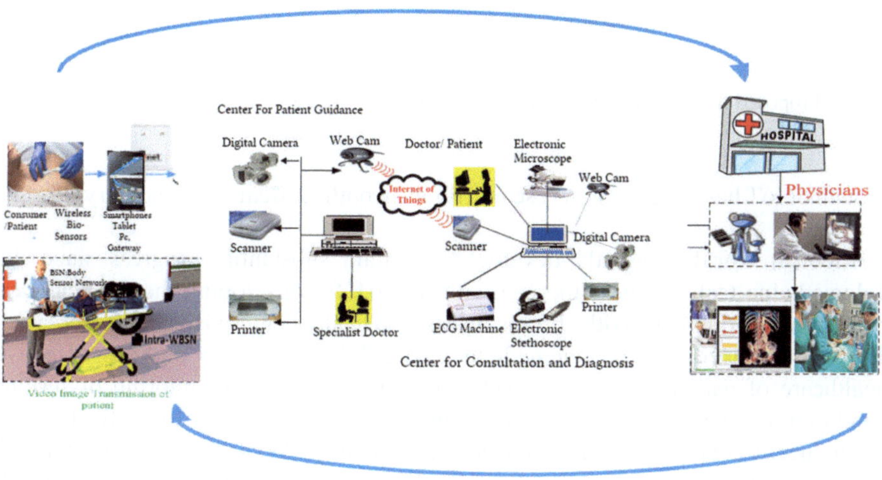

Fig. 1 Inter-operability between telemedicine and IoT

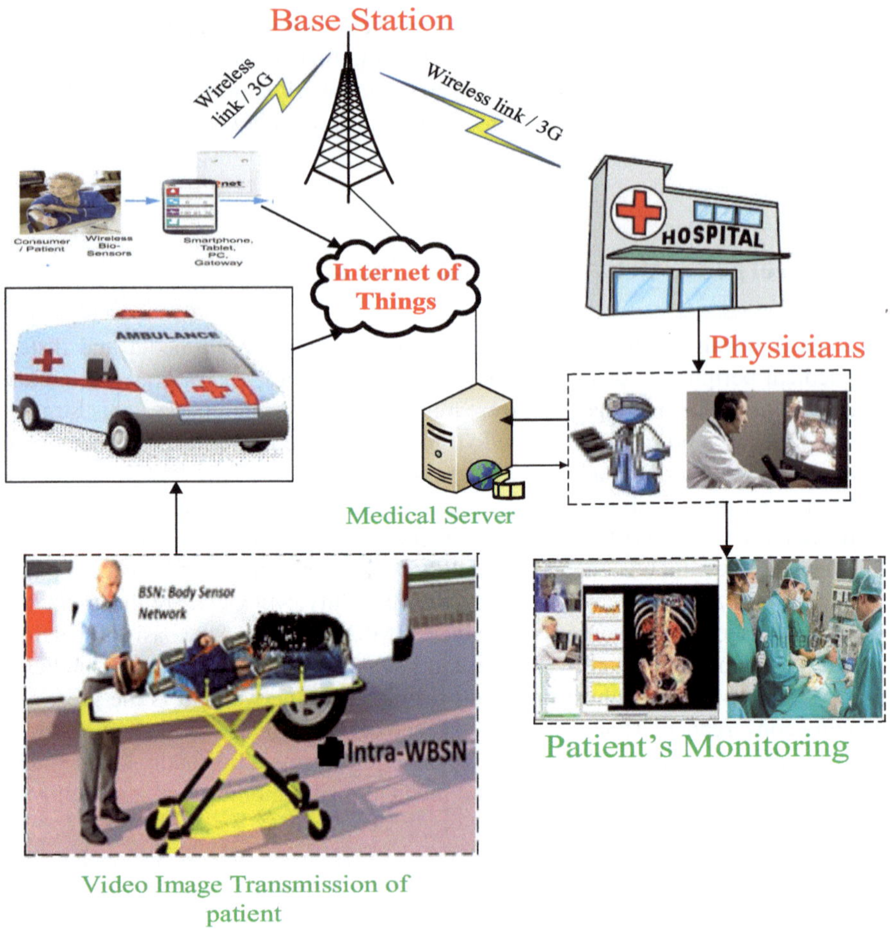

Fig. 2 Proposed framework of joint telemedicine and IoT

growth of IoT have opened new research doors in both academia and industry (Fig. 2).

Medical world is greatly developed to attain substantial progress since the Telemedicine technology is merged with the IoT for developing the medical care provision to individuals with chronic disease, so health center visits have decreased. There is a convergence of Telemedicine and IoT for the proper and cost-effective healthcare of patients in order to get timely diagnosis and monitoring of serious condition patients. The integration of Telemedicine services over IoT infrastructure has reduced care cost as well as enhanced care quality. People exhibit more concern for their health due to this convergence and convenience of Telemedicine and IoT. Advanced Telemedicine have integrated consultation over multimedia with different medical peripherals using IoT gateway. So integrated reference Telemedicine system identifies various important issues for research like inter-operability and streaming.

3.1 Inter-operability Between Telemedicine and IoT

Inter-operability between heterogeneous networks allows devices from different manufacturers to integrate and operate without any custom programming. Currently more and more people are becoming familiar and accepting the idea that real time data feeding should be greatly supported by Telemedicine. At the same time emerging sensor and digital world has a great boom of progress and prosperity in different areas.

However, a cumbersome problem appears that is inter-operability between IoT gateway and various Telemedicine devices [1]. This makes health data integration of various Telemedicine and Telehealth devices ambiguous and thus markdowns IoT health benefits. For instance, a perfect Telemedicine system is frequently connected to various medical devices like blood-pressure meter, video streaming, blood glucose meter, weighing meter etc. Generally, these device are manufactured by various vendors but the Telemedicine system provider have to work together with vendors to provide more features in these devices.

Therefore, inter-operability among different medical devices is a major issue for Telemedicine system provider. Inter-operability is required by users and service providers but vendor interests often obstruct it. It is of utmost importance due to various standards and technologies being employed in the smart network.

4 Rate Control Video Transmission Algorithm for High Definition Video Transmission

As enabled by emerging joint Telemedicine and IoT infrastructure, high definition (HD) video transmission is now a cornerstone for several applications and services. HD provides video quality 4$\times$ times larger than traditional PAL resolution, therefore recording more details and improving the video quality vividly. Here, we concentrate on applying the HD video for Tele-surgery application. The complete HD format video with 1920 $\times$ 1080 resolution needs 1.485 Gbps along with planes of three color (YUV), 10 bpcp (bits per color plane), interlacing 60 Hz field rate, 4:2:2 sampling and reducing 2/3rd data. It is called raw 1080i HD Video. Although, transmitting HD video streams in bandwidth constrained wireless medium is very challenging [41] (Fig. 3).

As a real time medical VBR compressed video contains N frames, each of size L bytes ($1 \leq i \leq N$). A video stream is stored at server side while at the client side that offline video stream can be decoded and played-out. First model of client's buffer

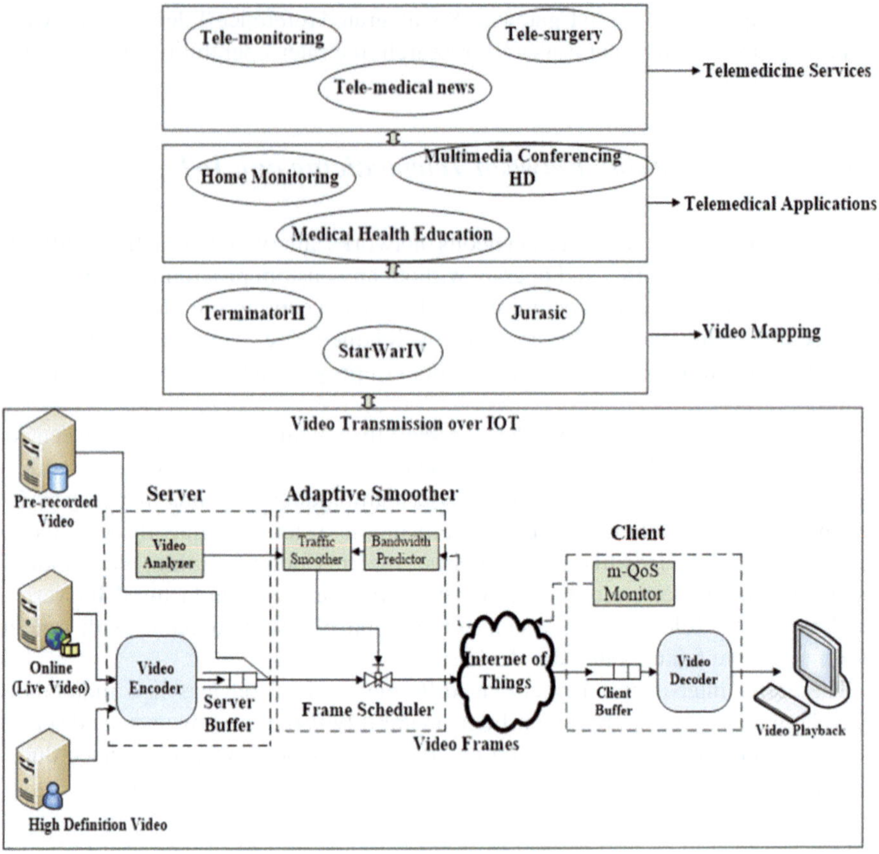

Fig. 3 Framework of video transmissions over IoT

is considered for discrete time slot t as fundamental time unit. A frame time is the time interval in which a medical video frame is transmitted (i.e., 1/15 s). To obtain a reasonable transmission schedule, the collective input data at receiving buffer in discrete time t, *S*(t) must arrive swiftly to avoid underflow of buffer. Mathematically, the curves of buffer underflow and overflow is given in Eqs. (2)–(4).

$$\mathrm{D(t)} = \sum_{\mathrm{i}=1}^{\mathrm{t}} \mathrm{L(i)} \tag{1}$$

$$\mathrm{B(t)} = \mathrm{b} + \sum_{\mathrm{i}=1}^{\mathrm{t}} \mathrm{L(i)} = \mathrm{D(t)} + \mathrm{b} \tag{2}$$

Thus,

$$D(t) \leq S(t) = \sum_{i=1}^{t} s \leq B(t) \tag{3}$$

Whereby, $s(i)$(i.e., $i = 1, 2, 3, \ldots, t$) gives smoothed stream bit rate at ith discrete time slot. The smoothed medical video transmission schedule is determined through monotonically growing and piece-wise linear path that is between $D(t)$ curve and $B(t)$ curve as depicted in Fig. 4.

Generally, after assign of number of frames (N) and window (W) it is hard to change video stream during transmission. A smoothening window size W is chosen of 600 frames or 20 s frame times for online optimal smoothening algorithm to maintain delay between server and client. Smooth transmission is performed once N unsmoothed frames are available to server. A window size (W) must be chosen as good compromise between small smoothed high peak rate, reduced rate variability. Our proposed algorithm smoothes VBR medical video stream taking into account both the feedback information on the real buffer fill level and total client buffer size. As the real buffer fill level information is periodically provided by real-time transmission control protocol (RTCP) packets in Δt_s seconds to the streaming server as a multiple of 64 KB. For each sliding window of W frames, online optimal smoothening algorithm generates a transmission plan that minimizes both the high peak data rate and rate variability by taking care of $D(t)$ and $B(t)$ curves as defined by (1) and (2).

When free buffer information from RTCP varies in time, than $B(t)$ curve is modified as in Eq. (4) and Fig. 5.

$$B(t) = D(t) + b(t) \tag{4}$$

Whereby $b(t)$ denotes function in discrete interval t which is supposed to be constant for each. Suppose, empty buffer information is saved during complete period till next

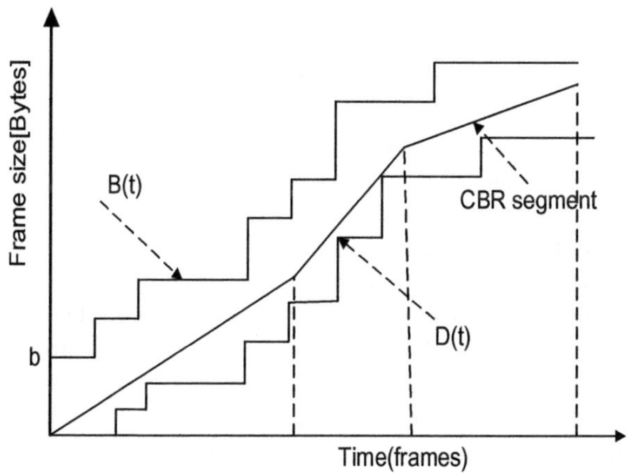

Fig. 4 Transmission schedule of rate control video transmission algorithm [5]

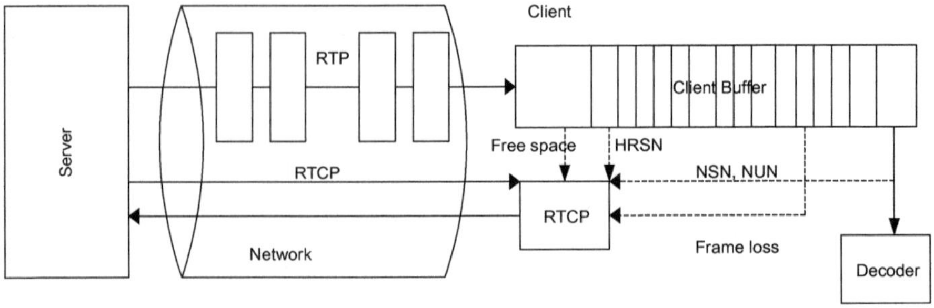

Fig. 5 Feedback information during VBR medical video streaming

information is received from client. This information is instantly reflected on server end, so smoothening is performed uninterrupted. Video streaming over joint IoT and BSN plays vital role in healthcare while in the meantime every patient needs good quality in terms of visualization, analysis etc. of the results from physician's side. In this connection we focused on the m-QoS optimization during video streaming over joint Telemedicine and IoT network.

5 Medical Quality of Service Mapping over Joint Telemedicine and IoT

We simulate in MATLAB a 5-min HD VBR medical video of MPEG-4 CIF resolution (352×288) [21]. Simulation is executed by considering client buffer size, initial delay, smoothening window (W) and server buffer size which is obtained by RTCP for HD video stream with MPEG-4 encoder. RCVTA algorithm for Tele-surgery over joint BSNs and IoT network is proposed to optimize the m-QoS in terms of Std dev, PMR, jitter, average jitter, delay and average delay. We establish trade-off amongst between client buffer size and standard deviation, peak-to-mean ratio, jitter, average jitter, delay and average delay by adopting start-up delay and window size. It is analyzed that if start-up delay, window size (W) and client buffer size for HD video increases then std. dev and PMR decrease which shows the smooth and uniform transmission of the real time VBR medical video.

On the contrary delay and jitter values increase with the increase of the client buffer size but their range is below the range of threshold values of obtained delay and jitter for Telemedicine medical applications. For example; threshold value of delay and jitter are 250 ms and 20 ms respectively, and our proposed algorithms provide (0.0475, 139 ms, 144 ms) of delay value and (2.48e−4, 0.60, 1.81 ms) of jitter value. We also analyzed from simulation results that if any value from start-up delay, client buffer size and window size varies not in proportion to each other than the performance of RCVTA will be affected (Table 1).

Table 1 Classification of services, applications, videos and m-QoS requirements

Service type	Video stream	Application	Algorithm	m-QoS requirement
Tele-surgery	Tracking and retargeting in GI endoscopy	IoT and BSNs	RCVTA	Throughput (bps), PMR, delay (ms), jitter (ms), average delay, average jitter

6 Experimental Results and Discussion

After experimental analysis it is observed that RCVTA performs better with acceptable delay and jitter values with window size of 300 frames or playback delay of 20 s and client buffer size of 2 MB over joint Telemedicine and IoT network (Tables 2 and 3).

Table 2 Simulation parameters

Parameter	Value
Tracking and retargeting in GI endoscopy	I-P-B Quant scale, 15-15-15
Encoding standard	MPEG-4
Peak-to-mean ratio	9.864
Mean bit rate (Mbps)	0.901
Peak bit rate (Mbps)	8.89
1-GOP (group of pictures)	12 frames
Time of one frame	5 ms
Total time of video	5 min = 5 × 60 = 300 s
Total number of frames (N)	900
Window size	600
Frame delay time	20 s
Bit rate (kbps)	2.691e+07
Average bit rate (kbps) un-compressed video	2.82e+05
Average bitrate (kbps) compressed video	2.88e+05
Client buffer size	64 KB, 512 KB, 2 MB

Table 3 Medical quality of service (m-QoS) requirements [8]

Application type	Throughput	Peak-to-mean ratio	Delay	Jitter
Tele-surgery	High	High	High	High

Experimental results in Table 4, reveals that RCVTA reduces rate variability and high peak rate dramatically at window size (W) of 600 frames and different client buffer size than the Baseline (unsmoothed video) during HD Tele-surgery video for m-QoS optimization in terms of network metrics such as, PMR, Std dev, jitter, delay, average jitter and average delay. As buffer size increases then PMR and Std dev are decreasing while jitter, delay, average jitter and average delay are increasing. But the performance of the RCVTA is better by using large window size and client buffer size. The delay and jitter values of RCVTA are large but are acceptable for the real time medical application such as, Tele-surgery, because the threshold values for real time video medical applications are 250 ms and 20 ms respectively. After deep analysis of different metrics of medical videos and m-QoS requirement, the major issue is addressed i.e., comparison between standard network metrics and simulation results. It is observed that delay and jitter values found through simulation are in microseconds which are less than their standard values as presented in Table 4.

Furthermore, Std dev and PMR are smaller and optimal in our simulation than the standard values. To summarize, it is inferred that simulation results of parameter values are appropriate for medical video requirements and m-QoS. Simulation results in Figs. 6, 7, 8, 9 and Table 4 shows the relationship between buffer size and number of frames for Tele-surgery video transmission over joint Telemedicine and IoT network. Proposed RCVTA recues the rate variability and peak data rates with high quality during HD video transmission with slightly more delay and jitter at the receiver, which is not suitable for emergency applications as compared to the pre-recorded

Table 4 Comparison between unsmoothed and HD videos

Video streams		Peak to mean ratio	Std. dev	Delay (ms)	Jitter (ms)	Average delay (ms)	Average jitter (ms)
Buffer size= 64 KB	Un-smoothed video	9.52	34.08	9.74 e−05	5.11e−07	1.082e−05	5.67e−08
	High definition video	2.37	1.76	2.0	0.005	0.23	3.39e−4
Buffer size= 512 KB	Un-smoothed video	1.46	0.13	0.0015	8.17e−06	1.67e−04	9.07e−07
	High definition video	2.30	1.73	17.92	0.053	1.951	0.0061
Buffer size= 2 MB	Unsmoothed video	1.39	0.013	0.0058	3.11e−05	6.44e−04	3.45e−06
	High definition video	2.25	1.64	69.01	0.35	9.82	0.025

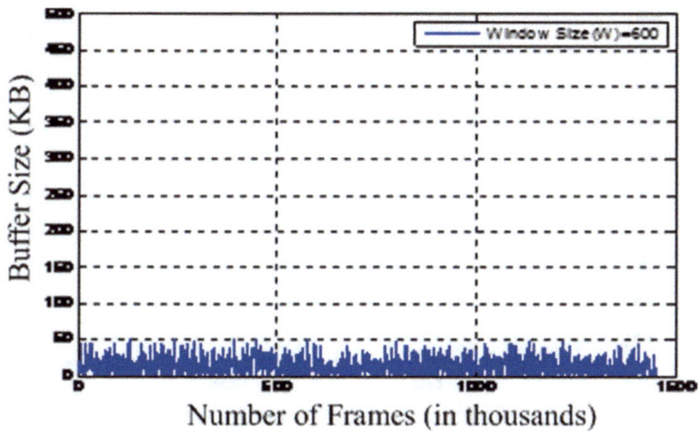

Fig. 6 Unsmoothed video stream

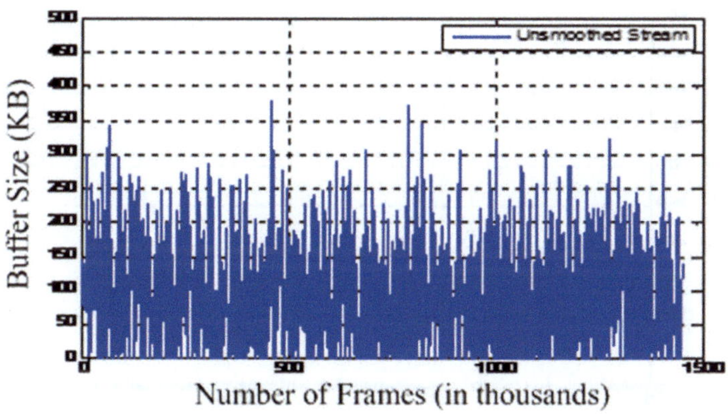

Fig. 7 Smoothed video stream with buffer size 64 KB

and online video transmission. The main reason of large delay and jitter values for RCVTA is that there is less prior knowledge of future frames and it depends upon video transmission.

7 Conclusion

In this paper we propose RCVTA for Tele-surgery HD video transmission through workahead transmission by delaying that video *W* time units to optimize m-QoS over the joint IoMT and BSNs. Besides, m-QoS optimization framework for joint BSNs and IoT networks is proposed. In addition the feedback information mechanism about real time buffer fill level from client to server with RTCP is established to

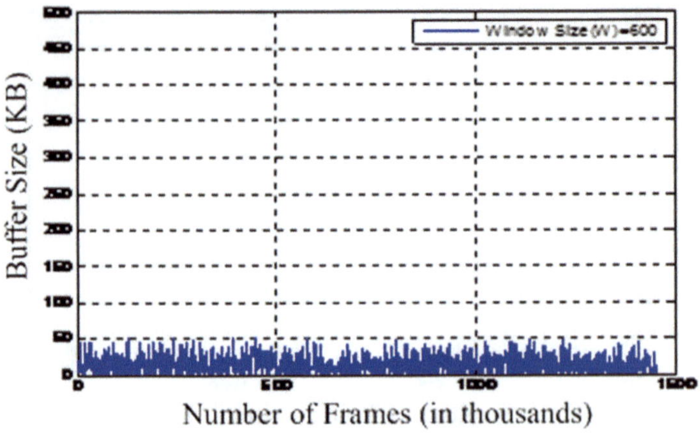

Fig. 8 Smoothed video stream with buffer size 64 KB

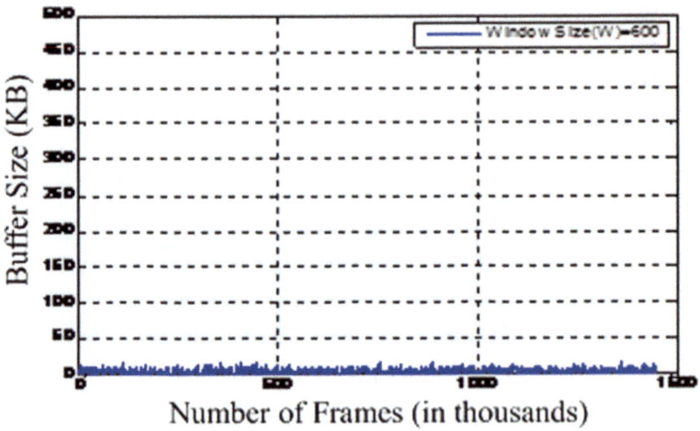

Fig. 9 Smoothed video stream with buffer size 2 MB

reduce high peak data rate and rate variability. A trade-off between window size and client buffer size is established with m-QoS metrics both mathematically and experimentally. By using large window size, client buffer size and knowledge of future frames through RTCP, server provides good transmission schedule for Telesurgery and also balances the use of processing, memory and network resources (i.e., standard deviation, peak-to-mean ratio, jitter, delay, average jitter and average delay).

In near future inter-operability in joint IoT and BSNs during video transmission will be center of attention and our proposed algorithm will be implemented for several real-time medical applications.

References

1. Alvi, S.A., et al.: Internet of multimedia things: vision and challenges. Ad Hoc Netw. **33**, 87–111 (2015)
2. Atzman, O., et al.: The convergence of eHealth and IoT in the future of healthcare. In: IoT Congress and Exhibition 2014. Hasharon Hotel Herzliya, Israel (2014)
3. Shah, G.A.: Enabling green video transmission over internet of things (2014)
4. Gubbi, J., Buyya, R., Marusic, S., Palaniswami, M.: Internet of Things (IoT): a vision, architectural elements, and future directions. Future Gener. Comput. Syst. **29**(7), 1645–1660 (2013)
5. Hassan, A., Li, Y.: Medical quality-of-service optimization in wireless telemedicine system using optimal smoothing algorithm. E-Health Telecommun. Syst. Netw. J. **2**(1) (2013)
6. Westerborn, J.: On particle-based online smoothing and parameter inference in general hidden Markov models. KTH Royal Institute of Technology (2015)
7. Sodhro, A.H., Li Y., Shah, M.A.: Green and friendly media transmission algorithms for wireless body sensor networks. Multimed. Tools Appl. **76**(19), 20001–20025 (2016)
8. Sandeep, P., Zhang, H., Wu, W., Zhang, Y.-T.: A comparative study of fuzzy vault based security methods for wireless body sensor networks. In: IEEE 10th International Conference on Sensing Technology (ICST), pp. 1–6 (2016)
9. Sodhro, A.H., Fortino, G.: Energy management during video transmission in WBSNs. In: 14th IEEE International Conference on Networking, Sensing an Control (ICNSC), pp. 1–6. Calabria, Italy (2017)
10. Zhang, X.M., Xu, C.: A multimedia telemedicine system in internet of things. In: 2nd International Conference on Information and Multimedia Technology (ICIMT 2010) IPCSIT, vol. 42. Singapore (2012)
11. Sodhro, A.H., Li, Y., Shah, M.A.: Green and battery-friendly video streaming in wireless body sensor networks. J. Multimed. Tools Appl. **67**(10), 1–25 (2016)
12. Sodhro, A.H., Shah, M.A.: Role of 5G in medical health. In: IEEE International Conference on Innovations in Electrical Engineering and Computational Technologies (ICIEECT), Indus University, Karachi, Pakistan (2017)
13. Aggarwal, C.C., Ashish, N., Sheth, A.: The internet of things: a survey from the data-centric perspective. In: Managing Cooperation Among Heterogeneous Multimedia Devices and Mining Sensor Data, Springer, pp. 383–428 (2013)
14. Hassan, A., Li, Y., Shah, M.A.: Energy-efficient adaptive transmission power control in wireless body area networks. IET Commun. **10**(1), 81–90 (2016)
15. Sodhro, A.H., Kumar, A., Sodhro, G.H.: 5G-based transmission power control mechanism in fog computing for IoT devices. MDPI Sustain. **10**(4), 1–17 (2018)
16. Jornet, J.M., Akyildiz, I.F.: The internet of multimedia nano-things in the Terahertz band. In: Wireless Conference (European Wireless), 2012 18th European. VDE (2012)
17. Sandivine, Sandvine Global Internet Phenomena Complete (2014)
18. Cisco VNI: The Zettabyte Era: Trends and Analysis, June 2014
19. Qualcomm. www.eeworld.com.cn/zt/wireless/downloads/qca4002-4004fin.pdf
20. Distefano, S., Merlino, G., Puliafito, A.: A utility 471–480 paradigm for IoT: the sensing cloud, Perv. Mob. Comput. (2014)
21. Sodhro, A.H., Kumar, A.: An energy-efficient algorithm for wearable electrocardiogram signal processing in ubiquitous healthcare applications. MDPI Sens. **8**(3), 923 (2018)
22. Fazio, M., Puliafito, A.: Cloud4sens: a cloud-based architecture for sensor controlling and monitoring. IEEE Commun. Mag. **53**(3), 41–47 (2015)
23. Distefano, S., Merlino, G., Puliafito, A.: Device-centric sensing: an alternative to data-centric approaches. IEEE. Syst. J. (2015)
24. Sodhro, A.H., Li, Y.: Battery-friendly packet transmission strategies for wireless capsule endoscopy. In: The International Conference on Health Informatics, International Federation for Medical and Biological Engineering (IFMBE) Proceedings, vol. 42, pp. 236–239 (2014)
25. Imperial College London. http://hamlyn.doc.ic.ac.uk/vision/

26. Gravina, R., Alinia, P., Ghasemzadeh, H., Fortino, G.: Multi-sensor fusion in body sensor networks: state-of-the-art and research challenges. Inf. Fusion **35**, 68–80 (2017)
27. Sodhro, A.H., Sangaiah, A.K., Pirphulal, S., Sekhari, A., Ouzrout, Y.: Green media-aware medical IoT system. In: Multimedia Tools and Applications, pp. 1–20. Springer. http://link.springer.com/article/10.1007/s11042-018-5634-0 (2018)
28. Fortino, G., et al.: A framework for collaborative computing and multi-sensor data fusion in body sensor networks. Inf. Fusion **22**, 50–70 (2015)
29. Fortino, G., et al.: BodyCloud: a SaaS approach for community body sensor networks. Future Gen. Comp. Syst. **35**, 62–79 (2014)
30. Sodhro, A.H., Sekahri, A., Ouzrout, Y.: Energy-efficiency comparison between data rate control and transmission power control algorithms for wireless body sensor networks. Int. J. Distrib. Sensor Netw. (IJDSN) **14**(1), 1–18 (2018)
31. Fortino, G., et al.: Enabling effective programming and flexible management of efficient body sensor network applications. IEEE Trans. Human-Mach. Syst. **43**(1), 115–133 (2013)
32. Sodhro, A.H., Sangaiah, A.K.: Convergence of IoT and product lifecycle management in medical health care. In: Future Generation Computer Systems: Special Issue on Emerging Edge-of-Things Computing: Opportunities and Challenges, vol. 86, pp. 380–391. Elsevier (2018)
33. Xu, X.J., Andrepoulos, Y., Xiao, Y., van der Schaar, M.: Non-stationary resource allocation policies for delay-constrained video streaming: application to video over internet-of-things-enabled networks (2014)
34. Hassan, A., Li, Y., Shah M.A.: Novel key storage and management solution for the security of wireless sensor networks. Ind. J. Electr. Eng. TELKOMNIKA **11**(6), 3383–3390 (2013)
35. Sodhro, A.H., Pirbhulal S.: Power control algorithms for media transmission in remote healthcare systems. IEEE Access **6** (2018)
36. Sodhro A.H., et al.: Convergence of IoT and product lifecycle management in medical health care "Future generation computer systems: Special issue on emerging edge-of-things computing: Opportunities and challenges. Elsevier **86**(2018), 380–391 (2018)
37. Sodhro A.H., Kumar A.: An energy-efficient algorithm for wearable electrocardiogram signal processing in ubiquitous 3 healthcare applications. MDPI Sensors **8**(3), 923 (2018)
38. Sodhro A.H., Sekahri A., Ouzrout Y.: Energy-efficiency comparison between data rate control and transmission power control algorithms for wireless body sensor networks. International Journal of Distributed Sensor Networks (IJDSN) **14**(1), 1–18 (2018)
39. Sodhro A.H., Pirphulal S., Sekhari A., Ouzrout Y.: Green media-aware medical IoT system, multimedia tools & applications. Springer, pp.1–20, 2018. http://link.springer.com/article/10.1007/s11042-018-5634-0
40. Sodhro A.H., Kumar A., Sodhro G.H.: 5G-based transmission power control mechanism in fog computing for IoT devices. MDPI Sustainability **10**(4), 1–17 (2018)
41. Dhumane, A., Prasad, R., Prasad, J.: Routing issues in internet of things: a survey. In: Proceedings of IMECS 2016, Hongkong, China (2016)
42. Pirbhulal, S., et al.: Analysis of efficient biometric index using heart rate variability for remote monitoring of obstructive sleep apnea, neuropsychiatry (2017)
43. Yun-qiang, L., Song-yu, Y., Xiang-wen, W., Jun, Z.: Segment-based traffic smoothing algorithm for VBR video stream. J. Zhejiang Univ. Sci. A **7**(4), 543–548

An Encryption Method for BAN Using the Channel Characteristics

Liangguang Peng, Jinzhao Lin, Tong Bai, Yu Pang, Guoquan Li, Huiquan Wang, Xiaoming Jiang, Junchao Wang and Zeljko Zilic

Abstract The protection of information security is a very important technique requirement in Body area networks (BANs). According to the limitation of energy, the traditional encryption methods which adopt the complex algorithm and the large consumption are not suitable for BANs. This paper proposes a new encryption method based on the channel model of BANs, which adopts the path loss from the BAN systems to form the initial key, utilizes the LFSR (Linear Feedback Shift Register) circuit to generate the key stream, and then encrypts the data in the coordinator of the BAN system. This new encryption method has the advantages of low energy consumption, simple hardware implementation, and dynamic key updating.

Keywords Body area network · Channel model · Path loss · LFSR Encryption method

1 Introduction

Body area network has caused a widely interest in academia and industry all around the world since it was born [1, 2]. BANs can obtain the human vital signs continuously, and support the remote clinical diagnosis, the emergency treatment and the health data service [3, 4]. Different from the common wireless sensor network, security is one of the main factors in the application of BAN [5–7], which requests to support the identity authentication, the privacy protection and the information completeness.

The researches on the security of BAN have carried out a series of methods. Paper [8] puts forward a multi-path reinforcement method with pre-distribution of

L. Peng · J. Lin (✉) · T. Bai (✉) · Y. Pang · G. Li · H. Wang · X. Jiang
Chongqing University of Posts and Telecommunications, Chongqing, China
e-mail: Linjz@cqupt.edu.cn

T. Bai
e-mail: baitong03@126.com

J. Wang · Z. Zilic
McGill University, Montreal, Canada

G. Fortino and Z. Wang (eds.), *Advances in Body Area Networks I*,
Internet of Things, https://doi.org/10.1007/978-3-030-02819-0_17

random number, which is effectively resisting the small attack. However, the algorithms require a large number of operations before the encryption and thus the power consumption would increase rapidly. The BAN system is a not suitable to use the key pre-distribution scheme according to the extremely low-power BAN nodes. Papers [9–11] generate the keys according to the periodicity physiological signs information. Different from the general sensor network nodes, the BAN nodes' parameters are the human physiological signs, which show different features according to the different individual. Even for the same individual, the values are not constant appearing to change slowly over time. Therefore, these parameters can be used as keys, especially the ECG (electrocardiograph) signal containing much information. However, this method is not suitable for the simple BAN systems without the detecting of periodicity physiological signs information. There are some other complex encryption methods such as AES (advanced encryption standard) encryption method [12] and ellipse curve cryptography. But these encryption methods cost high consumption and adopt the fixed key [13], and are not suitable for the BANs.

In this paper, we design a method whose keys are based on the channel characteristics of BANs, which has the following advantages: (i) the path loss of different individuals are discrepancy, and even for the same individual, the values of channel characteristic changes depend on the activities of the human body. Therefore, the parameters of the channel can be used as the initial keys. And the channel characteristics could be easily calculated based on the registers of the BAN sensors. Thus the generating of the initial keys not require the extra overhead; (ii) based on the initial keys, the encryption method generates the stream cipher by the LFSR circuit, and thus has the advantages of fast conversion speed, simple hardware implementation, lower consumption and safer than the software encryption; (iii) the initial keys is generated by the channel parameters and it changes over time, therefore, the initial keys are dynamic update. Even if the key is cracked by the eavesdroppers, only a small part of the data would be intercepted. Therefore, the new key agreement is a lightweight encryption method which is very suitable for BANs.

The remainder of this paper is organized as follows: Sect. 2 presents the design of the human body model. Section 3 provides the encryption method based on the LFSR circuit. The simulations and experiment results are validated in Sect. 4, and the numerical results are analyzed. Finally, Sect. 5 concludes the paper.

2 Human Body Model

This paper establishes a three dimensional human body model based on the VariPose simulation software, which is a kind of reconfiguration visualization tool of human grid, and it can be used to build the digital body model of various body postures. Since the simulation software calculates different body postures by adjusting the position of the joints, and we could obtain different positions of the human body model through the method of the human grid. As shown in Fig. 1, this paper adopts 36 kinds of dielectric constant and conductivity of human tissue to simulate the human

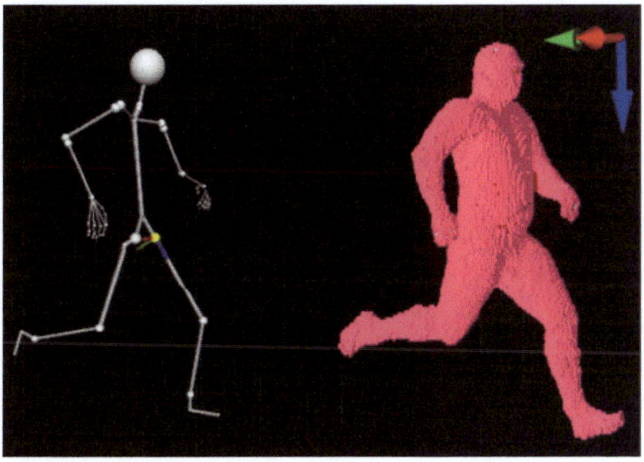

Fig. 1 Body posture model

body model and thus import the model to the XFdtd, which is a three-dimensional electromagnetic field simulator and adopt the FDTD (finite difference time domain) method.

2.1 Human Parameter Configuration

Through the simulation analysis, the corresponding path loss values could be obtained. The path loss of the 2.4 GHz human surface channel is discussed as follow: This paper adopts the half wave dipole antenna to simulate the transceiver, the half wave dipole antenna was placed 2 mm outside the human skin, and the impedance is set to 50 Ω. The dielectric constant and conductivity of the human tissue in the 2.4 GHz human surface channel are shown in Table 1. The formula for the length and radius of the antenna are shown as follows:

$$L = \frac{\lambda}{2} \tag{1}$$

$$r = \frac{\lambda}{200} \tag{2}$$

where λ represents the incident wavelength. L = 62 mm and r = 0.62 mm in 2.4 GHz.

The uniform grid structure could be established based on the FDTD algorithm. In order to get the stable numerical calculation in the grid dissection, the time step of the simulation Δt and the side length of the minimum unit grid Δx should be satisfied the formulas (3)–(5):

$$\Delta t \leq \frac{1}{c\sqrt{\frac{1}{\Delta x^2} + \frac{1}{\Delta y^2} + \frac{1}{\Delta z^2}}} \tag{3}$$

Table 1 The dielectric constant and conductivity of the human tissue in 2.4 GHz

Human tissue	Dielectric constant	Conductivity (S/m)
Body fluid	68.24	2.4392
Eyeball	44.681	1.4738
Fat	5.2853	0.10235
Protein	1.22	36.17
Muscle	52.791	1.705
Bladder	18.026	0.67265
Testis	57.629	2.1273
Cartilage	38.878	1.7172
Skin	38.063	1.4407
Tooth	47.7	1.92
Blood	58.347	2.5024
Marrow	5.3024	0.092843
Cornea	51.697	2.2588
Small intestine	54.527	3.1335
Heart	54.81	2.26

$$\lambda_{\text{min}} = \frac{c}{f\sqrt{\varepsilon}} \tag{4}$$

$$\Delta x \leq \frac{\lambda_{\text{min}}}{10} \tag{5}$$

where $\triangle x$, $\triangle y$ and $\triangle z$ is the side length of the grid, c is the speed of light in the medium, ε is the maximum dielectric constant of the human tissue in the model, and λ_{min} is the minimum wavelength of the human tissue. The dielectric constant and conductivity of the human tissue in 2.4 GHz are shown in Table 1.

2.2 Channel Path Loss

According to the limitation of the computation complexity, this paper makes $\triangle x = 5$ mm, the total number of girds are $60 \times 120 \times 360$, and the simulation boundary condition is PML (perfectly matched layer).

After the human body model is established, we study for the channel characteristics influence by the distance between transmitters and receivers in the body surface channel. As shown in Fig. 2, the simulation adopts half wave dipole antenna instead of transceiver antenna, the antenna length is 62 mm, the impedance is 50 Ω, and works in 2.4 GHz. The transmitter antenna is fixed on the right hand wrist, and the receiver antenna is not fixed. The BAN channel model is mainly the path loss according to the characteristics of electromagnetic field distribution in human body. S_{21} in FDTD represents the relation between the input and the output of the

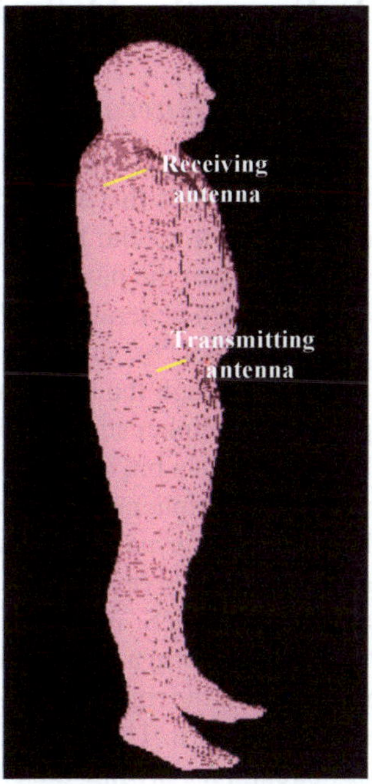

Fig. 2 Standing body model

Table 2 S_{21} values of body surface channel

d (mm)	S_{21} (dB)	d (mm)	S_{21} (dB)
25	−28.74	300	−63.97
50	−30.12	325	−65.38
75	−32.47	350	−66.04
100	−36.1	375	−67.16
125	−37.61	400	−68.92
150	−44.86	425	−69.34
175	−50.02	450	−69.87
200	−55.43	475	−70.16
225	−56.53	500	−70.69
250	−58.97	525	−71.25
275	−61.25	–	–

signal in the body surface channel, and meanwhile reflects the electromagnetic wave transmission characteristics of the human channel. Table 2 is the S_{21} values of body surface channel with different distance of the antennas.

3 Encryption Method

3.1 Key Generation

Shannon proved that if the key is a random sequence, which is only used once, and the length of the key sequence is equal to or longer than that of the plaintext sequence, then the password system is unbreakable. As shown in Fig. 3, the stream cipher system generates random sequence, and uses the sequence as the key.

The symbol "⊕" in the diagram indicates the addition of module 2. The initial key λ passes through the key generator to generate the key stream k_j. In the encryption process, the cipher stream c_j is obtained by adding up the plaintext stream m_j and the key stream k_j in bit, and the decryption unit takes the addition operation of module 2 in bit for the cipher stream c_j and the key stream k_j to get the plaintext stream m_j, The key generator usually adopts LFSR [14, 15], and the n-level LFSR consists of the shift register $R = (r_n, r_{n-1}, \ldots, r_1)$, the tap sequence $T = (t_n, t_{n-1}, \ldots, t_1)$ and the enter digital signature S. When generating a new key, one of bits in digital signature and the feedback bit form a new bit though a operation, then the new bit would insert the shift register, and the right side output a bit at the same time, the other bits moves one bit to the right in turn. The structure of the key generator could output the self-circulation key stream, meanwhile the circuit construction of the key generator is relatively easy and some logic gate would provide a high security. But if the eavesdroppers master a certain number of corresponding plaintext and ciphertext, the tap sequence would be deduced and then crack the encryption system. Therefore, the structure need to the optimization design. This paper adopts a 16-order nonlinear feedback shift register to extend the period, and its maximum value is 2^{16}. In addition, the dynamic key update is essential, because the key is refreshed every T seconds leading to lower probability of cracking.

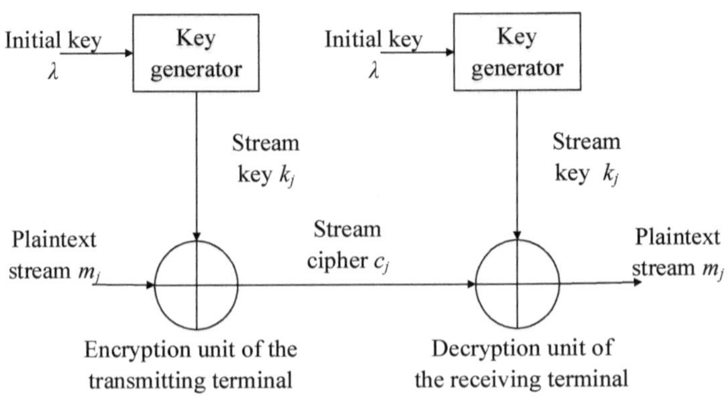

Fig. 3 Stream cipher system

In this paper, λ is the path loss value of S_{21}. After detecting the path loss value of t seconds S_{21}, the characteristic value λ can be calculated as:

$$\lambda_i = S_{21}^i \tag{6}$$

$$\lambda = \frac{\sum_{i=1}^{n} \lambda_i}{n} \tag{7}$$

where S_{21}^i is the amplitude of the ith S_{21} of the sensor nodes and coordinator within the t seconds. Due to the path loss have a significant difference on different individuals and a single individual has slight differences over time. Calculate the characteristic values of S_{21} which reflect the path loss differences. Therefore, the characteristic value λ is dynamic updated, and then the key is refreshed based on λ. Meanwhile, in the actual applications of BAN systems, S_{21} is the receiving sensitivity minus transmitting power. The receiving sensitivity and the transmitting power are easy to obtain from RF (radio frequency) chips without any additional costs, therefore, this method is conducive to low power implementation of the sensor nodes in BANs.

Then the characteristic value λ is quantified as the initial key for the LFSR with N-order to generate the stream key. The key stream broadcasts to the nodes in the BAN system, and the nodes use the key to decrypt the data. Meanwhile, the encryption method would re-generate the key and repeats the above steps every T seconds.

3.2 *Method of LFSR Circuit Encryption*

In this paper, the value of T could be 200–400 s, and the value of t could be 4 s. Using the LFSR circuit to produce the stream cipher has advantages of the fast conversion speed, simple hardware implementation, lower consumption and safer than the software encryption. Here we use the 16-order LFSR to encrypt the plaintext, whose maximum value is allowed $2^N - 1$. Moreover, the dynamical update of the key from S_{21} every T seconds can realize real-time encryption, which can extremely reduce the possibility of attack.

As shown in Fig. 4, the common bits move by the arrow, and the bits before the XOR symbols move after the exclusive-or operation.

In the BAN initialization, the coordinator extracts S_{21}, calculates the key after the reception of the preamble sequence, and then broadcasts the key to all nodes.

Receiving the encrypted data generated by the node, the intelligent terminal calculates the characteristic value according to S_{21}. In this paper, path loss has large difference among individuals. The flow chart of the proposed encryption method is shown in Fig. 5.

Since BAN system requires the node of physiological sign acquisition which should has extremely low consumption, the encryption algorithm should be as simple as possible. As the method mentioned before, the additional function for the node does not need extra cost except a simple LFSR circuit. This is very suitable to the implementation of low power sensor nodes.

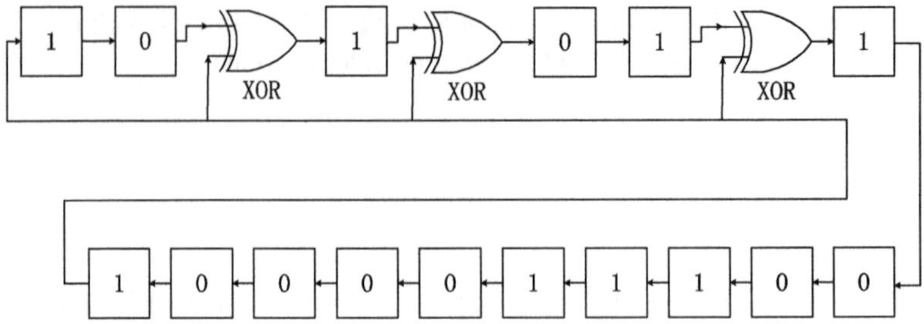

Fig. 4 16-LFSR flow chart

Fig. 5 The flow chart of the proposed encryption method

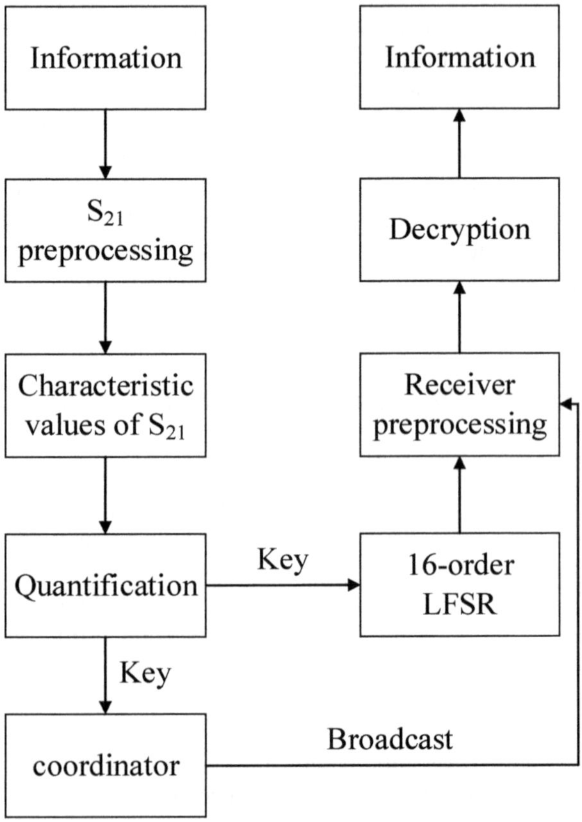

3.3 *Key Distribution*

BANs require the low consumption sensor nodes. Therefore, this paper minimizes the computing tasks in the nodes when considering the data encryption. The key interaction scheme is: firstly, the sensor node sends the transmitting power information and the leader sequence during initialization, in which the sensor node does not make any calculations; secondly, the intelligent terminal calculates the characteristic values of S_{21} based on the received data of the transmitting power information, the leader sequence and its receiving sensitivity, and then generates the keys and broadcasts them; thirdly, according to the above method, the nodes encrypt the data and interact with the intelligent terminal after receiving the keys; lastly, every T time, the system randomly selects a node to send the leader sequence to the intelligent terminal, and the intelligent terminal re-calculate the characteristic values based on the transmitting power and receiving sensitivity. Therefore, the task of calculating the characteristic values is completed by the intelligent terminal, the other sensor nodes do not have to take heavy computational effort leading to the low power consumption for the whole system. Even if the eavesdropper captures the key at a time, it could only get a period of information due to the key changes over time.

4 Simulation Results

4.1 Human Model Simulations

This paper breaks a full walking state into 9 frames, as shown in Fig. 6. The 62 mm length half-wave dipole antennas are simulated to the transceiver antennas on the right waist and the left ankle, as shown in the Fig. 7. The antennas are 2 mm from the human skin and the transmitter frequency is 2.4 GHz. And then calculates the path loss values of each frame by the simulation.

As shown in Table 3, it can be seen that the path loss of each frame is changed when the human posture changes in the communication link between the right waist and the left ankle. The average fluctuation is 2.89 dB, the maximum path loss value 72.679 dB appears in the 9th frame, and the minimum path loss value 62.246 dB appears in the 5th frame (in this frame, the left foot of the human body is straight forward and the right foot is backward, and thus the antennas are basically unobscured by the human body). The difference between the maximum path loss and the minimum path loss is 10.43 dB.

Fig. 6 The 9 frames of human walking state

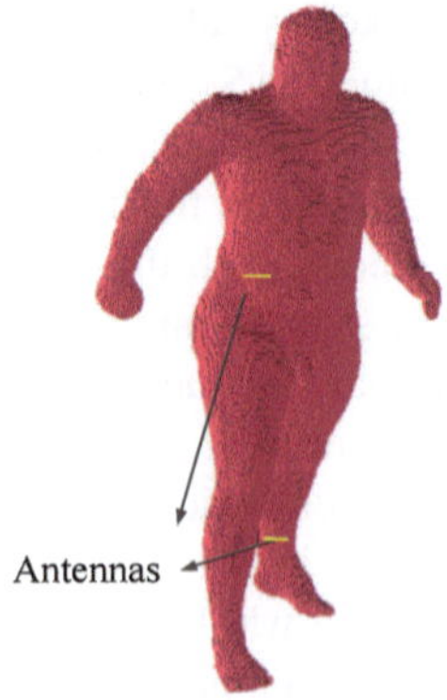

Fig. 7 The position of antennas in walking state

Table 3 The human surface channel path loss of each frame in 2.4 GHz

Frame	S_{21}(dB)	Frame	S_{21}(dB)
1	−66.275	6	−65.623
2	−70.146	7	−68.546
3	−72.068	8	−72.429
4	−65.567	9	−72.679
5	−62.246	–	–

4.2 Experiment Results and Analyses

Figure 8 shows the visual result of the encryption proposed in this paper using the picture Lena.

The 2-dimensional correlation of the original image and the encrypted image can be represented by the 2-dimensional correlation coefficient of the two corresponding matrices, as shown in formula (8).

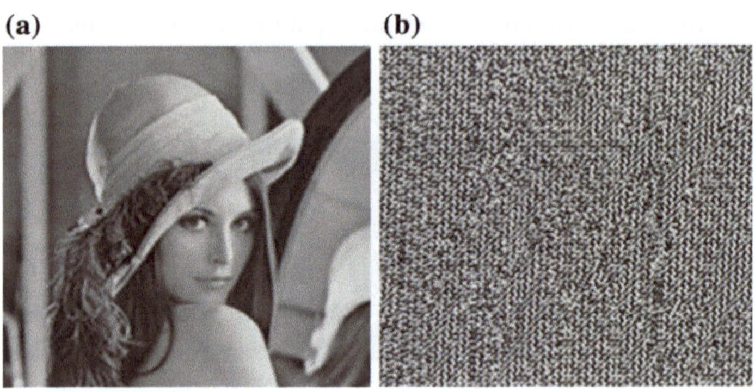

Fig. 8 Visual result of Lena: **a** The original image. **b** The encrypted image

$$C_{AB} = \frac{\sum_{i=1}^{M}\sum_{j=1}^{N}(A_{ij}-\overline{A})(B_{ij}-\overline{B})}{\sqrt{\left(\sum_{i=1}^{M}\sum_{j=1}^{N}(A_{ij}-\overline{A})^2\right)\left(\sum_{i=1}^{M}\sum_{j=1}^{N}(B_{ij}-\overline{B})^2\right)}} \tag{8}$$

where A_{ij} and B_{ij} represent the grey values of the original image and the encrypted image in the location (i, j) of the data matrix, respectively. $\overline{A}$ and $\overline{B}$ represent the average of the image pixel values of the original image and the encrypted image, respectively. This experiment selects four test images: Lena, Camera, Plane and Barbara, and uses the S_{21} from Table 3 to generate the initial key and encrypts the four images. The results are shown in Table 4, C_{AB} *is* close to zero which means the correlation between the original image and the encrypted image is very **small**.

Respectively selected 100 neighboring pixel pairs from three different directions (horizontal, vertical and diagonal) from the original image and the encrypted image, and calculate the correlation coefficient of the adjacent pixels with the formulas (9)–(12):

$$\gamma = \frac{\text{cov}(x, y)}{\sqrt{D(x)}\sqrt{D(y)}} \tag{9}$$

$$\text{cov}(x, y) = \frac{1}{N}\sum_{i=1}^{N}(x_i - E(x))(y_i - E(y)) \tag{10}$$

$$D(x) = \frac{1}{N}\sum_{i=1}^{N}(x_i - E(x))^2 \tag{11}$$

$$E(x) = \frac{1}{N}\sum_{i=1}^{N}x_i \tag{12}$$

where x_i and y_i represent the gray values of two adjacent pixels and N is the pixel number. The correlation coefficients of the three directions are shown as the Table 5. The correlations in three directions are all beyond 0.9 that denotes adjacent pixels have very similar gray values before encryption, while they decrease to under 0.1 after encryption, which means that the much smaller correlation can make better resistance of attack.

Table 4 2-dimensional correlation of the original image and the encrypted image

Images name	Lena	Camera	Plane	Barbara
C_{AB}	1.1465×10^{-9}	1.2432×10^{-9}	-1.2756×10^{-8}	2.5801×10^{-9}

Table 5 The correlation coefficient of the three directions

Correlation coefficient	Vertical	Horizontal	Diagonal
Original image	0.9388	0.9633	0.9417
Encrypted image	0.0672	0.0218	0.0735

5 Conclusions

In the BAN systems, protecting security of vital signs information plays a role of important technique requirement. This paper proposed a new encryption method which adopts the characteristic value of BAN channel as the initial key, introduces a 16-order LFSR circuit to encrypt the plaintext, and effectively hide the private information with the advantages of dynamic key, low cost, rapid operation and easy implementation. The investigation of encryption method is very helpful to design low-power sensor nodes, which provides theoretical support and engineering implementation.

Acknowledgements This work is supported by the National Science Foundation of China (Grant no. 61471075, 61671091), National Key Technology Research and Development Program of the Ministry of Science and Technology of China (2014BAI11B10), Key National Science Foundation of Chongqing (cstc2017jcyjB0203), Chongqing Integrated Demonstration Project (CSTC2013jcsf10029), Wenfeng Innovation Foundation of CQUPT, University Innovation Team Construction Plan Funding Project of Chongqing (Smart Medical System and Key Techniques, CXTDG201602009), Chongqing Key Laboratory Improvement Plan (Chongqing Key Laboratory of Photoelectronic Information Sensing and Transmitting Technology, cstc2014pt-sy40001), Chongqing Research Program of Basic Research and Frontier Technology (cstc2017jcyjA0982), Science and Technology research project of Chongqing Education Commission (KJ1704073), Research and Innovation Project of Graduate Students in Chongqing (CYS17239), the Scientific Research Foundation of CQUPT (A2016-73), the Priority Academic Program Development of Jiangsu Higer Education Institutions (PAPD) Fund, and Jiangsu Collaborative Innovation Center on Atmospheric Environment and Equipment Technology (CICAEET) Fund.

References

1. Seyedi, M, et al.: A survey on intrabody communications for body area network applications. IEEE Trans. Bio-Med. Eng. **60**(8), 2067 (2013)
2. Barakah, D.M., Ammad-Uddin, M.: A survey of challenges and applications of wireless body area network (WBAN) and role of a virtual doctor server in existing architecture. In: International Conference on Intelligent Systems, pp. 214–219. IEEE (2012)
3. Monton, E., et al.: Body area network for wireless patient monitoring. IET Commun. **2**(2), 215–222 (2008)
4. Ullah, S., et al.: A comprehensive survey of wireless body area networks: on PHY, MAC, and Network layers solutions. J. Med. Syst. **36**(3), 1065 (2012)
5. Lim, S., et al.: Security issues on wireless body area network for remote healthcare monitoring. In: IEEE International Conference on Sensor Networks, Ubiquitous, and Trustworthy Computing, pp. 327–332. IEEE Computer Society (2010)

6. Warren, S., et al.: Interoperability and security in wireless body area network infrastructures. In: International Conference of the Engineering in Medicine & Biology Society, p. 3837 (2005)
7. Mainanwal, V., Gupta, M., Upadhayay, S.K.: A survey on wireless body area network: security technology and its design methodology issue. In: International Conference on Innovations in Information, Embedded and Communication Systems, pp. 1–5. IEEE (2015)
8. Chan, H., Perrig, A., Song, D.: Random key predistribution schemes for sensor networks. In: Proceedings of 2003 Symposium on Security and Privacy. IEEE (2003)
9. Ali, A., Khan, F.A.: An improved EKG-based key agreement scheme for body area networks. In: Proceedings of the International Conference on Information Security and Assurance, ISA 2010, DBLP, Miyazaki, Japan, June 23–25 2010, pp. 298–308 (2010)
10. Mana, M., Feham, M., Bensaber, B.A.: SEKEBAN (Secure and efficient key exchange for wireless body area network). Int. J. Advanc. Sci. Technol. 12.12 (2009)
11. Yao, L., et al.: A biometric key establishment protocol for body area networks. Int. J. Distrib. Sens. Netw. 1550–1329, 1063–1067 (2011)
12. Gangadari, Bhoopal Rao, Ahamed, Shaik Rafi: Design of cryptographically secure AES like S-Box using second-order reversible cellular automata for wireless body area network applications. Healthcare Technol. Lett. **3**(3), 177–183 (2016)
13. Al-Janabi, S., et al.: Survey of main challenges (security and privacy) in wireless body area networks for healthcare applications. Egyptian Informat. J. (2016)
14. Sharma, K., et al.: Use of LFSR for sensor network security: a new approach. In: International Conference on Information Security and Assurance, pp. 61–66. IEEE Computer Society (2008)
15. Sharaf, M., et al.: A complex linear feedback shift register design for the A5 keystream generator. National IEEE Xplore 395–402 (2005)

Experimental Performance Evaluation of BLE 4 Versus BLE 5 in Indoors and Outdoors Scenarios

Heikki Karvonen, Carlos Pomalaza-Ráez, Konstantin Mikhaylov, Matti Hämäläinen and Jari Iinatti

Abstract This paper focus on an experimental performance evaluation of the recently published Bluetooth Low Energy (BLE) 5 technology. Measurements have been conducted both in indoors and outdoors scenarios. Performance of BLE 5 is compared to a previous release of BLE 4 which is currently the most used technology in commercial wireless healthcare and medical devices. This new improved BLE version may continue fostering the success of BLE use in those application scenarios as well as enable novel Internet of Things solutions. The main goal of this work was to evaluate, experimentally, the communications range and throughput performance of BLE 5 coded version which claims to provide fourfold improvement to the previous version of BLE. Measurement results obtained using the Nordic Semiconductor nRF52840 chipset are reported for indoor and outdoor cases relevant to healthcare and medical scenarios. Results show the practical communications range and throughput of the BLE 5 coded version, giving insight about the possible application space improvements for BLE technology. Specifically, our measurements showed that BLE 5 coded mode provides approximately 9 dB radio link budget gain compared to BLE 4, which leads to more than twofold communications range improvement in line-of-sight outdoor scenario and 10–20% improvement in non-line-of-sight indoor scenario.

Keywords BLE 5.0 · Communications range · Unlicensed band Healthcare and medical applications · Internet of Things

H. Karvonen (✉) · K. Mikhaylov · M. Hämäläinen · J. Iinatti
Centre for Wireless Communications, University of Oulu, Oulu, Finland
e-mail: heikki.karvonen@oulu.fi

C. Pomalaza-Ráez
Department of Electrical and Computer Engineering,
Purdue University, Fort Wayne, IN, USA

G. Fortino and Z. Wang (eds.), *Advances in Body Area Networks I*,
Internet of Things, https://doi.org/10.1007/978-3-030-02819-0_18

1 Introduction

Over recent years, the use of wireless body area networks (WBAN) has been continuously increasing [1, 2]. Today these devices are used in the context of smart healthcare applications, operating in hospitals and other similar environments, as well as for versatile sport and fitness activities. Due to the very nature of these use cases, wireless links are the most efficient way to enable connection of WBAN devices between themselves as well as to the external world and backbone systems.

Among all wireless standards employed for medical applications, Bluetooth Low Energy (BLE) stands out as the most widely used in current commercially available products [3]. The range of today's products communicating using BLE spans from simple pulse oximeters to complete body networks measuring the wide sets of physical parameters, such as body temperature, electrocardiogram (ECG) and electroencephalogram (EEG). BLE operates in the industrial, scientific and medical (ISM) 2.4 GHz band, which is available worldwide enabling interoperability in different regions. Nonetheless, today this band is rapidly becoming congested due to the presence of several other wireless technologies such as IEEE Std. 802.11x (Wi-Fi) [4], IEEE Std. 802.15.4 [5] (ZigBee [6]), IEEE Std. 802.15.6 [7], ETSI SmartBAN [8], ANT/ANT+ [9] and most recently, the upcoming unlicensed LTE solutions (LTE-U) [10]. Despite the competition for the spectrum and the corresponding interference and coexistence issues [11], Bluetooth has shown a considerable resilience and continuous improvements since it was introduced already in the late 1990s.

The low-power version of Bluetooth, BLE 4, has been in use since June 2010 and today it can be found in almost every smartphone, tablet, and laptop in the market in addition to a large set of other wireless devices. The most recent version, Bluetooth 5 was introduced in December 2016 [12] with the first commercial development kits being available in early 2017. Upcoming state-of-the-art smartphones are expected to support Bluetooth 5. The first smartphone supporting Bluetooth 5 is Samsung Galaxy S8, which is however lacking the support for the long-range LE coded mode. The long-range feature has made BLE 5 very suitable for versatile Internet of Things (IoT) applications. The official announcement of BLE 5 states that the increase in range is up to 4 times compared with BLE 4.2 [13].

To the best of our knowledge, the currently available literature doesn't provide results of any experimental investigation of the BLE 5 performance. In this paper we aim at fixing this omission by reporting the results of a comprehensive experimental measurement campaign, studying the actual range and throughput of BLE 5 in real indoor and outdoor scenarios. The measurements are conducted using one of the first available BLE 5 commercial chipsets. Importantly, we conduct our measurements and report the results not for BLE 5 only, but also for BLE 4.2. The latter is used as the reference, enabling the mapping of our results against the ones of previous works and allowing to easier estimate the benefits of using BLE 5 in an application currently employing BLE 4.2. Another contribution of this paper, which is especially important in the context of medical/health care WBANs, is that in our experiments we also address the cases of multi-floor buildings and outdoor scenarios, implying the cases when mobile patients might want to spend time such places as gardens and well delineated sidewalks.

The structure of the rest of the paper is as follows. Section 2 briefly describes the specifics of the BLE 5 technology. Section 3 details the hardware and software used in our measurements, Sect. 4 and Sect. 5 details the indoor and outdoor measurements, respectively. Section 6 provides a discussion of the results. Conclusions and future work can be found in Sect. 7.

2 Features of the BLE 5 Technology

The BLE 5 represents a further evolution of the BLE technology, which comprehensively addresses the major problems of its predecessor, namely the limited range, long transmission times and limited functionality of the broadcasting modes. In the rest of this section we will focus on the most important changes of BLE 5 compared to BLE 4.2.

The problem of improving the communications range and the maximum throughput has been addressed in BLE 5 specification by introducing three new physical layer (PHY) options. In addition to the 1 Mbit/s Gaussian frequency shift keying (GFSK) of BLE 4 (addressed in Bluetooth v 5.0 core specification as LE 1M), the BLE 5 specifies a 2 Mbit/s GFSK PHY (named LE 2M) for short range high-speed transmission and two coded PHY (referred to as LE Coded) with payload coded at 500 kbit/s or 125 kbit/s. The LE coded PHYs are modulated using GFSK at 1 Msym/s rate, but the payload data are coded in two stages: first by forward error correction convolutional encoder and then spread by the pattern mapper. In theory, this enables to improve the link budget of a coded transmission by over 5 dB and 12 dB compared to LE 1M for LE coded at 500 kbit/s and 125 kbit/s respectively. Note, that only support of LE 1M PHY is mandatory.

Another change introduced to improve the communications range is the increase of the maximum transmit power of a BLE from 10 dBm (10 mW) to 20 dBm (100 mW). Unfortunately, due to the transmit power restrictions imposed by the frequency regulations, this higher transmit power does not provide any benefit for some regions (namely, EU, Japan and Korea). The maximum link layer protocol data unit (PDU), increased in BLE 4.2 from 39 to 257 octets, stayed at this level also in BLE 5. The problem of coexistence of devices in the 2.4 GHz band has been addressed in BLE 5 by introducing the special interface proving signaling and messaging mechanisms between collocated Bluetooth and other mobile wireless standard radios.

In addition to these changes, the functionality of the broadcasting channels in BLE 5 has been substantially enriched by the introduced extended advertising feature. First, the concept of the secondary advertising channels which are co-allocated with the BLE data channels was introduced. The format of the advertising packets used in the secondary channels has been reworked enabling them to carry up to 255 octets of PDU (compared with 37 octets allowed in the primary channels of BLE 4) and even to support fragmentation. Another interesting feature enabled in the

Table 1 Summary of BLE 5 specification characteristics

PHY	Error control	Range multiplier	PDU length (B)	Packet duration	Max. throughput (kbps)
1M	CRC	1×	0–257	80 μs–2.12 ms	800
2M	CRC	0.8×	0–257	44 μs–1.06 ms	1438
Coded, S = 2	CRC and FEC	2×	0–257	462 μs–4.54 ms	382
Coded, S = 8	CRC and FEC	4×	0–257	720 μs–17.04 ms	112

BLE 5 is the periodic advertisements. Hopping between the secondary channels in a predefined pseudo-random sequence, a periodic advertiser broadcasts the packets, with PDU of up to 255 octets, at regular intervals of time ranging from 7.5 ms to almost 82 s. Importantly, a scanner device may synchronize with one or even several non-overlapping (in time) periodic advertisers and get the data from all of them. This equips BLE 5 with a more efficient and reliable solution for data broadcast than the one possible with BLE 4. Note, that the support of periodic advertisements and extended advertising features is optional.

Importantly, the BLE 5 is backward-compatible with the earlier versions of BLE—all the discussed features are optional and are not necessarily needed to be supported.

Nonetheless, as one can easily see, they can substantially increase the communications range or throughput or enable new modes of operation. Due to this fact, in the marketing materials of Bluetooth SIG [13], the BLE 5 is claimed to provide double bandwidth, up to four times higher range and up to 8 times broadcasting capacity compared to BLE 4.2 as summarized in Table 1. However, it must be noted that the improved data rate and communication range cannot be achieved at the same time since they are provided by different PHY modes.

3 Hardware and Software

In our experiments we have used one of the first commercial chipsets that support BLE 5.0, the nRF52840 [14] from Nordic Semiconductor. The nRF52840 is a system on chip (SoC) integrating a multiprotocol 2.4 GHz transceiver with an ARM Cortex-M4F based microcontroller. The chipset was programmed with S140 v5.0.0 (alpha) SoftDevice, which is a precompiled and linked binary software implementing BLE protocol developed by Nordic Semiconductor. Note that in the time of our experiments S140 was the only SoftDevice providing support for some of the BLE 5 features.

Fig. 1 Devices used for measurements

In the experiments we have used two nRF52840 Preview DK development kits shown in Fig. 1 with the laptop running a measurement software. The firmware for them was developed in this work based on the ATT_MTU Throughput Example of the nRF5 software development kit (SDK) v13.0.0-1.alpha. Executing this firmware, the two devices first use an LE 1M advertising channel to discover each other, establish the connection and set the desired connection parameters (including the used PHY).

After the start of the experiment one of the boards, connected to a computer via serial interface, starts spamming the Attribute Protocol (ATT) notifications to the other board at maximum rate. The amount of data sent are periodically reported via the serial interface. Once the pre-defined amount of data bytes has been sent, the experiment is ended and the average throughput is calculated.

Especially for this tests the firmware has been modified. First, we made the boards send data indefinitely while the link is active. Second, we have enabled measurements of the received signal strength indicator (RSSI) by the board with serial interface enabled. Third, we have enabled the control over the transmit power of the board with serial interface enabled. Fourth, we have included in the periodic reports the precise number of bytes sent in notifications and the time passed since the beginning of the experiment, as well as the average RSSI of the last eight radio packets received. These results were logged down and further statistically processed. The first two values have also been later used to calculate the average throughput.

Note, that at a time of our experiments the support for BLE 5 functionality by nRF52840 chipset was limited to enablement of LE 2M and 125 kbit/s LE Coded PHYs in data channels only. Prior to this, the connection has to be established using LE 1M PHY. This limitation has sufficiently affected the procedure of our measurements. Namely, in the initial phase the two devices were placed close enough to each other to enable them establishing the connection using the desired parameters (i.e., maximum transfer unit size, data channel PHY, transmit power, connection interval, etc.). After this, one of the devices (the one receiving the notifications—referred further in the paper as Rx node) was kept static and the other one (Tx node) was moved away from it until the connection was broken. This enabled us to detect the maximum communications range for particular set of parameters. Within this range we have selected few points and measured the average throughput and RSSI in order to investigate how the propagation environment affects the performance of BLE communications.

4 Indoor Measurements

The measurements were conducted in the Tietotalo Building of the University of Oulu. The first scenario was on the fourth floor of the north-west wing of the building illustrated in Fig. 2. The node receiving the notifications is represented by a red dot.

Blue dots represent the RSSI and throughput values measured for LE 1M PHY mode using 0 dBm transmit power. The green dots represent the RSSI and throughput values measured when in BLE 5 long-range (125 kbps data) mode, 0 dBm Tx power and S = 8 coding were used. The orange dot presents a point for BLE 5 long-range mode with 9 dBm transmit power. Most of the interiors' walls are standard plaster boards except for the bathrooms' walls (represent as green lines) which are made of concrete. The bathrooms and their walls have significant number of metallic pipes and fixtures. We carried out measurements for a plaster wall and concrete wall by using the same devices. It was found out that for a plaster wall the attenuation was only 1–2 dB and for the concrete wall the attenuation was 4–5 dB.

The second scenario was from a fixed Rx node location in the south-east corner of the fourth floor to the corresponding Tx node location in the third, second, and first floor of the building. In this case, we measured how much is the floor attenuation in this Tietotalo building. This measurement we did at first using BLE 4 PHY with 0 dBm transmit power. The reference RSSI value was −13 dBm when the Tx and Rx node are close to each other. Received RSSI value from 3rd floor (one floor between Tx and Rx) was −60 dBm and throughput was 210 kbps. Received RSSI value from 2nd floor (two floors between Tx and Rx) was −81 dBm and throughput was 40 kbps. By using BLE 4 mode we were not able to reach the 1st floor. Therefore we changed to coded (S = 8) BLE 5 mode and when using 0 dBm transmit power, we still were not able to reach first floor. With 9 dBm transmit power, the BLE 5 coded mode connection range reach the first floor with RSSI value −91 dBm and throughput 10 kbps.

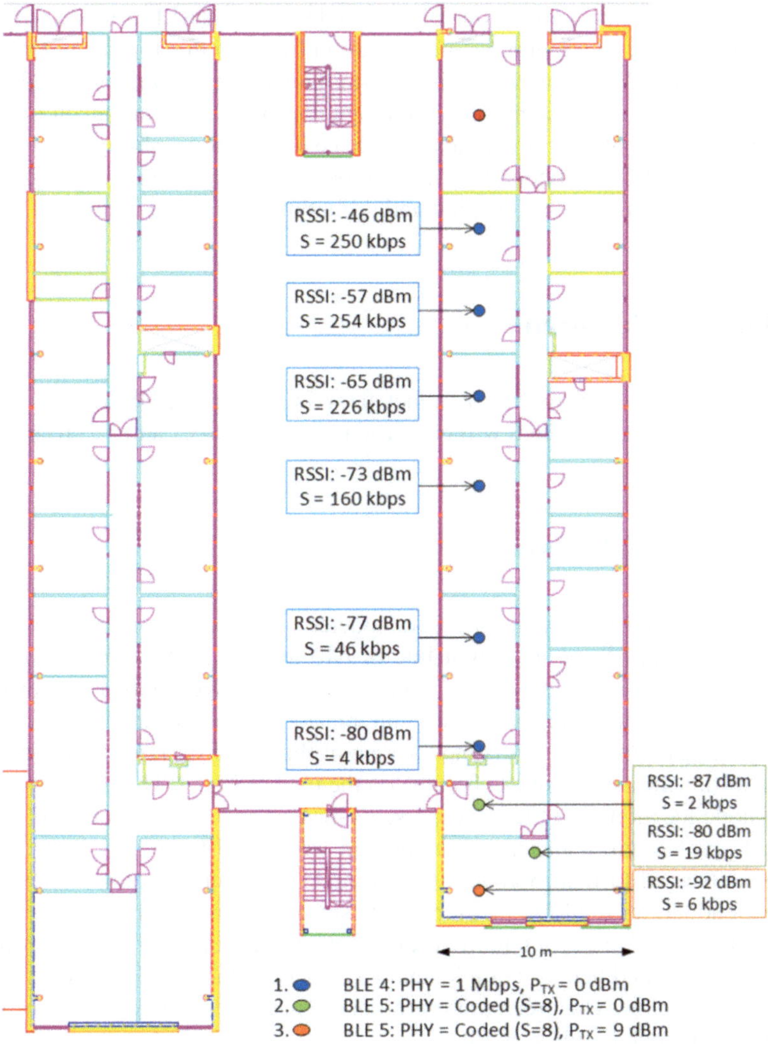

Fig. 2 Results of the indoor measurements

In these indoor measurements, the communications range improvement that was achieved by using the coded BLE 5 mode is around one extra room between the Tx and Rx nodes. The coding gain was found to be approximately 9 dBm. Measurements between different floors showed that by using BLE 4 mode with 0 dBm transmit power, the communication was successful from 4th floor to second floor. In order to reach the first floor, we had to change to BLE 5 coded (S = 8) mode and increase the transmission power. One floor addition to communications range was achieved when using coded mode and increasing the transmission power to 9 dBm. We can conclude that when changing from BLE 4 mode to coded BLE 5 mode with same transmit power, the range improvement is around 10% in this type of indoor office communications. When increasing the transmission power by 9 dBm, an additional 10% improvement can be achieved. The range improvement is not much in meters, but when operating at the communications range limits, the reliability is better when using the coded BLE 5 mode. This is an important factor when a patient carrying wearables is moving inside a hospital and is continuously being monitored.

5 Outdoor Measurements

Outdoor measurements have been conducted for line-of-sight (LOS) and non-line-of-sight (NLOS) scenarios which will be introduced in Sects. 5.1 and 5.2, respectively.

5.1 Line-of-Sight Case

The first LOS measurements were conducted outdoors on a bicycle path at Kaitoväylä, Oulu, having a straight line of 640 m. The purpose of this measurement was to find out the maximum communications range for BLE 4 and BLE 5 with coded PHY when using the maximum transmit power 9 dBm, since increase of the communications range is claimed as one of the major design targets of BLE 5. Results of this measurement can be seen on the map of Fig. 3. The red dot represent the Rx node location. The blue dots presents measurement points for BLE 4 mode, the final point where connection was broken was found to be at 430 m distance. Then we changed to BLE 5 coded (S = 8) mode and found out that this road was not long enough for the LOS measurements since the connection was still working at the end of bicycle road at 640 m distance between Tx and Rx node. However, the results are presented here to complement the other outdoor LOS measurement that are described below. The green dot represent the measurement point and results for BLE 5 coded mode in this scenario.

The second outdoor LOS measurements were conducted on a longer straight (~1 km) bicycle road to find out the maximum communications distance also for BLE 5 coded mode. In order to complement the set of previous measurements, we measured at first BLE 4 and BLE 5 coded mode performances using transmit power of 0 dBm and then used the BLE 5 mode with transmit power of 9 dBm. This very long straight bicycle path allowed for LOS measurement, as illustrated in Fig. 4 taken from the Rx point. Results are shown for each case in Fig. 5. The RSSI and throughput values just before connection break are shown for each case. Red circle is the RX point and blue circle is the point where BLE 4 connection was broken. Yellow circle is the point where BLE 5 mode with transmit power 0 dBm stopped working, and green circle is the point where connection was broken when using BLE 5 mode with transmit power of 9 dBm. From the results of Fig. 5 it can be observed that for

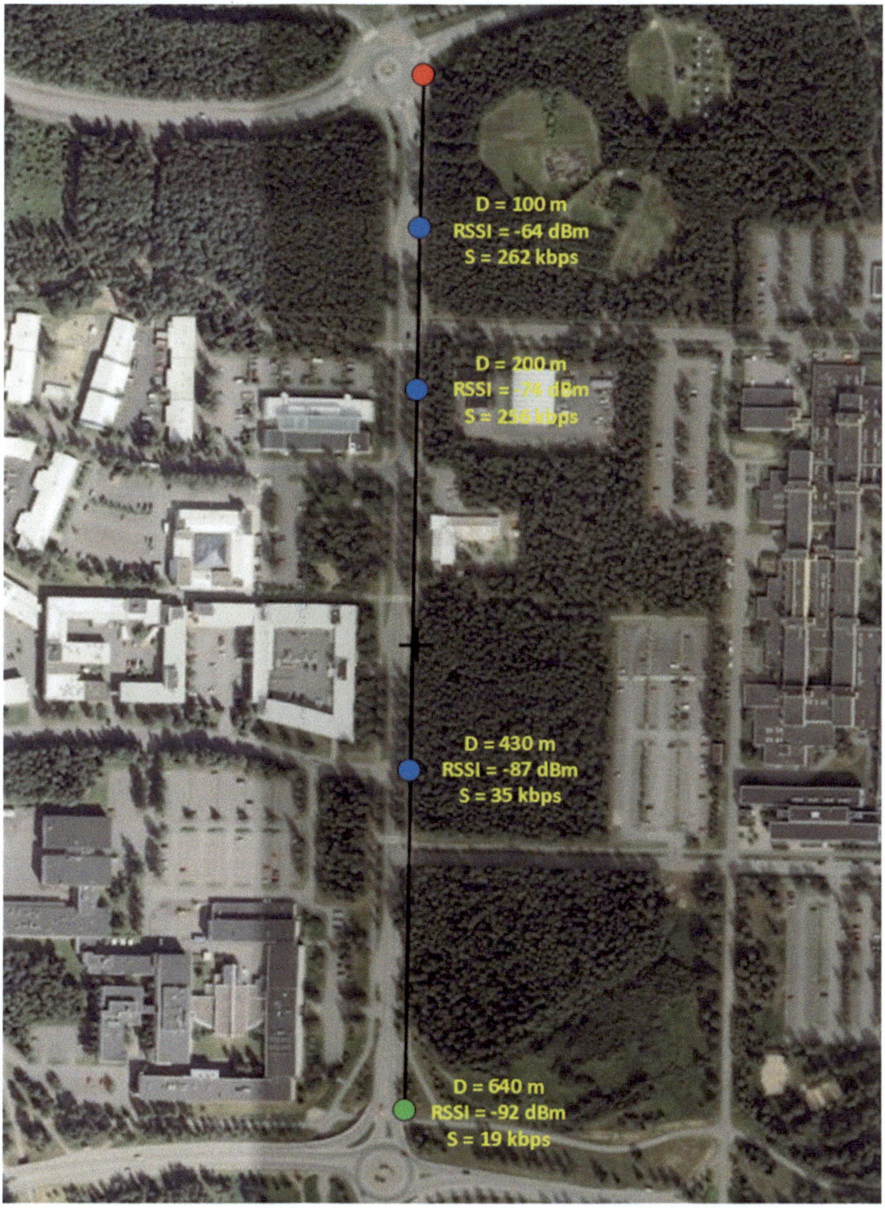

Fig. 3 Results of the first outdoors LOS measurements

BLE 4 with 0 dBm, the maximum range was 220 m. For BLE 5 case with 0 dBm transmit power, the communications range was found to be 490 m. In the previous outdoor measurements it was found that for BLE 4 with 9 dBm, the maximum range

Fig. 4 Straight bicycle road used for LOS measurements

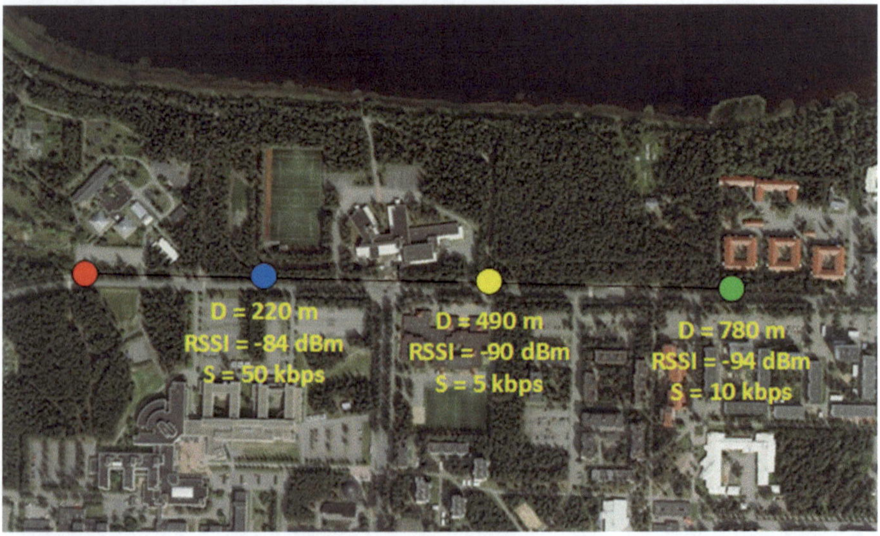

Fig. 5 Results of the second outdoors LOS measurements

was 430 m. Therefore, the BLE 5 coded mode can achieve longer range with 9 dB lower transmit power (0 dBm). Figure 5 shows that the communications range for BLE 5 coded (S = 8) mode with transmit power of 9 dBm is very impressive, being 780 m.

5.2 Non-line-of-Sight Case

We did measurements at University of Oulu Botanical Garden to find out the communications range of BLE 4 and BLE 5 in the case of non-line-sight situation where trees and bushes block the signal. Measurements were done in two different scenarios ('Scen1' and 'Scen2') with different type of obstructions created by vegetation.

The first measurement set (Scen1) was done by setting the Rx node on top of small hill (approx. 5 m height) and moving the Tx node around the garden to find out when the connection breaks in case of BLE 4 and BLE 5 coded mode. A view from the Rx node position is shown in Fig. 6 to illustrate the type of vegetation of the measurement environment. The walking route and connection breaking points for this measurement set are shown in Fig. 7. In this scenario the transmit power was set to 9 dBm for both BLE 4 and BLE 5 coded mode.

In BLE 4 case the connection was broken at the blue circle point shown in Fig. 7. Connection quality was good via the shown walking path at the garden, average being 190 kbps, even there are trees and bushes blocking the signal between Tx and Rx nodes. The final measured RSSI value before the connection break was −88 dBm and throughput was 2 kbps.

In BLE 5 case, the connection was working longer, as expected, and was broken at a point marked with green circle in Fig. 7. However, the range improvement is not

Fig. 6 View from the Rx node position in NLOS measurement Scen1

Fig. 7 Walking path and results of NLOS Scen1

drastic in this NLOS scenario. The NLOS path was very difficult at the point where BLE 5 connection was broken since there was a dense tree stand. The final measured RSSI value before connection break was −95 dBm and throughput was 494 bps. This behavior is somewhat similar to the NLOS indoor case since the relative range improvement is not very large.

In the second outdoor NLOS scenario (Scen2) the Rx node was 1 m above the ground level and the vegetation was dense as illustrated in Fig. 8. The walking path and results are illustrated in Fig. 9. In BLE 4 case, the connection was broken at the point which is shown in Fig. 9 by using a blue circle. We can see from the result that the achievable communications range through a heavy vegetation was 90 m. Average throughput for this walking path measurement was 170 kbps. Vegetation at the end point of this case is shown in Fig. 10 and from the starting point towards the end point is illustrated in Fig. 8. Then BLE 5 coded mode was measured using the same transmit power of 0 dBm. The point where the connection was broken is shown using a green circle in Fig. 9. We can see that this mode was able to increase the communications distance to 123 m. However, it must be noted that the vegetation is not homogenous through the measurement path as can be seen from the Fig. 9.

Fig. 8 View from the Rx node position in NLOS measurement Scen2

Outdoor measurements illustrate the improved communications distance when using BLE 5 coded mode especially in the LOS case. The coding gain was found to be approximately 9 dBm. We can conclude also that in general, BLE performance is quite good even with quite heavy vegetation at the communication path. When there is a heavy vegetation, the range improvement (BLE 4 vs. BLE 5) behavior is somewhat similar to indoor case, i.e., the relative improvement in distance is not drastic. BLE 5 coded ($S = 8$) mode with transmit power of 9 dBm enables a very long communications range in LOS conditions, being 780 m in our real-life measurements.

6 Discussion of the Results

Based on the introduced indoor and outdoor results, which are summarized also in Table 2, we can compare the communications range performance of BLE 4 versus BLE 5 coded version. It is evident that BLE 5 coded version can improve the communications range, in comparison to BLE 4 version, which has been its design goal. However, based on our measurement results it can be seen that most of the time the range improvement is not fourfold as has been advertised in BLE 5 marketing material. Based on the results introduced in this paper, and our additional measurement

Fig. 9 Walking path and results of NLOS Scen2

experiments with the same development kit, we can conclude that the gain of the coded mode is approximately 9 dB. Namely, in the indoor and outdoor measurement we have found that the BLE 5 coded mode can achieve approximately similar communications range than BLE 4 mode, by using 9 dB lower transmit power. When using the same transmit power for both modes, the range improvement has been found to be around 10–20% for indoor scenario. In NLOS outdoor scenario, the range improvement was found to be approximately 20–37% in our measurement scenarios. However, it must be noted that in outdoor scenarios the vegetation was not homogenous, as well as in indoor scenarios the construction material is varying depending on the signal path, therefore in order to acquire exact range improvement percentages and coding gain, more measurement are needed to find out statistically reliable results. Outdoor LOS measurements showed that the BLE 5 coded mode can

Fig. 10 Vegetation at the point where BLE 4 connection was broken in NLOS Scen2

improve the communications range more than twofold when using the same transmit power than BLE 4 mode. When using the transmit power of 9 dBm, the communications range of BLE 5 coded mode was found to be very impressive, 780 m, which is closer to fourfold improvement in comparison to BLE 4 range with 0 dBm transmit power. Therefore, our finding is that the fourfold communications range improvement requires the BLE 5 coded mode combined with increased transmit power.

7 Conclusions and Future Work

This paper described experimental communications range and throughput measurement results for new the BLE 5 coded mode. Reference measurements are also conducted using BLE 4 (1 Mbps) PHY settings. The goal was to find out the performance of the BLE 5 coded (S = 8) version, since with improved communications range it can open up more options and enable new uses for BLE in medical/healthcare scenarios. Results show that BLE 5 coded mode improves the communications range, with lower throughputs that are acceptable for low-power sensor devices. However, our measurements show that the promised fourfold communications range improvement for BLE 5 requires both coded mode and increased transmission power. The coding gain of BLE 5 coded version (S = 8) was found to be approximately 9 dB, i.e., the coded version was able to achieve same (or longer) communications range than BLE 4 by using 9 dB lower transmission power. The maximum BLE 5 communications range which we have witnessed in our experiments outdoors was almost 800 m, using

Table 2 Summary of measurement results

Mode	Tx power (dBm)	Range (m)	Max throughput (kbps)	Min throughput
Indoor scenario				
BLE 4	0	43	262	1761 bps
BLE 5	0	48	26	521 bps
BLE 5	9	51	26	3.5 kbps
Outdoor NLOS Scen1				
BLE 4	9	138	262	2.7 kbps
BLE 5	9	165	26	494 bps
Outdoor NLOS Scen2				
BLE 4	0	90	262	8.3 kbps
BLE 5	0	123	26	2.1 kbps
Outdoor LOS				
BLE 4	0	220	262	1.1 kbps
BLE 4	9	430	262	7.5 kbps
BLE 5	0	490	26	1.7 kbps
BLE 5	9	780	26	1.1 kbps

the transmit power of 9 dBm. The presented results indicate that BLE 5 is an efficient technology that can be used in remote patient monitoring via wireless connectivity between wearable nodes carried by the mobile patients. Larger coverage is accomplished by using strong coding which reduces the amount of required infrastructure at the hospital buildings.

To the best of our knowledge, this paper reports the first measurement campaign conducted to find out the performance of BLE 5 coded mode and having it compared to BLE 4.2. In future work, we plan to perform additional measurements to study the effect of different PHY settings that the new BLE 5 version offers compared to previous BLE version. Specifically, since the increase of the range by using the coded PHY brings along the increase of the on-air time, we plan to study the interference nature of other wireless technologies on BLE 5. In addition, we plan to perform measurements around the human body since it is a difficult communications environment.

Acknowledgements This work has been partially funded by the European Regional Development Fund (ERDF) through the WILLE project.

References

1. Lamkin, P.: Wearable Tech Market to Be Worth $34 Billion by 2020 (2016). https://www.forbes.com/sites/paullamkin/2016/02/17/wearable-tech-market-to-be-worth-34-billion-by-2020/#4cbe76e83cb5
2. Tractica: Healthcare Wearable Device Shipments to Reach 98 Million Units Annually by 2021 (2016). https://www.tractica.com/newsroom/press-releases/healthcare-wearable-device-shipments-to-reach-98-million-units-annually-by-2021/
3. Karvonen, H., Hämäläinen, M., Iinatti, J., Pomalaza-Ráez, C.: Coexistence of wireless technologies in medical scenarios. In: European Conference on Networks and Communications (EUCNC), Oulu, Finland (2017)
4. IEEE 802.11 LAN Working Group. http://www.ieee802.org/11/
5. IEEE Standard for Low-Rate Wireless Networks, IEEE Std 802.15.4-2015 (Revision of IEEE Std 802.15.4-2011) (2016)
6. ZigBee Alliance. http://www.zigbee.org/
7. IEEE Std. 802.15.6: IEEE Standard for Local and Metropolitan Area Networks—Part 15.6: Wireless Body Area Networks. Standard, The Institute of Electrical and Electronics Engineers, Inc. (2012)
8. Hämäläinen, M., et al.: ETSI TC SmartBAN: overview of the wireless body area network standard. In: International Symposium on Medical Information and Communication Technology (ISMICT) (2015)
9. ANT/ANT+. https://www.thisisant.com/
10. Nokia White Paper: LTE evolution for IoT connectivity (2017)
11. Karvonen, H., Pomalaza-Ráez, C., Mikhaylov, K., Hämäläinen, M., Iinatti, J.: Interference of wireless technologies on BLE based WBANs in hospital scenarios. In: IEEE International Symposium on Personal, Indoor and Mobile Radio Communications (PIMRC), Montreal, Canada (2017)
12. Bluetooth SIG: Bluetooth Core Specification v 5.0 (2016). https://www.bluetooth.com/specifications/bluetooth-core-specification
13. Bluetooth SIG: Bluetooth Core Specification 5.0 FAQ (2016). https://www.bluetooth.com/~/media/files/specification/bluetooth-5-faq.ashx?la=en
14. Nordic nRF52840. https://www.nordicsemi.com/eng/Products/nRF52840

References

Electrode Impedance Modeling for Channel Characterization for Intra-body Communication

Ahmed E. Khorshid, Ibrahim N. Alquaydheb and Ahmed M. Eltawil

Abstract This paper discusses techniques for modeling the electrode/human contact impedances for Intra-body communication applications. Factors that affect the electrode impedance are considered and tuned in order to study their impact on the channel model (gain/attenuation profile). Finally, an explanation is provided for the relation between the different basic impedances and blocks that are considered in the channel model, and the sensitivity of the channel gain to the variation in such parameters.

Keywords Body area networks · Intra-body communications
Electrode impedance · Channel modeling · Circuit model · Galvanic coupling

1 Introduction

Wearable devices are rapidly being adopted as means of augmenting and improving health care services. To provide a cable-free biomedical monitoring system, new wireless technologies associated with sensor applications have been promoted as the next biomedical revolution, promising a significant improvement in the quality of health-care applications. Yet the size and power requirements of wireless sensors which are typically dominated by the RF section of the associated transceivers, has limited their adoption. To overcome such concerns, system architects proposed designing the system in a way that would allow more than one sensor to share the same wireless gateway, providing a distributed solution, with less power consumption.

A. E. Khorshid (✉) · I. N. Alquaydheb · A. M. Eltawil
Electrical Engineering and Computer Science Department,
University of California, Irvine, USA
e-mail: Khorshia@uci.edu

I. N. Alquaydheb
e-mail: Alquaydi@uci.edu

A. M. Eltawil
e-mail: aeltawil@uci.edu

G. Fortino and Z. Wang (eds.), *Advances in Body Area Networks I*,
Internet of Things, https://doi.org/10.1007/978-3-030-02819-0_19

That approach paved the way for adopting Intra-Body Communication (IBC) systems where data transmission is carried out through the skin, rather than through air [1, 2]. Sensors and actuators inter-communicate through skin, relaying messages to a centralized wireless hub that could be a smart watch for instance. This emerging technology would ultimately lead to body area networks (BANs) that operate at extremely low power, with minimal foot print by replacing expensive, power consuming Radio Frequency (RF) front ends, for each individual node with simpler interfaces. Furthermore, while the skin operates as an interference channel for the communicating nodes, it is relatively protected from the higher levels of interference expected when broadcasting via an air interface.

A basic block in Intra-body communication systems is the electrode used. Electrodes are the main interface between the circuits/systems blocks and the human body (the channel), responsible for coupling the signal into the body at the transmitter side and picking it up at the receiver one. Accordingly, accurate modeling of the electrode impedance is crucial for IBC system performance estimation. This paper presents different approaches for modeling the electrode and electrode-body contact impedances. Each approach is thoroughly investigated and the channel model and overall gain characteristics for each case are presented. Different physical factors (material, size, spacing…etc.) are also taken into consideration to study their impact on the overall attenuation profile. The paper concludes with a study of the relation between the different impedances constituting the channel model, how electrode parameters affect them, how they change with respect to each other, and their role in shaping the final channel gain characteristics.

2 Channel Model

Modeling the body as a communication channel has always been a major concern for BANs, in order to design reliable communication systems accordingly. Based on a previous study performed by the authors [3], it was proven that electromagnetic waves are the best candidate to act as the data carrier through the human body for the range of frequencies starting from 100 kHz till 100 MHz. Trials performed in [4, 5] present a circuit model for the signal path through the body. Improved models then followed, attempting to account for the electrode contact impedance [6]. In [7], the authors presented a model for intra body communications using galvanic coupling technique, where both the electrodes (signal and ground) of each node (transmitter and receiver) are connected to the body, so that the whole signal is confined into the human body, thus making it more immune to noise and interference. In the proposed model, biological parameters are assumed to be variable, taking into consideration the impact of important factors; such as age and weight, on these parameters and thus on the overall gain/attenuation profile. The model was built to represent the main five layers constituting the human body; skin, fat, muscle, cortical bone and bone marrow.

2.1 Electrode Impedance Modeling

A basic element in the human body communication channel is the electrode used at both the transmitter and the receiver sides. Modeling the electrode and electrode-body compact impedance has a significant impact on the overall channel model. Parameters as the electrode size, material, shape, width, thickness ...etc. should all be taken into account. Other factors that further determine the value of the electrode-contact impedance include the operating frequency, spacing between each of the electrode pairs (ex: separation between the two electrodes of the transmitter), location of the electrodes on the human body and finally the transmitted power level (operating voltage and induced current). Providing an accurate impedance model that would take in as many from the above-mentioned factors as possible is a complicated task. Several approaches had been considered for accurately modeling the electrode contact impedances. One approach is through measuring the impedance of the electrodes experimentally, as followed in [5], where the impedance measurement for different materials and different electrode areas, as well as different locations on the body were carried out at a number of frequencies. Collected data is then used to calibrate the circuit model. In [6], a set of measured impedances was also used to try to generate a simple equation that relates electrode impedance to the operating frequency and electrode size, using curve fitting techniques. Yet another approach is to use 3D Finite Element Models (FEM) [8], where more complex human body models are proposed, taking into consideration more biological features and factors, to accurately model the signal path through and around the human body. Although such models accurately solve the EM wave simulations, providing more insight on how the signal travels in various body tissues, diffusion of power into deeper tissues, power dissipation through the human limbs, and other vital information for the channel modeling, it still fails in accurately modeling the electrode-body contact impedance. The main reason for that is failing to represent the physical/biological interactions at the electrode interface; the ionization of the electrode/gel, movement of the ions and the interaction at the interface.

The same shortcomings apply to simplified circuit models used to represent IBC channel models [6, 7, 9–11]. Although some of these models were able to accurately take into account the electrodes properties, it still suffered from drawbacks when it comes to the electrode-body impedance modeling, since in most cases very simple models were assumed, limiting the fidelity of the model when representing electrode-body interface. In [12], a new method was proposed for a better representation for the electrodes impedance, where a hybrid approach (circuit model, FEM technique and experimental measurements) was adopted for modeling the IBC channel. The proposed simulation model uses measurement-derived S-parameters for modeling the electrode-body contact impedances and uses lumped circuit models to describe the electrode properties. Although this approach provided more accurate results, it still had some limitations. Mainly, the measurement-derived S-parameters are very unique to the measurement setup/conditions, including; the electrodes material and dimensions, subjects participating in the measurement and the operating frequency.

If any of those parameters are changed, a new set of measurements will be needed, which would be a tedious process with expected variations due to the measurement setup conditions. Clearly there is still a need for an approach that would accurately model the electrode and electrode-contact impedances, taking into account the different factors and electrode properties, yet is not case specific (takes into account the electrodes material, size, spacing, operating frequency …etc.).

Electrodes are responsible for connecting electronic systems to the human body through transducing ionic currents from the human body into electric currents and vice versa. Chemical reactions (oxidation-reduction) at the interface between the metal electrode and the electrolyte gel solution, as well as that between the electrolyte and the skin are responsible for carrying out this process, through the transfer of the charges between the electrons in the metal and the ions in the electrolyte solution/the human body building up a potential difference at the interface [13]. Modeling this interaction is the main concern in this paper. The complexity of this process resides in the fact that it is a function of variable parameters such as the metal of the electrodes, the electrolyte, electrodes' dimensions, quality of the electrode attachment to the skin and the skin electrode interface which may vary from one person to the other. The basic electrical representation for the skin electrode interface [13] is shown in Fig. 1. It shows the electrode, electrolyte, skin layers, and the potential differences developed at the interfaces; $E_{electrode}$ representing the potential difference at the electrode electrolyte interface and $E_{sweat\text{-}epidermis}$ representing that at the electrolyte skin interface. Model parameters are highly subjective, due to the complexity of subject dependent voltage sources, capacitances and ohmic resistances.

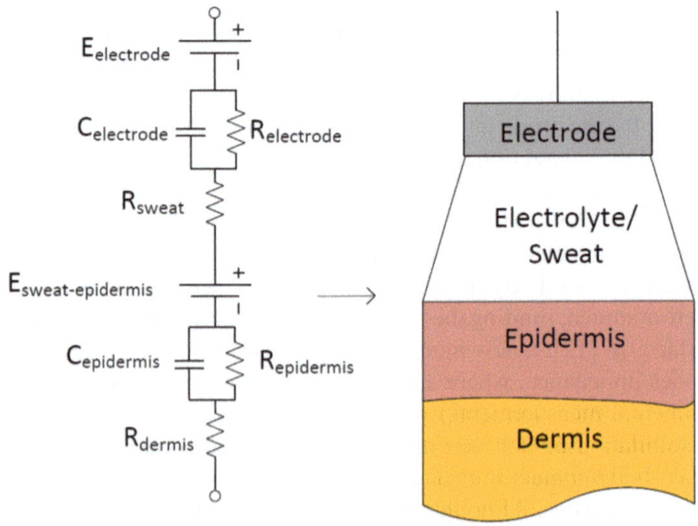

Fig. 1 The electrical representation of the skin-electrode interface, showing the electrode, electrolyte, the skin layers' impedances and the potential developed at the interfaces due to the chemical reactions

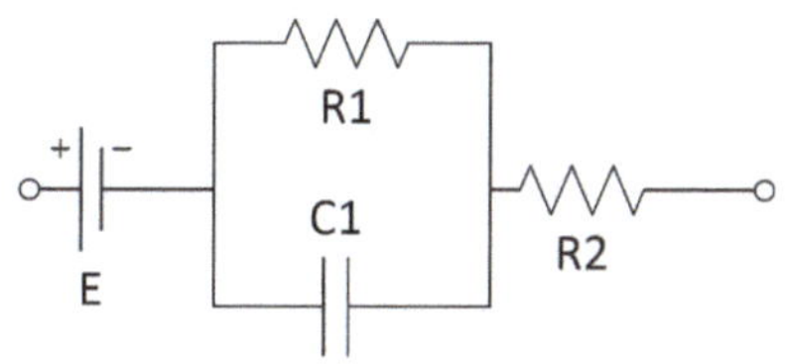

Fig. 2 Simple circuit model representation for the skin-electrode interface

A number of circuit modeling representations (with varying degrees of fidelity) have been proposed to capture the parameters shown in Fig. 1. The following section, compares three common models (under the same assumptions) to illustrate the variability in channel gain profile that can be attributed to electrode-body interface modeling.

2.2 *Single Order Model*

For simplicity, an alternative electrical representation [13] for the electrode skin interface is shown in Fig. 2. R_1 and C_1 represent the impedance associated with the electrode-electrolyte interface and the polarization at that interface, R_2 is the series resistance associated with interfacial effect as well as the resistance of the electrode material, and E is the potential developed at the interface due to the chemical interaction between electrons/ions. This simplified model is commonly used in circuit models for channel representation in the IBC field.

The amplitude of the electrode-interface impedance using the single order model is plotted in Fig. 3 over the frequency range of interest for galvanic coupling (100 kHz till 50 MHz). In order to generate the impedance shown in the figure, curve fitting was

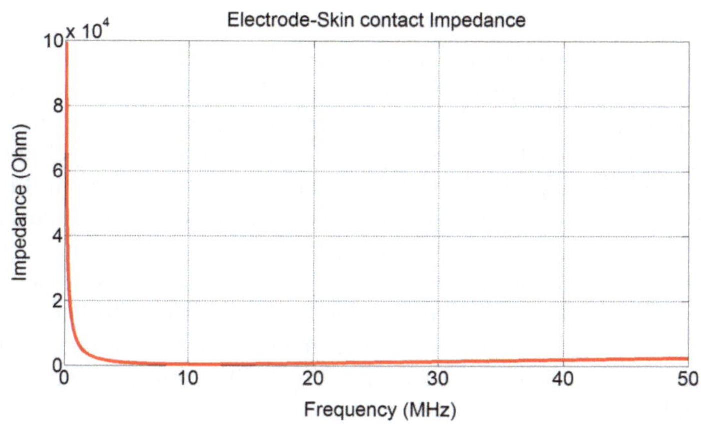

Fig. 3 Equivalent impedance for the first order electrode-skin contact model for copper electrodes of radius 5 cm

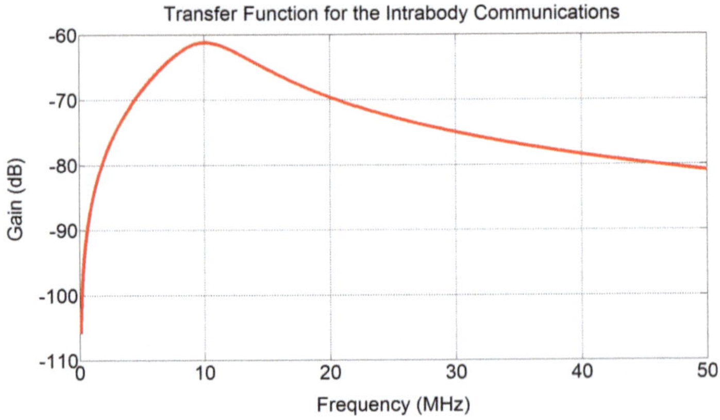

Fig. 4 Channel model for galvanic intra-body communications using the first order model for representing the electrodes contact impedance. Gain is plotted from 100 kHz till 50 MHz, distance between transmitter and receiver is 10 cm

used to find appropriate R, C values based on the experimental frequency dependent values of R, C presented in [5].

The first order electrode impedance is then integrated into a full channel model as presented in [7] to encapsulate all main five layers constituting the human body; skin, fat, muscle, cortical bone and bone marrow. The corresponding channel model (Gain profile) is shown in Fig. 4, where the gain of the channel through a human arm is plotted; the transmitter and receiver are 10 cm apart, electrodes of the same node are 3.5 cm apart, electrode radius is 5 cm and is made from copper.

2.3 *Double Order Model*

While the single order model is simple and fast in simulations, it lacks the details needed to accurately represent the details shown in Fig. 1. Thus, a double order model was introduced in [14, 15]. The proposed electrode-gel-skin model consists of two main parts; a circuit model representing the electrode—electrolyte (gel) impedance, and another representing the electrolyte-skin impedance, as shown in Fig. 5. The authors used curve fitting techniques to obtain the values of the circuit elements from experimental data that they conducted.

In order to compare the two approaches, we used the same channel model as presented in [7]. Impedance values (equivalent impedance for each of the two sections solely, then the overall combined electrode-skin impedance) were computed over the range of frequencies of interest, as shown in Fig. 6. The corresponding channel gain is plotted in Fig. 7.

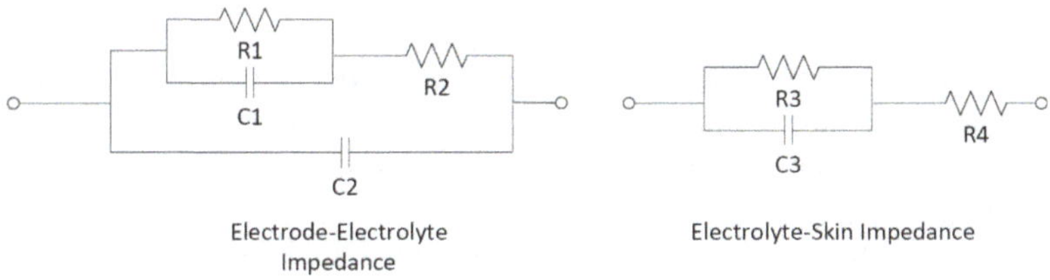

Fig. 5 A double order model, consisting of a section modeling the electrode-electrolyte impedance and another modeling the electrolyte-skin impedance

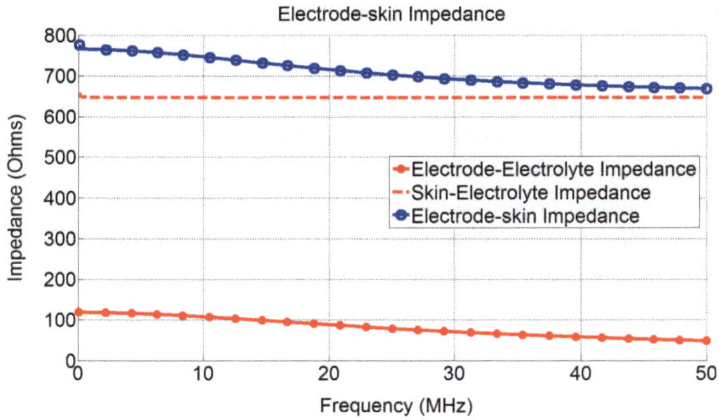

Fig. 6 Equivalent impedance of the double order model; electrode-electrolyte impedance, electrolyte-skin impedance and the overall electrode skin impedance

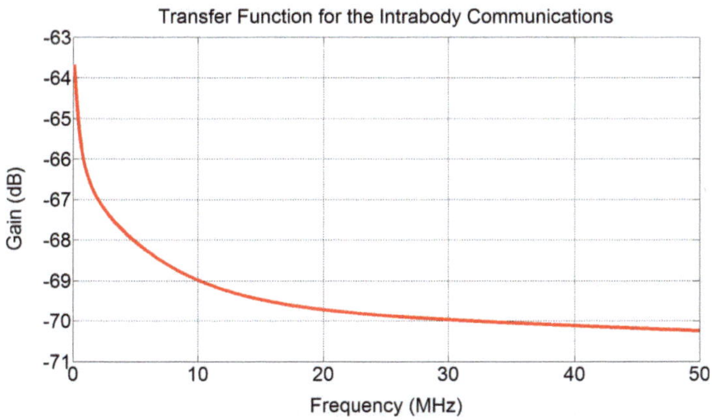

Fig. 7 Channel model for galvanic intra-body communications using the double order model for representing the electrodes contact impedance

It is important to note that the two channel gain profiles are not identical. This fact highlights that the channel gain profile is highly dependent on the electrode modeling approach used and hence a thorough understanding of the parameters affecting the model is needed. In subsequent section, we elaborate on such parameters and their impact on channel gain profiles.

In [16], the authors proposed a more generic model for the second order electrode-human body contact impedance, shown in Fig. 8, where the model included variables that represent the electrode dimensions and material. In the model, C_d represents the double layer capacitance between the electrode and the electrolyte solution in the body tissue, R_a is the activation polarization resistance, R_w and C_w represent the diffusion polarization impedance (Warburg impedance), Z is the reaction impedance and Z_t is the impedance of the tissue under the electrode (skin). Using this model, we computed the equivalent impedance for the electrode-skin contact for three different materials; copper, brass and stainless steel over the frequency range of interest, where material dependent parameters are as given in [16]. Results are shown in Fig. 9. The corresponding channel profile using results obtained from this model, for the three electrode materials, are then shown in Fig. 10, which again show a deviation from the two channel gain profiles presented earlier.

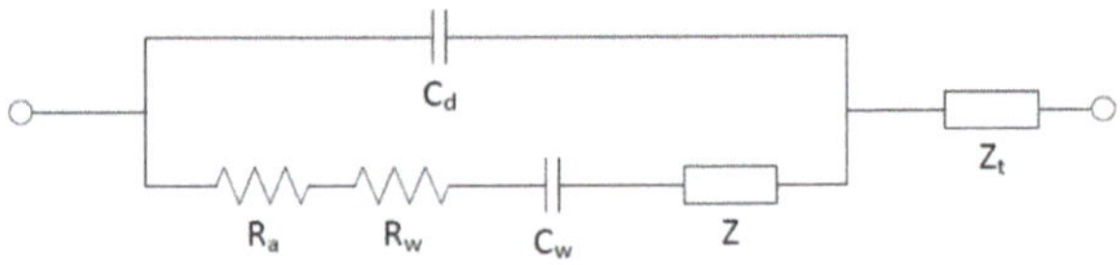

Fig. 8 Contact impedance circuit model proposed in [16], taking into consideration the activation polarization, diffusion polarization, reaction impedance and the body tissue impedance

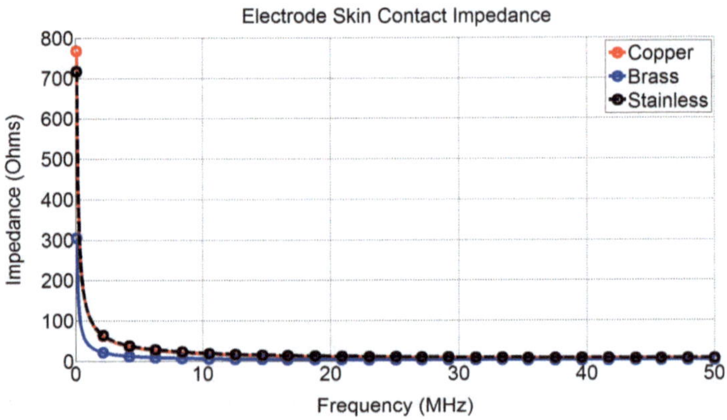

Fig. 9 Equivalent impedance using the model suggested in [16], for three different materials; copper, brass and stainless steel, with an electrode area of 100 mm^2 each

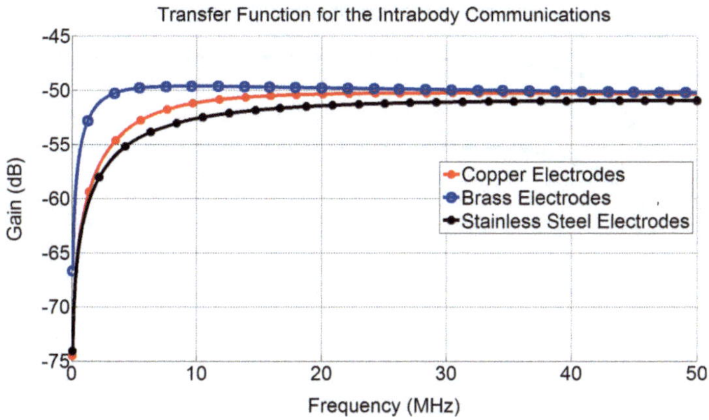

Fig. 10 Channel model for galvanic intra-body communications for three different electrode materials; copper, brass and stainless steel

3 Channel Model Variation with Respect to Electrodes Parameters

From the previous section, it is clear how sensitive the channel gain profile is to the adopted electrode skin contact impedance model. It is thus imperative to choose a model that captures as many of the biological processes as possible. The model presented in [16] considers the activation polarization, diffusion polarization, reaction impedance and the body tissue impedance. In the remainder of the paper, we use this circuit model to gain further insight in the impact of various parameters on the channel model.

Effect of distance between the transmitter node and the receiver node electrodes: In Fig. 11, the electrode area is fixed at 1 cm^2, the separation between each pair of electrodes at the same node is 6 cm. The distance between the two nodes is varied (10, 30 and 50 cm). As expected, the results show that as the distance between the transmitter and the receiver increases, the channel gain drops (more attenuation).

Effect of node specific parameters: In Fig. 12, the separation between electrodes of each node is varied; 1 cm, 6 cm and 10 cm, while the electrode area and distance between the transmitter and the receiver are kept fixed at 1 cm^2 and 30 cm respectively. In Fig. 13, the electrode area is varied; 1 cm^2, 10 cm^2 and 100 cm^2, while the distance between the nodes and separation between pair of each nodes are kept constant at 30 cm and 6 cm respectively. From the results, one can infer that as the area of the electrode increases, the impact its impedance has on the overall channel gain profile decreases, especially at the high frequencies (above 30 MHz). To further illustrate this point, consider the results shown in Figs. 14 and 15 where the electrode area is kept constant at 100 cm^2 and the distance between the transmitter and the receiver as well as the separation between each electrode pairs is varied. Although the same

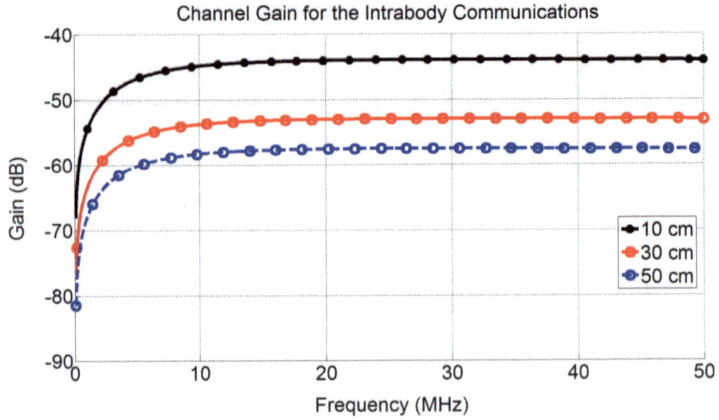

Fig. 11 Effect of the distance between the transmitter electrodes and the receiver on the channel gain profile, where the distance is varied, while the electrode area and the separation between each pair of electrodes are kept constant at 1 cm^2 and 6 cm respectively

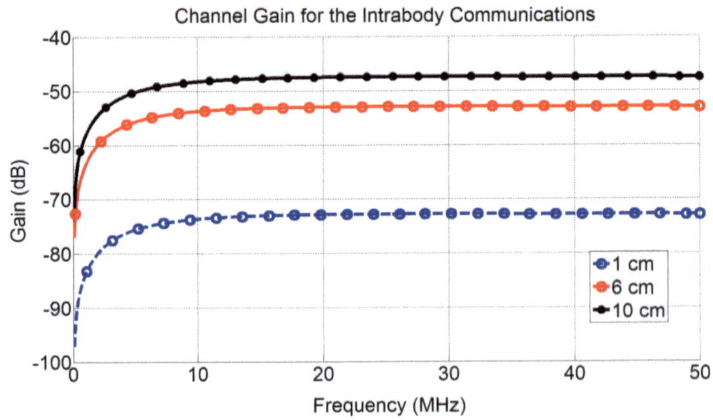

Fig. 12 Varying the separation between electrodes of each node; 1 cm, 6 cm and 10 cm, while both the electrode area and distance between the transmitter and the receiver are kept fixed at 1 cm^2 and 30 cm respectively

relations deduced earlier stays the same, the channel profile, especially at lower frequencies, follows a different trend.

Electrode Contact Impedance versus Body Impedance: In order to better understand the change in channel gain profiles for different electrode areas, it is important to understand the regions in which certain impedances dominate; namely the electrode contact impedance and the body impedance (input impedance seen between the two terminals of the node). To better understand the impact of electrode area, a simulation was run, where two extremes for the electrode area were used; namely 1 and 100 cm^2. The magnitude of the equivalent impedances is plotted in Figs. 16 and 17 respectively. From the figures, we notice that for smaller area electrodes,

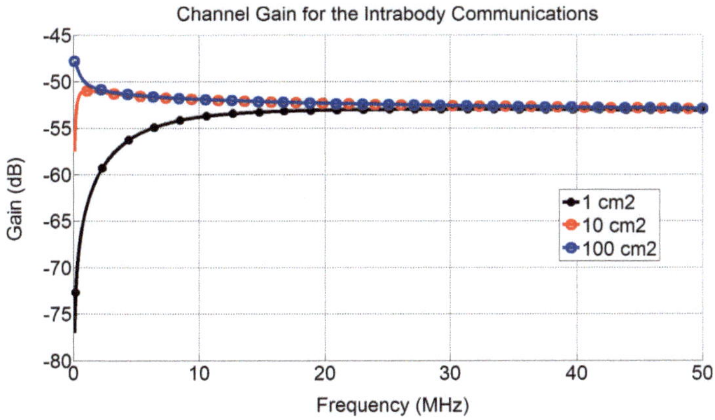

Fig. 13 Changing the electrode area; 1 cm^2, 10 cm^2 and 100 cm^2 while distance between nodes and separation between pair of each nodes are kept constant at 30 cm and 6 cm respectively

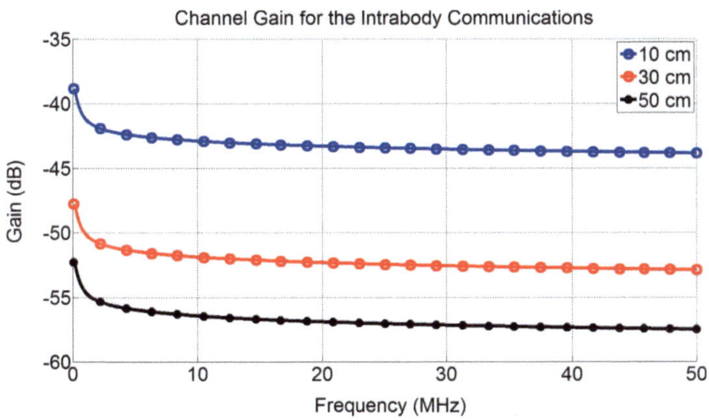

Fig. 14 Gain versus distance for electrode area of at 100 cm^2 and separation between each pair of electrodes of 6 cm

the electrode contact impedance is dominant over the body input impedance, which means the contact impedance has higher impact on the channel gain profile. As the electrode area increases, its impedance drops till the body input impedance becomes dominant over it, thus the body input impedance will have the higher impact on the channel profile. This also explains the results shown in Fig. 13, as the electrode area increases its impact on the channel gain profile, especially at higher frequencies, becomes negligible. Moreover, changing the separation between the electrode pairs at each node, changes the input impedance seen between these two electrodes according to the model proposed in [7], thus the final channel gain will be determined according to the relation between the various impedances (input impedance, electrode impedance, transmission path impedance ...etc.).

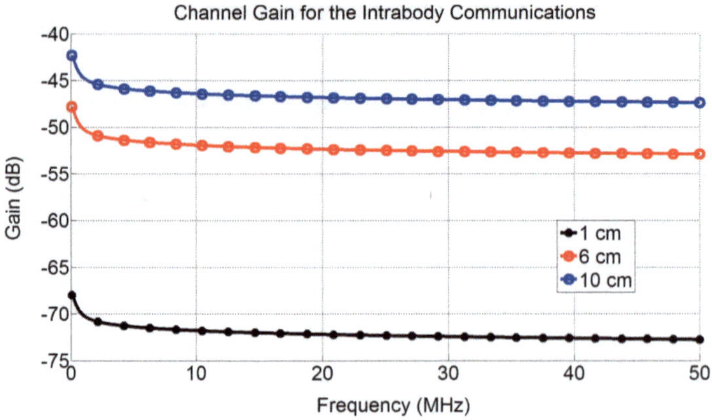

Fig. 15 Gain versus separation between electrode pairs of the same node, for electrode area of at 100 cm^2 and distance between transmitter and receiver of 30 cm

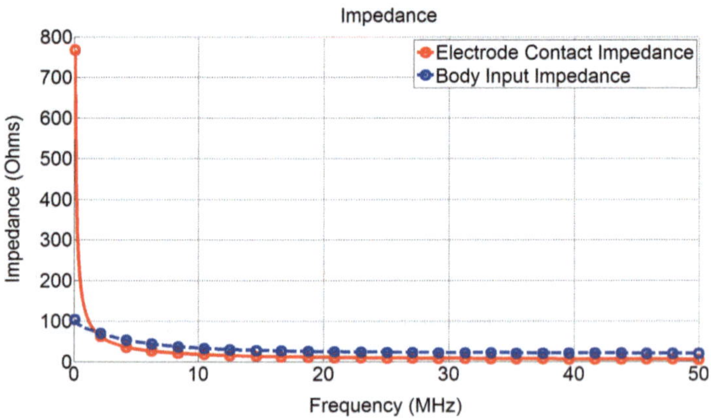

Fig. 16 Magnitude of the electrode contact impedance as well as the body input impedance seen between the two electrodes of each node, for electrode area of 1 cm^2

4 Conclusion

In this paper, we investigated various methods used to model the electrical circuit representation of the electrode-body contact. The impact of adopting each of these models on the characteristics of the galvanic intra-body communication channel, using the channel model that we developed in an earlier work [7], is presented. We then studied the impact of varying different parameters, related to the electrodes as the electrode area, distance between the transmitter and the receiver nodes and the separation between each pair of electrodes on the characteristics of the channel gain/attenuation profile. It is clear that the gain drops with increasing the distance between the two nodes, yet it increases with increasing the separation between the

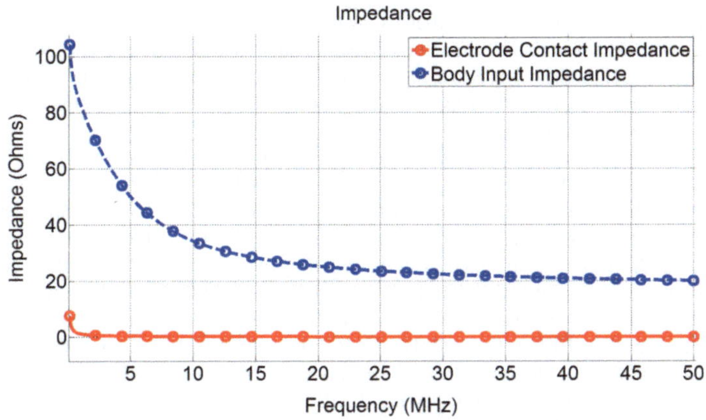

Fig. 17 Magnitude of the electrode contact impedance and the body input impedance for electrode area of 100 cm^2

electrodes of the same node. To further understand the obtained results, we studied the relation between the basic impedances forming our channel model; mainly the electrode contact impedance and the body input impedance. Through observing the values of each for different electrodes dimensions, it became clear the role and impact of each one of them on the channel gain characteristics, and how the relation between both of them dictates the behavior of the channel profile.

Acknowledgements The authors gratefully acknowledge that this work was supported by the National Institute of Justice (NIJ) grant number 2016-R2-CX-0014.

References

1. Zimmerman, T.G.: Personal Area Network (PAN). M.S. thesis. MIT Press, Cambridge, MA (1995)
2. Handa, T., Shoji, S., Ike, S., Takeda, S., Sekiguchi, T.: A very low power consumption wireless ECG monitoring system using body as a signal transmission medium. In: Proceedings of the International Conference on Transducers, Solid-State Sensors Actuators, vol. 2, pp. 1003–1006, June 1997
3. Khorshid, A.E., Eltawil, A.M., Kurdahi, F.: On the optimum data carrier for intra-body communication applications. In: Proceedings of the 11th EAI International Conference on Body Area Networks, pp. 137–140 (2016)
4. Hachisuka, K., Terauchi, Y., Kishi, Y., Hirota, T., Sasaki, K., Hosaka, H., Ito, K.: Simplified circuit modeling and fabrication of intrabody communication devices. In: Proceedings of the IEEE International Conference on Solid-State Sensor, Actuators, Microsystems, pp. 461–464 (2005)
5. Wegmueller, M.S., Kuhn, A., Froehlich, J., Oberle, M., Felber, N., Kuster, N., Fichtner W.: An attempt to model the human body as a communication channel. IEEE Trans. Biomed. Eng. **54**(10), 1851–1857 (2007)

6. Wegmueller, M.S., Oberle, M., Felber, N., Kuster, N., Fichtner, W.: Signal transmission by galvanic coupling through the human body. IEEE Trans. Instrum. Meas. **59**(4), 963–969 (2010)
7. Khorshid, A.E., Eltawil, A.M., Kurdahi, F.: Intra-body communication model based on variable biological parameters. In: 2015 49th Asilomar Conference on Signals, Systems and Computers, pp. 948–951, November 2015
8. Xu, R., et al.: Electric-field intrabody communication channel modeling with finite-element method. IEEE Trans. Bio-Med. Eng. **58**(3), 705–712 (2011)
9. Cho, et al.: The human body characteristics as a signal transmission medium for intrabody communication. IEEE Trans. Microw. Theory Tech. **55**(5), 1080–1085 (2007)
10. Callejón, M.A., et al.: Distributed circuit modeling of galvanic and capacitive coupling for intrabody communication. IEEE Trans. Bio-Med. Eng. **59**(11), 3263–3269 (2012)
11. Kibret, B., et al.: Investigation of galvanic-coupled intrabody communication using the human body circuit model. IEEE J. Biomed. Health Informat. **18**(4), 1196–1206 (2014)
12. Park, J., Garudadri, H., Mercier, P.P.: Channel modeling of miniaturized battery-powered capacitive human body communication systems. IEEE Trans. Biomed. Eng. **64**(2), 452–462
13. Neuman, M.R.: Biopotential electrodes. In: Bronzino, J.D. (ed) The Biomedical Engineering Handbook, 2nd edn. CRC Press LLC, Boca Raton (2000)
14. Saadi, H., Attari, M.: Electrode-gel-skin interface characterization and modeling for surface biopotential recording: Impedance measurements and noise. In: 2nd International Conference on Advances in Biomedical Engineering (ICABME), pp. 49–52, September 2013
15. Assambo, C., Baba, A., Dozio, R., Burke, M.J.: Determination of the parameters of the skin-electrode impedance model for ECG measurement. In: Proceedings of the 6th WSEAS International Conference on Electronics, Hardware, Wireless and Optical Communications, Corfu Island, Greece, pp. 90–95, February 2007
16. Kanai, H., Chatterjee, I., Gandhi, O.P.: Human body impedance for electromagnetic hazard analysis in the VLF to MF band. IEEE Trans. Microw. Theory Tech. **32**(8), 763–772 (1984)

Analysis and Estimation of Intra-body Communications Path Loss for Galvanic Coupling

Ibrahim N. Alquaydheb, Ahmed E. Khorshid and Ahmed M. Eltawil

Abstract The desire to have ultra-compact, low power patient monitoring techniques that include intercommunicating wearable and implanted sensors/actuators encourages researchers to develop new communication methods that can replace current Radio Frequency (RF) wireless communication links. RF links require power and area hungry analog circuitry that limits the usability of such systems. This paper evaluates different techniques for Intra-body communication (IBC) where the signal is coupled galvanically to the human tissue. Finite element method (FEM) technique is utilized to determine the path loss of the human channel (human arm model) and to examine the current density distribution in human tissues using both a full and a reduced order model. In addition, we investigate the effect of bone fracture internal fixation implant effect on the channel parameters.

Keywords Path loss · Galvanic coupling · Intra-body communications

1 Introduction

Intra-body communication (IBC) is a promising method that utilizes the human body as a waveguide or transmission medium to connect wireless implanted or wearable sensors and actuators to other receiving or transmitting devices, typically co-located on the human body. Developing new wireless real-time health monitoring and diagnosis sensors facilitates the process of acquiring vial data from patients with conditions that requires constant monitoring, such as in diabetic cases. Biosensors offer several medical and personal applications. For instance, implanting a biosensor inside

I. N. Alquaydheb · A. E. Khorshid (✉) · A. M. Eltawil
University of California, Irvine, CA, USA
e-mail: Khorshia@uci.edu

I. N. Alquaydheb
e-mail: Alquaydi@uci.edu

A. M. Eltawil
e-mail: Aeltawil@uci.edu

G. Fortino and Z. Wang (eds.), *Advances in Body Area Networks I*,
Internet of Things, https://doi.org/10.1007/978-3-030-02819-0_20

a certain human organ to measure a chemical reaction or an enzyme replaces the need of using more invasive approaches. Moreover, wearable biosensors which are placed on human skin are used to detect bio signals, e.g. Electrocardiography (ECG) or Electroencephalogram (EEG) signals. These sensors can communicate using conventional wired or wireless using RF techniques resulting in many useful applications, such as continuous glucose Monitors which utilizes an under skin sensor that measures glucose level and transfer data wirelessly to a monitoring device [1]. However, this technique is limited by the transmitting and receiving excessive power consumption which leads to frequent battery replacement and inflexible designs for battery powered sensors. This obstacle can be resolved by (IBC) that communicate locally using the human body as a communication medium without the need for complex RF subsystems [2].

Electrostatic or capacitive coupling is one technique of IBC systems. In this approach, the exciting and receiving electrodes are attached to the human skin while the reference electrodes (GND) of Tx and Rx are left exposed to the air to form a capacitive return path [3]. The main disadvantage of this setup is that it can be impacted by the surrounding medium in the form of interference. Galvanic coupling offers a solution to this issue by connecting both terminals of transmitting and receiving electrodes to the skin and injecting an alternating current that results in electromagnetic waves to initiate a potential difference between receiving electrodes.

The main goal of this paper is to investigate the human channel gain profile via a comparative study of full Finite Element Model (FEM) analysis versus a reduced state FEM, both simulated in ANSYS HFSS for an arm model. Results are then compared to a circuit model representation that is able to capture human tissues properties and determines the channel path loss. Finally, a bone fracture internal fixation implant is included in the structure of the arm model to investigate the effect on channel attenuation profile.

2 Related Work

Many research efforts have been directed at estimating the path loss of the human body channel along different tissues, limbs and organs [4]. One of the direct methods is by capturing in vivo measurement from live subjects. These efforts then drive modeling and simulation approaches to create a complete description of the channel. Finite element method (FEM) and Equivalent circuit model (ECA) are common methods for analysis of the human body channel. Most presented papers in the field of intra-body communication develop an equivalent circuit model (ECA) to derive a transfer function that can calculate and determine the instantaneous gain of the human channel and avoid using FEM [5]. This preference is due to the lengthy simulation time incurred by FEM approaches resulting from the complex geometry and interconnections of the human body [6]. Our work relates to conducting FEM simulation on a reduced order human arm model channel with focus on the human arm. The arm model is designed with simple constructed details yet capable to capture essential arm properties. A comparative study of the two approaches is presented and the channel gain profile plotted. An equivalent circuit model is derived and compared against both FEM approaches. Using the reduced order FEM model, the impact of an internal implant on the channel gain profile is further discussed.

3 FEM Simulation Setup and Analysis

For the purpose of FEM, ANSYS Electronics Desktop was used for accurate full wave electromagnetic simulation to find an estimation of the human channel path loss. The tool can solve for Electro-Magnetic (EM) fields every tetrahedron of the 3D meshed structure of the arm model for the transmitting and receiving electrodes [7].

A simplified geometry an arm model has also been constructed, as show in Fig. 1c, by building five main layers that include skin, fat, muscle, bone marrow and cortical bone. The dimensions of the cylinders mimic the average total thickness of the arm related to each tissue thickness [5, 8]. The dielectric properties, which is described in terms of conductivity σ and relative permittivity ϵ, of every tissue are measured and averaged by Gabriel et al. [9].

The full FEM model was based on the model created by NEVA Electromagnetic group which was subsequently imported on ANSYS HFSS environment as shown

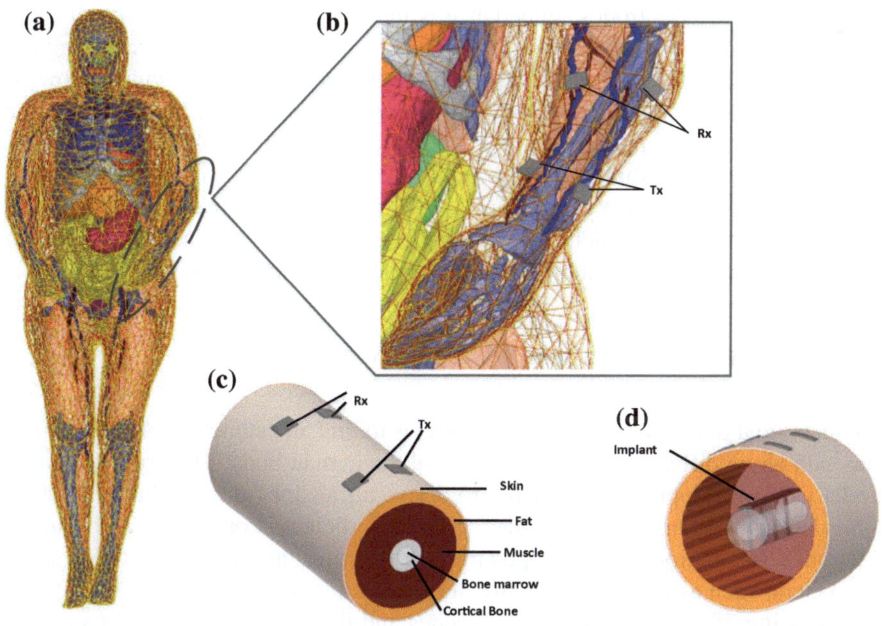

Fig. 1 Human arm model. **a** Complete human model including all tissues and blood streams. **b** Enlarged view of the human model arm. **c** An equivalent arm model with simpler structure. **d** An equivalent arm model including an implant

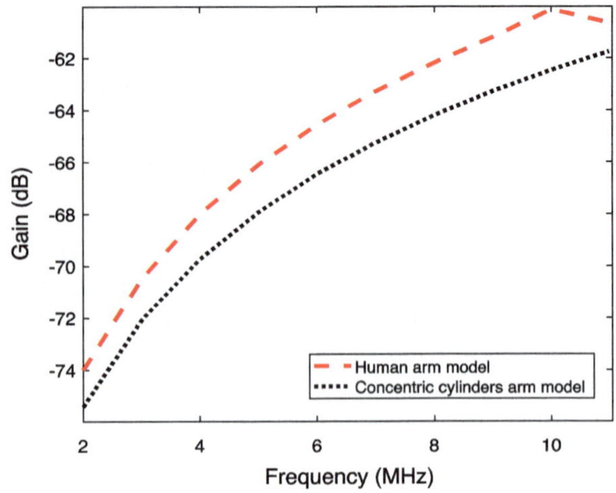

Fig. 2 Simulation results of cylindrical and human arm models

in Fig. 1a, b [10]. In addition to the five main layers of human body, blood streams or veins are included in the design of the arm model. Although the detailed model results in higher accuracy, the simulation time is much longer than the required time in case of the concentric cylinders.

The transmitting and receiving electrodes dimensions of both models were 15 × 15 mm. Since the electrodes' impedance is independent of the electrode material, Copper is used as the electrodes material [11]. The transmission path length is 10 cm. The excitation of transmitting side is set to a lumped port which allows us to embed internal ports of excitation with an input impedance of 50 Ω. The current source of the exciting ports is set to 1 mA which conforms to the safety standards of the International Commission on non-ionizing radiation protection (ICNIRP) [12]. The operating frequency range of 2–11 MHz is selected based on the obtained frequency of 10.5 MHz which showed the highest transmission gain in our previous work [13]. Finally, the arm model results are plotted as shown in Fig. 2.

4 Equivalent Human Body Circuit Model

The main goal of developing a circuit model is to find an estimation of the transmission gain along different paths and tissues of the human body. The proposed circuit model in [5] is adopted in this paper to derive a transfer function that is basically dependent on the human body tissues' geometry, dielectric properties and coupling electrodes. In addition, the developed transfer function can capture multiple varying parameters, such as the spacing between the transmitter and receiver, the interelectrode spacing of the Tx or the Rx electrodes and the electrodes' size. As shown in Fig. 3, the proposed model of the galvanic coupling represents a four-terminal circuit where V_i is the transmitter output voltage, R_o and Z_{ceo} are the transmitter output resistor and receiver input impedance respectively, Z_c is the coupling impedance

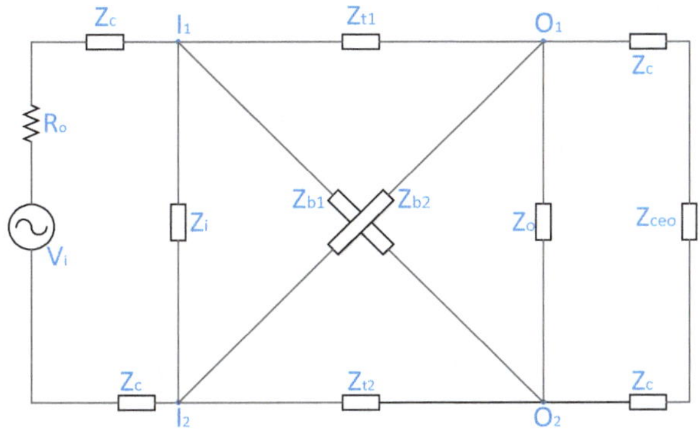

Fig. 3 Equivalent circuit model for galvanic coupling

between the human skin and the transmitting or receiving electrodes, Z_i representing the input impedance between the two input terminals I_1 and I_2, Z_o is the output impedance between the two output terminals 0_1 and 0_2, Z_{b1} is the cross impedance between I_1 and 0_2, Z_{b2} is the cross impedance between I_2 and 0_1. Z_{t1} and Z_{t2} are the transmission path transverse impedances which can be expressed in the following formula

$$Z_t = \frac{1}{\sum_{n=1}^{5}\left(\frac{1}{R_n} + j\omega C_n\right)} = \frac{L}{\sum_{n=1}^{5}\sigma_{nf} S_n + j\omega\epsilon_o \sum_{n=1}^{5}\epsilon_{rnf} S_n}$$

where L is the signal transmission path length between the transmitting terminals and receiving terminals, σ_{nf} and ϵ_{rnf} are the conductivity and relative permittivity, respectively, corresponding to the nth human body tissue and to the input signal frequency. S_n is the cross sectional area of the nth layer.

5 FEM and ECA Simulation Time

The required simulations have been completed on a common computing setup. Simulations were performed using a quad core processor running at 2.3 GHz with 8 GB of RAM.

Simulation time is calculated in terms of CPU time to determine the required time for the central processing unit to complete each simulation. The following table summarizes the estimated CPU time for each model simulation Clearly, the total CPU time of the complete human arm FEM simulation is $\approx$47 times the estimated time in case of cylindrical arm. As expected, ECA model shows the fastest simulation time (Table 1).

Table 1 CPU time of all simulations

Model	CPU time (hh:mm:ss)
ECA	(00:00:1.77)
FEM of cylindrical arm	(00:02:34)
FEM of human arm	(01:34:07)

6 ECA Model Validation

To verify the simulation results of cylindrical and human arm model, path loss readings of analytical model are included as shown in Fig. 4. The detailed geometry of human arm model which includes veins, uneven thicknesses of muscle and bone layers, the nonuniform surface of skin that affects the positioning of all electrodes structure play a main role in the observed difference of gain values between human and cylindrical model. However, shown results provide that using cylindrical model can be an alternative to using a model with detailed human geometry when targeting less accurate yet fast results of path loss profile. Analytical and cylindrical model show a tolerance of about ± 3 dB where analytical and human model exhibit an agreement in the range of 8–11 MHz. Factors that contribute to the difference between models include variations in contact impedance calculations which is not unified among IBC researches. Also, the meshing convergence procedure of FEM simulation and electrode internal impedance is a factor in data variations.

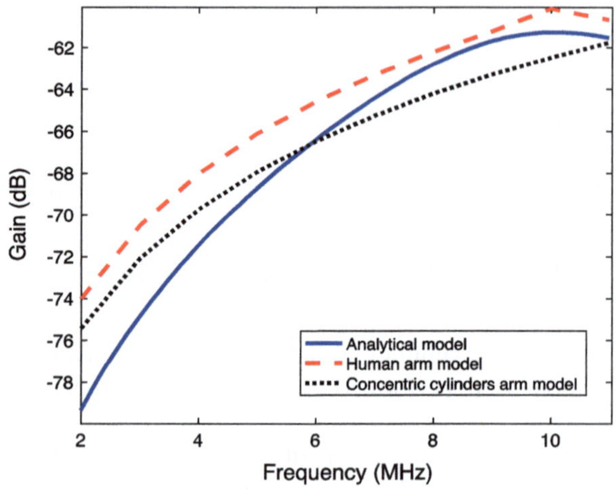

Fig. 4 Simulation results of analytical, cylindrical and human arm models

7 Current Density Distribution

In the galvanic coupling approach, the coupled electromagnetic signal exhibit dipole type field behavior [14]. This field can be represented in terms of current density vectors to show current paths along multiple tissues. Current paths are mainly composed of primary path that flows between the transmitting electrodes and secondary path the results in potential difference between the receiving electrodes as shown in Fig. 5. The source electrode current exciting the primary and secondary flows is as follows

$$I = \iint_s J \cdot ds = \iint_s \sigma \cdot E \cdot ds \tag{2}$$

where J $\left(A/m^2\right)$ is the current density and s is the surface area of Tx electrode and E is the normal electric field intensity component between the contact area of Tx and skin tissue [4].

Current density distribution within the five main human layers is affected by the relative permittivity and conductivity of these tissues. The skin layer has higher σ than fat tissue. Due to this fact, current density distribution values in skin is higher than observed values in fat, which can be seen in Figs. 6 and 7. In addition, the amount of movable charges between Tx and Rx electrodes are impacted by the described impedances in the previous sections. For instance, in case of longitudinal transmission for wearable sensors, transverse impedance Z_t is affected by the value of corresponding tissue's conductivity. As σ_n increases, Z_t decreases which leads to higher values of detected potential difference.

The coupled EM signal creates two current components described as conduction and displacement current [4, 15]. Conduction current at lower frequency which results from flow of charges is high. Consequently, coupled signal is confined within the human tissues and therefor leads to higher detected signal at Rx electrodes. However,

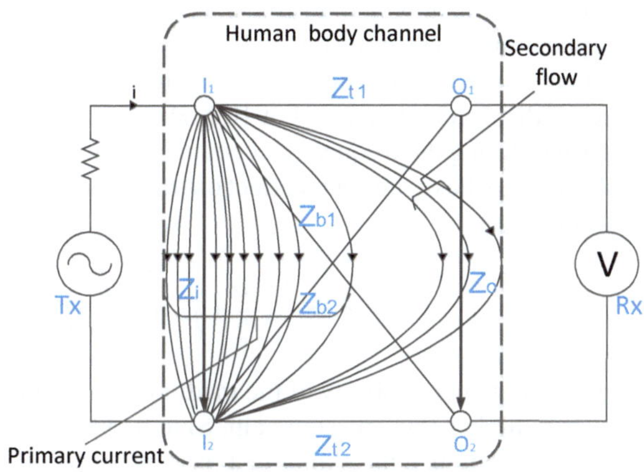

Fig. 5 Current flow paths of galvanic coupling

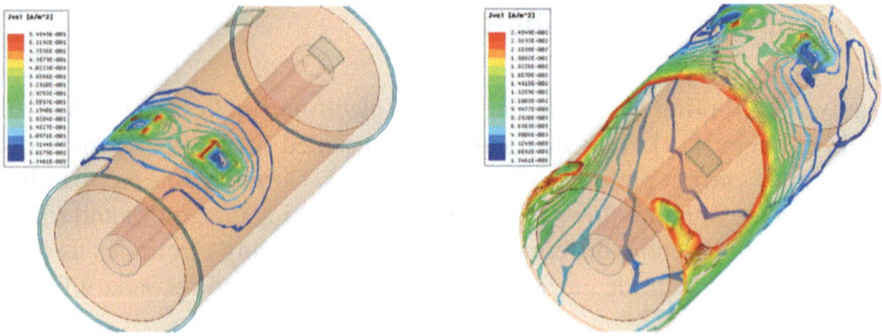

(a) Primary current distribution.

(b) Secondary current distribution.

Fig. 6 Skin tissue current density distribution

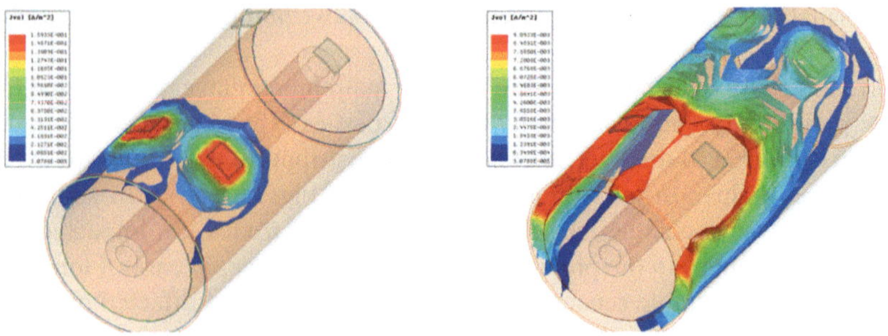

(a) Primary current distribution.

(b) Secondary current distribution.

Fig. 7 Fat tissue current density distribution

at higher frequencies, conduction current remains constant and the capacitive effect becomes dominant which results in a continuous increases in displacement current. This phenomenon leads to signal dissipation into the surrounding environment which eventually decreases the intensity of the received signal. From Fig. 4, we observe the effect of displacement current at 10 MHz.

By increasing the frequency of coupled signal, another factor that is characterized as skin depth δ has a dominant effect on transmission gain. The skin depth is described as follow

$$\delta = \frac{1}{\omega}[\frac{\mu\epsilon}{2}(\sqrt{1+\left(\frac{\sigma}{\epsilon\,\omega}\right)^2} - 1)]^{-\frac{1}{2}} \tag{3}$$

where μ is the permeability of the medium. Although higher skin depth value incorporates in higher transmission gain, Skin depth is a decreasing function of σ. From Fig. 4, we recognize the skin effect passing the 10 MHz frequency point.

Table 2 Properties of DCP

Plate material	Length	Width	Hole spacing
Stainless steel Titanium	25–145 mm	10 mm	12 and 16 mm

8 Internal Fixation Implant Effect

Bone fractures are one of the most common injuries that requires immediate care. The typical procedure for dealing with these fractures is by using an implant that can attach both broken parts of the bone for a long time that can take up to several weeks until both separated parts of the bone are healed and attached together. In this section, we investigate the effect of this implant by using the modified concentric cylinders as shown in Fig. 8. Dynamic compression plate (DCP) is a conventional choice by orthopedic surgeons to compress and stabilize bone fractures [16]. The plate has a spherical gliding holes which allows the spherical headed screws to have a self compression effect. Table 2 describes the characteristics of (DCP) [17].

By exciting the electrodes of the arm model which includes the implant, it has been noted that path loss increased by about 0.5 dB, which can be seen in Fig. 10. Since DCP is adjacent to cortical bone and all electrodes are on skin layer, the plat

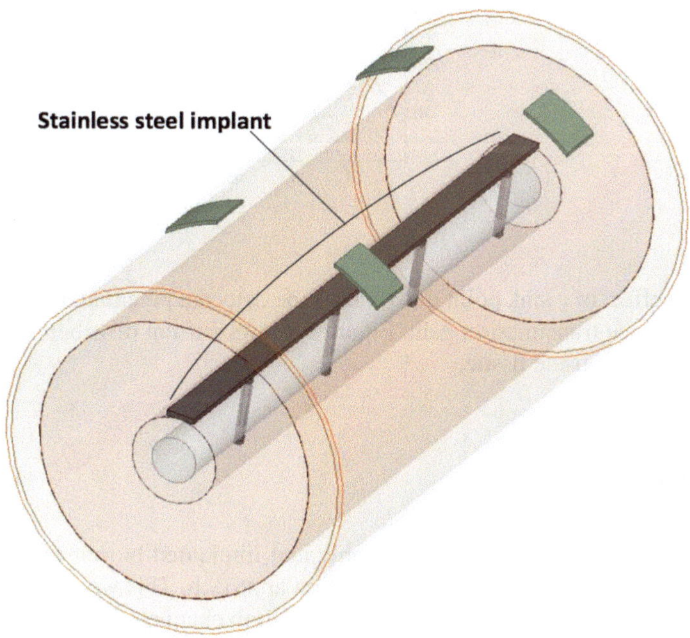

Fig. 8 Arm model including an internal fixation implant

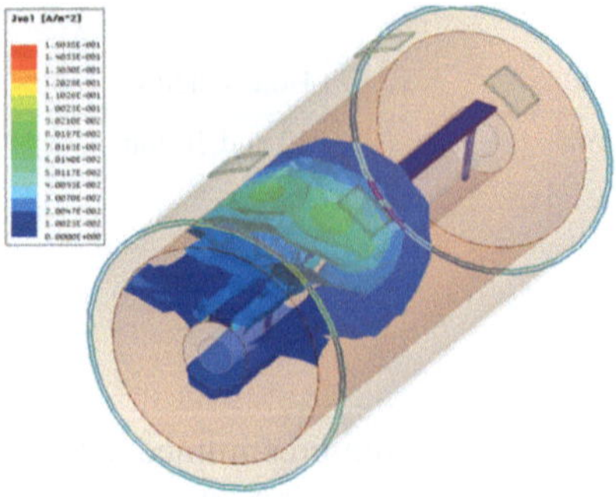

(a) Primary current distribution.

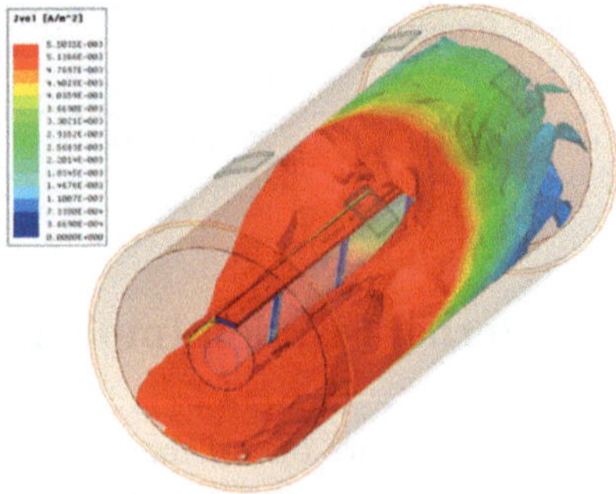

(b) Secondary current distribution.

Fig. 9 Implant and muscle current density distribution

Fig. 10 Simulation results of the arm model with and without DCP

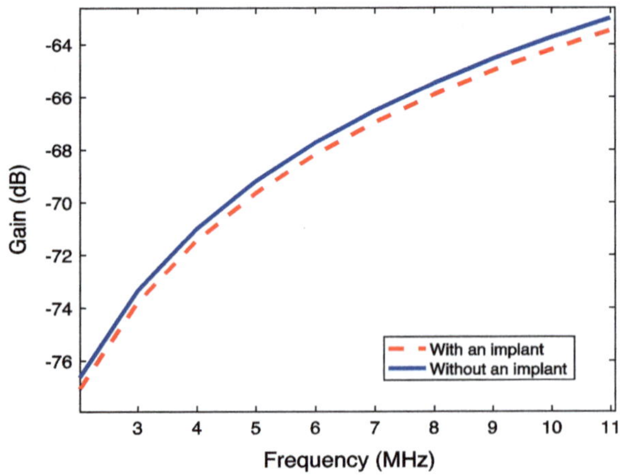

can have an effect of a sink point that reduces the detected potential difference by Rx electrodes, hence transmission gain. Figure 9 shows current distribution in implant and surrounding muscle tissue.

9 Conclusion

IBC will lead to the evolution of wearable and implanted biosensors industry by replacing the conventional RF communication approach. This work evaluates alternative means of modeling galvanic IBC human limb channels. The paper presents a full FEM model, a reduced order FEM model and an ECA circuit model of a human arm. The transfer function of the channel is presented for all three cases as well as the associated simulation time. Finally, analysis of the effect on current density due to an internal fixation implant is presented.

References

1. Ways to Get Started. https://www.medtronicdiabetes.com/treatments/continuousglucose-monitoring
2. Teshome, A.K., Kibret, B., Lai, D.T.: Galvanically coupled intrabody communications for medical implants: a unified analytic model. IEEE Trans. Antennas Propag. **64**(7), 2989–3002 (2016)
3. Callejon, M.A., Naranjo-Hernandez, D., Reina-Tosina, J., Roa, L.M.: Distributed circuit modeling of galvanic and capacitive coupling for intrabody communication. IEEE Trans. Biomed. Eng. **59**(11), 3263–3269 (2012)
4. Swaminathan, M., Cabrera, F.S., Pujol, J.S., Muncuk, U., Schirner, G., Chowd-bury, K.R.: Multi-path model and sensitivity analysis for galvanic coupled intra-body communication through layered tissue. IEEE Trans. Biomed. Circuits Syst. **10**(2), 339–351 (2016)
5. Song, Y., Hao, Q., Zhang, K., Wang, M., Chu, Y., Kang, B.: The simulation method of the galvanic coupling intrabody communication with different signal transmission paths. IEEE Trans. Instrum. Meas. **60**(4), 1257–1266 (2011)
6. Hachisuka, K., Terauchi, Y., Kishi, Y., Sasaki, K., Hirota, T., Hosaka, H., Ito, K.: Simplified circuit modeling and fabrication of intrabody communication devices. Sens. Actuators A: Phys. **130**, 322–330 (2006)
7. Ansoft Corporation. Ansoft High Frequency Structure Simulator v10 Users Guide. Ansoft Corporation (2005)
8. Wegmller, M.S.: Intra-body Communication for biomedical sensor networks. Doctoral Dissertation, ETH Zurich (2007)
9. Gabriel, S., Lau, R.W., Gabriel, C.: The dielectric properties of biological tissues: III. Parametric models for the dielectric spectrum of tissues. Phys. Med. Biol. **41**(11), 2271 (1996)
10. NEVA Electromagnetics. https://www.nevaelectromagnetics.com/
11. Hachisuka, K., Takeda, T., Terauchi, Y., Sasaki, K., Hosaka, H., Itao, K.: Intra-body data transmission for the personal area network. Microsyst. Technol. **11**(8–10), 1020–1027 (2005)
12. Ahlbom, A., Bergqvist, U., Bernhardt, J.H., Cesarini, J.P., Grandolfo, M., Hietanen, M., Swicord, M.L.: Guidelines for limiting exposure to time-varying electric, magnetic, and electromagnetic fields (up to 300 GHz). Health Phys. **74**(4), 494–521 (1998)
13. Khorshid, A. E., Eltawil, A. M., Kurdahi, F.: Intra-body communication model based on variable biological parameters. In: 2015 49th Asilomar Conference on Signals, Systems and Computers, pp. 948–951, Nov 2015. IEEE (2015)
14. Kibret, B., Seyedi, M., Lai, D.T., Faulkner, M.: Investigation of galvanic-coupled intrabody communication using the human body circuit model. IEEE J. Biomed. Health Inf. **18**(4), 1196–1206 (2014)
15. Hoyt, K., Castaneda, B., Parker, K.J.: 5C-6 muscle tissue characterization using quantitative sonoelastography: Preliminary results. In: In: IEEE Ultra-sonics Symposium, pp. 365–368, Oct 2007. IEEE (2007)
16. McCartney, W.T.. The design, manufacture and analysis of a new implant for fracture fixation in human and veterinary Orthopaedic surgery: the Bone Fastenerod. Doctoral Dissertation, Dublin City University (2002)
17. Dynamic Compression Plates, DCP, 3.5 mm. http://www.wheelessonline.com/ortho/

An Improved Mathematical Model for the Autonomic Regulation of Cardiovascular System

Yuanyuan Liu, Yingfei Sun, Zhipei Huang, Yu Meng, Jiankang Wu and Xinxia Cai

Abstract The activity of the autonomic nervous system is hardly measurable. This study presents an improved mathematical model to estimate the baroreceptor nerve firing rate, the efferent parasympathetic and sympathetic response in a scenario of postural change from sitting to standing, based on observed blood pressure and heart rate changes. An optimization step is then applied to find the model parameters best fitting to observed cardiac data of healthy people and hypertensive patients using unbiased estimation and Nelder-Mead method. The experimental results on 59 subjects have shown that the improved model can describe autonomic regulation mechanism well and the estimated system parameters have clear clinical meaning.

Keywords Autonomic nervous regulation · Heart rate
Baroreceptor reflex control · Parasympathetic and sympathetic activity
Cost function

Y. Liu · Y. Sun · Z. Huang (✉) · Y. Meng · J. Wu · X. Cai
School of Electronic, Electrical and Communication Engineering, University of Chinese Academy of Sciences, Beijing 101408, China
e-mail: zhphuang@ucas.ac.cn

Y. Liu
e-mail: liuyuanyuan115@mails.ucas.ac.cn

Y. Sun
e-mail: yfsun@ucas.ac.cn

Y. Meng
e-mail: mengy_2014@163.com

J. Wu
e-mail: jkwu@ucas.ac.cn

X. Cai
e-mail: xxcai@mail.ie.ac.cn

Y. Liu · Y. Sun · Z. Huang · Y. Meng · J. Wu · X. Cai
Institute of Electronics, Chinese Academy of Sciences, Beijing 101408, China

G. Fortino and Z. Wang (eds.), *Advances in Body Area Networks I*,
Internet of Things, https://doi.org/10.1007/978-3-030-02819-0_21

1 Introduction

Cardiovascular diseases remain the biggest cause of deaths worldwide [1], exceeding cancer and AIDS. Clinical study shows that the dysfunction of the autonomic nervous system (ANS) plays an significant role in cardiovascular disease (CVD) [2]. Therefore, studying the model for the autonomic regulation of cardiovascular system is very essential [3] with the respect to cardiovascular diseases management.

However, physiological systems are usually complex, non-stationary, since they are characterized by a network structure with a number of interacting elements. Several mathematical models have been proposed to describe the autonomic regulation of cardiovascular system. Karavaev et al. [4] presented a physiology-based mathematical model described by a set of three nonlinear, coupled differential equations, but the model is too complex to understand and engineer in Medicine and Biology. Chen et al. [5] simulated the role of the central neural control in a hypertensive patient using bond graph model, but the model only suits hypertensive patient. Calvo et al. [6] described an mathematical model for the control of the autonomic nervous system in the test of head-up tilt, however their study only cover Brugada syndrome patients.

Baroreceptor reflex control and efferent parasympathetic and sympathetic activity are important determinants of the autonomic regulation of cardiovascular system [7, 8]. Several methods have been proposed so far to assess them, however they still used an invasive method to give a quantification of baroreflex sensitivity and sympathetic activity [9], rather than an non-invasive method. We presented a model in [10] (P-model) using blood pressure (BP) measured in the finger as an input to non-invasively calculate the baroreceptor nerve firing rate, sympathetic and parasympathetic responses, and concentrations of norepinephrine and acetylcholine and the heart rate (HR) dynamics in response to the change of the BP. In this paper, in order to get a more effective model, we improve the P-model based on physiological significance by: (1) dividing the baroreflex firing rate into three types according to their ability to conduct the signal, assuming that three neuron types can be differentiated and self-governed; (2) establishing an new mathematical expression of efferent parasympathetic and sympathetic outflows to avoid the extra complexity aroused by the fixed time delay between the sympathetic signal and the parasympathetic signal; (3) using unbiased estimation to define the cost function (J). The improved model (M-model) are proved through analysing the system parameters, the value of cost function and running speed compared to the P-model. In the following part, we will describe the model-improved method, present the experimental results and discuss the results.

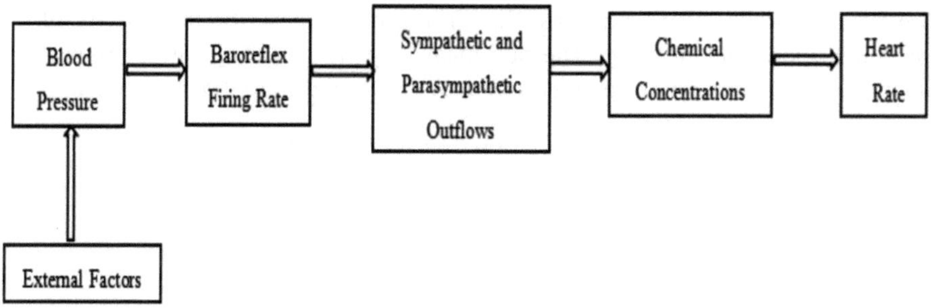

Fig. 1 Schematic of the model about the autonomic regulation of cardiovascular system

2 Method

2.1 *The Autonomic Regulation of Cardiovascular System*

Our mathematical model for the autonomic regulation of cardiovascular system responses to postural change is divided into five parts (see Fig. 1). Baroreflex firing rate, the sympathetic and parasympathetic outflows are the linchpins and considered as the system parameters of the model. Baroreceptors are stretch sensitive nerve fibres located at the carotid sinus and the aortic arch that senses changes in arterial BP [11], which is induced by an external factor in this model. During posture changes from sitting to standing, amounts of blood is poured into legs because of the gravitational force, leading to an obvious drop in pressure. The fall in BP results in the decreased firing of the afferent baroreceptor nerves, which gives rise to parasympathetic withdrawal and sympathetic activation, causing a rapid increase in heart rate ultimately. In the following part, the M-model will be discussed to analyze the system parameters of the autonomic regulation mechanism which changes HR to counteract the changes in BP quantitatively.

2.2 *Improved Mathematical Model*

2.2.1 Baroreflex Modeling

One of the major linchpin of the autonomic regulation of cardiovascular system is the baroreflex control, which consists of three parts: an affecter part, a control center, and an effector part. As the BP being the affecter, the average value is estimated via pulse transmission time which is determined by the measurement time and the physical properties of the blood vessel [12]. This part of the model describes the dynamics of the baroreflex effector signalling to a change in arterial blood pressure. We consider the three neuron baroreflex types independently and every type has threshold baroreflex firing rate (see Eq. 1) in order to describe the activity of the baroreflex more meticulously. The nonlinear response in firing rate of type i is a function of BP using the following system of ordinary differential equations.

$$\begin{aligned} \frac{dn_i}{dt} &= k_i \frac{dp}{dt} \frac{n_i(M - n_i)}{(M/2)^2} - \frac{n_i - N_i}{\tau_i} \\ n &= n_S + n_I + n_L \\ N &= N_S + N_I + N_L \end{aligned} \tag{1}$$

where i is the neuron type including short time-scale term S, intermediate time-scale term I, and long time-scale term L; n_i is the current baroreflex firing rate displayed by the neuron of type i respectively; n is the current total firing rate; k_i is a constant determining the sensibility of the neuron of type i responses to changes in blood pressure; $\frac{dp}{dt}$ denotes the time derivative of the continuous blood pressure; τ_i and N_i are the time constant and the threshold baroreflex firing rate associated with the neuron of type i; M is the maximum baroreflex firing rate and is set to 120 (Hz) according to the previous study [13]; N is the total threshold baroreflex firing rate and have to be lower than the maximum firing rate M. To enforce these limit, we use the following equation to express the relation of the two variables.

$$N = \frac{M}{2} + \frac{\eta^2}{1 + \eta^2} * \frac{M}{2} \tag{2}$$

where η is an unknown parameter.

2.2.2 Parasympathetic and Sympathetic Response Modeling

This submodel describes effects of the afferent baroreceptor nerve activity on the parasympathetic and sympathetic nervous system, which are often recognized, activated in the parasympathetic and sympathetic outflows [14]. The parasympathetic outflow Tpar is known to follow the "direct law", i.e.

$$Tpar = \frac{n(t)}{M} \tag{3}$$

But the sympathetic outflow is inversely related to the baroreflex firing rate. In addition, the previous studies [10, 13] revealed that there is a time delay between the activity of the sympathetic nervous system and the parasympathetic nervous system responding to the baroreflex firing rate. However, delay equations are difficult to solve numerically [15] and discrete time delay hardly appears in biological neuronal networks, thus, the sympathetic outflow is defined by the follow equation.

$$k = \frac{a^{\rho}}{(\rho-1)!}$$
$$Tpar\_d = k^* \int_{-\infty}^{t} Tpar(\eta)^*(t-\eta)^{\rho-1*}e^{-a(t-\eta)}d\eta \quad (4)$$

and

$$Tsym = \frac{1 - Tpar\_d + u(t)}{1 + \beta Tpar} \quad (5)$$

where α and ρ are integers that determine the shape of the gamma distribution used as weight of the signal, *Tpar_d* is the delayed parasympathetic outflow; η is an unknown parameter; β is a constant controlling the damping of the sympathetic outflow T_{sym} by The parasympathetic outflow T_{par}. $u(t)$ is the impulse function accounting for the regulation of the sympathetic nerve activity by the vestibulo-sympathetic system. The impulse function is modelled as

$$u(t) = -[b(t - t_{over})]^2 + u_0$$
$$b = \sqrt{\frac{4u_0}{(t_{over} - t_{start})^2}}$$
$$t_{over} = t_{start} + t_{per} \quad (6)$$

where t_{star}, t_{per}, u_0 are the start time, duration and magnitude of the response.

2.3 Parameter Estimation and Optimization

The M-model described in (1)–(6) and (8)–(11), contains a total of 17 variable parameters denoted by $\theta = (k_s, k_I, k_L, \tau_S, \tau_I, \tau_L, \eta, \alpha, \rho, \beta, u_0, t_{start}, t_{per}, \tau_{Snor}, \tau_{ach}, \zeta_S, \zeta_P)$. To estimate these parameters from experimental data, we use unbiased estimation to define the cost function (J), the mean square error between the calculated and the measured values of heart rate. We seek the parameters to minimize J by using the Nelder–Mead method (a simplex algorithm) [16], which uses the sampling of the parameter space to guide the minimization and gets the optimum parameters value through a simplex of points created at each iteration. Then the autonomic regulation model of cardiovascular system is established based on these optimized parameters. The cost function is defined as

$$J = \frac{1}{N-1}\sum_{i=1}^{N}(HR_i^m - HR_i^c)^2 \quad (7)$$

where HR_i^m is the measured value of heart rate, HR_i^c is the calculated value of heart rate by the model, N is the number of HR_i^m. It should be emphasized that, during the parameter estimation process, we consider the model as one integrated component. That is, all unknown parameters in the model are identified simultaneously.

3 Result and Discussion

3.1 Improved Mathematical Model

In this study, we have validated the model with 42 data sets from the healthy people and 17 data sets from the hypertensive patients. These data sets are collected according to the measurement in [10]. The subject stands up at 10th seconds. Figure 2 shows the baroreflex firing rate as described in Eq. 1. As the baroreflex firing rate is higher in the healthy people than in the hypertensive patient, the change in firing rate dynamics is also greater in the healthy people while the transition time of the firing rate is slightly slower in the hypertensive patient (see Fig. 2). Furthermore, the firing rate has never fallen below the minimum firing rate ($N = 60$), which solves the problem occurred in the P-model and the former model [13], proving M-model is more reasonable in physiology.

Figure 3 shows the sympathetic (*Tsym*) and parasympathetic (*Tpar*) outflows described in Eqs. (3)–(5). Since the sympathetic and parasympathetic outflows are affected by the baroreflex nerve activity, their dynamics and characteristics are similar to the baroreflex firing rate. The response of sympathetic increases while the response

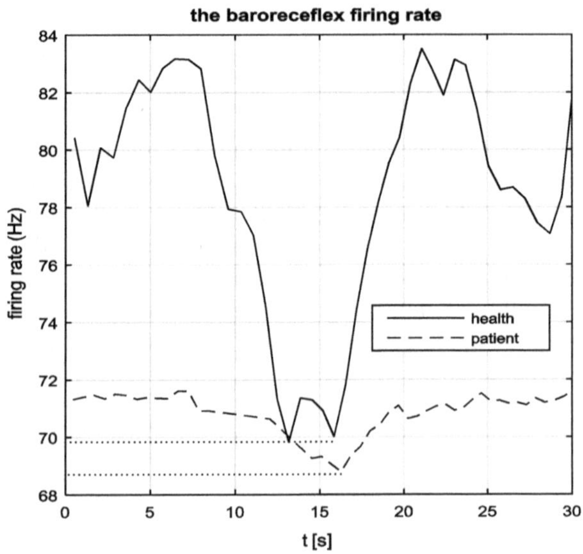

Fig. 2 The baroreflex firing rate as a function of time. Results are shown from a healthy person (solid) and a hypertensive patient (dash)

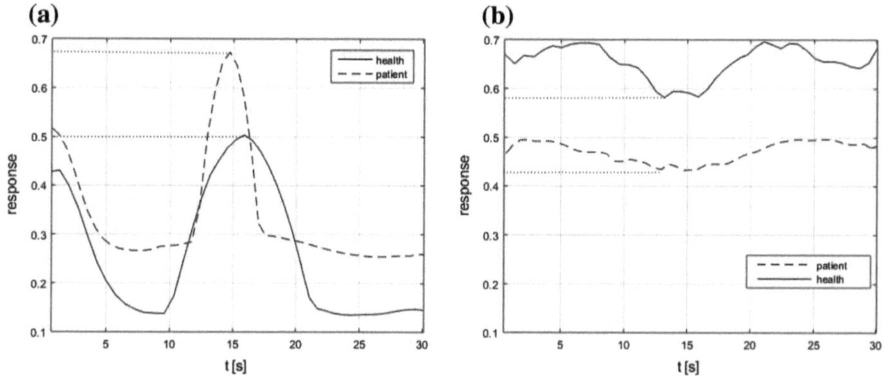

Fig. 3 **a** Shows the response of sympathetic, **b** shows the response of parasympathetic

Table 1 Relative values of the neural response

	Sympathetic maximum response	Parasympathetic minimum response	The time of sympathetic maximum response	The time of parasympathetic minimum response
Health	0.50 ± 0.07	0.58 ± 0.07	15.86 ± 0.72	13.17 ± 0.72
Patient	0.67 ± 0.07	0.43 ± 0.07	14.69 ± 0.13	12.92 ± 0.13
P values	0.000	0.000	0.43	0.41

of parasympathetic decreases both in healthy people and hypertensive patient when the stimulus appears at 10th seconds (see Fig. 3). The response of sympathetic is lower in the healthy people than in the hypertensive patient (see Fig. 3a), while the response of parasympathetic is significant higher in the healthy people than in the hypertensive patient (see Fig. 3b). Thus, the hyperactivity of sympathetic nerves and excessive inhibition of parasympathetic nerves may induce hypertension. We use ANOVA analysis to compare the relative values of the neural response between the healthy people and the hypertensive patient (see Table 1). The extremum of the neural response has an evident difference between healthy people and patient ($p < 0.00001$), while there is no difference in the time of the extremum generation ($p = 0.43$, $p = 0.41$).

3.2 *The Results After Optimization*

The Nelder-Mead nonlinear optimization is used to minimize the discrepancy between the calculated and the measured values of heart rate. The values in Table 2 are calculated according to the dynamic process of the autonomic nervous control (10–25 s). The calculated HR using the initial parameters shows significant difference

Table 2 The value of J

Group	Initial (J)	Optimized (J)	Decrease rate (J) (%)
Health	33.7	6.1	82.1
Patient	75.5	10.7	85.9

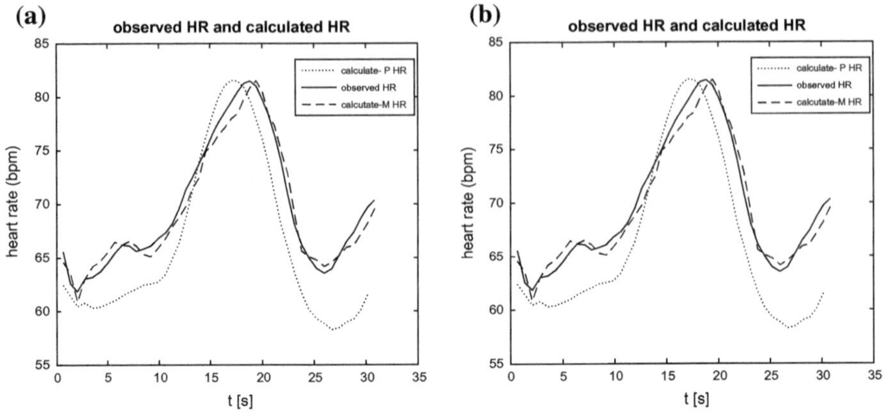

Fig. 4 **a** Shows the observed HR, calculated HR by our model (calculated-M HR) and the model in [10] (calculated-P model) of a healthy people. **b** Shows the result of a hypertensive patient

Table 3 Compare our model with the model in [8]

	The running speed	The cost function (J)
M-model	50 ± 10 (s)	6.1
P-model	5 ± 1 (min)	8.1
Decrease rate (%)	83.3	25.0

compared to the observed HR. Through optimization, the value of the cost function has reduced by 82.1% in healthy people and 85.9% in patient, thus the calculated HR by M-model is very close to the observed HR both in healthy people and patient (see Table 2 and Fig. 4).

After the modification of the baroreflex firing rate and the parasympathetic and sympathetic response, the HR calculated by M-model is closer to the observed HR than P-model [10] (see Table 3 and Fig. 4), with the value of cost function reduces considerably by 25.0%. Furthermore, the running speed of processing a set of data through M-model is obviously faster than P-model, with the running speed has reduced by 83.3%.

4 Conclusions

Improving model for the autonomic regulation of cardiovascular system based on physiological significance, then further analyzing the system parameters of the model, is an important approach to describe dynamic performance of cardiovascular system. The experimental results on 59 subjects have shown that the M-model can describe the autonomic regulation mechanism well and the estimated system parameters have clear clinical meaning. The analysis of the parasympathetic and sympathetic outflows shows that the hyperactivity of sympathetic nerves and excessive inhibition of parasympathetic nerves may cause hypertension. Compared to the P-model, experimental results have presented that the HR calculated by M-model is closer to the observed HR and the computational complexity has been reduced greatly. Furthermore, M-model is applicable both in the healthy people and the hypertensive patients.

Appendix

In this appendix we describe the part of chemical concentrations and heart rate modeling by the previous model and we use it.

Chemical concentrations: The concentration of neurotransmitters acetylcholine *Cach* and noradrenaline *Cnor* are defined using linear differential equations which depend on the sympathetic (*Tsym*) and parasympathetic (*Tpar*) responses.

$$\frac{dCnor}{dt} = \frac{-Cnor + Tsym}{\tau_{nor}}$$
$$\frac{dCach}{dt} = \frac{-Cach + Tpar}{\tau_{ach}} \tag{8}$$

where τ_{nor} and τ_{ach} are two time parameters.

Heart rate: HR potential (ψ) is linked with these chemical concentrations by an integrated model of the form.

$$\frac{d\varphi}{dt} = H_0(1 + M_S C_{nor} - M_P C_{ach}) \tag{9}$$

where

$$M_S = \frac{\xi_S^2}{1 + \xi_S^2}$$
$$M_P = \frac{\xi_P^2}{1 + \xi_P^2} \tag{10}$$

where H_0 denotes intrinsic HR, M_s and M_p represent the strength of the response to changes in the concentrations. ζ_s and ξ_p are the two parameters used to constrain M_s and M_p in the interval [0, 1], in order to bound HR within physiological values.

A heartbeat occurs when ψ reaches 1, then is reset to 0. So the HR is defined as the following equation:

$$HR = 1/(t_{\varphi=1} - t_{\varphi=0}) \tag{11}$$

References

1. Ahmad, N., Iqbal, D.: Cardiovascular Diseases and Its Treatment (2016)
2. Castrodiehl, C., Diez Roux, A.V., Redline, S., Seeman, T., Mckinley, P., Sloan, R., Shea, S.: Sleep duration and quality in relation to autonomic nervous system measures: the multi-ethnic study of atherosclerosis (MESA). Sleep **39** (2016)
3. Schultz, H.D., Li, Y.L., Ding, Y.: Arterial chemoreceptors and sympathetic nerve activity: implications for hypertension and heart failure. Hypertension **50**, 6 (2007)
4. Karavaev, A.S., Ishbulatov, Y.M., Ponomarenko, V.I., Prokhorov, M.D., Gridnev, V.I., Bezruchko, B.P., Kiselev, A.R.: Model of human cardiovascular system with a loop of autonomic regulation of the mean arterial pressure. J. Am. Soc. Hypertens. Jash **10**, 235–243 (2016)
5. Chen, S., Zhang, S., Gong, Y., Dai, K., Sui, M., Yu, Y., Ning, G.: The role of the autonomic nervous system in hypertension: a bond graph model study. Physiol. Meas. **29**, 473 (2008)
6. Calvo, M., Le, R.V., Romero, P.D., Behar, N., Gomis, P., Mabo, P., Hernandez, A.I.: Analysis of a cardiovascular model for the study of the autonomic response of Brugada syndrome patients. In: International Conference of the IEEE Engineering in Medicine & Biology Society, pp. 5591
7. Seravalle, G., Grassi, G.: Sympathetic nervous system, hypertension, obesity and metabolic syndrome. High Blood Press. Cardiovasc. Prev. Off. J. Ital. Soc. Hypertens. **23**, 175 (2016)
8. Charkoudian, N., Rabbitts, J.A.: Sympathetic neural mechanisms in human cardiovascular health and disease. Mayo Clin. Proc. **84**, 822–830 (2009)
9. Abboud, F.M.: In search of autonomic balance: the good, the bad, and the ugly. Am. J. Physiol. Regul. Integr. Comp. Physiol. **298**, R1449 (2010)
10. Wang, L., Huang, Z., Wu, J., Meng, Y., Ding, R.: A model-based method to evaluate autonomic regulation of cardiovascular system. In: IEEE International Conference on Wearable and Implantable Body Sensor Networks, pp. 1–6
11. Guyton, A.C.: Guyton and Hall Textbook of Medical Physiology. Saunders/Elsevier (2011)
12. Sheng, J., Ji, L., Wang, F., Sun, Y., Zhang, Z., Ji, F., Liu, B., Zhang, W., Jiankang, W., 蒋升, 冀连营.: Modeling human vascular networks and estimation of central aortic pressure. Scientia Sinica Informationis **43**, 964 (2013)
13. Ottesen, J.T., Olufsen, M.S.: Functionality of the baroreceptor nerves in heart rate regulation. Comput. Methods Programs Biomed. **101**, 208–219 (2011)
14. Parati, G., Esler, M.: The human sympathetic nervous system: its relevance in hypertension and heart failure. Eur. Heart J. **33**, 1058 (2012)
15. Bellen, A., Zennaro, M.: Numerical Methods for Delay Differential Equations. Clarendon Press (2003)
16. Nelder, J.A., Mead, R.: A simplex method for function minimization. Comput. J. **7**, 308–313 (1965)

Design of Fall Test System Based on Arduino 101

Nan Wang and Yaxia Liu

Abstract This paper designs a fall detection system based on Arduino 101. It is mainly composed of NNs (Neural Networks) and IMU (Inertial measurement unit). It is used to detect if an oldman falls down. When an old man falls, It can call the police to help the man get help in time. The core chip of Arduino 101 is the Intel Curie module. Intel Curie module microprocessor is Intel x86 Quark SE. It also carries GPRS wireless communication and GPS satellite positioning module. It analyzes and studies the characteristic parameters of the old man when he falls and does daily activities. It mainly uses the RBF (Radial Basis Function) algorithm to identify if a fall occurs. Experimental results show that: the system is able to identify most of the motion states correctly, with low reporting and false alarm rate. And it can quickly distinguish between daily activities and falls. For the old man, the detection accuracy rate can reach 95.5%. It has a high recognition rate, reliability and stability.

Keywords Fall detection · Arduino 101 · GSM · NNs · RBF

1 Preface

As China's aging population increases gradually, elderly people are more likely to die from falls, especially those aged 65 or older. Ministry of Health has released a report which called the China Shanghai prevention report. It showed that in our country residents over the age of 65, more than 20% of men and 45% of women had happened in the fall. The older you are, the more likely you are to fall [1]. When an

N. Wang (✉) · Y. Liu
Information Engineering, Beijing Institute of Fashion Technology, Beijing, China
e-mail: Wangnan1117@yeah.net

Y. Liu
e-mail: gxylyx@bift.edu.cn

G. Fortino and Z. Wang (eds.), *Advances in Body Area Networks I*,
Internet of Things, https://doi.org/10.1007/978-3-030-02819-0_22

old person falls, it can lead to a variety of wounds and diseases, such as scratches, fractures, fainting, and cerebrovascular accidents [2]. It may lead to paralysis. The worse the health assessment, the probability of falling is the higher, especially for older people who are less likely to walk [3]. Therefore, it is necessary to design a fall detection system to reduce the old because of sin and the damage has an important significance.

At present, the method of falling detection is broadly divided into four categories: (1) Autonomous alarm system. This method requires the wearer to press the alarm button. The advantage: there is no miscarriage of justice. The disadvantage is that the wearer can't start when he's unconscious. (2) Falls test system based on video image [4]. By installing a camera in a specific area, It films the area's human activities in the mainland. Then, the characteristics of the human body are extracted through related techniques. It is feasible to determine whether or not to fall by comparing these characteristics with the characteristics of normal human activities. Advantages: high accuracy, no need to wear, and can be detected when multiple falls occur simultaneously. Disadvantages: high manufacturing cost, large calculation and small coverage [5]. (3) Environment-type fall detection system [6]. Infrared sensors and microphones are installed in the area of human activity to obtain information such as motion status and attitude of human body. Then the information is analyzed to detect falls. Advantages: easy to use, no need to wear, relatively small amount of computation. Disadvantages: limited use area and high misdiagnosis rate [7]. (4) Wearable fall detection system [8]. To embed one or more sensor into a device suitable for human body wear. The fall detection algorithm can be detected according to the corresponding characteristic design. Advantages: relatively simple algorithm, small computation, convenient manufacturing; Disadvantages: need to wear, not easy and run jump and other daily sports division.

For the above problems, this paper designs a detection scheme based on NNs and IMU. It can separate ADL from falling. It carries the GPS and GPRS to make the old man get help in time when he falls.

2 Analysis of Detection Algorithm Based on Neural Network

2.1 NNs

2.1.1 Neurons

The core chip of the Arduino 101 is an Intel Curie module, with an Intel x86 Quark SE microprocessor containing 128 bytes (Neurons). A neuron is a cognitive and reactive memory which can autonomously evaluate the distance between an incoming pattern and a reference pattern stored in its memory. If this distance falls within a range called the influence field, the neuron returns a positive classification which consists of the

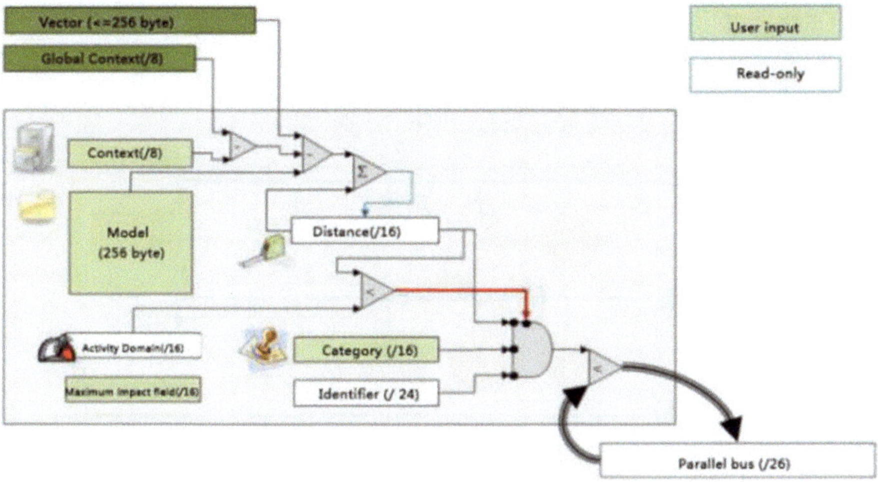

Fig. 1 The neuron receiving style and matching process

distance value and the category of its reference pattern [9]. Figure 1 shows the neuron receiving style and matching process.

2.1.2 Network Architecture

The fully parallel architecture of the NeuroMem chip is made possible because all the neurons are identical and do not require any controller or supervisor to interact with one another. They receive the same instructions and data over the neuron parallel bus and execute them at the same time. The execution of certain instructions requires that the Ready-To-Learn (RTL) and Committed neurons consult the response of one another over the neuron parallel bus. This interaction is necessary to build a consistent and adaptive knowledge.

The state of a neuron in the chain can be Idle, RTL or Committed. It is defined by the status of its daisy-chain-in (DCI) and daisy-chain-out (DCO) lines. The DCO of a neuron rises if its DCI is high and its category register is different from 0. As a result, the commitment of neurons is propagated automatically as examples are taught and retained. The RTL neuron also moves along until no more Idle neuron is available in the chain. Figure 2 shows the parallel lines of neurons.

The neural network is composed of N neurons:

- M committed neurons holding a reference pattern and a category value
- 1 Ready-To-Learn (RTL) neuron
- N-(M-1) idle neurons

The behavior of a neuron is function of its state in the chain (Table 1).

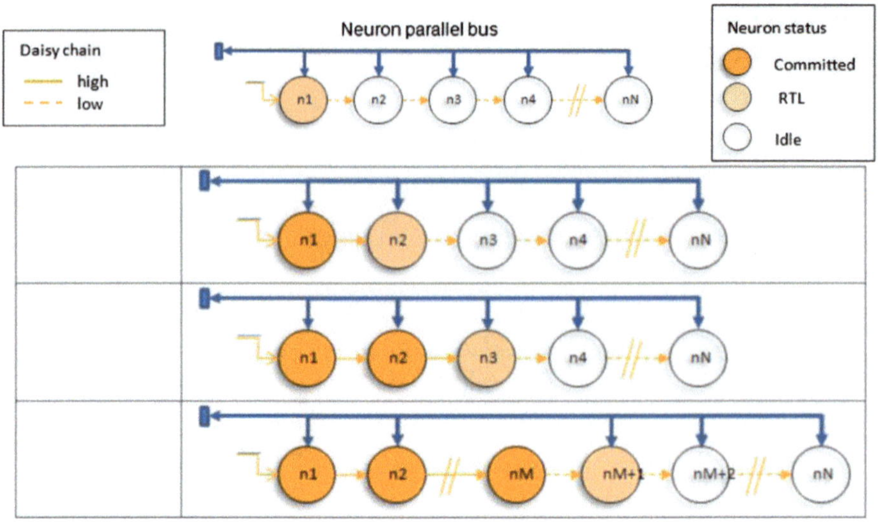

Fig. 2 The parallel lines of neurons

Table 1 Neuron chain function

Neuron state	Idle	Ready-To-Learn	Committed
Behavior	Does not respond to input patterns, but updates its global registers such as context, minimum influence field and maximum influence field	Holds the last input pattern. If a category is taught and not recognized by any of the committed neurons, the RTL neuron stores the category and becomes committed. The next idle neuron in the chain becomes the RTL neuron	All committed neurons attempt to recognize an incoming vector. Once committed, a neuron can only shrink its Influence Field and set its degenerated flag. Its status can return to Idle when it is instructed to forget
Memory	–	X	X
Distance calculator	–	X	X
Associative logic	–	–	X
Learning logic	–	X	X
Save and restore	–	X	X

Network segmentation into context: The neurons can be associated to different contexts and their use can be enabled or disabled by selecting a context value. For example, if an application uses two sensors such as a microphone to record a voice signal and a camera to record a video signal, two contexts will be used to toggle between two sub-networks: the neurons trained to recognize input vectors deriving from a voice signal and the ones trained to recognize input vectors deriving from a video signal. Furthermore, if the video camera is an outdoor camera, it might be useful to train it differently as a function of time of the day. Indeed the images will have nothing in common between day and night and the sensor might require different filter, gain and shutter adjustments. A context Day and context Night will allow training two sub networks based on the time of the day. In summary, usage of the context allows segmenting the network per family of input data. This segmentation can be based on the model of the input sensor, the settings of the input sensor, the feature extracted from the sensor data, the data length, the time of collection of the data and more.

A context is selected by writing a context value to the Global Context Register (GCR) of the chip. Any committed neuron with its own context register different from the global context register turns idle and does not participate in any learning or recognition operation. One exception: the context value 0 enables all the neurons without regards of their context. When a neuron gets committed, its Neuron Context Register (NCR) is set to the value of the Global Context Register (GCR). Whenever the GCR is changed, all the neurons with a different context value will not attempt to recognize any input vector, nor react to the learning of a new vector. They remain idle until the GCR is changed to a value matching their context value. A GCR equal to 0 activates all the neurons regardless of their context value.

The neurons belonging to a given context define a knowledge base. If the network has neurons belonging to N different contexts, it means that it contains N different knowledge bases. Finally if the Global Context is set to the value zero, all knowledge will be used in conjunction to recognize the input pattern.

2.2 *Algorithm Analysis*

2.2.1 RBF

RBF network can approximate any nonlinear function, can handle in the system is difficult to parse the regularity, has good generalization ability, and a fast convergent speed, has been successfully applied to nonlinear function approximation, time series analysis, data classification, information processing, pattern recognition, image processing, system modeling, control and fault diagnosis, etc.

This is a simple introduction to why RBF network learning converges faster. When one or more tunable parameters (weight or threshold) of the network affect any output, such networks are called global approximation networks. Due to each input, each weight of the network should be adjusted, which leads to the slow learning of the global approximation network. The BP network is a classic example.

If only a few of the connection weights affect the output in a local area of the input space, the network is called a local approximation network. The common local approximation network has RBF network, cerebellar model (CMAC) network, b-spline network, etc.

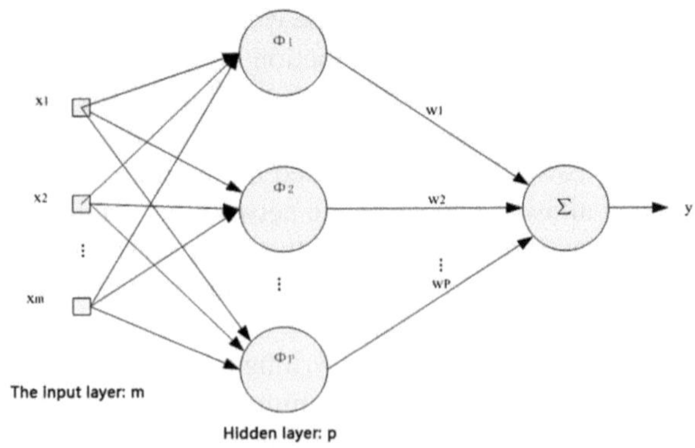

Fig. 3 Sample data processing

The radial basis function solves interpolation problem.

The interpolation function is required to pass each sample point, and the sample has a total of P.

$$F(X^p) = d^p \tag{1}$$

The method of RBF is to select the P basis function, each of which corresponds to a training data, in the form of each basis function.

$$\varphi(\|X - X^P\|) \tag{2}$$

Since the distance is radially homosexual, it is called the radial basis function. The magnitude of the difference vector, or 2 norm. The interpolation function based on the radial function is (Fig. 3).

$$\begin{aligned} F_X &= \sum_{P=1}^{P} w_p \varphi_P(\|X - X^P\|) \\ &= w_1\varphi_1(\|X - X^1\|) + w_2\varphi_2(\|X - X^2\|) + \cdots w_p\varphi_p(\|X - X^p\|) \end{aligned} \tag{3}$$

Input X is a vector of m dimensions, the sample size is P, P > m. You can see that the input data point is the center of gravity of the radial basis function. The role of the hidden layer is to map the vector from the low dimensional m to the higher dimension P, and the low dimension line is linearly separable in the case of high dimensions.

Plug in the interpolation conditions:

$$\begin{cases} w_1\varphi_1(\|X^1 - X^1\|) + w_2\varphi_2(\|X^1 - X^2\|) + \cdots + w_p\varphi_p(\|X^1 - X^P\|) = d^1 \\ w_1\varphi_1(\|X^2 - X^1\| + w_2\varphi_2(\|X^2 - X^2\|) + \cdots + w_p\varphi_p(\|X^2 - X^P\|) = d^2 \\ \vdots \\ w_1\varphi_1(\|X^P - X^1\|) + w_2\varphi_2(\|X^P - X^2\|) + \cdots + w_p\varphi_p(\|X^P - X^P\|) = d^P \end{cases}$$

$$\begin{cases} w_1\varphi_1(\|X^1 - X^1\|) + w_2\varphi_2(\|X^1 - X^2\|) + \cdots + w_p\varphi_p(\|X^1 - X^P\|) = d^1 \\ w_1\varphi_1(\|X^2 - X^1\|) + w_2\varphi_2(\|X^2 - X^2\|) + \cdots + w_p\varphi_p(\|X^2 - X^P\|) = d^2 \\ \vdots \\ w_1\varphi_1(\|X^P - X^1\|) + w_2\varphi_2(\|X^P - X^2\|) + \cdots + w_p\varphi_p(\|X^P - X^P\|) = d^P \end{cases} \quad (4)$$

In vector form for ϕ W = d. ϕ is a scale of the P symmetric matrix. It has nothing to do with the dimensions of X. When ϕ is reversible, there is a class of functions. When the input X is different, ϕ is inversible.

2.2.2 Regularization

The basic idea of regularization is to control the smoothness of the mapping function by adding a constraint containing the transcendental knowledge of the solution. Similar inputs correspond to similar outputs. Into the function F(x) achieves this by minimizing the target function below:

$$\min_F E(F) = \frac{1}{2}\sum_{P=1}^{P}\left[d^P - F(X^P)\right]^2 + \frac{1}{2}\lambda\|DF\|^2 \quad (5)$$

The first term is the mean squared error. Find the optimal approximation function minimize the mean square error. The second term is used to control approximation function smoothness. It's called the regularization term. λ is a regularized parameter. D is a linear differential operator. It represents the prior knowledge of F(x). F(x) of excessive curvature (smooth) usually have larger $\|DF\|$ DF. Therefore, there will be greater punishment.

Direct (5) solution:

$$F(X)\sum_{p=1}^{p} w_p G(X, X^p) \quad (6)$$

Weight vector:

$$W = (G + \lambda I)^{-1} d \tag{7}$$

$G(X, X^P)$ is called the Green function. G is called the Green matrix. The Green function is related to the form of operator D. When D has rotational invariance and invariance, $G(X, X^P) = G(\|X - X^P\|)$.

2.2.3 Supervised Learning

In general, the RBF function center, extension constant and output weight are trained by supervised learning algorithm. Experience an error correction learning process, It is similar to the learning principle of BP network. The definition objective function of gradient descent method is:

$$E = \frac{1}{2}\sum_{i=1}^{P} e_i^2 \tag{8}$$

e_i is the error signal when entering the *i*th sample

$$e_i = d_i - F(X_i) = d_i - \sum_{j=1}^{M} w_j G(\|X_i - c_j\|) \tag{9}$$

The upper-type output function ignores the threshold.

In order to minimize the objective function, the correction of each parameter should be proportional to its negative gradient:

$$\Delta c_j = \eta \frac{\partial E}{\partial c_j} \tag{10}$$

$$\Delta \delta_j = -\eta \frac{\partial E}{\partial \delta_j} \Delta \delta_j = -\eta \frac{\partial E}{\partial \delta_j} \tag{11}$$

$$\Delta w_j = -\eta \frac{\partial E}{\partial w_j} \tag{12}$$

The calculation formula is:

$$\Delta c_j = \eta \frac{w_j}{\delta_j^2} \sum_{i=1}^{p} e_i G(\|X_i - c_j\|)(X_i - c_j) \tag{13}$$

$$\Delta \delta_j = \eta \frac{w_j}{\delta_j^3} \sum_{i=1}^{p} e_i G(\|X_i - c_j\|) \|X_i - c_j\|^2 \tag{14}$$

$$\Delta w_j = \eta \sum_{i=1}^{p} e_i G(\|X_i - c_j\|) \tag{15}$$

The above objective function is the sum of errors caused by all training samples. The derived parameter correction formula is a batch type adjustment. That is, all samples are adjusted once in a round. The target function can also be the instantaneous value form, which is the error caused by the current input:

$$E = \frac{1}{2} e^2 \tag{16}$$

The revision of the parameters is as follows:

$$\Delta c_j = \eta \frac{w_j}{\delta_j^2} e \sum_{i=1}^{p} e_i G(\|X_i - c_j\|)(X_i - c_j) \tag{17}$$

$$\Delta \delta_j = \eta \frac{w_j}{\delta_j^3} e \sum_{i=1}^{p} e_i G(\|X_i - c_j\|)\|X_i - c_j\|^2 \tag{18}$$

$$\Delta w_j = \eta e G(\|X - c_j\|) \tag{19}$$

2.2.4 Algorithm Implementation

In the process of human body movement, different parts of the body have different movements. The different parts of the detection device will affect the data changes during the fall. In this paper, the center of gravity of the foot is selected. Choose rectangular coordinates. The vertical direction of the human body is the z-axis, the horizontal axis is the Y-axis, and the left side of the horizontal axis is the X-axis. The X, Y, and Z axes are perpendicular to each other. See Fig. 4.

The acceleration in the X-axis is ax, the acceleration in the y direction is ay, and the z-axis acceleration is az. The data collected by motion sensor is regarded as the acceleration value of the X-axis, Y-axis and z-axis.

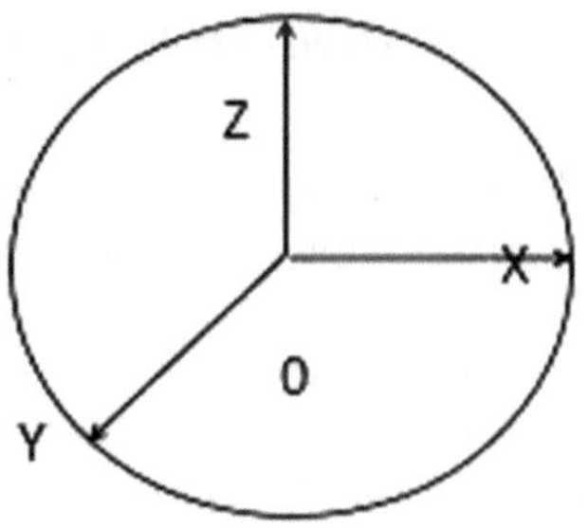

Fig. 4 The foot coordinates OXYZ

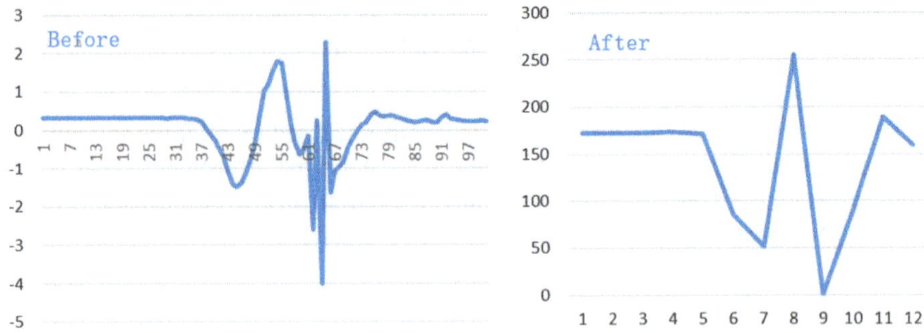

Fig. 5 The data before and after the feature extraction

The raw data is processed through the Curie simulator [10]. The simulator algorithm is mainly divided into the following steps: (1) Read data. (2) The data is segmented in the form of sliding window cutting. Each window has a length of 100. The length of each slide is 80. (3) Take the mean of the 100 data. And then you take the 100 Numbers and subtract them from the mean, and take the absolute value. You add up all the absolute values and you get a threshold. This threshold is compared with the standard threshold to determine whether the window is in a waveform. If the window is not in a waveform, the data will not be processed. If in a waveform, data dimension drop sampling is processed. To convert 100–12, convert the data between −4 to +4 and 0 to 255, which is the data planning. After the transformation is completed, the window continues to slide, repeating the above operations until all data is finished [10].

The following is an analysis of the X-axis data of a complete fall action. Figure 5 shows the data before and after the feature extraction.

It can be seen that after characteristic extraction software operation, the characteristics of data are basically not lost.

3 System Design Scheme

3.1 System Architecture

The system structure is shown in Fig. 6.

Preparation stage: The detected foot wears this system. The Arduino 101 is the first to collect movement information in real time through the acceleration sensor module. It uses neural networks to supervise learning and identification (Fig. 7).

The test phase: The Arduino 101 is analyzed and judged by the acceleration sensor module. When a fall is detected by a threshold algorithm, the buzzer makes a sound to prompt.

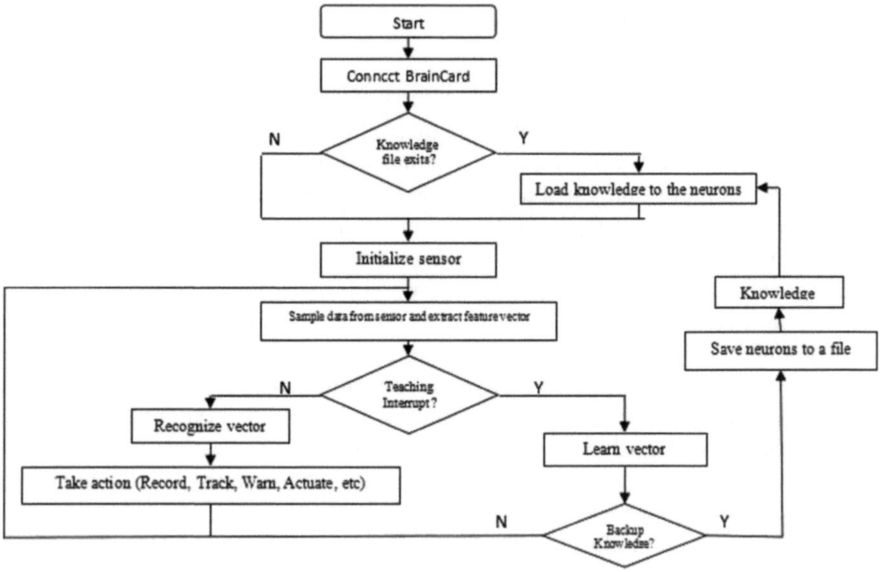

Fig. 6 Overall flow chart of the system

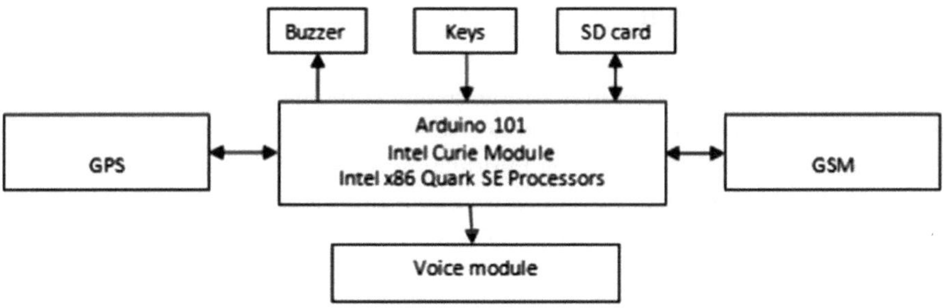

Fig. 7 General structure diagram of the system

The monitored system decides to make a fall, At the same time, the delay of the system, in the 30s, if support press cancel button, it is considered to be a miscarriage of justice or the person can help without help. Otherwise, the voice module will send a preset call for help to attract the attention of the surrounding people, and timely help the person who falls. Meanwhile, GPS modules are positioned to send messages or make calls to predetermined mobile phones through GPRS.

3.2 Microprocessor Selection

The microprocessor chose the ATMEL company's Arduino single-chip microcomputer. The series is based on an embedded application designed for high performance, low cost and low power consumption. Arduino 101 is an upgrade based on Arduino UNO, which is more functional and supports more functions and calls to libraries. It supports gesture recognition, acceleration sensors, and can handle the phone's bluetooth for the desired purpose.

The development board incorporates a low power high performance Curie module, which retains the interface and peripherals of Arduino UNO, as well as the bluetooth BLE module. Arduino 101 is the two core cpus for the asynchronous structure, which are x86 Quark SE and a 32 Bit arc architecture respectively. The two general a flash and ram, it comes with 14 digital pins, input pin and output pin, 6 analog pin to input pin, and used to download the USB interface and serial communication program. The neural network and 6-axis acceleration/gyroscope are collected in real time through the embedded 128 neurons. The SD card storage method is used to store the information in SD card. The computer is then analyzed and processed.

3.3 Sensor Selection

In order to detect the acceleration of the human body during the fall and the daily motion, this paper USES the 6-axis acceleration/gyroscope sensor embedded in Arduino to acquire the acceleration information of the XYZ triaxial direction. The sensor has the characteristics of low power consumption, high sensitivity and short response time. The seismic ability is strong and it is not easy to be damaged during the fall.

3.4 GPRS and GPS Selection

There are a lot of mature wireless networks such as Zigbee, WLAN, WIFI and so on, but these are not completely covered by our lives. In comparison, GPRS covers the widest range, requiring only cell phone signal retention for network communication.

SIM908 is a chip that integrates high-performance GSM/GPRS engines and GPS engines. The GSM/GPRS engines can work in GSM 850 MHz, EGSM 900 MHz, DCS 1800 MHz and PCS 1900 MHz. The GPS engine has the first-rate collection and tracking sensitivity, TTFF (time-to-first-fix) and accuracy. These features can well accomplish the task of locating and sending alarm signals. On SIM908 chip, GPS location and GSM communication can be completed simultaneously, which can greatly reduce the number and power consumption of system chips.

Table 2 Youth test data

Test events	Time	Alerts	Correct rate (%)
Standing	150	0	100
Walking	150	0	100
Running	150	0	100
Jumping	150	0	100
Lying down	150	15	90
Stopping	150	0	100
Falling	200	196	98

Table 3 Elderly test data

Test events	Time	Alerts	Correct rate (%)
Standing	100	0	100
Walking	100	0	100
Sitting	100	0	100
Lying down	100	4	96

4 The Experimental Results

This experiment selected 50 20–25 years old young man, respectively standing, walking, running, jump, lie down, stop, and fall (including forward fall down, fall back, side) test.

The experimental results are shown in Table 2: The system has high correct recognition rate. It can correctly identify most of the motion state, and the false report and low-rate. Among them, the high rate of false positives is the lie down behavior, mainly because lying down and backward falls the distinction of low.

The experimental results are shown in Table 3: For the elderly, the daily actions are mostly slow, and the daily behavior is extremely low.

5 Conclusion

In this paper, the fall-real-time detection device based on neural network and acceleration sensor is designed. The Arduino 101 processor with a Curie kernel is used to monitor the acceleration value of the three axes in real time and analyze the data. An effective detection method and a better detection alarm system are proposed. The location of the accident can be obtained by GPS module and the alarm information is automatically issued. The experiment shows that the device is easy to wear and has a high recognition rate, so that the elderly can be found to be saved in time when the accident occurs.

References

1. Dong, L., Zhou, J., Xu, G.: Progress in the research on the fall care of elderly residents. China Rehabil. Theor. Pract. (01), 30–32 (2012)
2. Cai, Q., Xiang, L.: Analysis of the current situation of rural aging in China and the analysis of development trends—based on the data analysis of the sixth censuses. J. Hubei Vocat. Tech. Coll. (01), 99–104 (2013)
3. Jiecheng, H.: The current situation, trend and suggestion of population aging in China. China Econ. Trade Guide (12), 59–62 (2017)
4. Yu, X.: Approaches and principles of fall detection for elderly and patient. In: 10th International Conference One-Health Networking, Applications and Services, 2008. Health Com. IEEE, pp. 42–47 (2008)
5. Vaidehi, V., Ganapathy, K., Mohan, K., et al.: Video based automatic fall detection in indoor environment. In: 2011 International Conference on Recent Trends in Information Technology (ICRTIT). IEEE, pp. 1016–1020 (2011)
6. Li, Y., Ho K.C., Popescu, M.: A microphone array system for automatic fall detection. IEEE Trans. Biomed. Eng. **59**(5), 1291–1301 (2012)
7. Scott, T.E.: Bed exit detection apparatus. US Patent 6,067,019 (2000)
8. Hwang, J.Y., Kang, J.M., Jang, Y.W., et al.: Development of novel algorithm and real-time monitoring ambulatory system using Bluetooth module for fall detection in the elderly. In: Engineering in Medicine and Biology Society, 2004. IEMBS'04. 26th Annual International Conference of the IEEE. IEEE, vol. 1, pp. 2204–2207 (2004)
9. XiaohuaQin, C., Yuan, K., et al.: A mobile health monitor system for the elderly. Chin. J. Med. Phys. **28**(1), 2407–2410 (2011)
10. Luo, Q., ZhenYan, X., Peng, Y., Peng, X.: A dynamic distance estimation method based on sliding window mode. J. Instrum. **36**(03), 499–506 (2016)

Data Reliability-Aware and Cloud-Assisted Software Infrastructure for Body Area Networks

Joseph Reeves, Carlos Moreno, Ming Li, Chengyu Hu and B. Prabhakaran

Abstract Cloud-assisted body area networks have been the focus of researchers in past years as a response to the development of robust wireless body area networks (WBANs). While software such as Signal Processing in Node Environment (SPINE) provide Application Programming Interfaces (APIs) to manage heterogeneous biomedical sensor networks, others have focused on data analysis within networks, laying the groundwork for a scalable cloud-assisted infrastructure. However, recent work in cloud-assisted architectures have revealed several issues, specifically pertaining to applications in the biomedical field. Data-reliability and context aware adaptations are paramount to the success of biomedical applications, due to the field's data quality needs when seeking in-depth analyses of the data sets. In addition, the cloud server must have a way to organize heterogeneous biomedical body sensor data and perform different types of biomedical body sensor research. The software infrastructure presented in this paper proposes several feedback mechanisms built off of dynamic variables within the system including data importance, data quality and network layout in order to provide researchers an optimal quality of service. The implementation of a domain specific language (DSL) will enable diverse biomedical

J. Reeves (✉) · C. Moreno (✉) · M. Li (✉)
Department of Computer Science, California State University,
2576 E. San Ramon, Fresno, CA 93740, USA
e-mail: josephreeves@mail.fresnostate.edu

C. Moreno
e-mail: mmxzbnl@mail.fresnostate.edu

M. Li
e-mail: mingli@mail.fresnostate.edu

C. Hu
School of Computer Science, China University of Geosciences, Wuhan, Hubei, China
e-mail: huchengyu@cug.edu.cn

B. Prabhakaran
Department of Computer Science, The University of Texas at Dallas, 800 W. Campbell Rd, Richardson, TX 75080, USA
e-mail: praba@utdallas.edu

G. Fortino and Z. Wang (eds.), *Advances in Body Area Networks I*,
Internet of Things, https://doi.org/10.1007/978-3-030-02819-0_23

data processing operations. Furthermore, a robust set of APIs will give researchers the ability to build flexible and unique biomedical applications.

Keywords Body sensor networks · Cloud-assisted · Wireless body area networks

1 Introduction

In the domain of wearable computing, WBANs have been the subject of research pertaining to data sensing, analytics, quality of service (QoS) and other interesting topics due to their exciting potential in a plethora of fields, namely medical. Researchers have addressed the problems of energy efficiency [1, 2], refined MAC protocol [3] and enhanced data analytics [4] in recent years, but off-the-shelf sensors and smartphones still do not provide sufficient support for deploying scalable and complicated biomedical systems.

The reason for this is threefold: (i) in addition to the basic operations of collection/display/visualization of sensor reading and data analysis, significant software support is required to allow development of complicated strategies, especially from the data analytics aspect; (ii) there is no data reliability-aware and cloud-assisted software infrastructure that facilitates design, implementation and evaluation of new algorithms and protocols in body sensor system research, making it difficult to meet the needs from the biomedical research community; and (iii) the current trend of running specific goal-based applications are lacking a systematic, data-analysis driven body sensor framework that incorporates transmission and analytics for maximum data reliability [5], which is imperative for heterogeneous biomedical sensor systems.

Several software tools attempted to address subsets of the issues laid out above; some attempts arising from Shimmer Sensing [6], Qualcomm [7], Intel [8] and Samsung [9]. However, these tools are centered around sensor connection/control, data receiving and visualization but are deficient in research driven components. Our proposed software infrastructure resolves these concerns by integrating various aspects of heterogeneous biomedical body sensor systems, from the sensor side's data preprocessing, transmission and critical event detection, to the data aggregator's data reliability evaluation/enhancement, data analysis and sensor control, and finally the cloud server's data storage, retrieval and extensive analysis.

Systems have been proposed to enable cloud assisted networks; specifically, BodyCloud [10] provides a general-purpose software that covers a wide range of sensors and provides APIs for the creation of new biomedical applications. Our system expands on the foundations [11] laid by other cloud assisted architectures, and addresses some of the key issues they have identified. In terms of flexibility, our system includes two types of data-aggregators including PC and Android (cell phones or tablets), which allows for a flexible research environment. Scalability is another desired feature of this architecture, with tools allowing new sensor hardware to easily be inherited into the system and a template enabled Domain Specific Language (DSL) supporting the addition of new data analytics algorithms into the cloud side. Finally, this system will focus on enhancing data-reliability and transmission in real-time with a context-aware approach.

The key components our system will provide include the following: (i) comprehensive APIs that enable the ability to conduct in-depth research at sensor, data aggregator and cloud sides for data processing, networking, storage, indexing/retrieval and analysis; (ii) built-in and configurable reliability enhancement mechanism especially for biomedical sensor systems; (iii) a function rich cloud platform that enables scalable, long term and collaborative research among members of various disciplines; and (iv) real-time analytics from multiple, multimodal sensor data. Among the projects directly benefiting from this system are Dietary Assessment, Body Sensor-based In-home Stroke Rehabilitation, Event Monitoring for Remote Patient Center and Computer-aided Fall Selection.

This paper is organized as follows. Section 2 reviews related works. Section 3 introduces and describes the proposed reliability-aware and cloud-assisted software infrastructure for body area networks. Section 4 describes the sensor side instrument. Section 5 describes the data aggregator instrument. Section 6 describes the cloud server. Section 7 discusses several design and implementation issues and, finally, Sect. 8 summarizes the work.

2 Related Works

Among many potential communication technologies, Zigbee [11] and Bluetooth are most widely deployed. Zigbee is a very low power, collision avoidance protocol optimized for lower power sensors. It has developed a health care specific protocol and is compliant with all IEEE 11073 devices as well as most other IEEE 802.15.4 wireless devices. Bluetooth supports high-bandwidth and many several existing devices and has a health care compliant version defined, but has very high power requirements and uptime for the radios. Bluetooth Low Energy [12] is a new proposed system from Bluetooth which will have lower energy requirements and still be interoperable with Bluetooth Classic.

SPINE [13], a TinyOS based platform, has enabled the implementation of a heterogeneous body sensor network by abstracting the hardware level of multiple sensors such as TelosB and MicaZ and creating an easy to use software. Furthermore, SPINE contains APIs for general-purpose processing functions for reduction of transmitted sample data such as average, median and RMS. Separately, Shimmer Research [14] delivers two sets of software tools: development drivers for LabView, MATLAB, Android and C#; standalone software such as Shimmer Connect, Shimmer Log and Shimmer Plot for easy connection, storage and visualization of sensor data. While these drivers and software provide a good combination of tools for easy direct use or further API developments, the research community needs a scalable, robust instrument to meet their fast research and development needs.

One WBAN architecture that incorporates SPINE is DexterNet [15]. The system takes a real-time approach and seeks to provide an open-source platform that allows indoor and outdoor persistent human monitoring. The Body Sensor Layer, Personal Network Layer and Global Network Layer provide a three-tier architecture to address the different API needs from the sensors, data aggregator and server. This system is both scalable and reconfigurable in real-time due to the versatility of the layered approach. CodeBlue [16], proposed by Fulford-Jones and Malan, presents an ad hoc infrastructure for emergent medical care. In this project, several types of body sensors (e.g., pulse oximeter, ECG/EKG sensor) are individually connected to Zigbee enabled radio transmitters. Due to the ad hoc architecture and the capability of self-organizing, CodeBlue yields scalability for network expanding and flexibility to connect various wireless devices. Jiang and Cao proposed CareNet [17], an integrated wireless environment used for remote health care systems. CareNet offers features such as high reliability and performance, scalability, security and integration with web based portal systems.

Combining Cloud Computing and Wireless Sensor Networks (CC-WSN) [18] and Open Sensor Web Architecture (OSWA) [19] propose architectures that integrate wireless sensor networks (WSN) with the cloud. CC-WSN has base services including sensor data management run on Google AppEngine and Microsoft Azure, filter chain and filter management and user services including visualization and notification. OSWA provides an architecture for integrating sensor networks with various distributed computing platforms like SOA and Grid-Computing. BodyCloud [10] proposes a SaaS approach that supports cloud-assisted BSN applications. The purpose of the software is to allow fast prototyping of cloud-assisted BSN applications and flexibility among architectural components utilizing web standards-based procedures and scalability aided by the Google App Engine PaaS infrastructure. Their paper discusses the issues of providing body-side and cloud-side context-aware sensing and adaptation, an issue our proposed architecture will address.

A survey of the architectures presented above found there are several issues still needing to be explored [20]. One such issue is providing an efficient collection of body senor data and implementing an adaptive sensing mechanism. The data transmission throughout the network, including sensor to aggregator and aggregator to cloud, still has potential to be optimized. Contextual adaptation is another area of concern, relating to dynamic adaptation in an array of services depending on the current network and sensing context. Finally, a general software that provides ease-of-use to programmers for developing specific healthcare related apps should be further investigated.

To address data reliability between sensor and data aggregator, an analysis driven adaptive framework [5] will be expanded upon. In this framework, a scheme to group data based on importance regarding the effect of data loss on analysis algorithms, and a hybrid transmission protocol developed to meet analysis needs are investigated. Furthermore, a multi-level importance ranking [21] can allow a fast organization of sensor data into priority and non-priority packets, which will then be transmitted progressively in response to data reliability demands and network constraints. Finally, the concatenation of small data frames from sensors with low sample rates like ECG/EKG, along with an adaptive transmission scheme can provide loss resilience among the sensor network [22]. By exploring these papers, a comprehensive system with adaptive, context aware feedback and real-time modifications can be created.

3 Data Reliability-Aware and Cloud-Assisted Software Infrastructure for Body Area Networks

3.1 System Model

The model seen in Fig. 1 consists of three basic parts: the sensors, the data aggregator and the cloud along with the network built between the adjacent parts. With the use of SPINE, a wide range of sensors with data acquisition from blood pressure to motion can be used to build a scalable WBAN. Data will be passed from the sensors to a data aggregator running on PC or Android, allowing for a flexible sensing environment. Finally, a cloud server receives data from the data aggregator through a streaming service from either a cellphone/tablet or PC, and provides a web portal so that users can access a pool of data analysis algorithms to analyze their data. This basic system model provides an integrated platform with context-aware adaptations and data-reliability built in as key features, providing enhanced scalability and QoS.

3.2 System Design

Each of the three components within the system contains a unique set of APIs designed to address the issues of data-reliability enhancement, data analysis and data transmission among others.

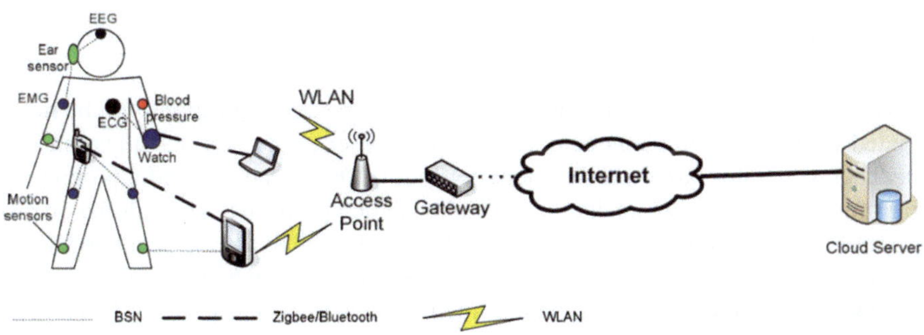

Fig. 1 System model

3.2.1 Sensor Side

Performs lightweight tasks such as: adjusting sample rate; monitoring critical activities from a preliminary screening of data samples; preprocessing data samples; and data transmission with either TCP (for reliable assurance) or UDP (for best efforts). These functions are reliability aware, meaning they incorporate the feedback messages received from the aggregator and adapt their behavior for enhanced data reliability.

3.2.2 Data Aggregator

Performs more intensive data analysis and critical event detection through mobile apps. Also, it provides data collection from various heterogeneous sensors, sends feedback messages on data reliability for specific sensors, sends control messages to turn on/off specific sensor data collection, streams data to the cloud server and enables data visualization. The data aggregator will also provide a context-aware streaming service that monitors battery life, network connection and data analysis.

3.2.3 Cloud Server Platform

Performs biomedical sensor data storage, indexing and analysis. Through a user-friendly web portal, researchers can test run representative data analysis algorithms for specific collected sensor data; submit their own analysis algorithms to test run sensor data of their interest; perform data visualization to view results. The cloud server also provides an API for real-time analysis that produces feedback for the data aggregator.

3.3 Contributions

During the development of this instrument, several research questions will be investigated through the implementation of APIs. For example, the actualization of a feedback generator based on Data Quality Index (DQI) readings and a progressive transmission formula, which has been detailed as an approach [21]. Building on progressive transmission research will allow data quality to be assessed and updated in real-time. Data quality assessment on the data aggregator will have access to the DQI of all sensors on the network, as well as their relative importance levels, battery life, desired DQI and analysis of current data. With this information, the data aggregator will dynamically adapt transmission schemes and pre-processing on sensors to provide the best possible data quality for researchers depending on the given network. Furthermore, the APIs within the sensor side and data aggregator will allow high levels of customization and flexibility in this process.

In the medical field, the instrument will improve sensor-based event monitoring for remote patient care. The scalable biomedical body area network computing platform can eliminate distance barriers and provide health care remotely. Another important result of this instrument will be the open source, user friendly and scalable cloud computing platform for sensor data storage, indexing, and analysis. As researchers contribute to the algorithm set, the platform will continue to grow in value over time. In addition, as data sets and analyses from various sensor networks and patient scenarios are consolidated, researchers will have the opportunity to compare their findings with the use of our web portal and cloud service.

4 Sensor Side Software Instrument

4.1 Body Sensor Design

The sensor side software instrument's tasks include data sampling, event detection, data analysis, pre-processing and data transmission, as well as a data receiver for receiving and interpreting feedback messages. The SPINE open-source software, built on TinyOS [23] provides a set of APIs that allow heterogeneous data collection, multiple forms of data analysis, threshold based event-detection and some forms of data reliability enhancement. This instrument will build off SPINE, seeking to improve the data-analysis and event detection APIs, while also addressing data reliability in a more comprehensive manner.

Figure 2 depicts the modules used to support the APIs the sensor side software provides, as well as gives an intuitive understanding of the scenario in which the system operates. Upon startup, the sensor will have several fields set either automatically, or by users, including sampling time, critical event thresholds, analytics functions and pre-processing tasks. Data is sampled through an abstraction that provides sensor independent functionality. Analysis is then performed as requested through the data analytics API. If a critical event is detected during analysis, an alert message is instantly created and forwarded to the data transmission module with top priority. Separately, data is pre-processed and packets are formed based on a data reliability metric decided users, as explained in Sect. 4.1. Throughout the run-time of the sensor network, feedback can be sent directly from the data-aggregator to sensors. The feedback message is parsed in the streaming adaptation manager, and the preprocessing module will accommodate the changes requested of the feedback message in a real-time effort.

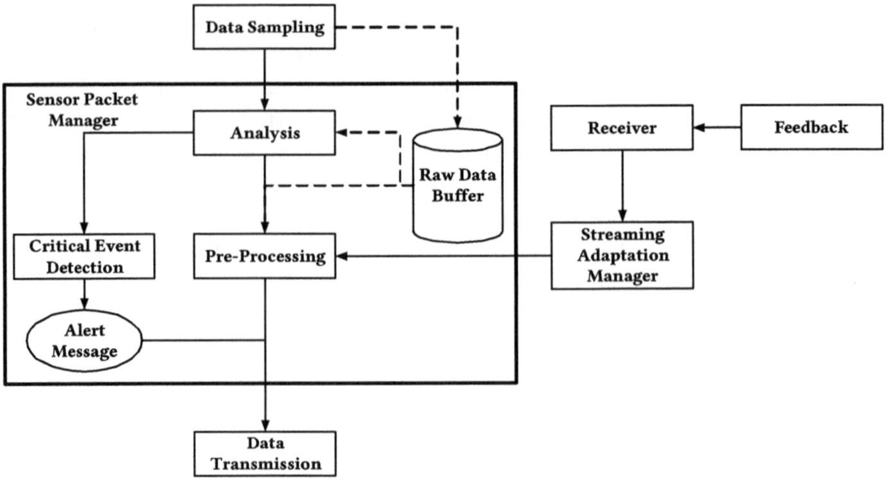

Fig. 2 Body sensor modules

4.2 APIs

4.2.1 Data Sampling API

Manually (through configuration) or automatically (through algorithm) manipulates the frequency of data sampling. This allows the system to find the appropriate sampling rate for a specific application dynamically and sensors to be emulated for experimental purposes. One such example is researchers using multiple TelOsb motes, changing the sampling rates of some motes to emulate EEG or EMG sensors based on their typical frequency.

4.2.2 Data Analytics API

Includes three subsets of APIs: (1) lightweight pre-processing API, which handles how sensor data samples can be processed before forwarding to the transport layer for delivery. Such processing includes extracting samples of certain properties (e.g. min, max, or importance) and reducing data samples to be transmitted based on some criteria (e.g. estimation accuracy by adjacent samples). (2) Lightweight feature extraction API, which provides native support to extract a variety of simple features from a (user-defined) time-series (i.e. sliding window or sensor data segmentation) of sensor data, such as average, standard deviation, time between peaks, binned distribution, etc. (3) lightweight event detection API, which handles lightweight analysis of data samples (e.g. threshold-based detection algorithms) to detect emergent events. The API consists of both general and application specific functions. General functions will detect abnormal sample values based on use configuration of sensors. Application specific functions will focus on identifying certain simple patterns indicating the occurrence of specific events.

4.2.3 Data Transmission API

Handles how sensor data samples are treated for network communication. This includes reliable delivery with re-transmissions, best-effort delivery or a combination thereof. In addition, this API provides a streaming adaptation manager. Several data reliability metrics will be investigated such as sample importance rankings and DQI value for the sensor. The data aggregator will access the feedback manager through feedback messages, changing packet priorities, sample importance and packet frequency, enabling real-time improvements of data quality.

5 Data Aggregator Software Instrument

5.1 Data Aggregator Design

The data aggregator is involved with data collection (from the sensor), data analysis, critical event alert, reliability enhancement, visualization and streaming (to the cloud server). The advantageous set of resources including processing power, energy and storage a PC or phone has relative to a sensor permits higher level tasks to be performed. Larger data sets can be analyzed with more sophisticated algorithms, critical event detection can provide more useful data-specific trained detection and data reliability can be effectively calculated and dynamically adapted through feedback messages sent directly to sensors.

The modules supporting the APIs for the data aggregator are shown in Fig. 3. A typical scenario is as follows: data is received from the sensor and is stored locally while also being passed to the processing module. The processing and analysis modules support the data analyzer API and perform its various tasks. Data will then get metadata attached so it may be streamed to the server with the appropriate form. The backup streaming module will adjust sending frequency and data load based on various factors such as battery life and network connection. In addition, event detection will check if a critical event has occurred, notifying the user immediately. Upon analysis, data reliability is calculated and the reliability enhancement module queries the sensor profile and current DQI status to determine the appropriate feedback response which it derives from the feedback table and sends directly to the sensor. Visualization can be obtained by the user at any time, and can show raw-data as well as any analytics acquired through the processing and analysis stages. Finally, the data aggregator may receive messages from the server and display them to users.

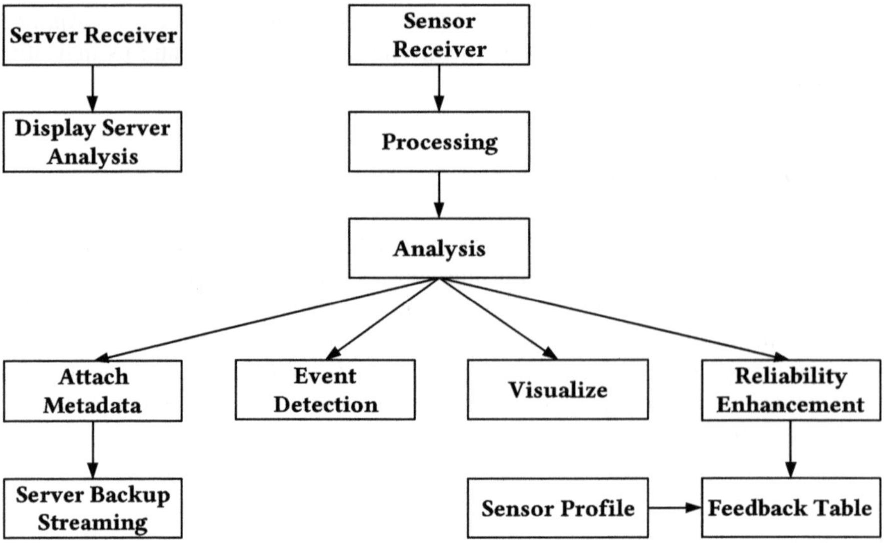

Fig. 3 Data aggregator modules

5.2 APIs

5.2.1 Data Collection API

Receives data from sensors, stores it locally, forwards it to the data analyzer and/or streams to the cloud server. In the case when a critical event occurs, it also triggers alarm messages for emergencies. Separate data analyzers and other modules may be implemented through additional mobile apps for flexibility and scalability, thus a system service will be created to move data into requesting apps' local storage.

5.2.2 Data Analyzer API

The data analyzer provides (1) a more comprehensive pre-processing API that can utilize complex processing methods such as Bayesian theorem-based data cleaning techniques and Wavelet Transform-based data reduction algorithms. Advanced processing like sample packetization will be included to further support quality of data analysis. (2) A comprehensive feature extraction API will provide optimized implementation for representative algorithms in time series mining (e.g. spectrum transform, wavelet transform, piecewise polynomials, symbolic mapping) and machine learning (e.g. regression, probabilistic generic models, discriminative models). Users may either configure the behavior of the algorithm by parameter configuration or modify the source code to implement their own algorithms. (3) A comprehensive event detection API. The event detection at the data aggregator is more capable of identifying scenario-specific events due to complex learning-based algorithms. (4) A visualization API offering reliable and easy accessible methods to visualize data at the data aggregator.

5.2.3 Data Streaming API

Sends sampled data to the cloud server. Data transmission is reliant on energy efficiency and resource constraints for mobile devices [24]. Functionality will be incorporated to adapt streaming based on battery power and network connectivity (i.e. offline or online with WiFi/3G/4G). Due to the excess storage on a phone, streaming will be kept as a background thread during optimum conditions (i.e. charging and on WiFi). Streaming adaptations in response to current resources can be controlled directly by users.

5.2.4 Feedback Based Reliability Enhancement

Analyzes the quality of data received at the aggregator and sends feedback if the quality needs to be improved. A sensor profile containing different sensor information and activities such as identity, algorithm used, reliability of data level, importance, etc. will be used to ensure a scalable and robust design for calculating feedback. The metric used to analyze reliability is the DQI from earlier work on data reliability in sensors [21]. In this work, a design was presented for progressive data-transmission based on sample layering, which will be implemented on the server side to allow for the adaptive implementation the data aggregator provides. The feedback generator will evaluate the current DQI in reference to the sensor profile, and create a feedback message from on a customizable feedback table such as Table 1. Feedback will not only be implemented on a per-sensor basis, but will also take into account the collective sensor network along with respective importance levels of sensors, and adapt accordingly.

Table 1 Sample action codes for feedback messages

Action code	Action
0	No adaptation required
1	Reliable transmission required
2	Unreliable transmission acceptable
3	More data samples
4	Less data samples
5	Stop data transmission
6	Resume data transmission

6 Cloud Server Software Instrument

6.1 *Cloud Server Design*

The cloud infrastructure will achieve large scale, distributed and parallel biomedical data management and analytics. The assistance given by the hardware resources in the cloud in conjunction with a robust set of APIs and a DSL will allow researchers a new opportunity to creatively compare data sets with algorithms tuned for varying fields of research. The main focus of the cloud infrastructure is to enable this with a template-based approach that will allow easy insertion of new algorithms, as well as a shared storage protocol to provide ubiquitous access to multiple data sets. Further, a web portal will present an entry point for professionals to manipulate data and run complex analyses while not being in direct contact with the patient wearing the body sensors, framing a discrete relationship between the two parties.

The modules presented in Fig. 4 provide the basis for the cloud server's flexible and scalable design. Data received from the data aggregator will be formatted and stored on the server. Through the web portal, a user can request real-time analysis as data is being received, triggering result messages as desired. Furthermore, users will have the opportunity to upload algorithms or select algorithms from the expansive existing libraries and make analysis requests. The results will then be displayed in the web portal and the user can retrieve the data in a standard format. In addition, the user can request raw data from the cloud server. This provides an opportunity not only for researchers to analyze multiple data sets at once, but for professionals to make real-time judgements based on patient data.

6.2 *APIs*

6.2.1 Analysis Algorithm API

The goal of this subcomponent is to provide APIs to both data-intensive and computational-intensive algorithms implemented in existing and new libraries focusing on biomedical data indexing, retrieval and analysis. The analysis APIs will draw from MPI [25] for computational-intensive algorithms and the MapReduce model [26] for data-intensive algorithms. Sample algorithmic solutions provided by the API include distributed clustering algorithms (e.g. K-means algorithm), parallel EM algorithms, parallel SVM algorithms, distributed random forest algorithms and distributed deep learning algorithms (e.g. Deep Convolutional Network, Auto Encoders, Restricted Boltzmann Machines). Another important feature is the API extension, giving users the ability to add new extensions using existing ones as templates. Categorized API adaptors specialized for different types of data analysis algorithm will make the process more expedient for users. The analysis API can further be broken down into real-time analysis and on-demand analysis. On-demand analysis presents

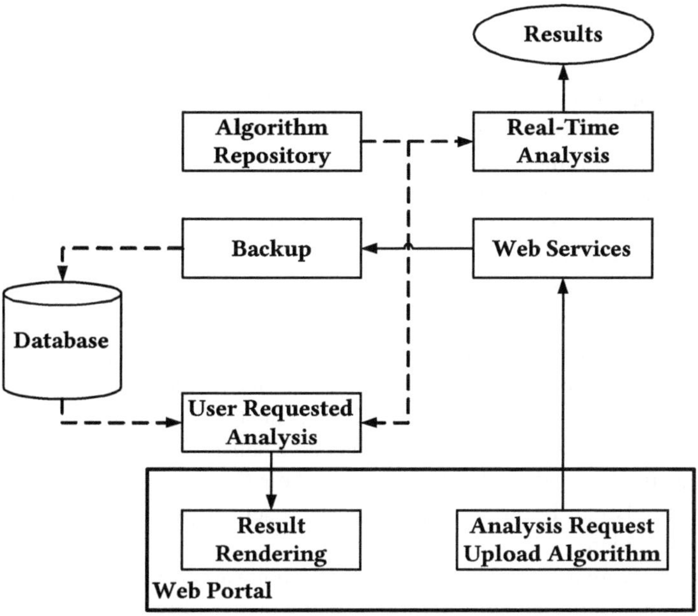

Fig. 4 Cloud server modules

an opportunity for users to manually configure algorithms and perform their analysis on data sets, with the results being returned with the aid of a visualizer. The real-time analysis runs user specified algorithms on data as it is received, directing results to the data aggregator or to a professional set to be monitoring the data.

6.2.2 Adapter Service API

This component provides two services to users, with the first being the retrieval of data from the cloud server using JSON. The second service is to orchestrate data processing workflow that is suited for biomedical body sensor applications. The system will be able to support computation-intensive, data-intensive and service-oriented applications in which the user can perform data processing operations modeled as abstract pipeline stages. A compiler will construct the pipeline of data processing workflow from the source program and schedule parallel and distributed computation operations on different platforms such as cloud services.

7 Discussions

The goal of a uniform language abstraction is to provide a DSL to support different types of biomedical body sensor research, heterogeneous biomedical body sensor data, and diverse biomedical data processing operations. A biomedical DSL with uniform language interface [27] will support multiple data sources such as time series data from accelerometers, image data from vision sensor networks, as well as different types of biomedical body sensors.

Regarding data reliability on a per sensor basis, specific thresholds must be determined to enable a feedback system to adjust the network appropriately. Several issues come into play, including sensor importance and sensor sampling time, with sensors like EKG and ECG having longer sampling intervals. In addition, each type of sensor data requires its own form of analysis, which must be weighed when determining sample importance. Due to these concerns being sensor and data focused, a domain expert from the biomedical field will aid in their fine-tuning. Once this process is done, thresholds can be established within the sensor profile, because most body sensors are standard.

A context-aware approach searches the dynamic variables within the network and adapts the system automatically to provide better data quality, analysis and streaming. Several variables will be considered, including the number of sensors within the network and their relative importance and data quality threshold, along with sensor battery life and connectivity to the data aggregator. Another point of discussion is data streaming from an Android device, with constraints being battery, network connection (WiFi/LTE/3G/4G), data importance and user requests. This system can be optimized by running a set of experiments with varying conditions to find the most intelligent adaptation mechanism, while also allowing user configuration.

Flexibility is a necessary component of systems that provide general purpose software to meet a range of research and real-time needs. Android has the largest market share for mobile phones and tablets and is supported by most existing platforms such as Shimmer and SPINE. Further, Android code (written in Java) can be ported to PC or Linux easily, and vice versa. These qualities make Android a more suitable choice over iOS devices for this project.

8 Summary

The cloud assisted WBAN architecture presented in this paper works to address several of the issues currently inhibiting similar architectures. It enhances data transmission within the network by providing a robust feedback mechanism for assuring optimized data quality among different sensor types based on several dynamic variables within the network along with a context-aware adaptation manager that controls streaming from the data aggregator to the cloud. In addition, each component, sensor side, data aggregator and cloud server contains a complete set of APIs allowing user friendly development platform focused on data quality and analysis. Lastly, the algorithm repository supported by a DSL on the cloud server will allow research focused applications a greater flexibility and scalability than they have seen in the past.

Acknowledgements This work was supported by the United States National Science Foundation, CNS division (Award No. 1626586) and NSF of China (Grant No. 61305087).

References

1. El-Hoiydi, A., Decotignie, J.D.: WiseMAC: an ultra low power MAC protocol for the downlink of infrastructure wireless sensor networks. In: 9th International Symposium on Computers and Communications, pp. 244–251. IEEE, Egypt (2004)
2. Polastre, J., Hill, J., Culler, D.: Versatile low power media access for wireless sensor networks. In: 2nd International Conference on Embedded Networked Sensor Systems, pp. 95–107. ACM, New York (2004)
3. Fang, G., Dutkiewicz, E.: BodyMAC: energy efficient TDMA-based MAC protocol for wireless body area networks. In: 9th International Symposium on Communications and Information Technology, pp. 1455–1459. IEEE, South Korea (2009)
4. Garg, M.K., Kim D.J., Turaga D.S., Prabhakaran B.: Multimodal analysis of body sensor network data streams for real-time healthcare. In: International Conference on Multimedia Information Retrieval, pp. 469–478. ACM, New York (2010)
5. Li, M., Cao, Y., Prabhakaran, B.: A data analysis driven streaming framework for body sensor area networks. In: 8th International Conference on Body Area Networks, pp. 205–208. ICST, Belgium (2013)
6. Shimmer Enabling Softwares. http://www.shimmersensing.com/research-and-education/applications/data-aquisition-software-sensor-systems/
7. The 2net Platform from Qualcomm Life. http://www.qualcommlife.com/wireless-health
8. Intel IoT Developer Kit. https://software.intel.com/en-us/iot/devkit
9. Samsung Architecture for Multimodal Interactions: An Open Data Platform for Innovation. http://www.samsung.com/us/globalinnovation/innovation_areas/digital-health
10. Fortino, G., Parisi, D., Pirrone, V., Fatta, G.D.: BodyCloud: a SaaS approach for community body sensor networks. Future Gener. Comput. Syst. **35**, 62–79 (2014)
11. Baronti, P., Pillai, P., Chook, V.W.C., Chessa, S., Gotta, A., Hu, Y.F.: Wireless sensor networks: a survey on the state of the art and the 802.15.4 and ZigBee standards. Comp. Commun. **30**, 1655–1695 (2007)
12. Bluetooth Low Energy Core Specification Version 4.0. http://www.bluetooth.com/English/Technology/Works/Pages/Bluetooth_low_energy_technology.aspx
13. SPINE Project. http://spine.deis.unical.it/
14. Shimmer: a small wireless sensor platform that can record and transmit physiological and kinematic data in real-time. http://www.shimmer-research.com/
15. Kuryloski, P., Giani, A., Giannantonio, R., Gilani, K., Gravina, R., Seppa, V.P., Seto, E., Shia, V., Wang, C., Yan, P., Yang, A.Y., Hyttinen, J., Sastry, S., Wicker, S., Bajcsy, R.: DexterNet: an open platform for heterogeneous body sensor networks and its applications. In: 6th International Workshop on Wearable and Implantable Body Sensor Networks, pp. 92–97. IEEE, California (2009)
16. Malan, D., Fulford-Jones, T., Welsh, M., Moulton, S.: CodeBlue: An ad hoc sensor network infrastructure for emergency care. In: International Workshop on Wearable and Implantable Body Sensor Networks. United Kingdom (2004)

17. Jiang, S., Cao, Y., Iyengar, S., Kuryloski, P., Jafari, R., Xue, Y., Bajcsy, R., Wicker, S.: CareNet: An integrated wireless sensor networking environment for remote healthcare. In: 3rd International Conference on Body Area Networks. ICST, Belgium (2008)
18. Kurschl, W., Beer, W.: Combining cloud computing and wireless sensor networks. In: 11th International Conference on Information Integration and Web-based Applications & Services, pp. 512–518. ACM, New York (2009)
19. Cheng, H., Guo, R., Su, Z., Xiong, N., Guo, W.: Service-oriented node scheduling schemes with energy efficiency in wireless sensor networks. Int. J. Distr. Sens. Netw. **10** (2014)
20. Fotino, G., Fatta, G.D., Pathan, M., Vasilakos, A.V.: Cloud-assisted body area networks: state-of-the-art and future challenges. Wirel. Netw. **20**, 1925–1938 (2014)
21. Li, M., Cao, Y., Prabhakaran, B.: Multi-level sample importance ranking based progressive transmission strategy for time series body sensor data. In: 16th International Symposium on a World of Wireless, Mobile and Multimedia Networks, pp. 1–3. IEEE, Massachusetts (2015)
22. Read, N., Liu, S., Li, M., Wilson, T., Cao, Y., Prabhakaran, B.: Loss resilient strategy in body sensor networks. In: 6th International Conference on Body Area Networks, pp. 99–102. ICST, Belgium (2011)
23. Levis, P., Madden, S., Polastre, J., Szewczyk, R., Whitehouse, K., Woo, A., Gay, D., Hill, J., Welsh, M., Brewer, E., Culler, D.: Tinyos: an operating system for sensor networks. Amb. Intell. **35** (2005)
24. Liu, F., Shu, P., Jin, H., Ding, L., Yu, J., Niu, D., Li, B.: Gearing resource-poor mobile devices with powerful clouds: architectures, challenges, and applications. IEEE Wirel. Commun. **20**, 14–22 (2013)
25. pGBRT: Parallel Gradient Boosted Regression Trees. http://machinelearning.wustl.edu/pmwiki.php/Main/Pgbrt
26. Dean, J., Ghemawat, S.: MapReduce: simplified data processing on large clusters. Commun. ACM **51**, 107–113 (2008)
27. Huang, X., Zhao, T., Cao, Y.: PIR: a domain specific language for multimedia information retrieval. Int. J. Multimed. Data Eng. Manage. **5**, 1–27 (2014)

Genetic-Algorithm-Based Feature-Selection Technique for Fall Detection Using Multi-placement Wearable Sensors

I Putu Edy Suardiyana Putra and Rein Vesilo

Abstract For a machine-learning-based fall-detection approach using wearable sensors, having a high number of features can not only cause a reduction in the detection rate because of irrelevant features, but it can also cause a high computational cost. Therefore, the number of features needs to be reduced through a feature-selection technique. However, current studies in fall-detection only consider features that can give an optimum detection rate without considering their computational cost. Having features with a high computational cost on wearable devices can cause their battery to drain fast. This paper presents a genetic-algorithm-based feature-selection technique that can search for a subset of low-computational-cost features from different sensor placements, where those features can give a relatively good detection rate in terms of F-score. The experimental results show that our technique is able to select a subset of low-computational-cost features that can achieve up to 97.7% of F-score on average. Compared to the SelectKBest and Recursive Feature Elimination (RFE) techniques (a filter and an embedded feature-selection technique, respectively), our approach is able to select features that can give a comparable F-score and significantly lower computational cost.

1 Introduction

Falls are the second leading cause of injury-related deaths worldwide [25]. Moreover, Age UK reported that falls are the main cause of disability and death for people aged over 65 in the UK [2]. Therefore, a fall-detection system becomes an urgent matter as it can help victims to receive help quickly.

I. P. E. S. Putra (✉) · R. Vesilo
School of Engineering, Macquarie University, Sydney 2109, Australia
e-mail: i-putu-edy-suardiyana.putra@hdr.mq.edu.au

R. Vesilo
e-mail: rein.vesilo@mq.edu.au

G. Fortino and Z. Wang (eds.), *Advances in Body Area Networks I*,
Internet of Things, https://doi.org/10.1007/978-3-030-02819-0_24

Igual et al. [12] categorized accelerometer-based fall-detection approaches into two: threshold-based and machine-learning. Vallejo et al. [30] show that threshold-based approaches are less effective than machine-learning-based approaches, because the threshold-based approaches produce a high number of false alarms. Those false alarms can cause users to reject the system [19].

Although machine-learning-based approaches give a better performance in terms of accuracy, these approaches have a high computational cost, especially in terms of feature extraction [15]. In theory, using more features should make the classifier more accurate. In fact, using more features can slow down the learning process or can even reduce the accuracy because there are some irrelevant or redundant features [34]. For an on-node-based classification system, extracting a high number of features can rapidly drain the battery of the device, because an extensive computation is required. A possible way to reduce the computational cost of the system is to reduce the number of features through a feature-selection technique.

Broadly speaking, feature-selection techniques are categorized into three: wrapper, filter, and embedded [10]. Both wrapper and embedded techniques involve a learning algorithm in selecting features. The wrapper methods [18, 32] choose a subset of features based on the machine-learning's performance, while the embedded methods select features during the training process of the learning algorithm [11]. On the other hand, the filter methods [3] have no dependency on learning algorithms.

Although the filter-based methods are computationally more efficient than the wrapper-based methods, the filter methods appear to ignore features that can give more information if they are used together [32]. This disadvantage can reduce the classifier's accuracy. The main disadvantage of wrapper methods is that they have a high computational complexity in searching for the optimum solution [32]. Doing an exhaustive search in a feature space of N features requires an evaluation of 2^N possible feature combinations [17]. Heuristic search is often used to reduce the computational complexity of an exhaustive search in wrapper methods.

Several studies in fall-detection and activity recognition used feature-selection techniques to reduce the number of features [3, 18, 28, 32]. However, those studies only consider the classifier accuracy as a criterion for selecting the subset of features. This is because most existing feature-selection techniques are not designed to handle multiple selection criteria [33]. In fact, there are some features that aid accuracy, but have a high computational cost.

The contribution of this paper is a genetic-algorithm-based feature-selection technique (one of the heuristic search techniques) that can select a subset of features based on three criteria: detection rate (in terms of F-score), computational cost (in terms of running time), and the number of sensors used. The genetic-algorithm has been shown to be effective for multi-criteria-based feature-selection in several other studies [21, 33].

We compared our approach with the SelectKBest and Recursive Feature Elimination (RFE) techniques. Our approach is able to find a subset of features that can give a comparable detection rate (in terms of F-score) than both SelectKBest and RFE, while having a significantly lower computational cost.

The rest of the paper is organized as follows. Section 2 presents related work. Our genetic-algorithm-based feature-selection is explained in Sect. 3. Section 4 provides the methodology. Section 5 discusses our experimental results and analysis while Sect. 6 gives conclusions as well as prospective future work.

2 Related Work

There have been several studies which have proposed a feature-selection technique for fall-detection using wearable sensors. Li et al. [18] implemented a Bayes framework and receiver operating characteristic (ROC) curve-based feature-selection. Their technique is able to achieve up to 86.08%, 94.31%, and 95.75% accuracy with 4 features, 8 features, and 12 features respectively. Wang et al. [32] proposed a game-theory-based feature-selection with k-Nearest Neighbor (k-NN) and a Support Vector Machine (SVM) as the machine-learning algorithms. Based on their experiment, their technique is able to achieve up to 74.42% accuracy with 21 features. On the other hand, Atallah et al. [3] used several feature-selection techniques to investigate the best sensor placement. However, there is no explanation of the performance of feature-selection techniques, as their main objective is to find the best sensor placement.

The main objective of the above fall-detection studies is to find a subset of features that can give the highest accuracy without considering the computational cost (especially for the feature extraction process). Moreover, those studies only focused on finding the best subset of features based on only one sensor placement. Gjoreski et al. [9] show that using multiple sensor placements can increase the detection accuracy. Therefore, this study develops a genetic-algorithm-based feature-selection for fall-detection (GA-Fade), aiming to find a subset of features from multiple sensor placements. These features are expected to give an optimum detection rate and have a low computational cost.

3 GA-Fade

This section describes our genetic-algorithm-based feature-selection for fall-detection (GA-Fade). Algorithm 1 shows the steady-state genetic-algorithm. The following sub-sections explain several main functions of our technique.

3.1 Chromosome Encoding and Initial Population

For a feature-selection using a genetic-algorithm, an individual consists of a sequence of features ($f_1, f_2, f_3, \ldots, f_n$) where each feature is represented by a binary number

Algorithm 1 A steady-state genetic-algorithm

```
initialize population P
non-increasingly sort individuals in P based on fitness value
while number of generation < desired generation do
    select two parents p1 and p2 from P
    child1, child2 = crossover(p1, p2)
    mutation(child1, child2)
    insert child1 and child2 into P
    sort P
    eliminate individual with the lowest fitness from P
end while
```

(either 1 or 0) [21]. Values 1 indicate selected features while values 0 show removed features. Figure 1 shows an illustration of the encoding of a chromosome.

Before searching for the optimal sequence of features using the genetic-algorithm, several initial chromosomes are generated. Algorithm 2 shows the process of chromosome generation of the initial population P. We ensure that there are not two (or more) similar chromosomes in P, to increase the diversity of individuals.

Algorithm 2 Controlled initial-population generation

```
for i = 1 to I do                          ▷ I : Initial size of population P
    Generate a chromosome C
    while C is exist in P do
        Re-generate C
    end while
    Put C into P
end for
```

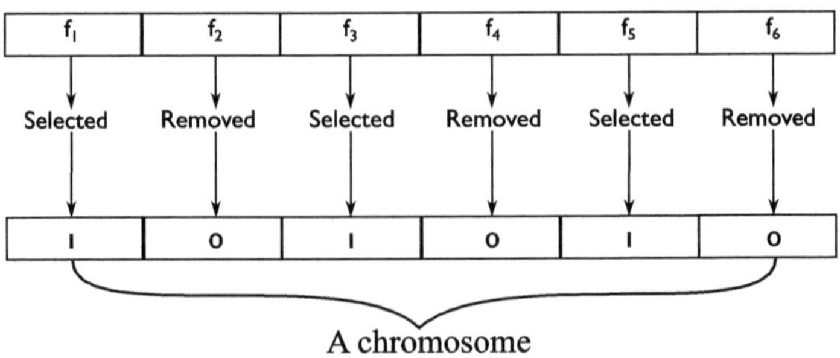

Fig. 1 An illustration of a chromosome encoding for features from multiple sensor placements

3.2 Fitness Function and Selection Process

To construct the fitness function of our technique, three aspects from Lara and Labrador [16] are considered: level of energy consumption (E), overall detection rate for fall activities (A), and number of sensors being used (N).

According to Dunkels et al. [7]:

$$\frac{E}{V} = I_m t_m + I_l t_l + I_t t_t + I_r t_r + \sum_i I_{c_i} t_{c_i}, \tag{1}$$

where V is the supply voltage, and I_x and t_x are the current draw and the running time, respectively, for x, where x can be:

- the microprocessor in normal mode m,
- the microprocessor in low-power mode l,
- the communication device in transmit mode t,
- the communication device in receive mode r, and
- other components c.

As this study is performed offline, all energy consumption of the microprocessor in all modes except the m mode were assumed to be zero ($I_l = I_t = I_r = I_c = 0$). Thus, the relative energy consumption between two sequences of features is

$$\frac{E_1}{E_2} = \frac{I_{m_1} t_{m_1} V_1}{I_{m_2} t_{m_2} V_2}. \tag{2}$$

Assuming that those two sequences of features are implemented using the same sensor device, the current draw and the supply voltage are the same ($I_{m_1} = I_{m_2}$ and $V_1 = V_2$). Then we conclude that the energy consumption is proportional to the running time: $E \propto t$. Therefore, we used running time to fulfill the level of energy consumption (E) aspect.

A fall-detection system must have low false positives (false alarms) and low false negatives (undetected falls) [19]. Therefore, the F-score was used to measure the detection rate of the classifier, as it considers both false positives and false negatives. The F-score is given by

$$F\text{-}score = \frac{2\text{TP}}{2\text{TP} + \text{FP} + \text{FN}}, \tag{3}$$

where TP is true positives, FP is false positives (false alarms), and FN is false negatives (undetected falls). Thus, we used F-score as the overall detection rate (A) aspect.

One of the characteristics of a successful human-activity recognition system is not letting the users wear many sensors [16]. Therefore, in the fitness function we considered the number of sensors used as N. Thus, the fitness function for this study is

$$fitness(x) = w_1 \cdot A - w_2 \cdot E - w_3 \cdot N, \quad (4)$$

where:

- x = a chromosome from the population
- A = overall detection rate in terms of F-score
- w_1 = weight for accuracy
- E = total computational cost of the system for a chromosome
- w_2 = weight for total computational cost
- N = number of sensors chosen
- w_3 = weight for number of sensors used.

To compute the total computational cost, Eq. (5) is used

$$E = w_c \cdot t_c + w_w \cdot t_w + w_t t_t \quad (5)$$

where w_x and t_x are the weight and running time of x, where x can be chest placement (c), waist placement (w), or thigh placement (t).

The units of A is percentage (%) and E is millisecond (ms). As N is number of sensors being used, it does not have a unit. Thus, for the implementation, the fitness function is changed into:

$$fitness(x) = (w_1 * A/100) - (w_2 * E/2.4) - (w_3 * N/3),$$

where 2.4 and 3 is the fastest time (in millisecond) of the machine (in this case a PC) to extract all features from all sensors and the total number of sensors used in this study, respectively. w_1 is set to be 1 while w_2 and w_3 is set to be 0.5. This is because accuracy (A) is more important than energy (E) and number of sensor (N). An equal value is given to w_c, w_w, and w_t, thus $w_c = w_w = w_t = 1$, as the importance of all sensor placements is assumed to be the same. Therefore, the optimum sequence is a sequence that has the highest detection rate with the lowest computational cost and the lowest number of used sensors. Those weights above can be adjusted to fit the needs of the application.

To select parents from the population, a roulette-wheel technique is implemented, adopted from Oh and Lee [21]. This technique is chosen to ensure that fitter chromosomes/individuals have a higher probability of being chosen. The probability $P(i)$ of selecting the ith item from a pool of n items is weighted more highly for lower numbered items according to:

$$P(i) = q(1-q)^{i-1} / \left(1 - (1-q)^n\right) \quad (6)$$

where $q = 1/4$ is the probability of selecting the first item for infinite n [21]. The form of this equation is set to ensure that it sums to unity $\sum_{1 \le i \le n} P(i) = 1$ while decreasing the probability geometrically for subsequent items $P(i+1) = (1-q)P(i)$.

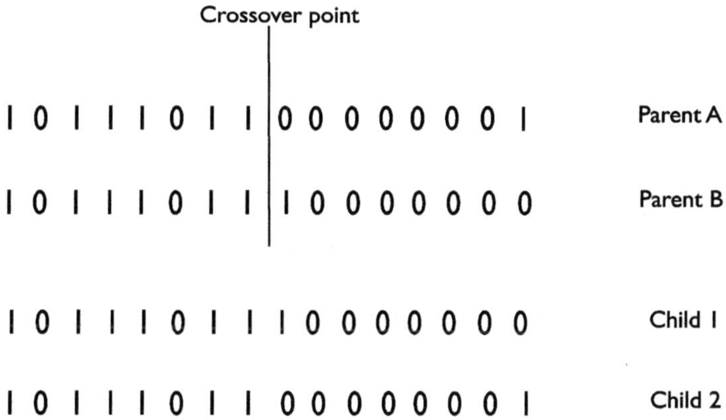

Fig. 2 An illustration of a crossover operator between parents for one sensor

It is possible to find an inverse of the cumulative probability distribution to help generate random numbers sampled from this distribution. Specifically, setting $A = \left(1 - (1-q)^n\right)$

$$Q(k) = \sum_{0 \le i \le k} q\,(1-q)^{i-1}/A = 1/A\left(1 - (1-q)^k\right) \tag{7}$$

which has a simple inverse of,

$$k(Q) = \lceil \log\left(1 - AQ\right) / \log\left(1-q\right) \rceil \tag{8}$$

3.3 Crossover and Mutation

We use a one-point crossover operator with a crossover probability of 1.0 [21]. An illustration of our crossover operator is shown in Fig. 2. This crossover operation produces two children. Then, the mutation operator is applied to those children using Algorithm 3 with the mutation rate p of 0.1 [21].

Algorithm 3 Mutation process

```
for each gene g in the chromosome do
    generate a random number r within [0, 1]
    if g = 0 and r < p then
        g = 1
    else if g = 1 and r < p then
        g = 0
    end if
end for
```

3.4 Features

The following features are used in this study:

1. Minimum, maximum, and average acceleration vector magnitudes [1, 22, 29].
2. Velocity [4, 22, 29].
3. Energy [22, 31].
4. Variance of the acceleration vector magnitude [5, 22, 29].
5. Root mean square (RMS) of the acceleration vector magnitude [4, 9, 29].
6. Acceleration exponential moving average [5, 22, 29].
7. Signal magnitude area [14, 22, 29].

The features above are extracted from the pre-impact, impact, and post-impact stages using a segmentation technique from our previous work [27]. Thus, each sensor produces 27 features. As the dataset consists of data from three different sensors, then the total number of features investigated is 81, meaning that the length of the chromosome is 81. Note that the order of the digits in a chromosome matters, as each digit represents a particular feature. An exhaustive search requires an investigation of approximately 2.4×10^{24} chromosomes, which needs a huge amount of time. Therefore, we developed GA-Fade to optimize the sequence searching process.

4 Experimental Methodology

This section describes a publicly-accessible dataset that is used in this study together with the setup for the experiment. This section consists of two sub-sections, dataset and experimental setup.

4.1 Dataset

A publicly accessible dataset[1] from Ojetola et al. [23] was used. The dataset was obtained with fully-approved ethics from Coventry University. The data were gathered from 46 healthy subjects (9 females and 37 males). Each of the subjects was required to perform 14 falls (forward fall, backward fall, and lateral fall) and some Activities of Daily Living (ADLs) for 23 mins on average. More detailed information on the protocol can be found in [23].

Several placements were considered when collecting the data: chest, waist, and thigh. However, not all subjects have data from a waist sensor. Therefore, a subset of the subjects that had sensors on all placements (18 subjects with 5 females and 13 males, age 22 ± 2.8 years old, height 172.4 ± 10.1 cm, and weight 66.1 ± 13.7 kg),

[1] The dataset can be downloaded at: http://skuld.cs.umass.edu/traces/mmsys/2015/paper-15/.

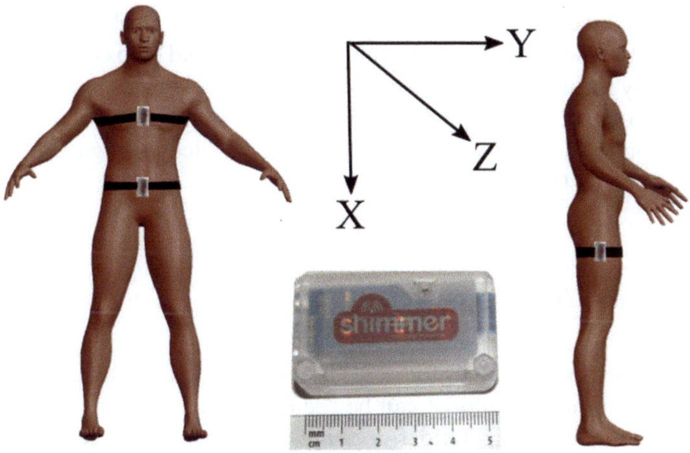

Fig. 3 Sensor placements (left and right) and the shimmer sensors (middle)

was used in this study. Using a publicly accessible dataset is important as it can give a fair comparison between techniques [13].

Three Shimmer sensors [6] were used for data collection with a sampling rate of 100 Hz. The sensor consists of a 3D Accelerometer, 3D gyroscope, Bluetooth radio, and an MSP430F1611 microcontroller. The data were transferred to a Personal Computer (PC) using Bluetooth then manually annotated with LabView. Figure 3 shows the sensor placements and and a Shimmer device.

4.2 Experimental Setup

The data were processed offline using a PC. For training and testing the classifier, Logistic Regression (LR) with the inverse regularization strength (C) of 10^9 from the Scikit-learn library [26] was used. To segment the data stream, an algorithm called EvenT-ML [27] is implemented. Because EvenT-ML extracts the features based on signal peaks, the number of instances produced from each placement is different for some activities. This means that fusion of the feature values between the three sensor placements for feature selection cannot be done directly. Therefore, a data imputation is done, so that all sensors have an equal number of instances. Since an activity is iterated more than one time for each subject, the imputation is done by using the median value of the samples from the same activity. The median value is used for the imputation process because the feature values are not normally distributed.

Leave-one-subject-out cross-validation was used as it is able to determine the ability of the classifier to generalize to unseen subjects and helps to identify the classifier variability from subject to subject [8, 24]. To avoid biased results, a validation step from Nowotny [20] is adopted. The significance of the improvement in detection rate and computational cost were evaluated using a Wilcoxon signed-rank test with significance level of 0.05 ($\alpha = 0.05$).

5 Results and Analysis

Table 1 shows the F-score values for each number of generation and number of initial population. From those results, it can be seen that, by increasing either the initial population or number of generations does not necessarily increase the F-score. The optimum result is achieved when the number of generations is 100 or 200 with initial population of 40 individuals, with a 97.7% F-score on average.

To evaluate our proposed technique, we compared GA-Fade with the SelectKBest (with score functions: chi-squared and analysis of variance (f_classif)) and Recursive Feature Elimination (RFE) techniques from the Scikit-learn library [26]. For the SelectKBest and RFE techniques, we use 30 as the number of selected features ($K = 30$), because the number of selected features by GA-Fade is around 30 (Table 2). Table 3 shows a detection-rate comparison between using all features, SelectKBest, and RFE. In terms of F-score, GA-Fade is able to select features that can give comparable results than using all features and features that are selected by both the SelectKBest-based technique (regardless of its score function) and RFE. These results show that using wrapper-based, filter-based, or embedded technique can give comparable results in this study. Also, reducing the features in this study does not significantly increase the performance.

Table 1 F-score of selected features for each generation

Initial population	Number of generations		
	50	100	200
40	96.9 ± 4.8	97.7 ± 3.4	97.7 ± 3.9
60	97.4 ± 4.9	96.8 ± 5.1	96.7 ± 5.5
80	97 ± 5.5	96.6 ± 5.3	97.1 ± 4.2

Table 2 Number of selected features

Number of generations	Size of the initial population (P)		
	40	60	80
50	33.1 ± 4.0	33.6 ± 4.5	33.4 ± 4.5
100	28.7 ± 4.3	31.5 ± 3.3	32.3 ± 4.5
200	25.6 ± 3.5	26.4 ± 5.1	25.6 ± 5.5

Table 3 Classifier performance on selected features

Feature-selection technique	F-score (%)
Full features	96.6 ± 5.2
SelectKBest + chi2	95.7 ± 5
SelectKBest + f_classif	96.5 ± 6.5
RFE + Logistic Regression-based classifier	97.3 ± 4

Table 4 Computational cost for selected features for each placement

Feature selection technique	Chest (ms)	Waist (ms)	Thigh (ms)
Full features	0.8 ± 0.0	0.8 ± 0.0	0.8 ± 0.0
SelectKBest + chi2	0.3 ± 0.0	0.3 ± 0.0	0.4 ± 0.0
SelectKBest + f_classif	0.5 ± 0.0	0.3 ± 0.0	0.1 ± 0.0
RFE + Logistic Regression-based classifier	0.7 ± 0.1	0.5 ± 0.1	0.4 ± 0.1

Table 5 Average and standard deviation of minimum running time (ms) selected features of GA-based feature selection with different initial population sizes and different numbers of generations

Initial population	Number of generations		
	50	100	200
(a) *Chest*			
40	0.3 ± 0.1	0.2 ± 0.1	0.2 ± 0.1
60	0.3 ± 0.1	0.2 ± 0.1	0.2 ± 0.1
80	0.3 ± 0.1	0.2 ± 0.1	0.2 ± 0.1
(b) *Waist*			
40	0.3 ± 0.1	0.2 ± 0.1	0.2 ± 0.1
60	0.3 ± 0.1	0.2 ± 0.1	0.2 ± 0.1
80	0.3 ± 0.1	0.2 ± 0.1	0.2 ± 0.1
(c) *Thigh*			
40	0.3 ± 0.1	0.2 ± 0.1	0.2 ± 0.1
60	0.3 ± 0.1	0.2 ± 0.1	0.2 ± 0.1
80	0.3 ± 0.1	0.2 ± 0.1	0.2 ± 0.1

Tables 4 and 5 show the computational cost of selected features from the feature-selection techniques for each sensor placement. GA-Fade is able to select features that can give the lowest computational cost on average for the chest, waist, and thigh placements (p-values $\leq 1.8 \times 10^{-4}$), except when SelectKBest with f_classif score function is used for the thigh placement.

From the results above, it can be seen that GA-Fade is able to select significantly lower-computational cost features in most of the cases, that can give a comparable F-score to all features and features that are selected by SelectKBest or RFE. We also found that all feature-selection techniques pick features from all placements (chest, waist, and thigh), which means that using more sensors can give a better F-score.

6 Conclusion and Future Work

This paper presents a genetic-algorithm-based feature-selection technique (GA-Fade) that can select a subset of features from multiple sensor placements for an accelerometer-based fall-detection system. This feature-selection technique is able to select a subset of features that can give a comparable accuracy with a significantly lower computational cost. To evaluate the performance of our approach, we implemented GA-Fade on a publicly-accessible dataset with Logistic Regression as the machine-learning algorithm. We compared our technique with SelectKBest with several score functions and RFE techniques.

The experiment shows that our technique can find a subset of features that can give a comparable F-score than subsets of features that are selected by both the SelectKBest and RFE techniques, but can give a significantly lower computational cost. Features that are selected by our approach can reach up to 97.7% of F-score on average. In terms of computational cost, our selected features have a significantly lower computational cost than selected features from the SelectKBest and RFE techniques, in most of the cases.

Future work will focus on improving the genetic-algorithm with the aim to increase its speed, so the optimal solution can be achieved in less time. Also, more features will be investigated.

Acknowledgements This work was supported by an International Macquarie University Research Excellence Scholarship (iMQRES) and a Coventry University studentship, under a co-tutelle scheme. The authors would like to thank Professor Elena Gaura and Dr. James Brusey (Coventry University) for their insightful inputs.

References

1. Abbate, S., Avvenuti, M., Bonatesta, F., Cola, G., Corsini, P., Vecchio, A.: A smartphone-based fall detection system. Pervas. Mobile Comput. **8**(6), 883–899 (2012). https://doi.org/10.1016/j.pmcj.2012.08.003
2. Age UK: Stop Falling: Start Saving Lives and Money. http://www.ageuk.org.uk/documents/en-gb/campaigns/stop_falling_report_web.pdf?dtrk=true (2013). Accessed 2 Mar 2017
3. Atallah, L., Lo, B., King, R., Yang, G.Z.: Sensor positioning for activity recognition using wearable accelerometers. IEEE Trans. Biomed. Circ. Syst. **5**(4), 320–329 (2011). https://doi.org/10.1109/TBCAS.2011.2160540

4. Bourke, A.K., O'Donovan, K.J., Nelson, J., OLaighin, G.M.: Fall-detection through vertical velocity thresholding using a tri-axial accelerometer characterized using an optical motion-capture system. In: 30th Annual International Conference of the IEEE Engineering in Medicine and Biology Society, pp. 2832–2835. Vancouver, BC (2008). https://doi.org/10.1109/IEMBS.2008.4649792
5. Brusey, J., Rednic, R., Gaura, E.I., Kemp, J., Poole, N.: Postural activity monitoring for increasing safety in bomb disposal missions. Meas. Sci. Technol. **20**(7), 075–204 (2009). https://doi.org/10.1088/0957-0233/20/7/075204
6. Burns, A., Greene, B.R., McGrath, M.J., O'Shea, T.J., Kuris, B., Ayer, S.M., Stroiescu, F., Cionca, V., Shimmer, A.: Wireless sensor platform for noninvasive biomedical research. IEEE Sens. J. **10**(9), 1527–1534 (2010). https://doi.org/10.1109/JSEN.2010.2045498
7. Dunkels, A., Osterlind, F., Tsiftes, N., He, Z.: Software-based on-line energy estimation for sensor nodes. In: Proceedings of the 4th Workshop on Embedded Networked Sensors, EmNets '07, pp. 28–32. ACM, New York, NY, USA (2007). https://doi.org/10.1145/1278972.1278979
8. Gaura, E., Kemp, J., Brusey, J.: Leveraging knowledge from physiological data: on-body heat stress risk prediction with sensor networks. IEEE Trans. Biomed. Circ. Syst. **7**(6), 861–870 (2013). https://doi.org/10.1109/TBCAS.2013.2254485
9. Gjoreski, H., Lustrek, M., Gams, M.: Accelerometer placement for posture recognition and fall detection. In: 2011 Seventh International Conference on Intelligent Environments, pp. 47–54. IEEE (2011). https://doi.org/10.1109/IE.2011.11
10. Guyon, I., Elisseeff, A.: An introduction to variable and feature selection. J. Mach. Learn. Res. **3**, 1157–1182 (2003). http://dl.acm.org/citation.cfm?id=944919.944968
11. Guyon, I., Weston, J., Barnhill, S., Vapnik, V.: Gene selection for cancer classification using support vector machines. Mach. Learn. **46**(1), 389–422 (2002). https://doi.org/10.1023/A:1012487302797
12. Igual, R., Medrano, C., Plaza, I.: Challenges, issues and trends in fall detection systems. Biomed. Eng. Online **12**, 66 (2013). https://doi.org/10.1186/1475-925X-12-66
13. Igual, R., Medrano, C., Plaza, I.: A comparison of public datasets for acceleration-based fall detection. Med. Eng. Phys. **37**(9), 870–878 (2015). https://doi.org/10.1016/j.medengphy.2015.06.009
14. Karantonis, D.M., Narayanan, M.R., Mathie, M., Lovell, N.H., Celler, B.G.: Implementation of a real-time human movement classifier using a triaxial accelerometer for ambulatory monitoring. IEEE Trans. Informat. Technol. Biomed. **10**(1), 156–167 (2006). https://doi.org/10.1109/TITB.2005.856864
15. Kau, L.J., Chen, C.S.: A smart phone-based pocket fall accident detection, positioning, and rescue system. IEEE J. Biomed. Health Informat. **19**(1), 44–56 (2015). https://doi.org/10.1109/JBHI.2014.2328593
16. Lara, O.D., Labrador, M.A.: A survey on human activity recognition using wearable sensors. IEEE Commun. Surv. & Tutor. **15**(3), 1192–1209 (2013). https://doi.org/10.1109/SURV.2012.110112.00192
17. Le, H.T., Van Tran, L.: Automatic feature selection for named entity recognition using genetic algorithm. In: Proceedings of the Fourth Symposium on Information and Communication Technology, SoICT '13, pp. 81–87. ACM, New York, NY, USA (2013). https://doi.org/10.1145/2542050.2542056. http://doi.acm.org/10.1145/2542050.2542056
18. Li, J., Li, M., Wang, Z., Zhao, Q.: An improved classification method for fall detection based on Bayesian framework. In: 2015 IEEE International Conference on Robotics and Biomimetics (ROBIO), pp. 237–242 (2015). https://doi.org/10.1109/ROBIO.2015.7418773
19. Noury, N., Fleury, A., Rumeau, P., Bourke, A., Laighin, G., Rialle, V., Lundy, J.: Fall detection—principles and Methods. In: 2007 29th Annual International Conference of the IEEE Engineering in Medicine and Biology Society, pp. 1663–1666 (2007). https://doi.org/10.1109/IEMBS.2007.4352627
20. Nowotny, T.: Two challenges of correct validation in pattern recognition. Front. Robotics AI **1**, 5 (2014). https://doi.org/10.3389/frobt.2014.00005. https://www.frontiersin.org/article/10.3389/frobt.2014.00005

21. Oh, I.S., Lee, J.S., Moon, B.R.: Hybrid genetic algorithms for feature selection. IEEE Trans. Pattern Anal. Mach. Intelligen. **26**(11), 1424–1437 (2004). https://doi.org/10.1109/TPAMI.2004.105
22. Ojetola, O.: Detection of Human Falls using Wearable Sensors. Ph.D. thesis, Coventry University (2013)
23. Ojetola, O., Gaura, E., Brusey, J.: Data set for fall events and daily activities from inertial sensors. In: Proceedings of the 6th ACM Multimedia Systems Conference, MMSys '15, pp. 243–248. ACM, New York, NY, USA (2015). https://doi.org/10.1145/2713168.2713198
24. Ojetola, O., Gaura, E.I., Brusey, J.: Fall detection with wearable sensors - Safe (SmArt Fall dEtection). In: IEEE 2011 Proceedings of 7th International Conference on Intelligent Environments, pp. 318–321 (2011). https://doi.org/10.1109/IE.2011.38
25. Organization, W.H.: Falls (2012). http://www.who.int/mediacentre/factsheets/fs344/en/. Accessed 2 Mar 2017
26. Pedregosa, F., Varoquaux, G., Gramfort, A., Michel, V., Thirion, B., Grisel, O., Blondel, M., Prettenhofer, P., Weiss, R., Dubourg, V., Vanderplas, J., Passos, A., Cournapeau, D., Brucher, M., Perrot, M., Duchesnay, E.: Scikit-learn: machine learning in python. J. Mach. Learn. Res. **12**, 2825–2830 (2011)
27. Putra, I.P.E.S., Brusey, J., Gaura, E., Vesilo, R.: An event-triggered machine learning approach for accelerometer-based fall detection. Sensors **18**(1) (2018). https://doi.org/10.3390/s18010020. http://www.mdpi.com/1424-8220/18/1/20
28. Shan, S., Yuan, T.: A wearable pre-impact fall detector using feature selection and support vector machine. In: IEEE 10th International Conference On Signal Processing Proceedings, pp. 1686–1689 (2010). https://doi.org/10.1109/ICOSP.2010.5656840
29. Suardiyana, I.P.E., Brusey, J., Gaura, E.: A cascade-classifier approach for fall detection. In: 5th EAI International Conference on Wireless Mobile Communication and Healthcare (2015)
30. Vallejo, M., Isaza, C., Lopez, J.: Artificial neural networks as an alternative to traditional fall detection methods. In: 2013 35th Annual International Conference of the IEEE Engineering in Medicine and Biology Society (EMBC), pp. 1648–1651 (2013). https://doi.org/10.1109/EMBC.2013.6609833
31. Wang, S., Yang, J., Chen, N., Chen, X., Zhang, Q.: Human activity recognition with user-free accelerometers in the sensor networks. In: International Conference on Neural Networks and Brain, vol. 2, pp. 1212–1217. Beijing, China (2005). https://doi.org/10.1109/ICNNB.2005.1614831
32. Wang, Z., Wu, D., Chen, J., Ghoneim, A., Hossain, M.A.: A triaxial accelerometer-based human activity recognition via eemd-based features and game-theory-based feature selection. IEEE Sens. J. **16**(9), 3198–3207 (2016). https://doi.org/10.1109/JSEN.2016.2519679
33. Yang, J., Honavar, V.: Feature subset selection using a genetic algorithm, pp. 117–136. Springer US, Boston, MA (1998). https://doi.org/10.1007/978-1-4615-5725-8_8. http://dx.doi.org/10.1007/978-1-4615-5725-8_8
34. Yu, L., Liu, H.: Efficient feature selection via analysis of relevance and redundancy. J. Mach. Learn. Res. **5**, 1205–1224 (2004). https://doi.org/10.1145/1014052.1014149

Link-Level Performance of FM-UWB in the Interfered IEEE 802.15.6 Channel

Harri Viittala, Matti Hämäläinen and Jari Iinatti

Abstract This paper provides simulation results of the frequency modulated ultra wideband (FM-UWB) system with channel coding in an interfered IEEE 802.15.6 channel. The paper discusses a FM-UWB receiver structure and presents the developed simulator model. The receiver applies a delay-line modulator followed by an amplitude modulation projection detection. A channel model used is the IEEE 802.15.6 channel model 3 for an on-body link. An interference is modeled as a colored Gaussian noise presenting an in-band IEEE 802.15.4 interferer. The simulation results show that the FM-UWB system can tolerate as low as −6 dB signal-to-interference power ratios (SIR) in the studied scenarios.

1 Introduction

Typically, wireless body area networks (WBAN) refer to a wireless system of devices carried by a human or animal. The communication architecture can be divided into three communication tiers: intra-WBAN (personal), inter-WBAN (local) and beyond-WBAN (global) based on the proximity of occurred communication around a body. Consequentially, the intra-WBAN can be separated to in-body, on-body and off-body communications.

Wireless communications around or vicinity of a human body is very challenging. A radio signal propagates around a body via diffractions and reflections from

H. Viittala · M. Hämäläinen (✉) · J. Iinatti
Centre for Wireless Communications, University of Oulu, P.O. Box 8000, 90014 Oulu, Finland
e-mail: matti.hamalainen@oulu.fi

H. Viittala
e-mail: harri.viittala@gmail.com

J. Iinatti
e-mail: jari.iinatti@oulu.fi

G. Fortino and Z. Wang (eds.), *Advances in Body Area Networks I*, Internet of Things, https://doi.org/10.1007/978-3-030-02819-0_25

extremities and surrounding environment. A direct signal propagating through a body is strongly attenuated. In addition, movement of extremities, variation in tissue permittivities and shape of a body involve [1–4].

The IEEE 802.15.6 standard defines narrowband, ultra wideband (UWB) and human body communication (HBC) physical (PHY) layers to be used in the intra-WBAN communication [5]. From these, UWB communications provides robustness in a dense multipath environment and a low power spectral density technology for WBAN [6]. The IEEE 802.15.6 standard defines two UWB PHY layers, i.e., impulse radio UWB (IR-UWB) and frequency modulated ultra wideband (FM-UWB) [5]. FM-UWB is a very simple UWB approach for low and medium data rates exploiting a double frequency modulation [7]. It applies a low-modulation index frequency shift keying (FSK) followed by a high modulation index frequency modulation (FM) to achieve a constant envelope UWB signal in the frequency domain. FM-UWB can be considered as an analog implementation of a spread-spectrum system, where a modulation index is equal to a spreading gain of a spread spectrum system. The performance of FM-UWB is studied, e.g., in [7–12]. As far as we know, the FM-UWB system is not investigated with Bose-Chadhuri-Hocquenghem (BCH) encoding in an interfered channel before. This paper contributes simulation results of the FM-UWB system with BCH encoding as defined in the IEEE 802.15.6 standard [5]. In addition, the system performance is tested against an in-band interference.

This paper is organized as follows. Section 2 introduces the concept of the FM-UWB PHY layer. The simulator model, channel model and interference are discussed in Sect. 3. It also describes the simulation parameters. The simulation results are given in Sect. 4, and the paper is concluded in Sect. 5.

2 Frequency Modulated UWB

FM-UWB is a simple low data rate UWB technique utilizing a double frequency modulation [7]. Basically, it constitutes an analog implementation of a spread spectrum system [13]. In FM-UWB, a data stream is modulated with a low-modulation index digital FSK followed by a high-modulation index FM. The modulation index is defined as $\beta = \Delta f / f_m$, where Δf is the frequency deviation and f_m is the frequency of a modulating signal. The modulation index can be chosen freely. A small index ($\beta < 1$) yields to a narrowband FM signal, whereas a large index ($\beta >> 1$) generates a UWB signal. A FM signal $s_{\text{FM}}(t)$ modulated by a signal $m(t)$ having a frequency f_m is expressed as [7]

$$s_{\text{FM}}(t) = A \sin\left(2\pi f_c t - \beta \cos\left(2\pi f_m t\right) + \varphi_0\right), \tag{1}$$

where f_c is the carrier frequency and φ_0 is the arbitrary but time-independent constant phase. Due to wideband frequency modulation, the processing gain PG_{FM} at a receiver can be expressed as the ratio of the RF bandwidth B_{RF} and subcarrier bandwidth B_{SUB} as

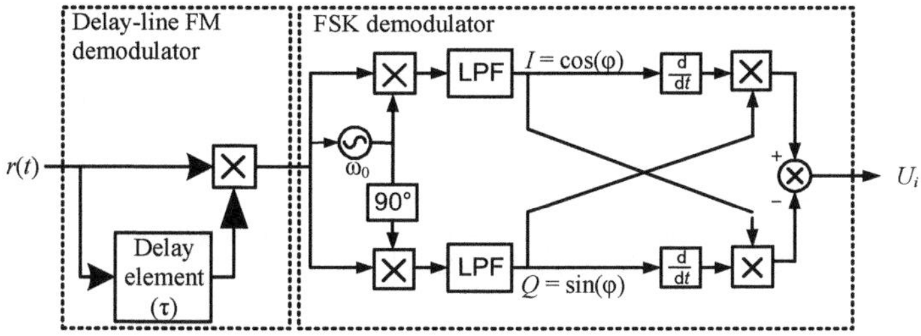

Fig. 1 Block diagram of FM-UWB receiver

$$PG_{\text{FM}} = 10\log_{10}\left(\frac{B_{\text{RF}}}{B_{\text{SUB}}}\right). \tag{2}$$

A received signal is demodulated with a delay-line FM demodulator followed by a FSK demodulator, as depicted in Fig. 1. The received signal is multiplied with the $\tau = N_d/(4f_c)$ delayed signal, where $N_d = 1, 3, 5, \ldots$ [7] is selected so that it optimizes the useful bandwidth of the demodulator B_{DEMOD} [7]. The useful bandwidth is the maximum frequency range where a demodulator transfer function is monotonic, defined as [7]

$$B_{\text{DEMOD}} = \frac{2}{N_d} f_c. \tag{3}$$

The output voltage of the FM demodulator $V_{\text{FM,out}}$ is given as [7]

$$V_{\text{FM,out}}(t) = (-1)^{(N_d+1)/2}\frac{A^2}{2}\sin\left(O\frac{\pi}{2}\sin\left(\omega_m t\right)\right), \tag{4}$$

where $O = N_d\,(\Delta f/f_c)$ is the FM demodulator overdrive and ω_m is the angular frequency of a modulating signal. The overdrive defines the ratio between a deviation of the FM signal and a useful bandwidth of the demodulator. From Eq. (4), it can be seen that when the overdrive is less than 1, the output voltage of the demodulator does not exploit the full available dynamic range. This occurs when a deviation of the FM signal is less than one half of the useful bandwidth of the modulator [7].

The output of the FM demodulator is followed by the FSK demodulator employing an amplitude modulation (AM) projection detection [14]. It converts an input to zero frequency and subsequently converts it to an AM wave with a differentiation stage and finally uses the AM—projection detection.

3 Simulator Model

The software simulator implemented in Matlab was applied to study performance of the FM-UWB system. The simulator consists of a transmitter, receiver and channel.

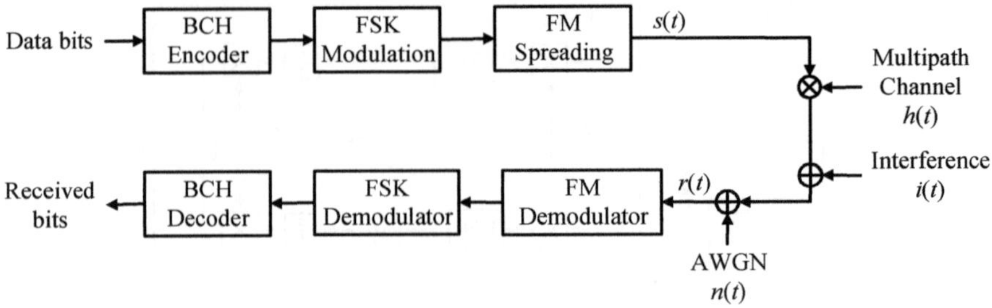

Fig. 2 Block diagram of the FM-UWB simulator

Random generated data bits are encoded by applying the BCH (63, 51) code followed by FSK and FM demodulators. A transmitted signal $s(t)$ propagates through a multipath channel having an impulse response $h(t)$. An interference $i(t)$ and additive white Gaussian noise (AWGN) $n(t)$ are added to the signal yielding to a received signal $r(t)$

$$r(t) = s(t) * h(t) + i(t) + n(t), \tag{5}$$

where $*$ denotes the convolution operator.

The receiver performs demodulation and decoding. Received bits are compared to the data bits to have a bit error rate (BER) and frame error rate (FER). Figure 2 gives a block diagram representation of the simulator model.

3.1 IEEE 802.15.6 Channel Model

The simulator model uses the IEEE 802.15.6 channel model for a body-to-body link [1]. The model is based on the measurements carried out in a hospital room where the applied frequency band was from 3.0 to 11.0 GHz [15]. A subject was laying down in a bed in a room of 5.0–7.0 m, a receive antenna was placed on the middle of a torso and a location of a transmit antenna varies from a head to an ankle. The distance between the antennas was from 176 to 984 mm. The S_{21}–parameter was measured 10 times for each location.

A single cluster model where a path amplitude a_l is modeled by an exponential decay Γ with a Rician factor γ_0 was extracted as [1].

$$10\log_{10}|a_l|^2 = \begin{cases} 0, & l = 0 \\ \gamma_0 + 10\log_{10}\left(\exp\left(-\frac{t_l}{\Gamma}\right)\right) + S, & l \neq 0, \end{cases} \tag{6}$$

where S is the normal distribution with zero-mean and a standard deviation of σ_S. A path arrival time t_l is modeled by the Poisson distribution [15]

Table 1 Parameters for the IEEE 802.15.6 CM3

Γ (ns)	γ_0 (dB)	σ_S	λ (1/ns)	$\overline{L}$
59.7	−4.60	5.02	0.541	38.1

$$p\left(t_l | t_{l-1}\right) = \lambda \exp\left(-\lambda(t_l - t_{l-1})\right), \tag{7}$$

where λ is the path arrival rate. The number of arrival paths L are modeled by the Poisson distribution, given as [15]

$$p(L) = \frac{\overline{L}^L \exp(\overline{L})}{L!}, \tag{8}$$

where $\overline{L}$ is the average number of L. ϕ_l is modeled by a uniform distribution over $[0, 2\pi)$. The parameter values for the channel model are tabulated in Table 1.

3.2 Interference

The simulator applies the colored Gaussian noise (CGN) noise to model an in-band interference, i.e., the white Gaussian noise process filtered by a bandpass filter. Using the band-limited characteristic of CGN, it is easy to implement specific interferences to the defined frequencies with the desired bandwidths. The interference generation procedure is presented in [16].

The bandpass filter uses the raised cosine waveform to have a baseband impulse response $h_{rc}(t)$ having spectrum $H_{rc}(f)$ as [17]

$$H_{\text{rc}}(f) = \begin{cases} T & 0 \le |f| < \frac{1-\alpha}{2T} \\ \frac{T}{2}\left(1 + \cos\left(\frac{\pi T}{\alpha}\left(|f| - \frac{1-\alpha}{2T}\right)\right)\right) & \frac{1-\alpha}{2T} \le |f| \le \frac{1+\alpha}{2T} \\ 0, & |f| > \frac{1+\alpha}{2T}, \end{cases} \tag{9}$$

where α is the roll-off factor and T is the symbol time related to the desired bandwidth BW. The filter is shifted to the desired center frequency f_c as

$$h_{\text{cgn}}(f) = h_{\text{rc}}(t)\cos\left(2\pi f_c t + \varphi\right) \tag{10}$$

An in-band interference $i(t)$ is generated by filtering the white Gaussian noise $n(t)$ with a filter h_{cgn} having a frequency response $H_{\text{cgn}}(f)$ as illustrated in Fig. 3. An interference power is adjusted according to a given signal-to-interference power ratio (SIR).

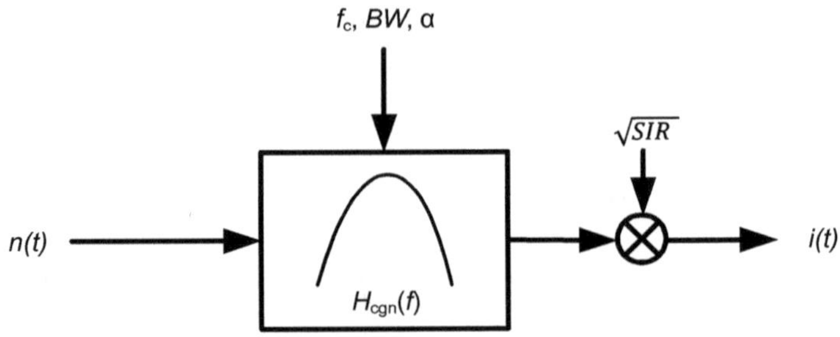

Fig. 3 Generation of the interference

Table 2 Simulation parameters

Parameter	Value
Operating frequency band	3
Channel center frequency (MHz)	6489.6
Bandwidth (MHz)	499.2
Uncoded bit rate (kbps)	250
BCH coding	(63,51)
Packet size, MPDU (octets)	50
FSK waveform	Triangular
Modulation index, FSK	1

3.3 Simulation Parameters

The simulation parameters are given in Table 2. The first frequency band from the high band group was selected as defined in the IEEE 802.15.6 standard, i.e., the channel number 3 at the 6489.6 MHz center frequency and the bandwidth is 499.2 MHz [5]. The IEEE 802.15.6 standard defines the uncoded bit rate of 250 kbps for FM-UWB as a mandatory rate [5]. When using the BCH code (63,51), the bit rate is 202.5 kbps. A modulating-carrier signal in FSK follows a triangular waveform. The selected parameters gives $N_d = 27$ and $O = 1.038$, which means that the FM demodulator is working almost an optimal range [7].

The applied channel model is the IEEE 802.15.6 channel model 3 as introduced in Sect. 3.1. An in-band interference is considered to be an IEEE 802.15.4 UWB system with the same bandwidth and center frequency as a victim IEEE 802.15.6 FM-UWB system, i.e., 6489.6 MHz and 499.2 MHz, respectively [18]. The roll-off factor for the raised cosine filter is 0.5.

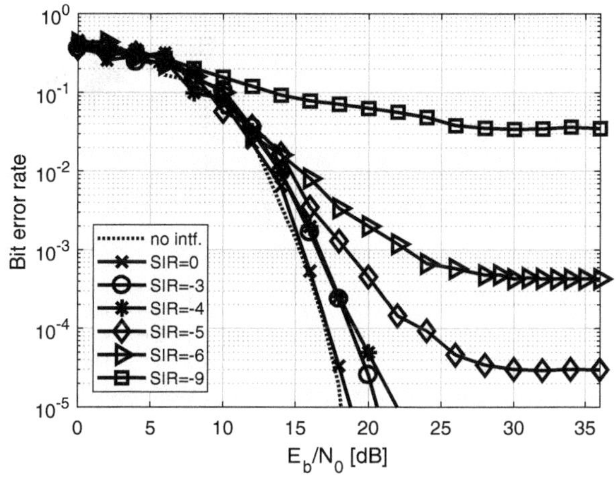

Fig. 4 BER of uncoded FM-UWB

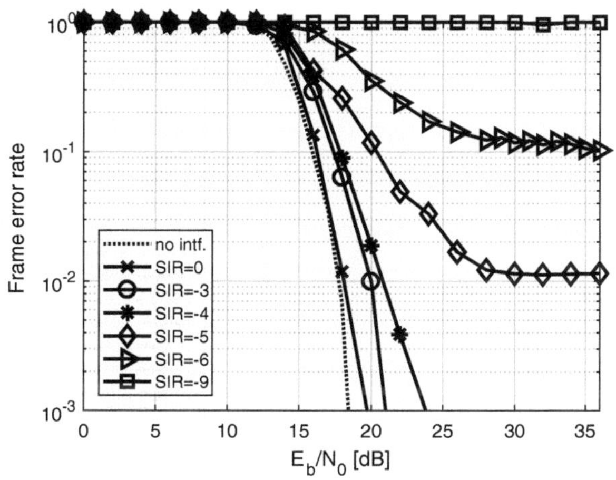

Fig. 5 FER of uncoded FM-UWB

4 Simulation Results

This section gives the simulation results of the uncoded and coded FM-UWB system. The performance of the system was simulated in the IEEE 802.15.6 channel model 3 without or with an IEEE 802.15.4 in-band interference. The channel model 3 models a body-to-body communication link.

Figure 4 and Fig. 5 represent the uncoded BER and FER performances, respectively. The dotted line without markers is the reference performance without interference. A SIR value is reduced from 0 to −9 dB, i.e., the interference power is increased. As it can be seen from the results, the performance saturates when SIR is equal or less than −5 dB. Otherwise, a reliable communication link can be established in the IEEE 802.15.6 channel model 3 even when an interferer is present.

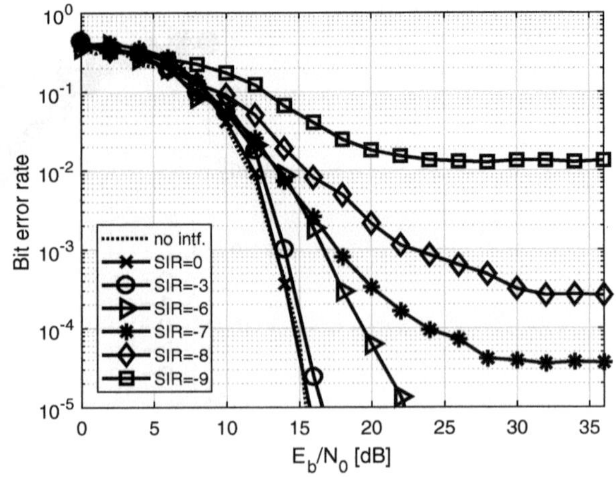

Fig. 6 BER of coded FM-UWB

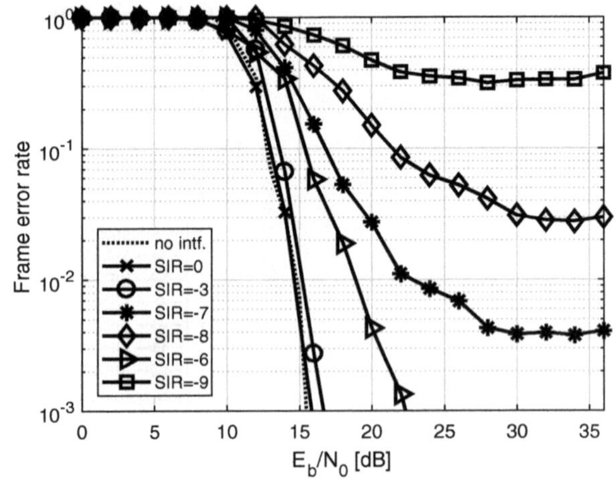

Fig. 7 FER of coded FM-UWB

The BCH code improves the system performance, and hence, it tolerates higher SIR level as showed in Figs. 6 and 7. The coding improves the system performance approximately 3 dB at the BER level of 10^{-5} without interference. The higher interference power can be tolerated with encoding, i.e., SIR of −6 dB. FM-UWB takes advantage of a large receiver processing gain to mitigate an in-band interference. After the FM demodulator, the part of the demodulated interference signal falls inside the subcarrier bandwidth. This explains a good performance with an in-band interference.

5 Conclusion

FM-UWB is a UWB PHY layer included in the IEEE 802.15.6 standard [5]. It is a simple technique where a double frequency modulation is applied to form an UWB waveform.

This paper presented the simulation results of the FM-UWB system with and without the BCH encoding in the interfered IEEE 802.15.6 channel model 3. The interference was modeled as CGN and it was intended to simulate an in-band IEEE 802.15.4 system. As the results showed, the FM-UWB system performs well in moderate interference environments, i.e., SIR is less than −6 dB.

Future work could include modeling of the IEEE 802.15.4 interference system in detail.

Acknowledgement The first author would like to thank Ulla Tuominen and Seppo Säynäjäkangas Foundations for financing the work with personal grants.

References

1. Yazdandoost, K.Y., Sayrafian-Pour, K.: Channel model for body area network (BAN). Technical Report IEEE P802.15-08-0780-09-0006, IEEE P802.15 Wireless Personal Area Networks (2009)
2. Fort, A., Desset, C., Doncker, P.D., Wambacq, P., Biesen, L.V.: An ultra-wideband body area propagation channel model: from statistics to implementation. IEEE Trans. Microw. Theor. Techniq. **54**(4), 1820–1826 (2006)
3. Taparugssanagorn, A., Raez, C.P., Isola, A., Tesi, R., Hämäläinen, M., Iinatti, J.: UWB channel modelling for wireless body area networks in a hospital. Int. J. Ultra Wideband Commun. Syst. **1**(4), 226–236 (2010)
4. Hämäläinen, M., Taparugssanagorn, A., Iinatti, J.: On the WBAN radio channel modelling for medical applications. In: Proceedings of the 5th European Conference on Antennas and Propagation (EUCAP), pp. 2967–2971 (2011)
5. IEEE: IEEE Standard for Local and Metropolitan Area Networks—Part 15.6: Wireless Body Area Networks. IEEE Std 802.15.6-2012, pp. 1–271 (2012)
6. Oppermann, I., Hämäläinen, M., Iinatti, J., (eds.): UWB: Theory and Applications. Wiley (2004)
7. Gerrits, J.F.M., Kouwenhoven, M.H.L., van der Meer, P.R., Farserotu, J.R., Long, J.R.: Principles and limitations of ultra-wideband fm communications systems. EURASIP J. Appl. Signal Process. **2005**, 382–396 (2005)
8. Gerrits, J.F.M., Farserotu, J.R., Long, J.R.: Low-complexity ultra-wide-band communications. IEEE Trans. Circuits Syst. II: Express Briefs **55**(4), 329–333 (2008)
9. Gerrits, J.F.M., Farserotu, J.R., Long, J.R.: Robustness and interference mitigation for FM-UWB BAN radio. In: 2011 5th International Symposium on Medical Information and Communication Technology, pp. 98–102 (2011)
10. Viittala, H., Hämäläinen, M., Iinatti, J.: Suitability study of DS-UWB and UWB-FM for medical applications. In: The 11th International Symposium on Wireless Personal Multimedia Communications, pp. 1–5 (2008)
11. Viittala, H., Nahar, B.N., Hämäläinen, M., Iinatti, J.: Medical applications adapting ultra wideband: a system study. Int. J. Ultra Wideband Commun. Syst. **1**(4), 237–247 (2010)

12. Viittala, H., Hämäläinen, M., Iinatti, J.: Impact of difference in WBAN channel models on UWB system performance. In: 2010 IEEE 11th International Symposiumon Wireless Personal Multimedia Communications, pp. 175–180 (2010)
13. Farserotu, J., Gerrits, J., Rousselot, J., van Veenendaal, G., Lobeira, M., Long, J.: CSEM FM-UWB Proposal. Technical Report IEEE P802.15-09-0721-00-0006, IEEE P802.15 Wireless Personal Area Networks (2009)
14. Kouwenhoven, M.H.: High-performance frequency-modulation systems. Ph.D. thesis, Technische Universiteit Delft, The Netherlands (1998)
15. Aoyagi, T., Ichi Takada, J., Takizawa, K., Sawada, H., Katayama, N., Yazdandoost, K.Y., Kobayashi, T., Li, H.B., Kohno, R.: Channel Models for WBANs—NICT. Technical Report IEEE P802.15-08-0416-04-0006, IEEE P802.15 Wireless Personal Area Networks (2008)
16. Tesi, R.: Ultra wideband system performance in the presence of interference. University of Oulu, Department of Electrical and Information Engineering, Licentiate Thesis (2004)
17. Proakis, J.: Digital Communications. Electrical Engineering Series, McGraw-Hill (2001)
18. IEEE: IEEE Standard for Local and Metropolitan Area Networks–part 15.4: Low-rate Wireless Personal Area Networks (LR-WPANs). IEEE Std 802.15.4-2011 (Revision of IEEE Std 802.15.4-2006) pp. 1–314 (2011)

Impedance Characteristics of the Skin-Electrode Interface of Dry Textile Electrodes for Wearable Electrocardiogram

Fan Xiong, Dongyi Chen, Zhenghao Chen, Chen Jin and Shumei Dai

Abstract Long-term dynamic Electrocardiogram (ECG) monitoring is considered as one of the main methods of preventing heart diseases. Ag/AgCl wet electrodes, although used clinically, are not suitable for long-time wearing. Dry textile electrodes, however, have won much attention for surmounting these drawbacks. This essay explains the impedance characteristics of the skin-electrode interface of wearable dry textile electrodes for measuring ECG. Specifically, through analyzing the characteristics of dry textile electrodes, the skin-electrode interface equivalent circuit models were built, the textile electrodes were made and the electrochemical impedance spectroscopy (EIS) for the skin-electrode interface was measured. Finally, the influence of each parameter to the interface was assessed. The research illustrated that interface of dry textile electrodes were more complicated than that of standard Ag/AgCl electrodes. The interface impedance $|Z|$ and the interface phase were relevant to the signal frequency and the key of descending the interface impedance was to lower the polarization resistance. The textile electrodes have the Constant Phase Angle Element (CPE) behavior due to the dispersion effect of the time constant within the Frequency of ECG measuring.

Keywords Wearable electrocardiogram · Textile electrode
Skin-electrode interface · Constant phase angle element

1 Introduction

Electrocardiogram (ECG) records much significant information on human heart such as the health status and the heart rate variability, which is one of the main evidences on discovering and diagnosing heart diseases [1–3]. It is of great importance to

F. Xiong · D. Chen (✉) · Z. Chen · C. Jin · S. Dai
School of Automation Engineering, University of Electronic Science and Technology of China, No. 2006, Xiyuan Ave, West Hi-Tech Zone, Chengdu 611731, China
e-mail: dychen@uestc.edu.cn

F. Xiong
e-mail: xiongfan2004@163.com

G. Fortino and Z. Wang (eds.), *Advances in Body Area Networks I*,
Internet of Things, https://doi.org/10.1007/978-3-030-02819-0_26

capture the short-term abnormal ECG changes so as to diagnose with heart diseases and save lives [2–4]. Hence, long-term monitoring of ECG is an effective method to discover and observe heart diseases. However, the Ag/AgCl wet electrodes for clinical use will get dry after some time, causing measurement errors, the gel also causes skin allergies, making them inappropriate to monitor ECG daily [4–6]. Dry textile electrodes are able to perform long-term dynamic monitoring without gel and are closer to clothing or textile [6–9], thus have drawn much attention.

Lately, a variety of textile electrodes with different textile structures (such as weaving [10], embroidery [4, 11, 12], non-woven [13], Knitting [14], etc.) have been developed. One of the obstacles to the clinically useful dry textile electrodes is that the skin-electrode interface possesses great impedance, which also varies in a wide range due to the lack of electrolyte, thus affect the quality of the ECG signal. In [12], the author used embroidery technique to make the textile electrodes, then he compared the noise and interface impedance under dry, wet and gel conditions. The result showed that the interface impedance under the dry condition was too high to measure the biopotential, while the impedance of wet electrodes was close to that of gel ones. The skin-electrode interface impedance also differed under different electrolyte conditions. In [4], the author used Ag/AgCl conductive ink on the non-woven fabrics to construct a dry textile electrode at first. And then he researched on the influence of the electrode area to the interface impedance. It indicated that the electrode area had an effect on the interface impedance. Besides the QRS wave crest amplitude relied on the area of the textile electrode and its location on the body. In [15], the author exerted pressure on the textile electrode and measured the interface impedance under different pressure. The result showed that the resistance decreased and the capacitance increased as the pressure increased. The author also stated that high interface impedance would impact the quality and waveform of the ECG signal, causing the ECG signal amplitude lower than actual ECG signal on human body. In [14], the author built the equivalent circuit model for the dry textile electrodes. Compared to the standard Ag/AgCl wet electrodes, the textile electrodes possessed strong capacity behavior as the electrolytic capacitor. Because they did not have gel, the moisture from skin and sweat would have extra resistance parallel connection to the electrolytic capacitor. According to those essays, dry textile electrodes skin-electrode interface impedance has great influence on the quality of the signal, the amplitude and accuracy of the ECG signal and the comfort level to the users. However, the problems have not been solved and require further research on the electrochemical theory and stability of the interface impedance to make dry textile electrodes available to long-term monitor the dynamic or static ECG signal.

Therefore, this essay focused on impedance characteristics of the skin-electrode interface of dry textile electrodes. Specifically, by analyzing the electrochemical theory of the skin-electrode interface of dry textile electrodes and the dielectric property of skin, we built the equivalent circuit model for the skin-electrode interface. Our textile electrodes were made of conductive silver fiber knitted fabric. Then the EIS of the skin-electrode contact interface was measured, calculated the model parameters using fitting method and analyzed the influence of each parameter to the interface. Eventually, we designed the dry textile electrodes with high stability and low interface impedance guided by these theories.

2 Skin-Electrode Interface

Assuming the design area of the electrodes is S_d; the geometric area is S (the smooth area that electrode contact with skin); the effective surface area is A; the ratio of effective area to surface area is $\gamma(\gamma \geq 1)$, which is also called effective ratio, and it is relative to the unevenness of the electrode surface [16] and the electrolyte. λ is the ratio of geometric area to design area ($0 \leq \lambda \leq 1$) called contact ratio, and it is relative to the contact condition of the electrodes and the skin. So the electrode effective area is

$$A = \gamma S = \gamma \lambda S_d \quad (1)$$

When dry textile electrode touches the skin, due to lack of electrolyte or the gel and the roughness between electrode and the skin, a part of the electrode interface touches the skin (λS_d). This part touches the skin through sweat and water, while the other part does not touch the skin ($(1 - \lambda)S_d$) because of the filled air. This is why dry electrodes can be regarded as a parallel connection of non-contacting electrode and contacting electrode. The equivalent circuit of the interface is shown in Fig. 1.

Electrode-electrolyte (sweat and water) interface and skin surface impedance track is shown as depressed semicircle, which can be represented by a parallel connection of impedance with constant phase angle and a charge transfer resistance [17, 18]. In Fig. 1a, electrode-electrolyte interface impedance can be represented by a parallel connection of constant phase impedance $Z_{CPEe(a)}$ and a resistance $R_{CT(a)}$. $Z_{CPEe(a)}$ denotes the non-faradic characteristics of the interface mainly caused by the double layer capacitor. $R_{CT(a)}$ is charge transfer resistance which represents the Faraday effect happened at the interface. Skin impedance is represented by a parallel connection of constant phase impedance $Z_{CPEs(a)}$ and a resistance $R_{s(a)}$. $Z_{CPEs(a)}$ represents the capacitance characteristics of the cuticle mainly caused by non-conductivity of

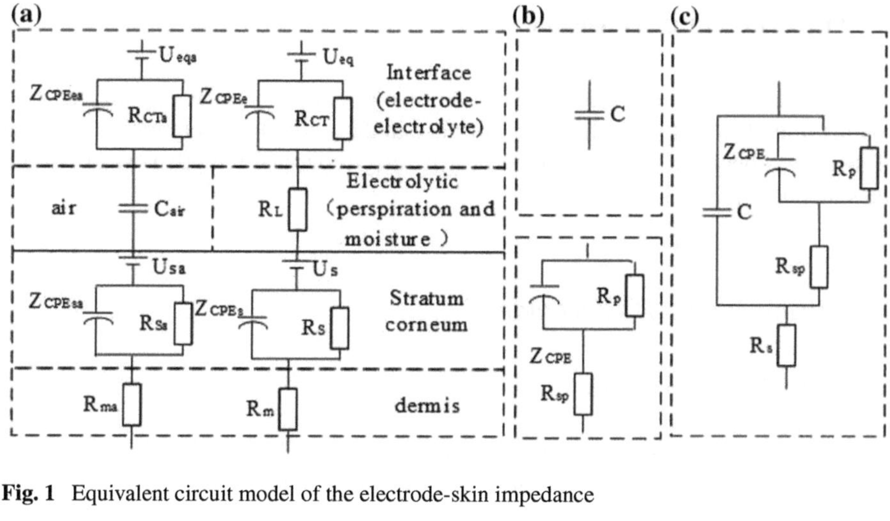

Fig. 1 Equivalent circuit model of the electrode-skin impedance

the cuticle, and $R_{s(a)}$ is an ionic current flow resistance which represents ionic current flow characteristics [18–22]. $U_{eq(a)}$ expresses electrode-electrolyte (sweat and water) electrode electric potential. $U_{s(a)}$ represents electric potential produced by the different ionic densities of different skin levels. In textile electrode perspective, R_L represents the resistance of the sweat and water in the electrode-skin interface, C_{air} represents the capacitance of air clearance in the electrode-skin interface. $R_{m(a)}$ represents the resistance of the corium layer [20–24] and subscript (a) represents the part corresponding the air medium.

Due to the lack of electrolyte medium between skin and dry electrodes, capacitance (C_{air}) is high. Non-contacting part of the electrode shows non-faradaic process, which can be regarded as a capacitor, so the equivalent circuit can be simplified as capacitance C, as shown in Fig. 1b. Contacting part of the electrode can be regarded as wet electrode so it can be simplified as shown in Fig. 1b, among which R_{sp} is the resistance of the electrolyte, R_p is the polarization resistance of electrode-electrolyte interface; The constant phase impedance, $Z_{CPE}(Z_{CPEe(a)}, Z_{CPEs(a)})$, can be calculated according to Eq. 2 as follows:

$$Z_{\mathrm{CPE}} = \frac{1}{(j\omega)^n Q} \tag{2}$$

In this equation, j is imaginary unit, ω is angular frequency, Q and n$(0 \le n \le 1)$ are constants irrelevant to frequency. If $n = 1$, Q is the capacitance of the contact area; otherwise if $n \ne 1$, Q can be regarded as parameter caused by the roughness of the surface or the continuous distribution of charge transfer time constant.

So, the whole skin-electrode interface equivalent circuit can be simplified as shown in Fig. 1c, $U_eq(a)$ and $U_s(a)$ is ignored because of AC measurement and R_s represents the sum of body resistance, wire and lead resistance.

3 Materials and Methods

In this research, textile electrode with laminated construction was built and two contact impedance of skin-electrode interface in the static state were measured. One of them was between human skin and textile electrode, and the other was between the skin and Ag/AgCl electrode.

3.1 Preparation of Electrodes

3.1.1 Ag/AgCl Electrodes

Dry electrode for ECG normally uses disposable Ag/AgCl gel electrode as the standard to compare the performance of electrodes. Here we used disposable 3M

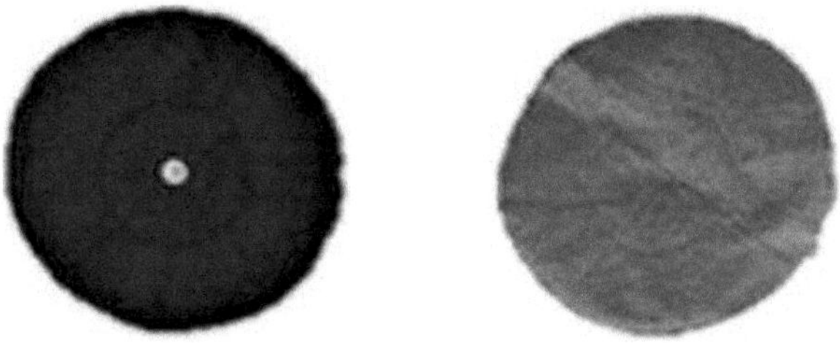

Fig. 2 Textile electrode (the radius: 2 cm)

Table 1 The radius and the area of the electrode

Electrode type	Area (cm^2)	Diameter (cm)	R (cm)	Symbols
Textile electrode	12.56	4	2	S_2
Ag/AgCl	3.14	1	0.5	$S_{Ag/AgCl}$

Ag/AgCl gel electrode (2223M, from 3M Medical Devices and Materials Manufacturing (Shanghai) Co. Ltd.) as the reference standard electrode.

3.1.2 Preparation of Textile Electrodes

In this research, we prepared 2 cm radius round dry textile electrode to study the interface impedance characteristics of the electrodes. Figure 2 illustrates the above textile electrode, the left one is front side and the right one is back side. Radius and symbol are listed in Table 1.

Dry textile electrode used concentric circular laminated structure, as depicted in Fig. 3. This electrode consists of joint snap, shield layer, buffer layer and sensor layer.

The shield layer was composed of joint snap, insulation woven fabric (obtained from Qi Feng textile co., LTD) and conductive fabric (average surface resistance $\leq 0.04\ \Omega$, obtained from Da Zhi Yuan radiation protection fabric business). Joint snap used stainless steel snap. During construction, we first cut the insulation woven fabric and conductive fabric into 2 cm radius round shape. Then a hole at the center was excavated, and put the insulation woven fabric (close to mother side) and conductive fabric (close to male side) between the snap. Finally embroidered on the 1/2 of radius and the edge to sew the fabrics together.

The buffer layer was the Polyurethane foam sponges with three-dimensional hydrophilic structure and great elasticity (Polyurethane foam sponges, surface resistance $\leq 1\ \Omega$, obtained from Shenzhen Hua Yue Yuan electronic co., LTD). The layers

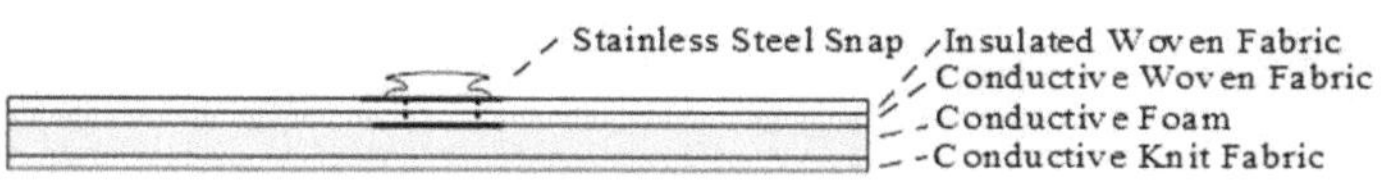

Fig. 3 Multilayered structure of textile electrode

Table 2 The radius and the area of the electrode (M^1 (male), F^2 (female))

Subjects no.	Age	BMI	Subject code
$1(M^1)$	24	24.1	A
$2(M)$	22	18.6	B
$3(M)$	21	25.6	C
$4(F^2)$	21	16.8	D
$5(F)$	24	18.2	E

were connected and packed with conductive and abrasive paste, which was obtained from SPES MEDICA.

3.2 Experiment and Data Acquisition

3.2.1 Subjects

Five healthy volunteers (3 males, 2 females) aged between 20 and 25 years old were enrolled in this study. They had BMIs of 16.8 and 25.6 kg/m^2. The results are listed in Table 2.

3.2.2 Bioimpedance Spectroscopy (BIS) Measurements

In this research, commercial BIS equipment (anba AT810A LCR meter from Applent Instrument Inc. China) with the double-electrode structure and frequency ranging from 10 Hz to 20 kHz was used to measure the skin-electrode contact impedance spectroscopy of both the standard wet electrode and the dry textile electrode ($Z_{bipolar} = Z_{body} + 2 * Z_{skin\text{-}electrode}$). Equivalent circuit and measuring schematic diagram are shown in Fig. 4.

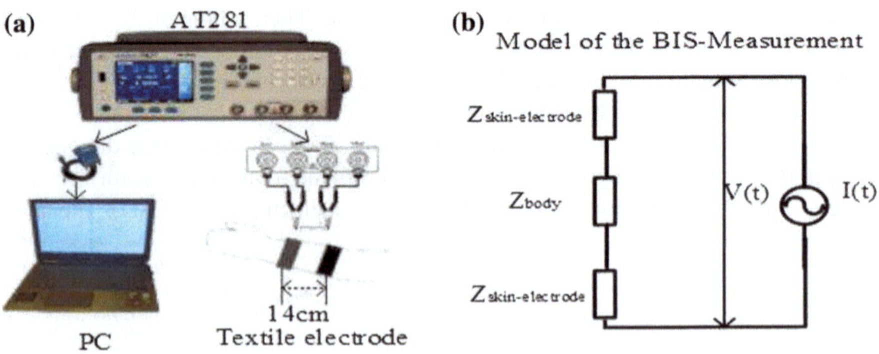

Fig. 4 Bipolar electrode configuration (**a**) and equivalent circuit (**b**) for the measurement of the skin-electrode impedance

Elastic band was used to fasten the textile electrode on the inside of left forearm. Besides, a pressure sensor (Flexiforce A201-1, obtained from Tekscan, Inc. USA) was placed tightly between skin and the band to measure the pressure skin taken so as to keep the pressure to be 0.8 N/cm^2. The center of two electrodes was 14 cm apart. In the course of measuring, volunteers were asked to sit and relax. The record began when the signal was stable. The amplitude of the signal was set to be 100 mV and the frequency accommodation mode was manual adjustment. The skin-electrode impedance of dry electrodes was measured initially and then that of standard Ag/AgCl wet electrode.

3.3 Data Analysis

We simultaneously measured the impedance and the phase of both dry textile electrode and standard Ag/AgCl electrode with 3 objectives: (1) To measure skin-electrode interface impedance and phase; (2) To calculate parameters for the equivalent circuit model; (3) To analyze the relationship between these parameters and the interface characteristics.

4 Results and Discussion

To better analyze skin-electrode interface impedance characteristics of the dry textile electrodes and compare it with that of standard Ag/AgCl electrodes. In this research, measured the skin-electrode interface impedance and the phase angle of both dry textile electrodes and standard Ag/AgCl electrodes by having fixed pressure of 0.8 N/cm^2 to inside front arms (A, B, C, D, E) of five volunteers with dual electrodes of commercial LRC measuring instrument.

Judging from the research, the interface impedance $|Z|$ (Fig. 5 drawn by MATLAB semilogx function) within the measuring frequency range (15 Hz–1 kHz) of both textile electrode and standard Ag/AgCl electrode decreases as the signal frequency increases. Standard Ag/AgCl electrodes have less interface impedance than dry textile electrodes do under the same frequency and the impedance itself differs in various interfaces (Fig. 5). The results indicate that just as the equivalent circuit, when the frequency approaches 0, the equivalent resistance relationship is $R_e = R_s + R_{sp} + R_p$, which means that the resistance was the highest, including the interface polarization resistance, when the frequency approaches the infinite, and that equivalent resistance equals to R_s. When the frequency is not very high, it can be neglected because C is very low, thus the equivalent resistance is $R_e = R_s + R_{sp}$, meaning that the equivalent circuit is mainly voltage drop loss, not including interface polarization resistance. In different electrodes, the amplitude characteristics differ much in low frequency and tend to coincide in high frequency (Fig. 6, drawn by MATLAB plot function). This implies that to lower the interface impedance while designing the dry textile electrodes, polarization resistance in low frequency should

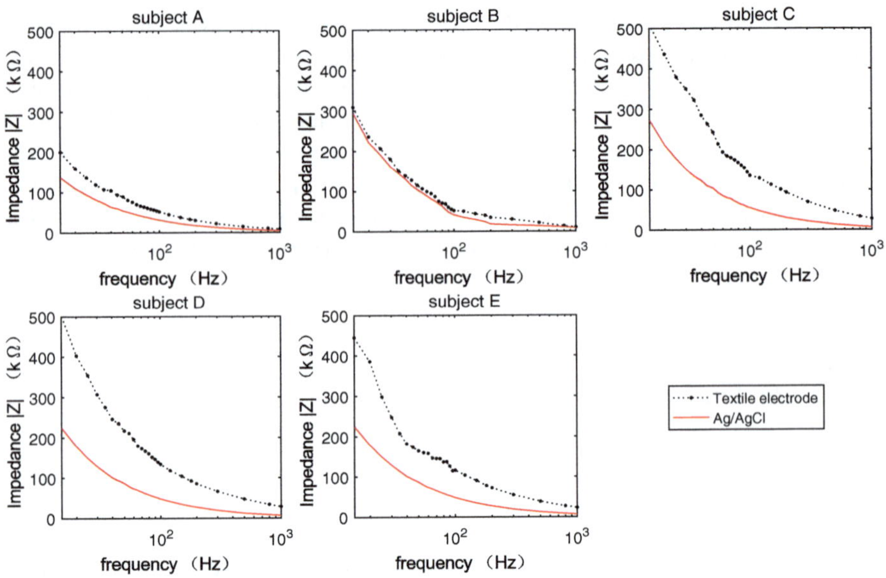

Fig. 5 Skin-electrode impedance $|Z|$ Bode of electrode

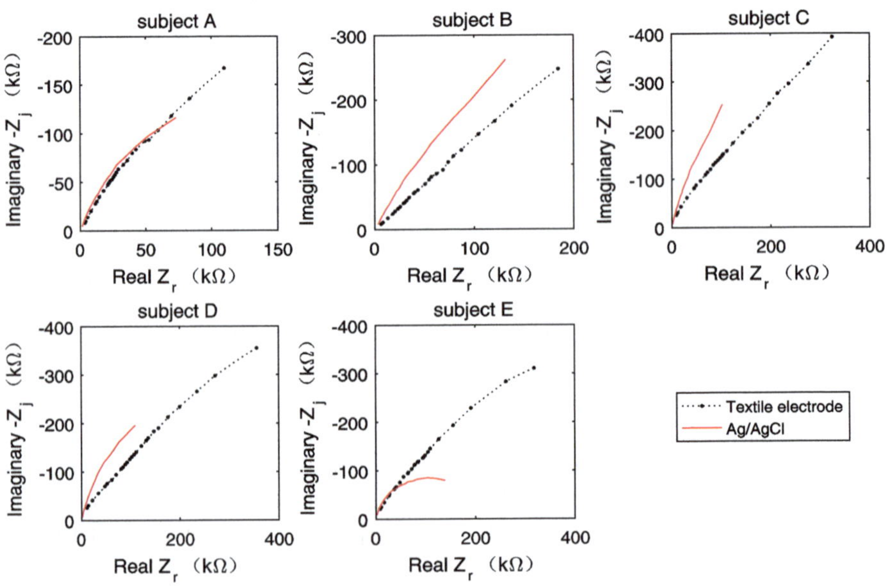

Fig. 6 Skin-electrode impedance Nyquist of electrode

be reduced, and that how to lower the resistance of the electrolyte interface (sweat and water) should also be considered, so the textile electrode should be hydrophilic and kept wet.

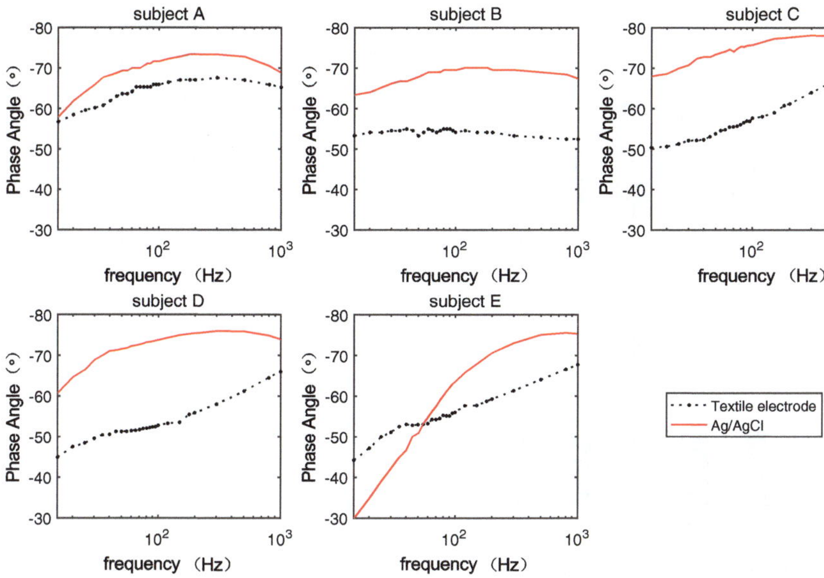

Fig. 7 Skin-electrode impedance Phase Angle Bode of electrode

According to the phase angles of the interface, the phase of skin-electrode interface is negative and it changes along frequency with the tendency of increasing and then decreasing within the range of the 15 Hz–1 kHz. The average phase angle value of the standard Ag/AgCl electrode is higher than that of fabric dry electrode (Fig. 7). It is possibly because the electrolyte conductivity is higher between skin and standard Ag/AgCl electrode, or the dielectric constant is low within the measuring frequency range, or the surface is smoother.

To better understand the impedance characteristics of the skin-electrode interface of dry textile electrodes, skin-electrode interface impedance and phase angle are used to match the skin-electrode interface equivalent circuit model parameters and compare it with the standard Ag/AgCl interface equivalent circuit model parameters. To improve fitting accuracy, the fitting frequency range is 15 Hz–20 kHz.

From the resistance parameters gotten from model fitting (Fig. 8, drawn by using MATLAB errorbar and bar function), resistance parameter R_s of dry textile electrodes skin-electrode interface equivalent circuit (Fig. 1b) is the sum of bulk resistance, wire resistance and electrode resistance, which is smaller than 1.5 k after these 5 measurements. R_s is smaller with dry textile electrodes than that with standard Ag/AgCl electrodes in B and E, but in A, C and D it leads to opposite result. The above result can be influenced by the area of dry textile electrode. R_{sp} represents electrolyte resistance of the interface. Dry textile electrodes have higher R_{sp} than standard Ag/AgCl electrodes in all 5 cases and the statistics differs greatly with great fitting error dispersion. But the standard Ag/AgCl electrodes have similar R_{sp} and less fitting error dispersion. This is because electrolyte between dry textile electrode

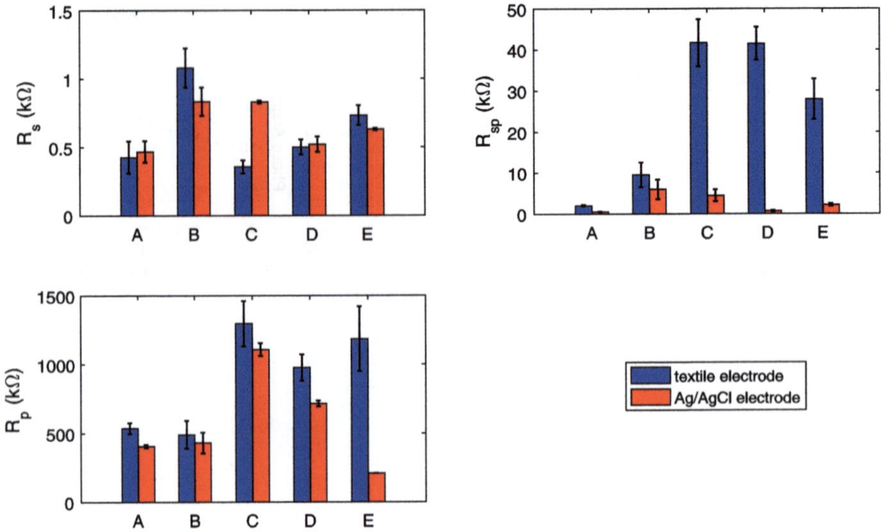

Fig. 8 Equivalent circuit parameters ($R_sR_{sp}R_p$) of skin-electrode impedance

and skin consists partly of sweat and water and also because the situation is different under different skin condition, while the standard Ag/AgCl electrodes are filled with conductive gel which has high and evenly distributed conductivity. R_p represents the polarization resistance of the interface. Apart from standard Ag/AgCl electrodes in E, R_p is all higher than 400 k, and dry textile electrodes have higher R_p because it is affected by electrode material, electrolyte, evenness of the electrode surface. Above all, interface polarization resistance is one-magnitude-order larger than any other resistance parameter, since which lowering the polarization resistance is one of the keys to lower the interface impedance.

From the CPE parameters and capacitance parameters gotten from model fitting (Fig. 9 drawn by MATLAB errorbar and bar function), the parameter C of dry textile electrode skin-electrode interface equivalent circuit model (Fig. 1a) is the plate capacitance formed between skin and electrode part which does not touch the skin or the air medium, and it is relevant to the electrode area, roughness of surface and oxidation blocking etc. Fitting data is within 20 nf using 2 cm radius round dry textile electrodes, which can be neglected within the frequency range of ECG measuring. However, Fig. 9 shows that standard Ag/AgCl electrodes whose parameter C is high than dry textile electrodes get unusual results, which is possibly because of the different distances and dielectric constants of the medium between skin and electrodes. Parameter Q and n are irrelevant to constant phase impedance and frequency (Eq. 2), Q represents the capacitance of ideal interface capacitor when $n = 1$. When $n \neq 1$,

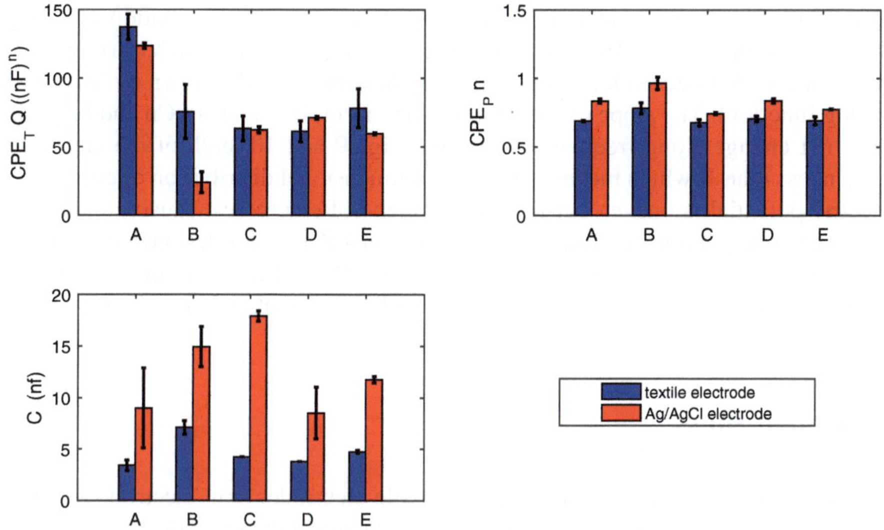

Fig. 9 Equivalent circuit parameters (Q n C) of skin-electrode impedance

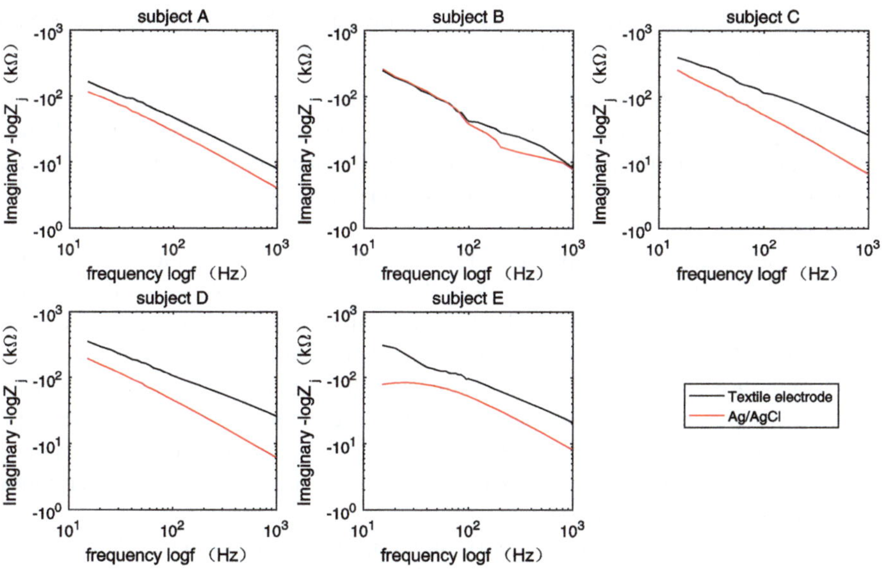

Fig. 10 Imaginary part Z_j Bode of skin-electrode impedance

Q is formed by uneven interface surface or the continuous distribution of charge transfer time parameter. Parameter n represents CPE parameter which is irrelevant to frequency within certain frequency range as shown in Fig. 10 (drawn by MATLAB loglog function). The slope of the straight line in Fig. 10 equals to Cn and basically does not change along frequency. From the Figs. 9 and 10, both of the electrodes have n less than 1, which means that the two-dimensional distribution of current and electric potential is uneven because of the shape of electrodes, causing dispersion effect on time parameter. However, standard Ag/AgCl electrodes have n high than dry textile electrodes, meaning worse dispersion effect of time parameter and more uneven on the two-dimensional distribution of both current and electric potential.

5 Conclusions

This essay systematically studied on dry textile electrodes skin-electrode interface impedance characteristics. Based on the electrochemical theory of the skin-electrode interface and dielectric property of the skin, skin-electrode interface equivalent circuit model was built. Combined with analysis of skin-electrode interface EIS, to theorize on designing dry textile electrodes and to provide ways to build dry textile electrodes whose performance was as good as standard wet electrodes.

(1) According to EIS, impedance $|Z|$ of the same electrode decreases as the frequency increases, and also different electrodes have different magnitude characteristic in low frequency and similar magnitude characteristic in high frequency. It means that polarization resistance should be lower while designing dry textile electrodes.
(2) Skin-electrode interface phase differs with different kind of electrodes. In the frequency range 15 Hz–1 kHz, skin-electrode interface phase is negative and differs along frequency with the tendency of increasing and then decreasing, standard Ag/AgCl electrodes has phase angle higher than dry textile electrodes.
(3) Dry textile electrodes skin-electrode interface has higher polarization impedance, lowering polarization resistance is one of the key to lower interface impedance. Therefore, while designing dry textile electrodes, materials with lower polarization resistivity should be used to increase effective area and to lower polarization resistance under certain area.
(4) Combining electrochemical theory and skin-electrode interface characteristic with interface equivalent circuit model analysis reveals that CPE behavior is caused by dispersion effect of time parameter which exists on dry textile electrodes within ECG measuring frequency. Thus CPE behavior should be averted, which means increasing Q and n.

However, designing high-quality textile ECG monitor still remains a challenge. Further work will continue to research on the effects of textile electrodes area and geometrical shape, especially the influence of CPE behavior. Follow-up studies should also focus on long-term stability and motion artifact while monitoring ECG signal under dynamic condition with dry textile electrodes.

Acknowledgements This work is supported by National Natural Science Foundation of China (no. 61572110) and National Key Research and Development Plan of China (no. 2016YFB1001401).

References

1. Sahoo, S., Biswal, P., Das, T., Sabut, S.: De-noising of ECG signal and QRS detection using Hilbert transform and adaptive thresholding. Procedia Technol. **25**, 68–75 (2016)
2. Weder, M., Hegemann, D., Amberg, M., Hess, M., Boesel, L.F., Abcherli, R., Meyer, V.R., Rossi, R.M.: Embroidered electrode with silver/titanium coating for long-term ECG monitoring. Sensors **15**, 1750 (2015)
3. Chi, Y.M., Jung, T.P., Cauwenberghs, G.: Dry-contact and noncontact biopotential electrodes: methodological review. IEEE Rev. Biomed. Eng. **3**, 106–119 (2010)
4. Yokus, M.A., Jur, J.S.: Fabric-based wearable dry electrodes for body surface biopotential recording. IEEE Trans. Biomed. Eng. **63**, 423 (2016)
5. Meziane, N., Yang, S., Shokoueinejad, M., Webster, J.G., Attari, M., Eren, H.: Simultaneous comparison of 1 gel with 4 dry electrode types for electrocardiography. Physiol. Meas. **36**, 513 (2015)
6. Meziane, N., Webster, J.G., Attari, M., Nimunkar, A.J.: Dry electrodes for electrocardiography. Physiol. Meas. **34**, R47–R69 (2013)
7. Dai, M., Xiao, X., Chen, X., Lin, H., Wu, W., Chen, S.: A low-power and miniaturized electrocardiograph data collection system with smart textile electrodes for monitoring of cardiac function. Australas. Phys. Eng. Sci. Med. **39**, 1–12 (2016)
8. Andreoni, G., Fanelli, A., Witkowska, I., Perego, P., Fusca, M., Mazzola, M., Signorini, M.G.: Sensor validation for wearable monitoring system in ambulatory monitoring. In: International Conference on Pervasive Computing Technologies for Healthcare, pp. 169–175 (2013)
9. Xu, P.J., Zhang, H., Tao, X.M.: Textile-structured electrodes for electrocardiogram. Text. Prog. **40**, 183–213 (2008)
10. Mottaghitalab, V., Haghi, A.K., Haghdoost, F.: Comfortable textile-based electrode for wearable electrocardiogram. Sens. Rev. **35**, 20–29 (2015)
11. Pola, T., Vanhala, J.: Textile electrodes in ECG measurement. In: International Conference on Intelligent Sensors, Sensor Networks and Information, pp. 635–639 (2007)
12. Puurtinen, M.M., Komulainen, S.M., Kauppinen, P.K., Malmivuo, J.A.V.: Measurement of noise and impedance of dry and wet textile electrodes, and textile electrodes with hydrogel. In: International Conference of the IEEE Engineering in Medicine and Biology Society, p. 6012 (2006)
13. Beckmann, L., Neuhaus, C., Medrano, G., Jungbecker, N., Walter, M., Gries, T., Leonhardt, S.: Characterization of textile electrodes and conductors using standardized measurement setups. Physiol. Meas. **31**, 233 (2010)
14. Marozas, V., Petrenas, A., Daukantas, S., Lukosevicius, A.: A comparison of conductive textile-based and silver/silver chloride gel electrodes in exercise electrocardiogram recordings. IEEE Trans. Instrum. Meas. **63**, 1412–1422 (2014)
15. Taji, B., Shirmohammadi, S., Groza, V., Batkin, I.: Impact of skin-lectrode interface on electrocardiogram measurements using conductive textile electrodes. J. Electrocardiol. **44**, 189 (2011)
16. Norlin, A., Pan, J., Leygraf, C.: Investigation of interfacial capacitance of Pt, Ti and TiN coated electrodes by electrochemical impedance spectroscopy. Biomol. Eng. **19**, 67 (2002)

17. Mcadams, E.T., Jossinet, J., Lackermeier, A., Risacher, F.: Factors affecting electrode-gel-skin interface impedance in electrical impedance tomography. Med. Biol. Eng. Comput. **34**, 397–408 (1996)
18. Mcadams, E.T., Jossinet, J.: Tissue impedance: a historical overview. Physiol. Meas. **16**, A1–13 (1995)
19. Rosell, J., Colominas, J., Riu, P., Pallas-Areny, R., Webster, J.G.: Skin impedance from 1 Hz to 1 MHz. IEEE Trans. Biomed. Eng. **35**, 649–651 (1988)
20. Mcadams, E.: Biomedical Electrodes for Biopotential Monitoring and Electrostimulation, pp. 31–124. Springer, US (2011)
21. McAdams, E.T., Jossinet, J.: DC nonlinearity of the solid electrode-electrolyte interface-impedance. Innov. Technol. Biol. Med. **12**, 329–343 (1991)
22. Kerner, Z., Pajkossy, T.: On the origin of capacitance dispersion of rough electrodes. Innov. Technol. Biol. Med. **46**, 207–211 (2015)
23. Webster, J.G.: Medical instrumentation-application and design. J. Clin. Eng. **3**, 306 (1998)
24. Webster, J.G., Clark, J.W.: Medical Instrumentation: Application and Design, pp. 197–221. Mifflin, Houghton

Compact Antipodal Vivaldi Antennas for Body Area Communication

Xiao Fang, Mehrab Ramzan, Qiong Wang and Dirk Plettemeier

Abstract In this paper, compact on-body antipodal Vivaldi antennas (AVAs) are proposed for body area network (BAN), which work at the lower ultra-wide-band (UWB). The antennas are modified by employing the tapered slot edge (TSE) to lower the resonance frequency and improve the radiation characteristics. In order to increase the gain of antenna further so that improve the transimission efficiency from in-body to on-body, two compact on-body AVA arrays are designed operating at the lower UWB. The polarization of the two antenna arrays are linear and mutually perpendicular with each other. In addition, each of antenna arrays are composed of four single elements which are fed by optimized 1x4 Wilkinson power divider. The performance of S11 and far field patterns show that the operating band of the antenna arrays covers the lower UWB band, which signifies that the benefits of introducing TSE to lower the resonance frequency is obvious. Furthermore, the influence of the human body on antenna array is discussed and the results satisfies the requirement of in-body to on-body communication.

1 Introduction

With the increasing research interest in BAN for medical and healthcare devices, the demand for miniaturized and high-efficiency electronic devices is increasing. Many kinds of BAN applications have been proposed, especially, one of the most interesting area is wireless capsule endoscope (WCE). In order to achieve high data rate and real-time video imaging transform for communicating between implanted and wearable devices, the UWB is the feasible operating frequency band for the transmitter and receiver systems. The co-authors have designed a wearable dual-polarized antenna array for in-body to on-body UWB communication [1]. To reduce

X. Fang (✉) · M. Ramzan · Q. Wang · D. Plettemeier
TU Dresden, Dresden, Germany
e-mail: xiao.fang@tu-dresden.de

M. Ramzan
e-mail: mehrab.ramzan@tu-dresden.de

G. Fortino and Z. Wang (eds.), *Advances in Body Area Networks I*,
Internet of Things, https://doi.org/10.1007/978-3-030-02819-0_27

the dimensions and increase the gain of antenna array, a modified AVA array has been proposed.

In this paper, the polarization of the designed two antenna arrays are linear and mutually perpendicular with each other. And each of antenna arrays are composed of four single elements which are fed by optimized 1x4 Wilkinson power divider. TSE is employed to reduce the dimensions and increase the gain and radiation efficiency. To obtain strong broadside radiation, a metallic reflector is positioned above the antenna array, and the distance between antenna and reflector is optimized. Based on the proposed idea, a full EM wave simulation has been implemented to get the retain loss and far field patterns of antenna arrays. In addition to that, the influence of human body on the antenna is also investigated. Different separation distance between the antenna array and human body dramatically influence the performance of the antenna. Finally, the optimized parameters of antenna array are proposed to improve the transmission efficiency for in-body to on-body communication.

2 Modified Antipodal Vivaldi Antenna

As the first step, conventional AVA is designed as shown in Fig. 1a. For this type of Vivaldi antenna, the two layers are positioned on the opposite sides of the substrate. The antenna is fed by a simple microstrip line which is conveniently achieve 50 Ohm matching. In order to reduce the dimensions and make the antenna independent of the influence of the human body, large permittivity substrate is employed, which is Rogers RO 3010 (thickness = 0.25 mm, $\varepsilon_r = 10.2$). The dimensions of the antenna is 40 mm × 45 mm, and the width of the feed line is 0.28 mm.

Figure 1b shows the modified AVA. On each patch surface, a pair of TSE is etched, each TSE is composed of two circular arcs, this idea is extension to the work of the report of Peng and Yong [2, 3]. The configuration takes full use of the path area and eliminates the current discontinuity on the edges and the parameters of arcs are optimized to satisfy the lowest-frequency requirement. Through employing TSE on

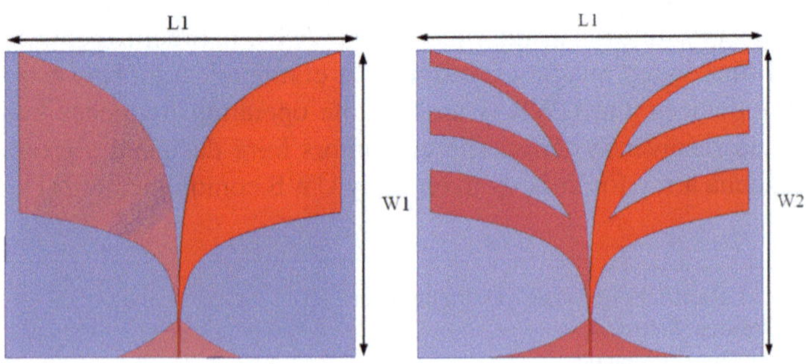

Fig. 1 Designed antipodal Vivaldi antenna. Left: original AVA, Right: modified AVA with tapered slot edge

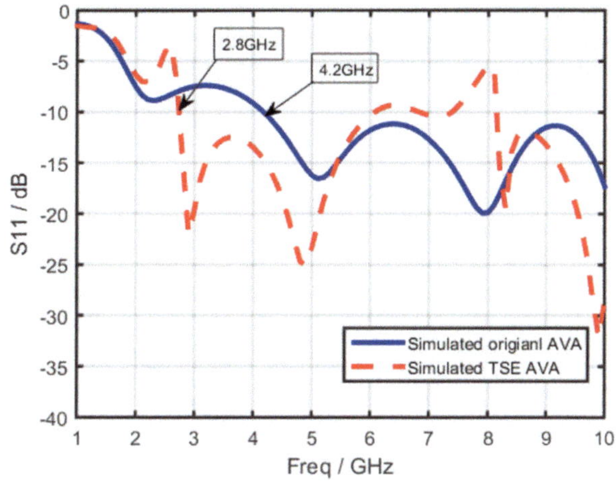

Fig. 2 S11 performance of original AVA and TSE AVA

Table 1 Parameters of AVA and modified AVA

Parameters	Length (mm)
L1	45
W1	40
L2	45
W2	40

the surface of the antenna to miniature the size of antenna and improve the current distribution at the edges of the antenna so that the front to back ratio is improved and the radiation efficiency can be increased and resonance can be shift to lower frequency.

Figure 2 illustrates the S11 variation of conventional AVA and TSE AVA. As shown in Fig. 2, the lower-end resonance frequency of conventional AVA is 4.2 GHz, which doesn't satisfy the requirement of UWB band. For the TSE AVA, without enlarging the dimension of the antenna, TSE is employed to shift the frequency to 2.8 GHz, which fits/satisfies the S11 lower UWB band.

As Table 1 shows, the whole dimensions of the original AVA and TSE AVA remain the same. Taking the electric size into account, for the original AVA, the dimensions are set to be 40×45 mm, which is approximately $0.56\lambda \times 0.63\lambda$, which is the wavelength of 4.2 GHz (the low end). In addition, for the TSE AVA, is set to be the wave wavelength of 2.8 GHz, thus the dimension is approximately $0.3\lambda \times 0.42\lambda$. Evidently, TSE is able to miniaturize the size of the antenna by lowering the minimum operating frequency.

The pattern of AVA can be seen from Fig. 4, which shows that the direction of pattern is along the plane of antenna. The realized gains varying with frequency of original AVA and TSE AVA are shown in Fig. 3. During the band between 3 and 5 GHz, the minimum realized gain of TSE AVA is 3.8 dB, and the maximum is 5.3 dB. Compared with the realized gain of original AVA, which is from 1.1 to

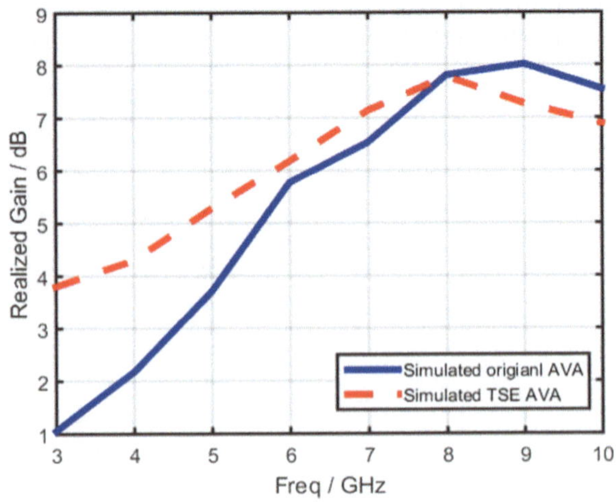

Fig. 3 Realized gain performance of original AVA and TSE AVA

3.7 dB between 3 and 5 GHz, the benefit of TSE to increase gain is prominent. The deteriorative behavior presenting in Fig. 3 is possibly caused by the irregular current along the edge of slot, especially at the high frequency (above 8 GHz), which can be accepted as the compensation for the low-end performance.

For the wireless capsule endoscope, the on-body antenna communicates with the implanted antenna, turning the direction of field pattern to be perpendicular to the plane of the antenna could further the transmission efficiency. A reflector can be employed to lead to turn the direction of the pattern and gain enhancement of the antenna by reflecting the antenna backlash energy back. In theory, the antenna gain can be increased by 3 dB, but the antenna distance from the reflector cannot be too close, otherwise, the mirror current radiation will offset the antenna radiation so that the radiation characteristics of the antenna are deteriorated. Since the plane wave perpendicularly incident on the surface of the ideal conductor producing an 180° phase difference, the reflector is usually placed at least a quarter of the wavelength

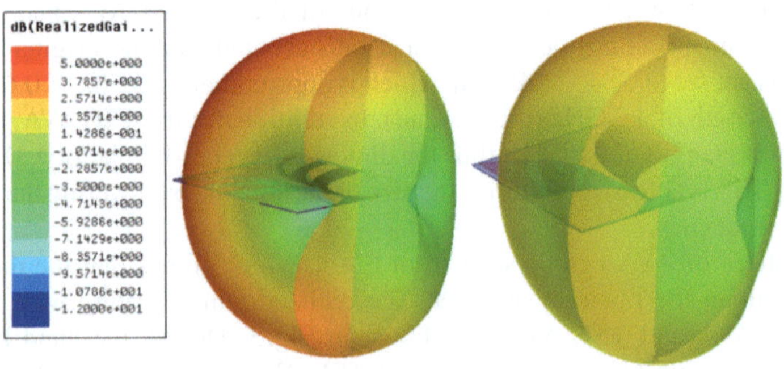

Fig. 4 Far field pattern of AVA and TSE AVA at 4 GHz

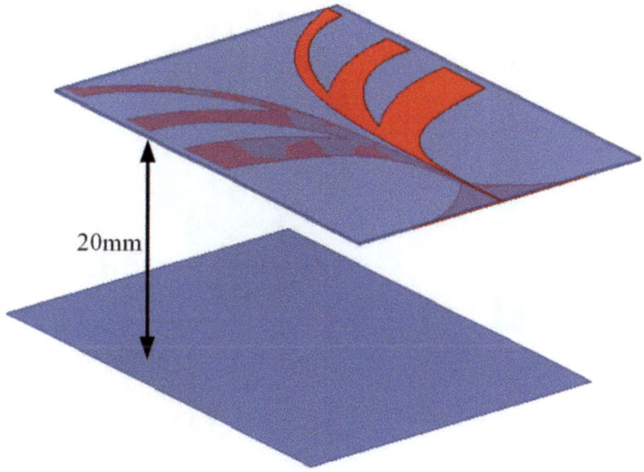

Fig. 5 TSE antipodal Vivaldi antenna with the reflector

from the antenna, then the phase difference of 180° by the antenna radiation to the reflector and then reflected to the antenna path difference can offset the 180° phase difference generated by the reflector.

Figure 5 is TSE AVA with metallic reflector. And the distance between antenna and reflector is 20 mm. From Fig. 6, after introducing the reflector, the S11 performance becomes deteriorated, especially between 3.2 and 3.7 GHz, which is possibly due to the reflected wave coupling with slot to disturb the current distribution. Despite all this, the deteriorated S11 can be accepted as compensation of gain enhancement.

As shown in Fig. 7, the direction of far-field pattern is perpendicular to the plane of the antenna, and the realized gain raises up to 7.6–10.6 dB at the lower band of UWB (3–5 GHz) comparing with the antenna without reflector whose realized gain is from 3.8 to 5.3 dB.

Moreover, in case 4 and case 5 the relative permittivity was varied from free space to 45 and a comparison was made between the dipole being direct contact with the surface and dipole at a distance of 1 mm from the surface. Figures 11 and

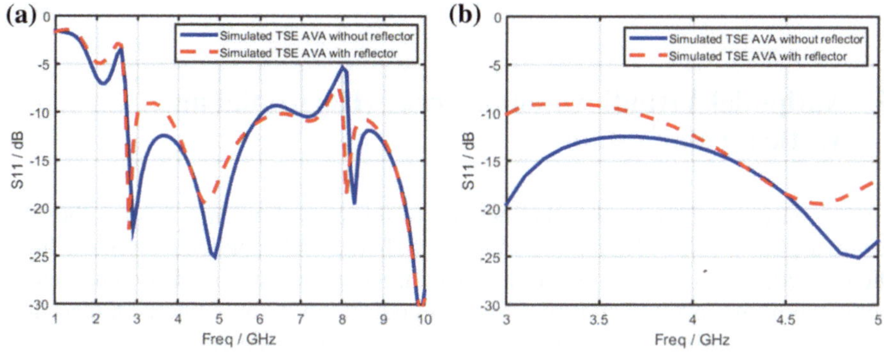

Fig. 6 S11 performances of TSE AVA with and without reflector. **a** 1–10 GHz, **b** 3–5 GHz

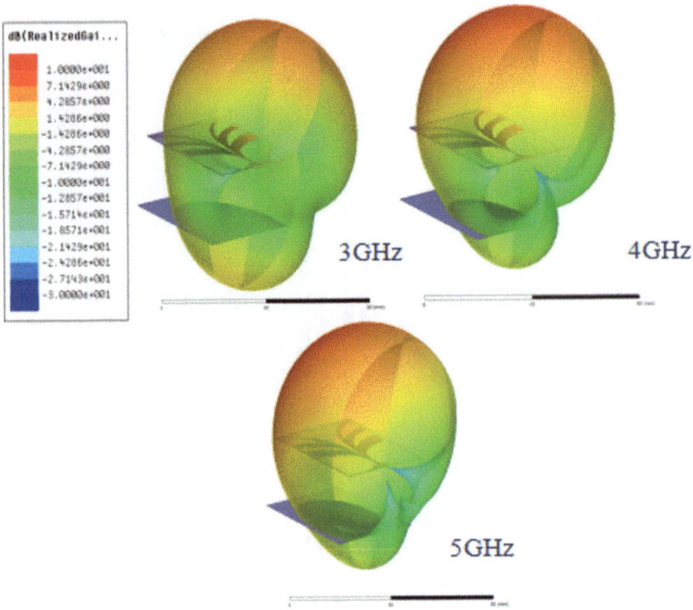

Fig. 7 Far-field pattern of TSE AVA with reflector (realized gain 3 GHz: 7.6 dB, 4 GHz: 9.8 dB, 5 GHz: 10.6 dB)

12 show the reflection coefficient results of the two cases, respectively. In case of direct contact the shift in the reflection is more prominent at different values of relative permittivity, however, in case of 1 mm distance from the surface the shift is insignificant. Figures 13 and 14 show the simulated gain pattern of the dipole in direct contact and 1 mm away from the surface, respectively. In case of direct contact, as the permittivity is increasing the side lobes starts to form more prominently in the medium. Whereas in the case of 1 mm distance from the surface the main lobe is preserved inside the medium and its directivity increases more as value of relative permittivity is increased.

3 Antipodal Vivaldi Antenna Array (Horizontal and Vertical Polarization)

For the purpose of facilitating the increase of the gain of antenna further, the AVA arrays are designed, which are shown in Fig. 8. In the WCE application senior, due to the unfixed of position and direction of the implanted antenna in abdomen, thus the polarization of wave transforming from in-body to on-body is unascertained. Due to every polarizations can be decomposed to horizontal and vertical polarizations,

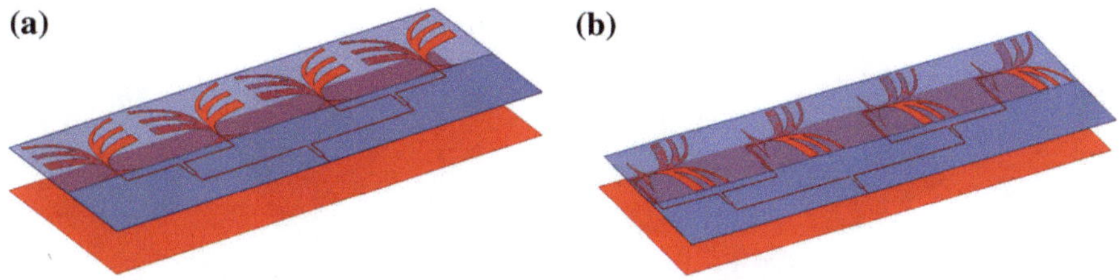

Fig. 8 Antipodal Vivaldi antenna arrays **a** horizontal polarization, **b** vertical polarization

Fig. 9 S11 performance of antipodal Vivaldi antenna arrays

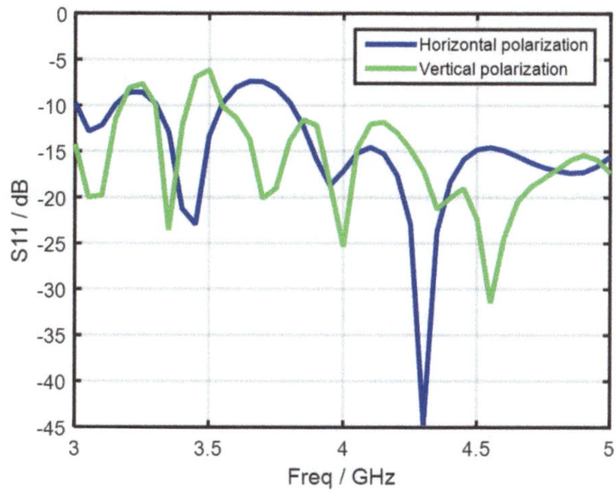

two mutually perpendicular linear polarizations can be utilized as the receiving to maximize the received power.

As Fig. 8 shows, Wilkinson power divider is designed to combine the four antennas. In addition, the dimension of the power divider is miniaturized to conform the requirement of on-body antennas. The width of the Wilkinson power divider is 35.62 mm, and the length is determined by the interval of each antenna. For the horizontal and vertical polarization array, the interval is separately 40 and 70 mm, thus the lengths are 200 and 280 mm. For the vertical polarization array, in order to eliminate the coupling of each antenna, the interval is set at 70 mm which is the approximate wavelength of center frequency 4 GHz. Moreover, two 100 Ω isolation resistances are soldered to the power divider to further isolation and enlarge antenna gain.

The S11 performances of the two arrays are shown in Fig. 9. In the band of 3–5 GHz, the S11 approximately satisfies the requirement of less than −10 dB, which can be accepted as the competence of gain enlargement and turning pattern direction. Obviously from Fig. 10, a reflector is employed to give the antenna backlash energy back and turn the direction of the pattern. The realized gain of the horizontal polarization array is from 9.5 to 12.3 dB, and the pattern is perpendicular to the antenna array. Meanwhile, for the vertical polarization array, the realized gain is from 12.5

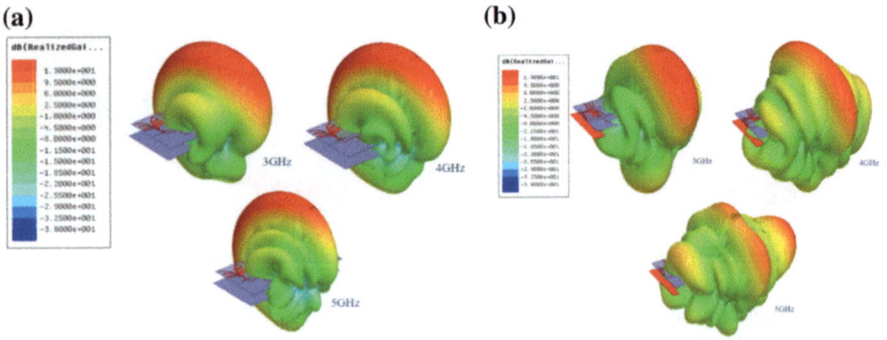

Fig. 10 Far-field patterns of AVA arrays **a** horizontal polarization, **b** vertical polarization

to 7.8 dB, the relatively decreasing trend of the gain is due to the large interval of each antenna with respect to the wavelength of high frequency.

4 The Influence of Human Body on Antipodal Vivaldi Antenna

The permittivity of the human body is high with respect to the free space so that it will affect the antenna a lot with the distance varying between the antenna and the human body, as Fig. 11 shows. The dielectric characteristics of the human body torso in lower UWB band can be approximated using an average muscle tissue. The average dielectric parameters for human torso tissue in the lower UWB band can be approximated as an average dielectric constant of 51.5 and average conductivity of 3.2 S/m. As shown in Fig. 11, the simulation configuration is that AVA with reflector is put on the human body and the distance between the human body and AVA is varied to acquire the different S11 results. The S11 performance is shown in Fig. 12.

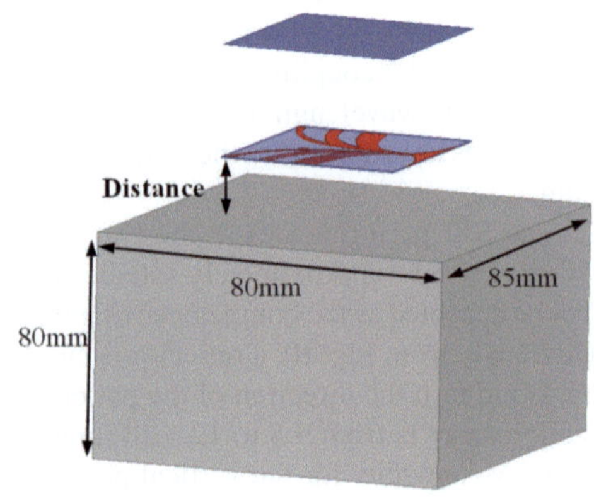

Fig. 11 AVA with reflector on human body

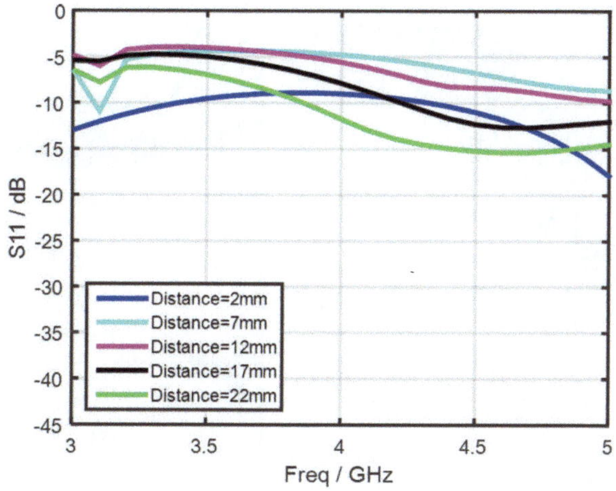

Fig. 12 S11 performance of AVA with reflector varying with the distance between antenna and human body

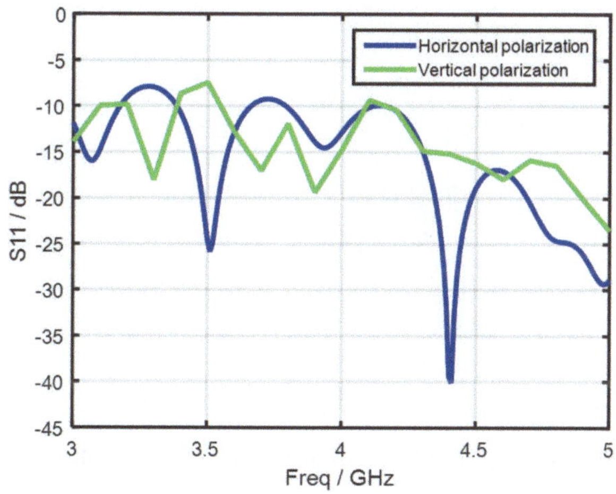

Fig. 13 S11 performance of horizontal and vertical polarization antenna arrays

In conclusion, the AVA array with reflector should be as close as possible to the human body to minimize the S11. In addition, for wireless capsule endoscope application, the on-body antenna should be close to the human body to maximize the receiving power. Thus, the distance between the array and human body is set as 2 mm. From Fig. 13, the S11 performances of horizontal and vertical polarization array fulfill the requirement.

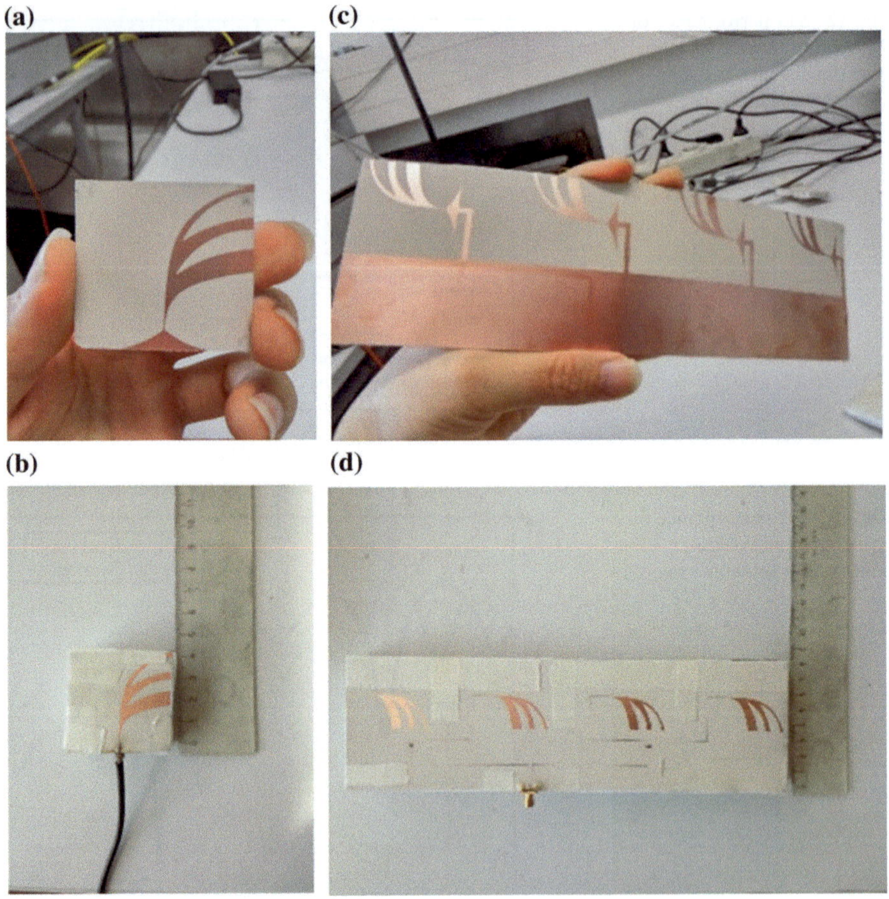

Fig. 14 Fabricated TSE AVA. **a**, **b** Signal antenna, **c**, **d** antenna array

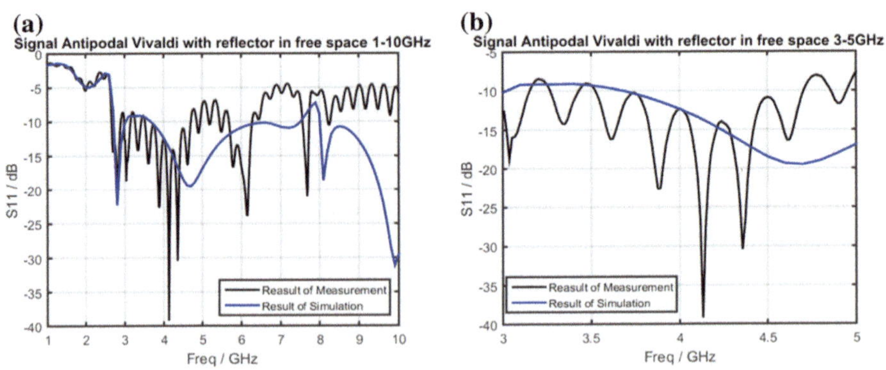

Fig. 15 Measurement and simulated S11 results of signal AVA array in free space. **a** 1–10 GHz, **b** 3–5 GHz

5 Measurement Results

The antennas are fabricated on 10-mil-thick Rogers 3010 substrate, for fixing the metallic reflector on the back, a piece of polyurethane foam is attached on, as Fig. 14. The measurement and simulation S11 in free space is shown in Fig. 15. It can be seen that the measurement and simulation results are similary with each other, especially in the lower UWB band. The ripples of the measurement results are introduced by the connector. And the antenna is measurement on the chest, the results are shown in Fig. 16. The lowest resonance frequency of antenna is extended to 1.5 GHz as the measurement results. Therefore, the antenna with optimized taper slot edge can dramatical improve the impedance bandwidth without increasing the dimensions of antenna.

The measurement results of antenna arrays in free space and on human body are shown in Figs. 17 and 18. It can be seen that the S11 performance on human body

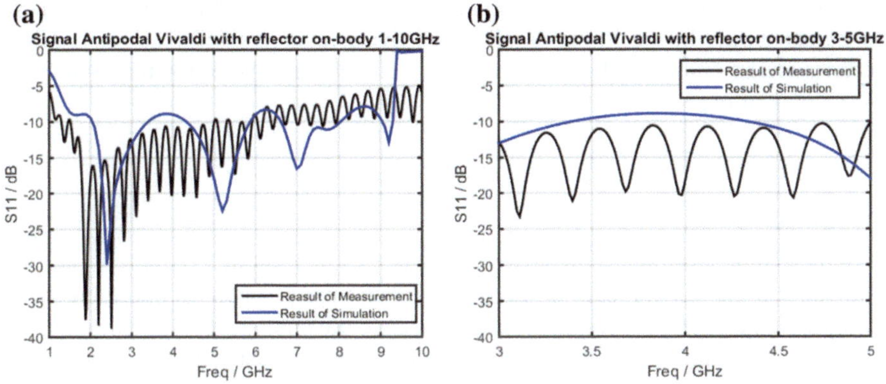

Fig. 16 Measurement and simulated S11 results on body. **a** 1–10 GHz, **b** 3–5 GHz

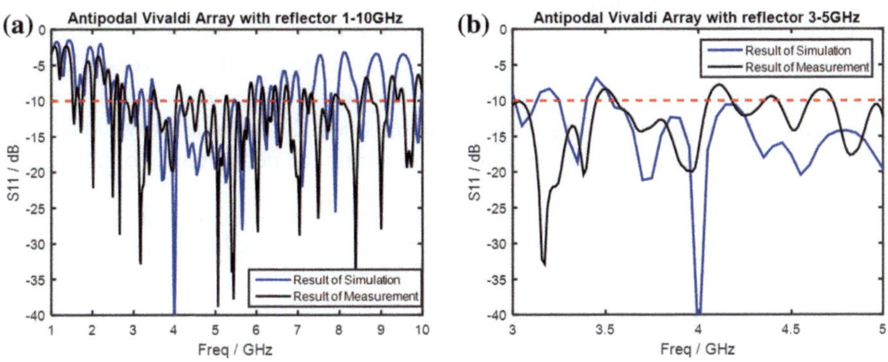

Fig. 17 Measurement and simulated S11 results of antenna array in free space. **a** 1–10 GHz, **b** 3–5 GHz

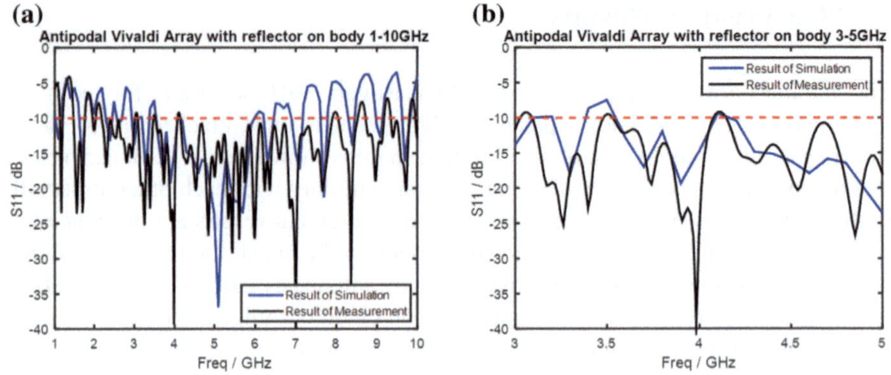

Fig. 18 Measurement and simulated S11 results of antenna array on human body. **a** 1–10 GHz, **b** 3–5 GHz

is better than that in free space, this is because that the antenna array is optimized on human body. In the lower UWB band, the measuremented S11 lower UWB performancing on human body is less than −10, which satisfies the requirement of BAN application.

6 Conclusion

The paper proposes compact on-body antipodal Vivaldi antennas operating at the lower UWB. By introducing the taper slot edge to the antenna, the lowest frequency can be lowered and the antenna achieve compact geometry, meanwhile the radiation efficiency is improved. In order to improve the transmission efficiency from in-body to on-body, two AVA antenna arrays are designed. The polarizations of the two antenna arrays are linear and mutually perpendicular with each other. And the metallic reflector has been added to turn the direction of patterns and reflect the antenna backlash energy back. Moreover, the influence of human body on antenna is investigated to optimize the performance of antenna arrays as on-body receiving antenna. It shows that when be put on the human body, the antenna arrays work well. The measurement results also are shown to illustrate the performance of antenna.

Acknowledgements This research is supported by "WiBEC" (Wireless In-Body Environment) project which has received funding from the European Union's Horizon 2020 research and innovation programme under grant agreement No. 675353.

References

1. Wang, Q., Hahnel, R., Plettemeier, D.: Wearable dual-polarized antenna array for in-body to on-body UWB communication. In: Bodynets 2014. Great Britain, London (2006)
2. Fei, P., Jiao, Y., Hu, W., Zhang, F.: A miniaturized antipodal Vivaldi antenna with improved radiation characteristics. IEEE Antennas Wirel. Propag. Lett. **10**, 127–130 (2011)
3. Bai, J., Shi, S., Prather, D.: Modified compact antipodal Vivaldi antenna for 4–50 GHz UWB application. IEEE Trans. Microw. Theory Tech. **59**(4), 1051–1057 (2011)

Study of a Dipole Antenna in the Vicinity of Lossless and Lossy Medium for On-body Antenna Analysis

Mehrab Ramzan, Xiao Fang, Qiong Wang and Dirk Plettemeier

Abstract In this paper, a detailed study of a dipole antenna in the close proximity of lossy and lossless human modeled structure is discussed. The main goal of the analysis is that which factors should be taken into account to design better on-body antennas and highlight the challenges associated with these kind of antennas in the vicinity of the lossy and lossless medium as compared to free space designs. The analysis is based on dipole separation distance from the equivalent human body structure. The antenna is analyzed in terms of shift in the frequency, reflection coefficient variation, input impedance and gain of the antenna. This analysis is very beneficial in terms of designing on body antennas and reliable wireless wearable devices. A full wave EM solver is used to demonstrate this study.

1 Introduction

From last few decades body area networks (BANs) have gained a lot of interest due their attracting applications in health monitoring system, military, sports and entertainment [1–4]. The development of reliable wireless wearable devices is rapidly growing. In these wearable systems, the position of the antenna with respect to the human body plays a vital role for in-body and on-body communication. Antenna designing for body area communication is much challenging as compared to free space antennas. It requires proper consideration of the permittivity of the surrounding material and how much the wave is slowed down and shortened [5]. In reality the

M. Ramzan (✉) · X. Fang · Q. Wang · D. Plettemeier
Chair of RF Engineering, Communication Laboratory, TU Dresden, Dresden, Germany
e-mail: mehrab.ramzan@tu-dresden.de

X. Fang
e-mail: xiao.fang@tu-dresden.de

Q. Wang
e-mail: qiong.wang@tu-dresden.de

D. Plettemeier
e-mail: dirk.plettemeier@tu-dresden.de

G. Fortino and Z. Wang (eds.), *Advances in Body Area Networks I*,
Internet of Things, https://doi.org/10.1007/978-3-030-02819-0_28

human biological tissue is lossy so it should also be taken into account for the performance analysis of body area antennas.

In this paper, a detailed study of a half wave dipole antenna in the proximity of the human body is investigated. The human body is modeled with equivalent average tissue characteristics. Initially the analysis is based on high permittivity material excluding the losses simplify the problem by giving an insight that how the waves are slowed down and how shortened wavelengths affect the performance of the antenna. Moreover, taking into consideration the losses gives a clearer analysis of the performance of the wearable antennas used for body area communication. In this study, different dipole antenna characteristics such as reflection coefficient, gain patterns, and input impedance are analyzed with various separation of the antenna with respect to lossless and lossy human body tissue. A full EM wave solver FEKO is used to demonstrate the different cases of the study.

2 In the Vicinity of Lossless Medium

The simulation setup in FEKO environment is shown in Fig. 1. The dipole is a half wavelength dipole at the designed frequency of 7 GHz. The dipole is analyzed at a height of h from the infinite half space dielectric medium having permittivity equivalent to the average value of lossless human body tissue.

Table 1 shows different cases of the study. Initial five cases are related to lossless infinite half space medium and last two cases are related to lossy medium. In case 1, the dipole height is varied in terms of λ with respect to the plane of dielectric substrate. Figure 2 shows the reflection coefficient results of the dipole with different separation distances from the surface of the medium for case 1. As dipole gets closer, the resonant frequency shifts to lower frequencies. A significant resonance shift is

Fig. 1 Dipole in the vicinity of lossless half space homogeneous dielectric medium with equivalent body tissue permittivity

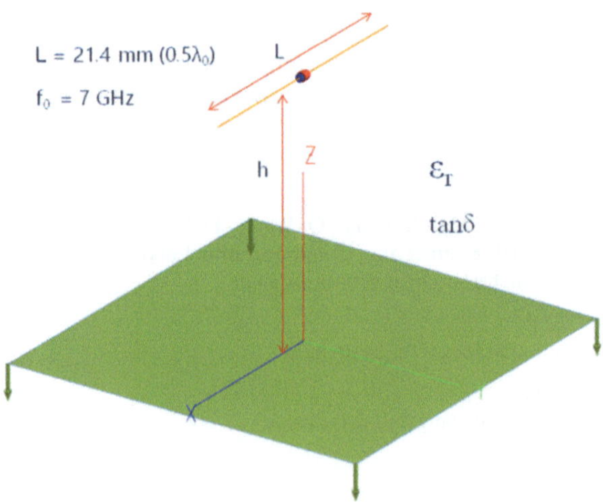

Table 1 Different cases of dipole antenna in the vicinity of lossless and lossy medium

Number	Permittivity (ε_r)	tanδ	Separation distance (h)
Case 1	50	0	$1 \rightarrow \lambda$
Case 2	50	0	$1 \rightarrow 3$ mm
Case 3	$1 \rightarrow 4$	0	0
Case 4	$1 \rightarrow 45$	0	0
Case 5	$1 \rightarrow 45$	0	1 mm
Case 6	5	$0 \rightarrow 0.1$	0
Case 7	5	$0 \rightarrow 0.1$	1 mm

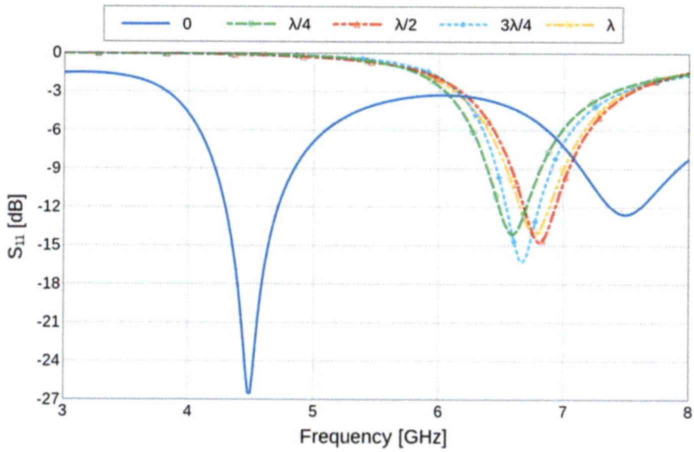

Fig. 2 Simulated reflection coefficient results for case 1

observed when the dipole becomes in contact with the medium. Figures 3 and 4 show the real and imaginary part of the input impedance of the dipole, respectively, with different separation heights from the surface of the lossless medium. In the case of full contact with the body, the imaginary values of dipole input impedance show multiple resonances due to shortening effect. Figure 5 shows the gain pattern of the dipole antenna. In the free space it is observed that patterns are showing constructive interference when the height is $\lambda/4$ and $3\lambda/4$ and destructive interference are observed when it is $\lambda/2$ and λ. Due to shortening effect in the medium much of the dipole gain is reduced and distributed among the side lobes. As the dipole gets far away from the body surface the gain of the dipole shows an increasing pattern. In case 2, the permittivity is kept same, however, the separation distance is varied in a smaller step near the surface of the lossless medium. Figures 6, 7 and 8 show the reflection coefficient, gain and gain pattern of the dipole antenna with different distance near the surface of the body. Even near to the surface the antenna resonance frequency shifts only too much when the dipole is in the direct contact with the

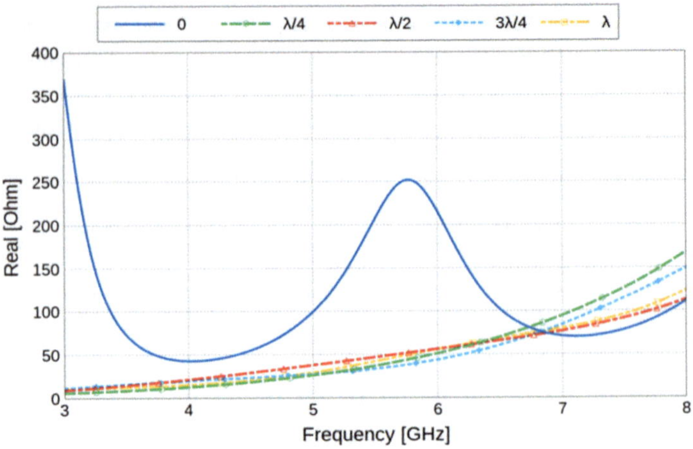

Fig. 3 Simulated real part of input impedance results for case 1

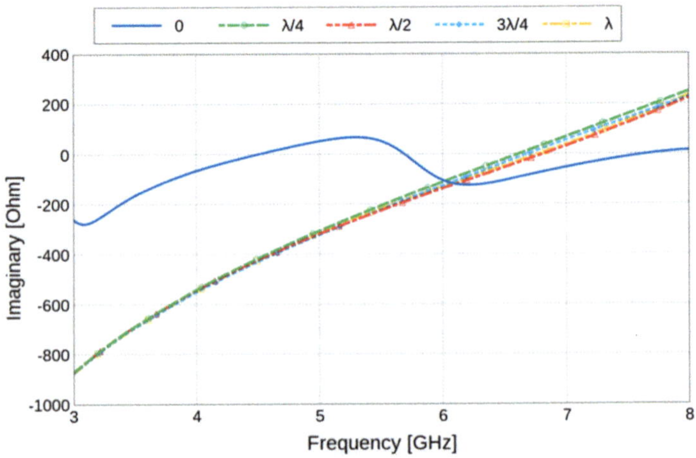

Fig. 4 Simulated imaginary part of input impedance results for case 1

surface. The side lobes start to form when the antenna is directly connected to the surface and gain starts to increase as it starts to move away from the surface.

In case 3, the dipole is in contact with the lossless half space homogeneous dielectric medium whose permittivity is varied from 1 to 4 and its influence was observed on the performance of the antenna in terms of shift in the frequency, gain, and impedance, respectively. Figure 9 shows the result of this case that as the permittivity value is increasing, the resonance frequency of antenna shifts to lower frequency

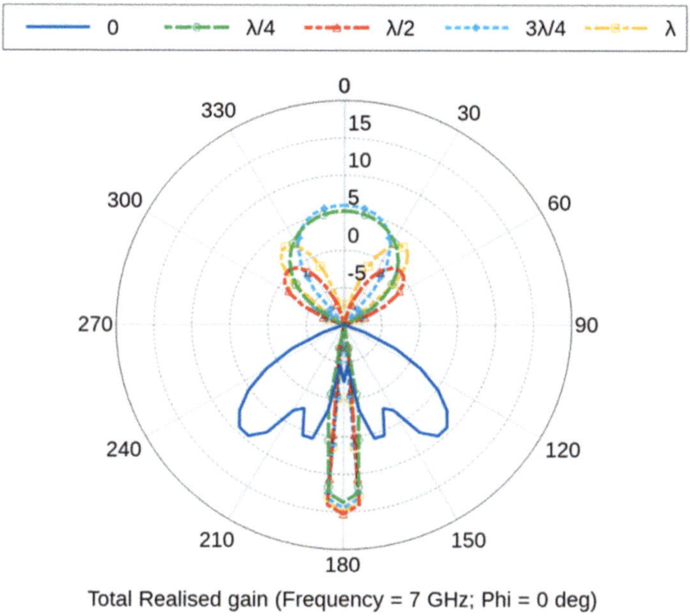

Fig. 5 Simulated realized gain pattern results for case 1

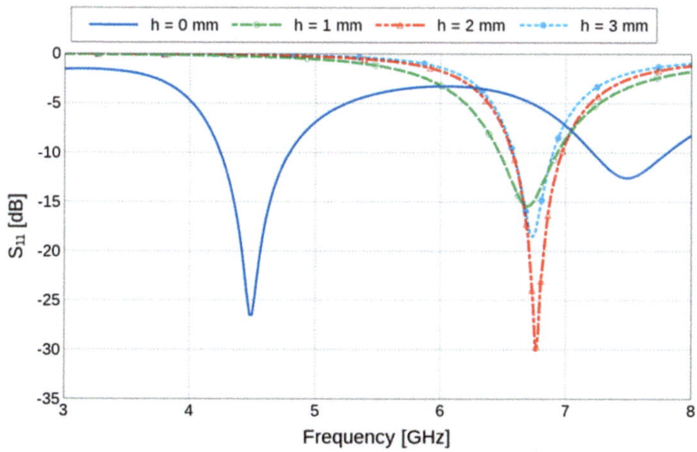

Fig. 6 Simulated reflection coefficient results for case 2

due to wave shortening effect. The effect of permittivity variance on the gain of the dipole antenna is shown in Fig. 10, which implies an increasing gain inside the medium as the permittivity value is increasing.

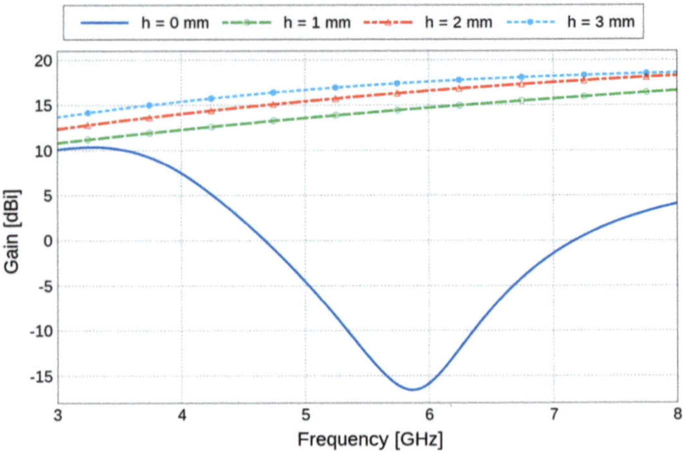

Fig. 7 Simulated gain of dipole antenna at different frequencies for case 2

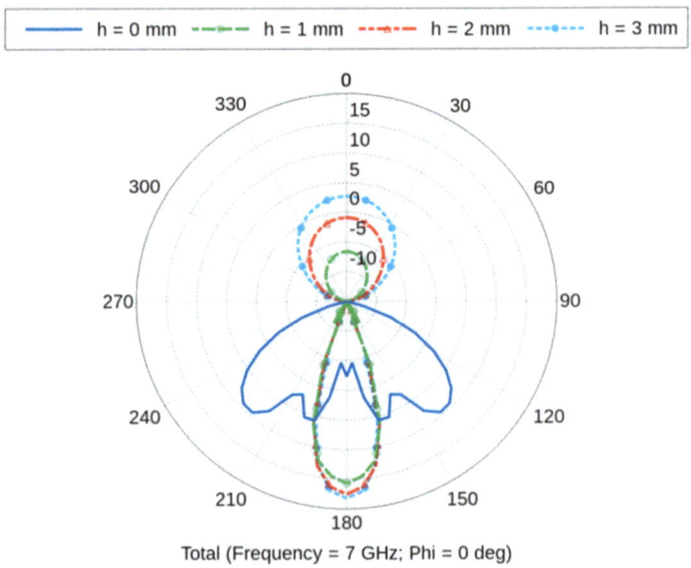

Fig. 8 Simulated gain of dipole antenna for case 2

Moreover, in case 4 and case 5 the relative permittivity was varied from free space to 45 and a comparison was made between the dipole being direct contact with the surface and dipole at a distance of 1 mm from the surface. Figures 11 and 12 show the reflection coefficient results of the two cases, respectively. In case of direct contact the shift in the reflection is more prominent at different values of relative permittivity, however, in case of 1 mm distance from the surface the shift is insignificant. Figures 13 and 14 show the simulated gain pattern of the dipole in

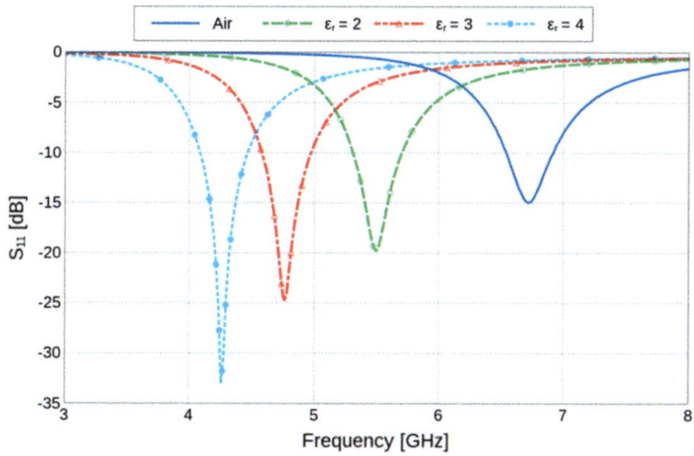

Fig. 9 Simulated reflection coefficient of dipole antenna for case 3

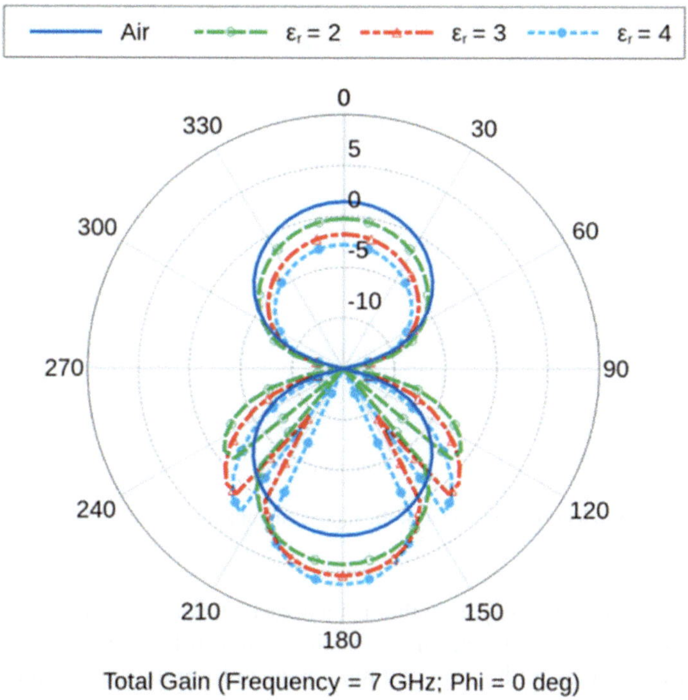

Fig. 10 Simulated gain of dipole antenna for case 3

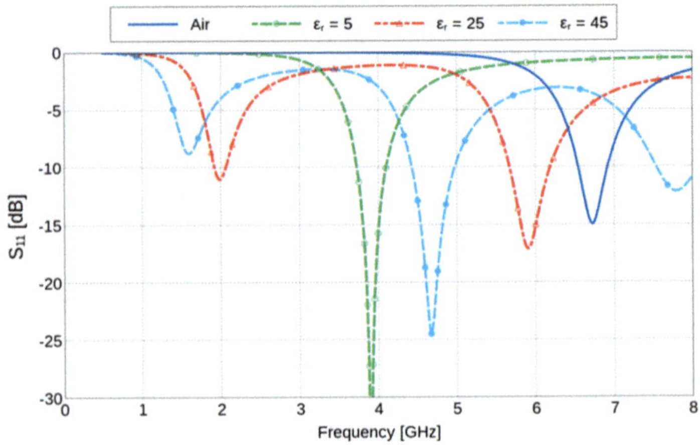

Fig. 11 Simulated reflection coefficient of the dipole antenna for case 4

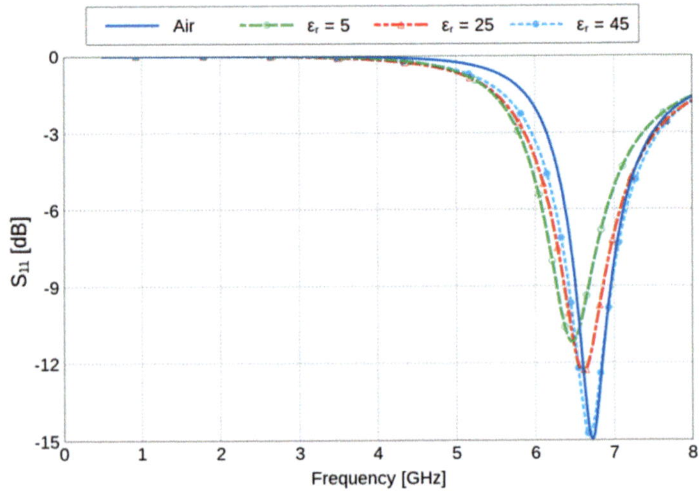

Fig. 12 Simulated reflection coefficient of the dipole antenna for case 5

direct contact and 1 mm away from the surface, respectively. In case of direct contact, as the permittivity is increasing the side lobes starts to form more prominently in the medium. Whereas in the case of 1 mm distance from the surface the main lobe is preserved inside the medium and its directivity increases more as value of relative permittivity is increased.

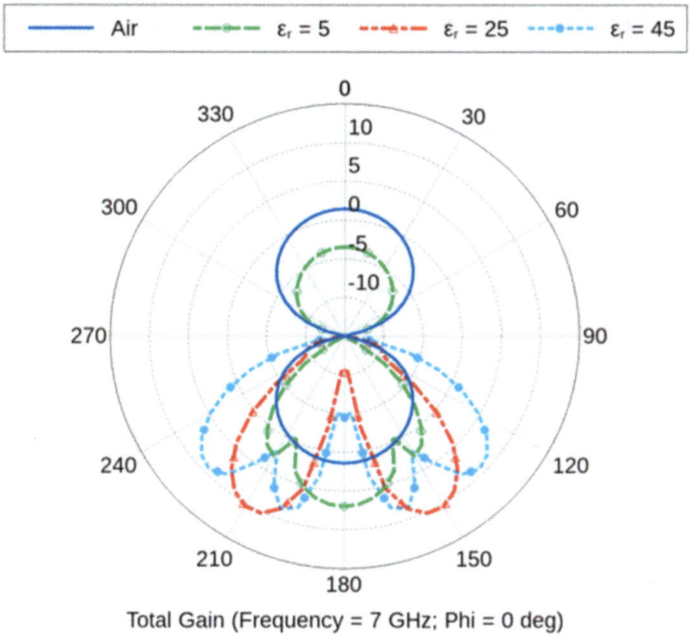

Fig. 13 Simulated gain of the dipole antenna for case 4

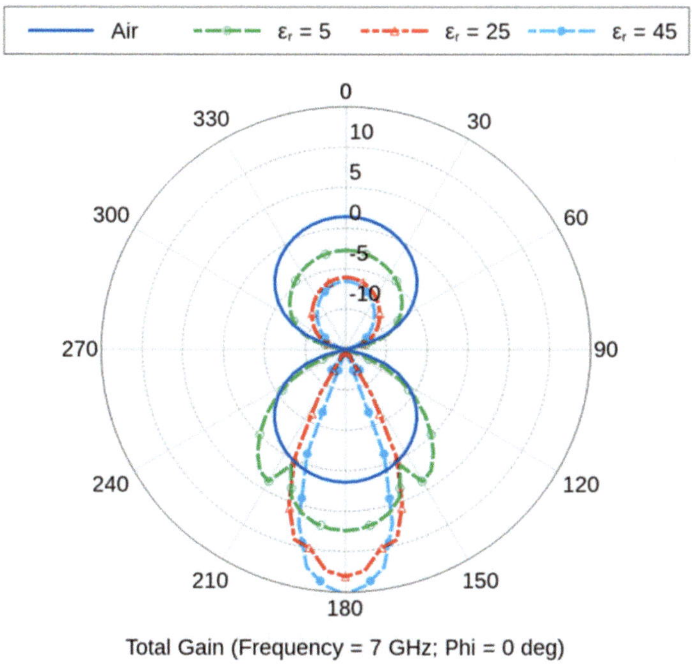

Fig. 14 Simulated gain of the dipole antenna for case 5

3 In the Vicinity of Lossy Medium

In the previous section, all the results were without losses. In reality, the human body is a lossy medium. In this part, losses are introduced in the medium and its different impacts on the antenna are analyzed. Table 1 highlights case 6 and case 7 which are associated with the losses of the medium and it shows that εr is kept constant with the value of 5 and $\tan\delta$ is varied from 0 to 0.1 to account for the losses. Figure 15 shows the reflection coefficient results of the dipole antenna in direct contact with the lossy medium. In case of direct contact, the losses influence the matching of the antenna and bandwidth is increased a little bit, however, it doesn't shift the resonance

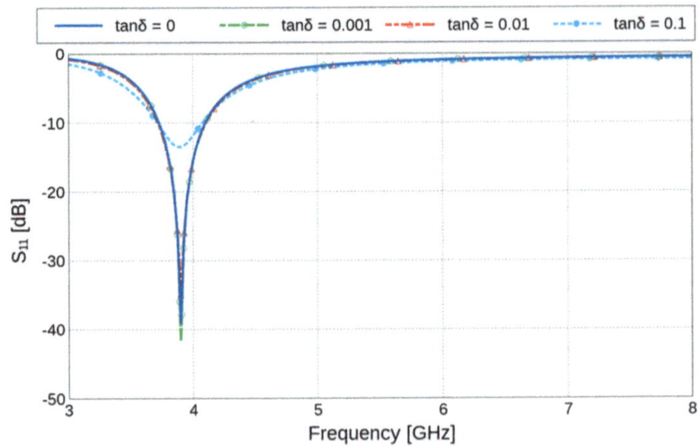

Fig. 15 Simulated reflection coefficient of the dipole antenna for case 6

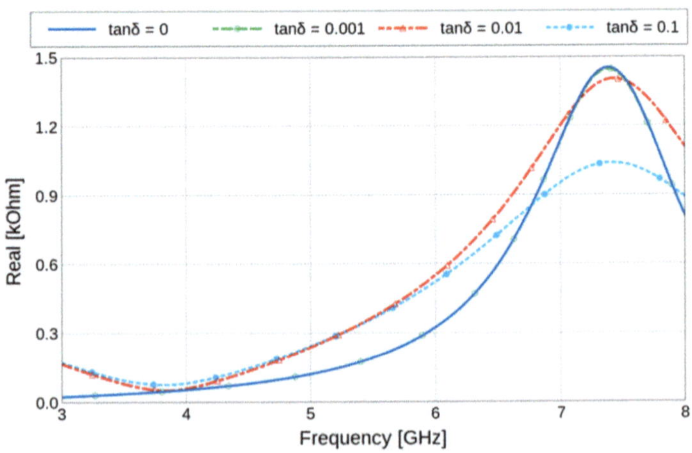

Fig. 16 Simulated real part of the input impedance of the dipole antenna for case 6

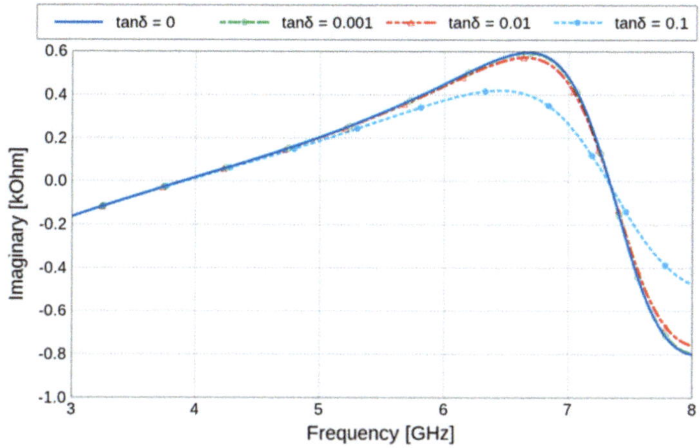

Fig. 17 Simulated imaginary part of the input impedance of the dipole antenna for case 6

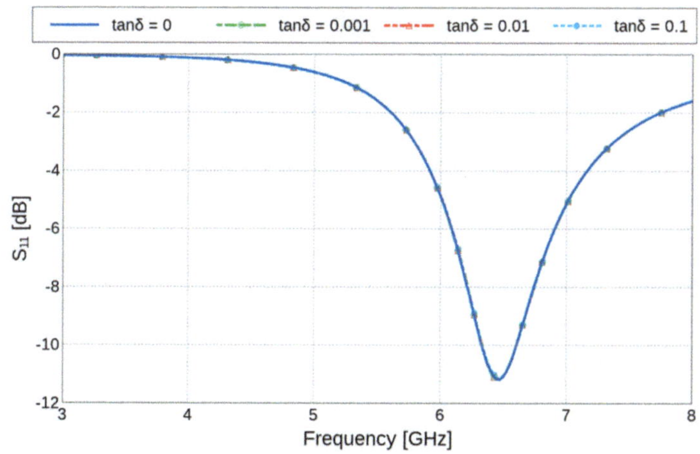

Fig. 18 Simulated reflection coefficient of dipole antenna for case 7

frequency of the antenna. Figures 16 and 17 show the results of real and imaginary part of the input impedance of the dipole antenna in direct contact with the lossy medium, the results show a decrease in the maximum value of real and imaginary part of the impedance with the increasing tanδ values, however, the zero crossing of the reactance values are not influenced by the loss. Figure 18 shows the reflection coefficient of the dipole at a 1 mm distance from the surface of the lossy medium. The lossy medium has no impact on the results if the dipole is kept at a distance from the surface. And Figs. 19 and 20 show that results are nearly equal for the real

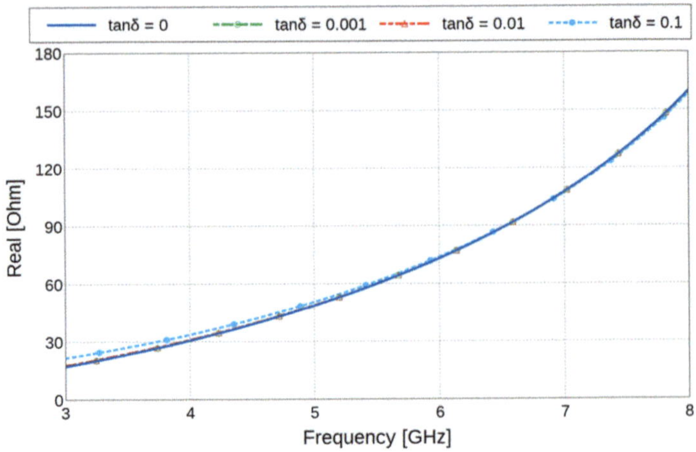

Fig. 19 Simulated real part of the input impedance of the dipole antenna for case 7

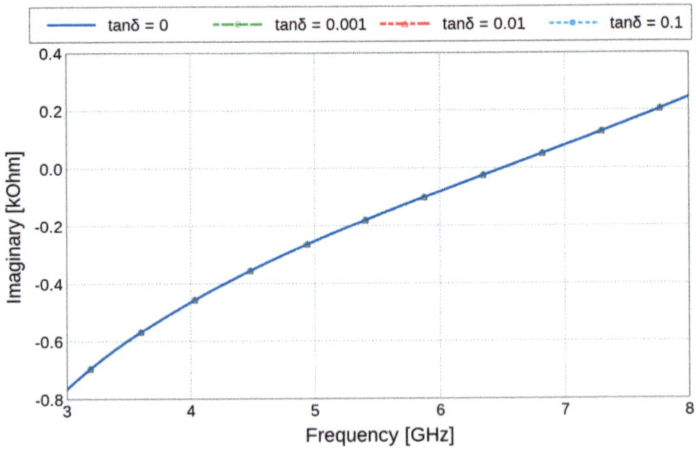

Fig. 20 Simulated imaginary part of the input impedance of the dipole antenna case 7

and imaginary part of the dipole antenna kept at a distance of 1 mm from the lossy medium. Which implies an isolation layer between the antenna and lossy medium could sustain the performance of the antenna in body area communication.

4 In the Vicinity of Lossless and Lossy Layered Medium

In this section, different characteristics of the dipole antenna are analyzed when it is kept in contact with the lossy and lossless layered structure. The layered structure is shown in Fig. 21. The structure consist of three layers, in the first case layer 1, layer 2

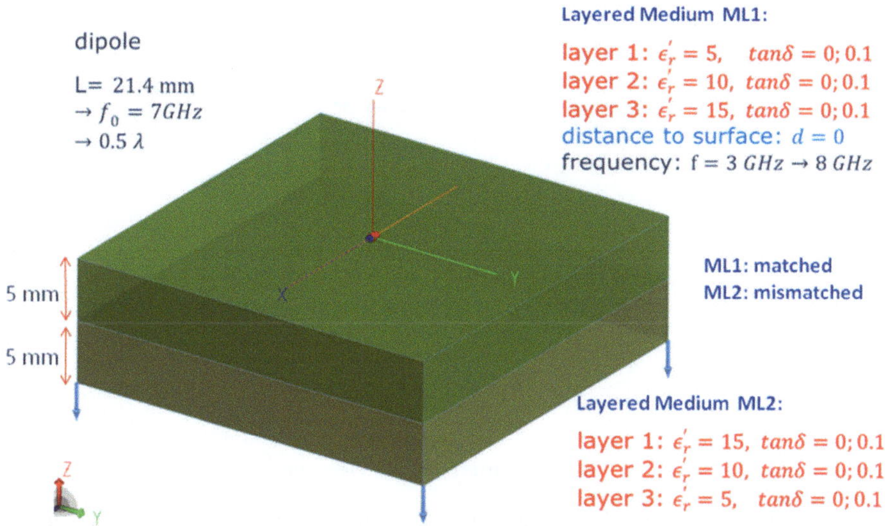

Fig. 21 Dipole in direct contact with the lossless and lossy multilayer dielectric medium 7

Fig. 22 Simulated gain pattern of the dipole antenna in contact with the lossy and lossless layered half space dielectric medium with different tangent loss values

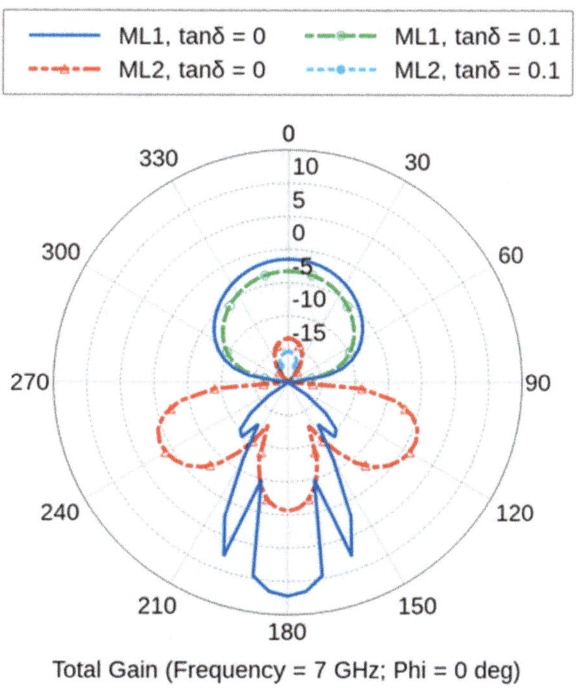

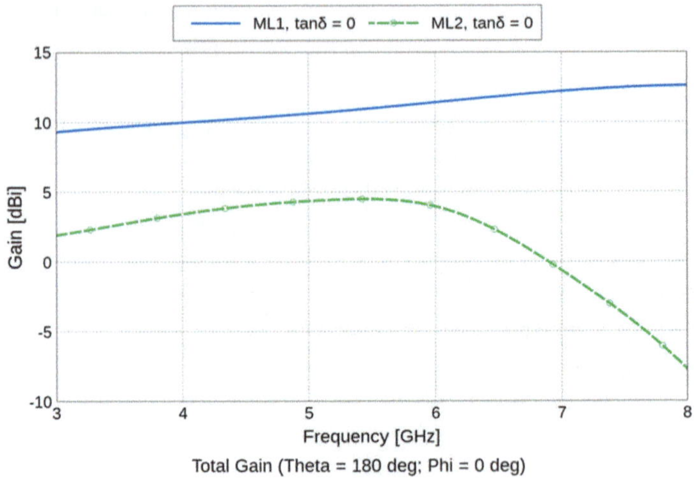

Fig. 23 Simulated gain of the dipole antenna in contact with the lossless layered half space dielectric medium

and layer 3 are assigned with the relative permittivity of 5, 10 and 15, respectively and named as multilayer 1 (ML1). In the second case the multilayer named as multilayer 2 (ML2), the permittivities of layers were reversed such that layer 1, layer 2 and layer 3 are assigned with the relative permittivity of 15, 10, and 5, respectively. The simulated gain pattern results in Fig. 22 show that in case of lossless ML1, the inside radiation is having high gain and main lobe is maintained, whereas in case of ML2 the gain is reduced and side lobes start to appear. In case of lossy ML1 and ML2 the inside gain deteriorates a lot in both of the cases. Figure 23 shows the gain of the antenna at different frequency values for the lossless case. In lossless ML1 the gain is nearly constant at different frequencies, however, in case of lossless ML2 the dipole gain drops tremendously at higher frequencies.

5 Conclusion

Different dipole cases in the close proximity of lossless and lossy medium are analyzed in depth. A matching condition was observed in lossless multilayer case when the dipole is in direct contact with the medium. The losses deteriorate the matching conditions of the dipole, but have negligible effect on the resonance frequency of the antenna. In order to maintain a good performance of the antenna for outside body communication, the body antenna must be isolated from the human body. The study is very beneficial in designing on body antennas.

Acknowledgements This research is supported by "WiBEC" (Wireless In-Body Environment) project funded by the European Union Horizon 2020 research and innovation program under grant agreement No. 675353.

References

1. Pellegrini, A., Brizzi, A., Zhang, L., Ali, K., Hao, Y., Wu, X., Constantinou, C.C., Nechayev, Y., Hall, P.S., Chahat, N., Zhadobov, M., Sauleau, R.: Antennas and propagation for body-centric wireless communications at millimeter-wave frequencies: a review. IEEE Antennas Propag. Mag. **55**(4) (2013)
2. Yilmaz, T., Foster, R., Hao, Y.: Detecting vital signs with wearable wireless sensors. Sensors **10**, 10837–10862 (2010)
3. Hao, Y., Foster, R.: Wireless body sensor networks for health-monitoring applications. Physiol. Meas. **29**(11), R27–R56 (2008)
4. Cotton, S.L., Scanlon, W.G., Madahar, B.K.: Millimeter-wave soldier-to-soldier communications for covert battlefield operations. IEEE Commun. Mag. **47**(10), 72–81 (2009)
5. Wang, Q., Fang, X., Plettemeier, D.: Impedance characteristics and field separation of body implanted antennas. In: Bodynets, Italy, Turin (2016)

Estimating Eavesdropping Risk for Next Generation Implants

Muhammad Faheem Awan and Kimmo Kansanen

Abstract Implanted medical devices are expected to be wireless in near future. Wireless nature of sensing, controlling and transmission brings along different security threats. In this work, an analysis of eavesdropping risk is performed for an unencrypted data transmissions from an implanted medical device such as cardiac leadless pacemaker. This work utilizes statistical attenuation model along with measures of capacity, information rate and outage probability. Results show that eavesdropping risk depends on pathloss with shadow fading, distance and information rate (*R*). In addition, probability of successful eavesdropping increases if legitimate nodes transmits at lower rate. Thus, a proper tradeoff between information rate (*R*) and eavesdropping risk should be made. Numerical results show that at an information rate of 650 kbps, an IMD has a 5% risk of successful eavesdropping at a distance of 500 mm. This work also consider different transmission parameters like heart rate, blood pressure, ECG and EMG with their information rates and find probability of successful eavesdropping at different distances. This study provide basis for designing secure implantable cardiac leadless pacemaker with associated risks involved due to wireless nature of transmission.

1 Introduction

Rapid development in personal health systems based on implantable and wearable medical devices results in better quality of diagnosis. This along with automatically treating different medical conditions, improves quality of life. Implantable medical devices (IMD's) continuously monitor and treat physiological conditions inside the body. Notable among these devices are cardiac pacemakers and implanted cardiac defibrillators (ICD's). They help treating different types of cardiac arrhythmia's. In 2001, there were approximately 25 Million patients with IMD's [1] in the USA. With

M. F. Awan (✉) · K. Kansanen
NTNU, Trondheim, Norway
e-mail: faheem.awan@ntnu.no

K. Kansanen
e-mail: kimmo.kansanen@ntnu.no

G. Fortino and Z. Wang (eds.), *Advances in Body Area Networks I*,
Internet of Things, https://doi.org/10.1007/978-3-030-02819-0_29

help of sensors these devices sense different conditions and provide proper actuation e.g. by maintaining heart rhythm.

IMD's are mostly equipped with on-board transceivers to transmit wirelessly over certain distance. These transmissions facilitate remote monitoring of a patient. The wireless nature of modern IMD's is a significant source of security risks. Security risk for an IMD increases, when patient is in non-medial or in an open environment. Wireless nature of an IMD makes it more visible and can facilitate an eavesdropper to listen. Thus an insecure communication channel makes it easier for an eavesdropper to perform attacks on an implant similar to attacks on other computing devices. Successful eavesdropping may result in retrieval of patient information (medical and non medical) or performing attacks like forging and data altering. In addition, it may enable the modification of implant configuration without knowledge of the patient or physician. Security and privacy of implantable medical devices received a lot of attention when it was revealed that US Vice President Dick Cheney disabled the wireless telemetry on his ICD [2, 3]. The work of Halperin et al. [4] is considered as pioneer work in security analysis of IMD's, followed by different research activities providing security for IMD devices. Most of research is focused on mitigating the security risks via providing different encryption mechanisms in order to protect data between sender and legitimate node [5–7].

Cardiac pacemakers are designed to facilitate proper functioning of human heart. Currently these pacemakers are mostly implanted using wired electrodes. The next generation of these pacemakers are expected to be wireless. Data transmitted by these devices include transmission of real time patient data, offline patient data and device information along with different indicators. Heart rate, blood pressure etc are considered to be real time data whereas patient name, patient date of birth, patient history lies under offline data transmission. Pacemakers also transmit device information like battery status, device number etc. Furthermore, it may transmit indicators with different diagnosed conditions along with therapies performed. These data transmissions are vital in order to keep pacemaker with right configurations and continuous remote monitoring of a patient. They also help physician to have an updated report regarding heart condition of a patient. Frequency of transmissions depends upon the configurations used for a pacemaker. In case of wireless capsule inside a heart and subcutaneous node under the skin as shown in Fig. 1, the data is transmitted every heart beat whereas wireless subcutaneous node with leads, logs an activity to a bed side external device on daily basis or data is retrieved during a visit to a physician. Table 1 shows examples of different types of traffic and their typical information rate for an IMD [8].

In order to perform the security analysis the first thing to do is analyzing different types of threats [9]. Providing a threat model is always considered to be starting point in security analysis. Among different threats one of the important threat is eavesdropping which provide basis for further attacks. Thus for the aforementioned purpose, this paper focus on finding probability of successful eavesdropping for legitimate transmission between leadless capsule inside human heart and subcutaneous implant. This work considers that an implant is using Medical Implant Communication system (MICS) band for telemetry between legitimate nodes with an assumption

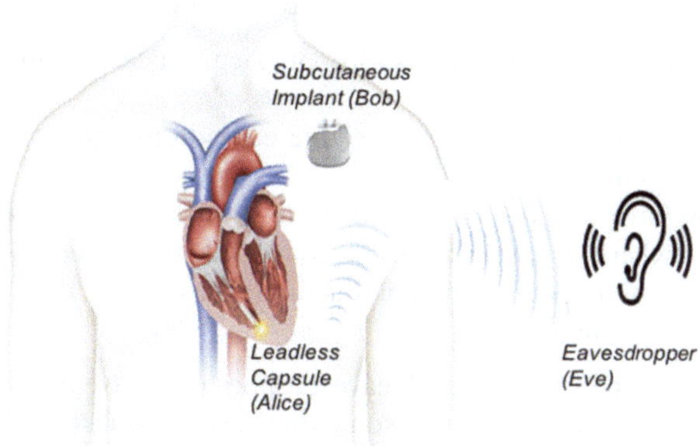

Fig. 1 Pacemaker scenario with an Eavesdropper

Table 1 Data rate for IMD traffic

Sensing parameter	Required data rate
Heart rate	1 sample/s or 600 bps
Medical image	2.4 Mbps
Blood pressure	1.2 kbps
EMG	600 kbps
EEG	4.2–32 kbps
ECG	1.2–250 kbps

that encryption is not used in data communication. We find probability of successful eavesdropping by using a concept of outage probability. We show that eavesdropping risk depends upon information rate, distance, and antenna gains. We utilize statistical attenuation model [10] to find path loss between legitimate nodes and path loss between legitimate node and an eavesdropper. This helps finding the distribution of signal to noise ratio (SNR γ) at different distances (d). We can then set different cut-off thresholds for SNR (γ_{th}) depending upon various information rates (R) and find the corresponding outage probability. We finally provide the design details which help setting an appropriate information rate (R) between legitimate nodes.

The rest of the paper is formulated as follows: Sect. 2 gives system model followed by eavesdropper model in Sect. 3, Sect. 4 provides numerical results, and finally we conclude and provides future work in Sect. 5.

2 System Model

Our system consists of an eavesdropper, an IMD, and a subcutaneous node as shown in Fig. 1. We consider implanted leadless capsule inside a right ventricle of human heart. This leadless capsule transmits unencrypted data to subcutaneous implant at a distance of 150 mm. We consider the famous **Alice**, **Bob** and **Eve** model which can be seen in Fig. 1 for our case scenario.

In case of IMD's, **Eve** can be categorized as

- An eavesdropper, eavesdropping legitimate transmission can be an active or passive node.
- There is possibility of single eavesdropper or it may be part of well organised group.
- Eavesdropper is most likely to be an external intruder but there can be the case where an eavesdropper is from inside the system e.g. physician, hospital administration, equipment manufacturers etc because they have the benefit of being close to patient.

It can be seen that each eavesdropper may have different intensions or goals. There may be possibility that some one eavesdrop just to get the private information of a patient or there may be a case where there is competition between manufacturers and they would like to obtain information on an IMD equipment. Eavesdropper may have higher capabilities than that of legitimate node like higher computational power, higher antenna gains, etc. In this work we consider single passive eavesdropper having same capabilities as a legitimate node trying to eavesdrop the communication. We also consider a case where the eavesdropper has high gain antenna.

3 Eavesdropper Model

This section focus on providing the probability of successful eavesdropping by utilizing path loss model for MICS band. We need to predict the received power from a transmitted node along with associated path loss at a certain distance (d). This helps finding distribution of signal to noise ratio (SNR). Then by considering different values of information rate (R), we can set different cutoff thresholds and find probability of successful eavesdropping at a certain distance (d).

3.1 Received Power/Attenuation

We consider that IMD (leadless pacemaker) is transmitting with power P_p to subcutaneous node at a distance of 150 mm. The eavesdropper is trying to listen the transmission and receives power P_e at a distance (d) away from the IMD. The power received by the eavesdropper can be given as

$$P_e = \frac{G_p G_e P_p}{w d^{\alpha}} \tag{1}$$

In (1), G_p and G_e are antenna gains of implant and eavesdropper antenna whereas α is path loss exponent and w is a random variable given by

$$w = 10^{\frac{S_h}{10}} \tag{2}$$

Here S_h is Gaussian random variable with zero mean and standard deviation σ ranging from 4 to 12 dB [10, 11]. This randomness is because of propagation through different materials (e.g. bone, muscle, etc.).

In order to compute the numerical results, we generally work with signal attenuation from sender to receiver, and then can utilize it into any case scenario. Here we consider sender as a legitimate node (IMD), where as the eavesdropper is considered as a receiver. Thus instead of received signal strength, attenuation between IMD and the eavesdropper can be given as

$$\delta(d) = \frac{P_p}{P_e} = \frac{wd^{\alpha}}{G_e G_p} \tag{3}$$

In this paper, we utilizes path loss model provided by [10] in order to find attenuation at distance (d). Thus, a pathloss/attenuation in logarithmic scale (dB) can be given as

$$PL(d)_{dB} = PL(d_o)_{dB} + G_{e(dB)} + G_{p(dB)} + 10\alpha \log_{10}(d/d_o) + S_h, \quad d \geq d_o \tag{4}$$

In (4), $PL(d)_{dB}$ is the path loss at distance d between the implant and the eavesdropper, d_o is the reference distance and is equal to 50 mm. Path loss exponent α is equal to 4.26 in [10]. Furthermore $PL(d_o)$ is the path loss at reference distance and is equal to 47.14 dB. Similarly, S_h is random scattering around the mean and is equal to 7.85 dB.

3.2 *Outage Probability, Information Rate and Detection Threshold*

The main focus of this section is outage probability (OP), information rate (R) and detection/cutoff threshold. This section also provide relation between these quantities for main link (link between legitimate nodes) and show how they provide eavesdropping risk.

Outage probability is the probability that "When signal to noise ratio (SNR γ) at the input of a receiver chain fall below certain threshold level [12]". It can be expressed as

$$p_{out}(\gamma_{th}) = Pr[\gamma < \gamma_{th}] \tag{5}$$

where γ is signal to noise ratio (SNR) and γ_{th} is the threshold SNR. For a communication between legitimate nodes, the value of SNR (γ) should be greater than threshold otherwise the communication link will be in outage. Based on this threshold, which can also be referred as cutoff threshold, we define the probability of successful eavesdropping at any distance (d). In (5) SNR (γ) can be expressed as

$$\gamma = \frac{P_e}{N_o B} = \frac{P_p}{\delta(d) N_o B} \tag{6}$$

where Pe, is received power at input of receiver chain, P_p is power transmitted, $\delta(d)$ is attenuation at a distance (d), B is bandwidth and N_o is noise spectral density. In logarithmic (dB) scale (6) can be expressed as

$$\begin{aligned} \gamma_{dB} &= P_{pdBm} - PL(d)_{dB} - (N_o B)_{dBm} \\ \gamma_{dB} &= P_{pdBm} - (PL(d_o)_{db} + G_{e(dB)} + G_{p(dB)} \\ &\quad + 10\alpha \log_{10}(d/d_o)) - (N_o B)_{dBm} + S_h, \end{aligned} \tag{7}$$

where, a constant per channel transmit power P_{pdBm} and constant $(N_o B)_{dBm}$ is considered. In (7), $PL(d)_{dB}$ is mean path loss obtained from (4) at any distance (d). Furthermore S_h is Gaussian random variable with zero mean and $\sigma_{dB} = 7.85\,\text{dB}$. Thus, (7) show that SNR (γ_{dB}) has Gaussian distribution with mean as

$$\begin{aligned} \mu_{\gamma_{dB}} &= P_{pdBm} - (PL(d_o)_{db} + G_{e(dB)} + G_{p(dB)} \\ &\quad + 10\alpha \log_{10}(d/d_o)) - (N_o B)_{dBm}, \end{aligned} \tag{8}$$

and standard deviation of $\sigma_{\gamma_{dB}} = 7.85\,\text{dB}$.

Our approach basically uses the concept of channel capacity as a measure. According to the concept of capacity, for a given information rate R, there exist a minimum received power to successfully decode the transmission based on certain threshold SNR. Using this concept, the capacity of a link is given by the famous Shannon capacity formula

$$C = B \times \log_2(1 + \gamma_{th}) \tag{9}$$

where B is per channel bandwidth and is equal to 300 kHz in case of MICS band and C is capacity. In order to find threshold SNR γ_{th}, (9) can also be expressed as

$$\gamma_{th}(R) > 2^{\frac{R}{B}} - 1 \tag{10}$$

In logarithmic scale (10) can be expressed as

$$\gamma_{th_{dB}}(R) = 10 \log_{10} \gamma_{th} \tag{11}$$

Equation (10) shows that threshold SNR is function of information rate (R). With variation in information rate (R), cutoff threshold (γ_{th}) varies and have different information rate R due to different traffic transmitted by an IMD. If we solve (10), by taking $B = 300\,\text{kHz}$ and $R = 600\,\text{bps}$, we have $\gamma_{th_{dB}}$ as $-28\,\text{dB}$ which shows that we have successful, error free decoding at that threshold. By reducing information rate (R), the cutoff threshold reduces where as by increasing, it increases the detection or cutoff thresholds. Thus, at a same distance d outage probability increases because of higher cutoff threshold and decreases because of lower cutoff threshold (γ_{th}) (5).

3.3 Probability of Successful Eavesdropping

In order to find the probability of successful eavesdropping, we use the concept of outage probability as shown in (5). We have to take certain cutoff threshold represented by $\gamma_{th_{dB}}$. Eavesdropper is able to eavesdrop communication with certain probability, when SNR (γ_{dB}) between Alice and Eve is greater than set cutoff threshold $\gamma_{th_{dB}}$. Here we express both SNR(γ) and cutoff threshold (γ_{th}) in dB scale, i.e. $\gamma_{dB} > \gamma_{th_{dB}}$(7) and (11). The probability of successful eavesdropping can be given as

$$P_{SE} = 1 - p_{out}(\gamma_{th_{dB}}) \tag{12}$$

As SNR (γ_{dB}) is Gaussian distributed (7) with mean $\mu_{\gamma_{dB}}$ (8) and standard deviation $\sigma_{\gamma_{db}}$, we can express (12) by Q-function as

$$P_{SE} = 1 - \left(1 - Q\left(\frac{\gamma_{th_{dB}} - \mu_{\gamma_{dB}}}{\sigma_{\gamma_{dB}}}\right)\right) \tag{13}$$

where, P_{SE} is probability of successfull eavesdropping, $\gamma_{th_{dB}}$ is cutoff threshold depending upon R (10) and $\mu_{\gamma_{dB}}$ is mean of signal to noise ratio (γ_{dB}) (8) at any distance (d). Equation (13) can also be expressed as

$$\begin{aligned} P_{SE} &= Q\left(\frac{\gamma_{th_{dB}} - \mu_{\gamma_{dB}}}{\sigma_{\gamma_{dB}}}\right) \\ P_{SE} &= 1 - \varphi\left(\frac{\gamma_{th_{dB}} - \mu_{\gamma_{dB}}}{\sigma_{\gamma_{dB}}}\right) \\ P_{SE} &= 1 - \frac{1}{2}\left(1 + \text{erf}\left(\frac{\gamma_{th_{dB}} - \mu_{\gamma_{dB}}}{\sigma_{\gamma_{dB}}\sqrt{2}}\right)\right) \end{aligned} \tag{14}$$

The eavesdropper can increase probability of successful eavesdropping by using high gain antennas. An antenna with high gain, have a reception from greater distances and have higher SNR. But nothing comes without cost, higher gain results from larger dimensions of an antenna. In case of MICS band the transmitting antenna gain is considered to be part of path loss model [10]. Different antennas has different antenna gain and aperture relation but in general it can be expressed as

$$A_e = \frac{\lambda^2 G_e}{4\pi} \tag{15}$$

From the eavesdropper perspective, she wants to eavesdrop without noticing. But with large aperture antennas it can not happen easily. So, the gain of an antenna can not be increased above certain limits e.g. if we are using MICS band and want to have an antenna gain of 20 dBi then the effective aperture of an antenna will be around 4.4 m^2 which can not go without noticing within short distance of a patient.

Table 2 Cutoff threshold SNR for IMD traffic

Sensing parameter	Required data rate	Threshold SNR ($\gamma_{th_{dB}}$)
Heart rate	600 bps	−28.5 dB
Blood pressure	1.2 kbps	−25.56 dB
EMG	600 kbps	−4.77 dB
ECG	250 kbps	−1.1 dB

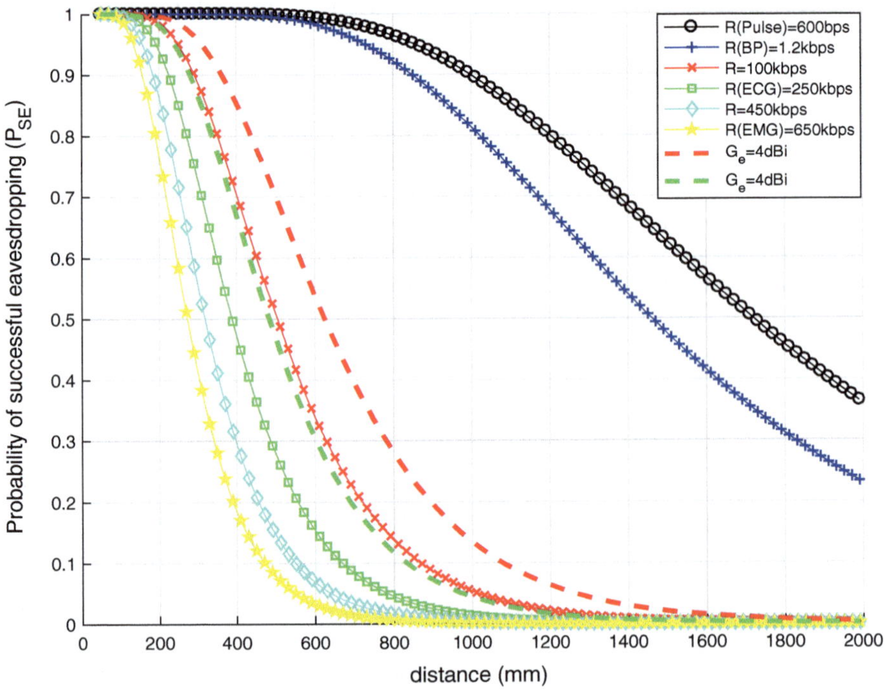

Fig. 2 Probability of successful eavesdropping at different distances (d)

4 Numerical Results

This section focus on numerical results by considering scenario as shown in Fig. 1 which focus on communication between implanted leadless capsule (pacemaker) and a subcutaneous node under the skin. For communication between the implant and subcutaneous node, we are interested to find the probability of successful eavesdropping. First, channel attenuation is predicted by utilizing pathloss model in [10], which give an SNR γ_{dB} distribution (7). Then different cutoff thresholds $\gamma_{th_{dB}}$ for an SNR are considered based on information rates (10) and (11).

We take different cutoff values by considering information rates from Table 1, whereas Table 2 provides different information rates with corresponding threshold SNR. Now, by looking into different threshold values, we utilize (14) and find probability of successful eavesdropping P_{SE}. Figure 2 shows probability of successful eavesdropping for different information rates (R) at different distances.

The results in Fig. 2, shows that if an eavesdropper is exactly at same distance as that of subcutaneous node i.e. 150 mm for an information rate of 650 kbps, then in practice the probability of successful eavesdropping is about 100%. But when an eavesdropper distance is increased the probability of successful eavesdropping reduces with distance e.g. for the same case at 400 mm, the probability of successful

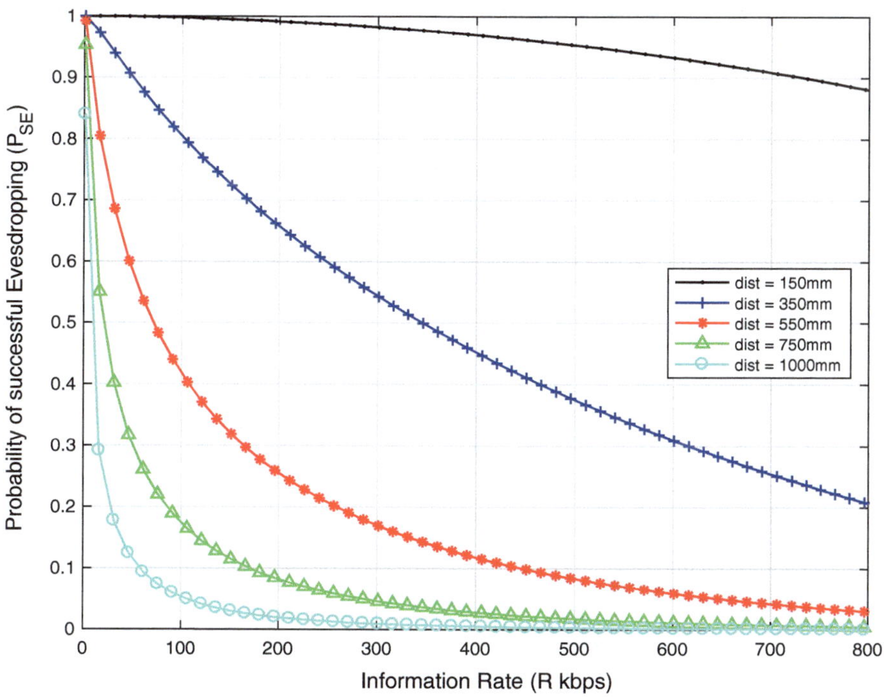

Fig. 3 Probability of successful eavesdropping at different information rates (R)

eavesdropping is reduced to around 19%. Each curve shows probability of successful eavesdropping for certain information rate (R) over different distances. An eavesdropper can also increase its receiver sensitivity against path loss by using high gain antennas, resulting in successful eavesdropping over larger distance because of higher SNR e.g. as shown in Fig. 2 the dashed lines show that probability of successful eavesdropping is increased over a distance by using an antenna gain of 4 dBi for the same respective cases without antenna gain (green and red curve). Here it is notable to mention that from a design perspective, a proper tradeoff is required between information rate and eavesdropping risk. We refer to the same example by considering 650 kbps information rate, which shows successful eavesdropping probability of 19% at 400 mm. But in case, if information rate is reduced to 0.6 kbps, at a same distance of 400 mm probability of successful eavesdropping is close to 100%.

Figure 3 gives probability of successful eavesdropping for different information rates, against certain fixed distance. Alice and Bob link is considered as 150 mm whereas Alice and Eve link is changed. For our case, we consider Alice-Eve distance of 150, 350, 550, 750 and 1000 mm. The eavesdropping risk is calculated for these distances with different information rates (R). For each information rate, we calculated corresponding cutoff SNR ($\gamma_{th_{dB}}$) and then find P_{SE} by (14). Figure 3 shows that if we have Alice-Eve distance of 1 m and an Information rate of 1 kbps, probabil-

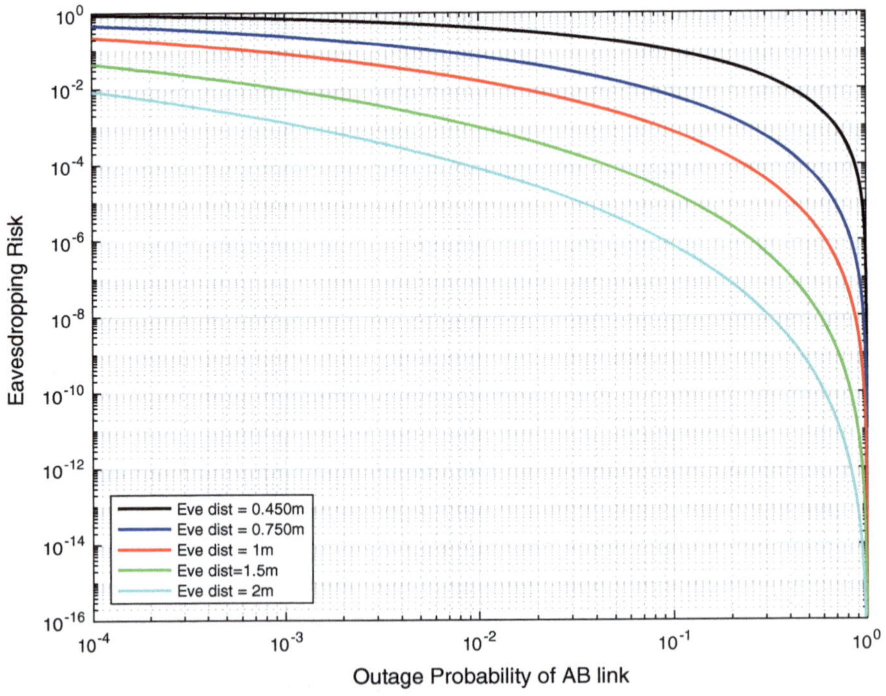

Fig. 4 Probability of successful eavesdropping with varying outage probability of Alice-Bob link at fixed distances

ity of successful eavesdropping is 85% and reduces to 10% at a rate of 60 kbps. Thus by increasing information rate, probability of successful eavesdropping reduces.

We also provide results of eavesdropping risk against outage probability of main link (Alice-Bob). Figure 4 is in logarithmic scale, which shows that as outage probability of Alice-Bob link increases, probability of successful eavesdropping reduces. It can also be seen that for Alice-Bob link outage probability of 0.01, the minimum distance should be approximately 1.5 m for having the eavesdropping risk of 0.001.

5 Conclusion and Future Work

In this paper we perform analysis of eavesdropping risk for a communication between an implant and subcutaneous node. We analyze probability of successful eavesdropping for different distances with different information rates. We analyze that proper tradeoff is required between information rate and corresponding eavesdropping risk. Our numerical results show that for the information rate (R) of 650 kbps, there is 5% probability of successful eavesdropping at a distance of 500 mm. If the information rate is reduced, eavesdropping risk increases over a larger distance. Our results also show that path loss reduces eavesdropping risk whereas antenna gains increases the risk. In future, it will be interesting to take different scenarios where we can consider transmission from subcutaneous node to external programmer or directly from leadless capsule to a programmer. We would also like to perform risk analysis using UWB band and finding probability of successful eavesdropping.

Acknowledgements This work was supported by the Marie Curie Research Grants Scheme under EU Horizon 2020 research and innovation network program, with project grant no 675353 WIBEC ITN (Wireless In-Body Environment).

References

1. Pope, A., Bouxsein, P., Manning, F.J., Hanna, K.E., et al.: Innovation and Invention in Medical Devices: Workshop Summary. National Academies Press (2001)
2. Rushanan, M., Rubin, A.D., Kune, D.F., Swanson, C.M.: Sok: security and privacy in implantable medical devices and body area network. In: 2014 IEEE Symposium on Security and Privacy (SP), pp. 524–539. IEEE (2014)
3. Times, N.Y.: Vice president news. October (2013). http://www.nytimes.com
4. Halperin, D., Heydt-Benjamin, T.S., Ransford, B., Clark, S.S., Defend, B., Morgan, W., Fu, K., Kohno, T., Maisel, W.H.: Pacemakers and implantable cardiac defibrillators: software radio attacks and zero-power defenses. In: IEEE Symposium on Security and Privacy, SP 2008, pp. 129–142. IEEE (2008)
5. Zhang, M., Raghunathan, A., Jha, N.K.: Medmon: securing medical devices through wireless monitoring and anomaly detection. IEEE Trans. Biomed. Circuits Syst. **7**(6), 871–881 (2013)
6. Son, S., Lee, K., Won, D., Kim, S.: U-healthcare system protecting privacy based on cloaker. In: 2010 IEEE International Conference on Bioinformatics and Biomedicine Workshops (BIBMW), pp. 417–423. IEEE (2010)

7. Gollakota, S., Hassanieh, H., Ransford, B., Katabi, D., Fu, K.: They can hear your heartbeats: non-invasive security for implantable medical devices. ACM SIGCOMM Comput. Commun. Rev. **41**(4), 2–13 (2011)
8. Islam, M.N., Yuce, M.R.: Review of medical implant communication system (MICS) band and network. ICT Express **2**(4), 188–194 (2016)
9. Hong, J.I., Ng, J.D., Lederer, S., Landay, J.A.: Privacy risk models for designing privacy-sensitive ubiquitous computing systems. In: Proceedings of the 5th Conference on Designing Interactive Systems: Processes, Practices, Methods, and Techniques, pp. 91–100. ACM (2004)
10. Sayrafian-Pour, K., Yang, W.-B., Hagedorn, J., Terrill, J., Yazdandoost, K.Y.: A statistical path loss model for medical implant communication channels. In: 2009 IEEE 20th International Symposium on Personal, Indoor and Mobile Radio Communications, pp. 2995–2999. IEEE (2009)
11. Wang, J., Wang, Q.: Body Area Communications: Channel Modeling, Communication Systems, and EMC. Wiley (2012)
12. Tse, D., Viswanath, P.: Fundamentals of Wireless Communications (2004)

A Study of Implant Antenna for FSK-Based Impulse Radio System in Human Body Communication Band

Qiancheng Liang, Jingjing Shi, Atomu Nakashima and Jianqing Wang

Abstract This paper aims to develop a miniaturized implant antenna for a Frequency Shift Keying (FSK)-based impulse radio system in Human Body Communication (HBC) band. The proposed implant antenna is realized with two layers of helical radiating elements formed on magnetic sheets, which contributes to a compact cylindrical shape with 1-cm diameter and 3-cm length for medical capsule endoscope application. The double-resonant return loss with 2 MHz bandwidth indicate the suitability of this antenna for a FSK-based impulse radio system. Moreover, the transmission characteristics and required transmit power using this antenna has also been investigated to clarify the feasibility of realizing real-time image transmission in HBC band.

Keywords Implant communication · FSK-impulse radio · Implant antenna
Body area network

1 Introduction

In recent years, body area network (BAN) has attracted a lot of attention in medical, health care, and monitoring supports due to its good properties in information acquisition, controlling and management [1, 2]. It can be defined as a miniature network consisting of sensors operating mainly on or in the human body. Classification of BAN on the basis of the device-operation environment can be divided in two main categories: wearable BAN and implanted BAN. Applications of wearable

Q. Liang · J. Shi (✉)
Northeastern University, Shenyang 110169, China
e-mail: shijj@bmie.neu.edu.cn

A. Nakashima · J. Wang
Nagoya Institute of Technology, Nagoya 466-8555, Japan
e-mail: wang@nitech.ac.jp

G. Fortino and Z. Wang (eds.), *Advances in Body Area Networks I*,
Internet of Things, https://doi.org/10.1007/978-3-030-02819-0_30

BAN include thermometer, electrocardiograph (ECG), pulse meter, etc. On the other hand, implant BAN has very typical application examples of pacemaker and capsule endoscope. Compared to conventional endoscopic examination with a gastrocamera, the capsule endoscope has huge advantage to reduce the burden on small bowel disease patients because of its pill-like swallowable feature. That is the reason why capsule endoscope has been drawing so much attention in recent years.

At the present stage, the available frequency band of the capsule endoscope mainly concentrated on Human Body Communication (HBC) band, Medical Implant Communication Service (MICS) 400 MHz band, Industrial, Scientific and Medical (ISM) 2.4 GHz, and ultra-wide band (UWB) [3]. However, signals at higher frequencies suffer from serious large scale attenuation when penetrating biological body tissues, especially in UWB band . For example, the average path loss reaches up to 60–100 dB in the case of in-body to on-body wireless UWB link distance at 10 cm [4]. A recent research [5] reveals that data rate of 1 Mbps was realized in a living animal within 10 cm transmission distance using an impulse radio (IR)-UWB transceiver. Nonetheless, it still be faced with severe challenges of reliable transmission larger than 10 cm and data rate higher than several Mbps inside the body. Therefore, to improve the transmission distance in body tissues, we tried to utilize HBC band possessing relatively lower frequency domination, and FSK-based impulse radio scheme to construct an in-body to on-body communication link. On the other hand, since the antenna size depends on wavelength (frequency), solutions of developing swallowable size implant antenna are in urgent demand in this band.

In this research, we propose an impulse radio system based on FSK modulation method in HBC band, and develop a compact implant antenna with double-resonant frequencies which can be effectively utilized in the proposed FSK-based impulse radio system. Moreover, we evaluate the transmission performance and required transmit power using our developed antenna to clarify the feasibility of in-body high-speed transmission in HBC band for capsule endoscope application.

2 FSK-Based Impulse Radio System

This study focused on the FSK impulse radio scheme as a communication method to improve the transmission performance in HBC band. Figures 1 and 2 show the transmitter and receiver block diagram of FSK-based impulse radio system with non-coherent detection. A Gaussian pulse signal expressed by the following equation was used as a transmission pulse.

$$v(t) = \frac{A}{\sqrt{2\pi}\sigma} e^{-\frac{t^2}{2\sigma^2}} \tag{1}$$

Here, σ was set to 2×10^{-7} so that the bandwidth of transmission pulse becomes 2 MHz or more. In order to generate two different center frequencies used in FSK method, the Gaussian pulse of Eq. (1) was multiplied by a cosine wave with frequen-

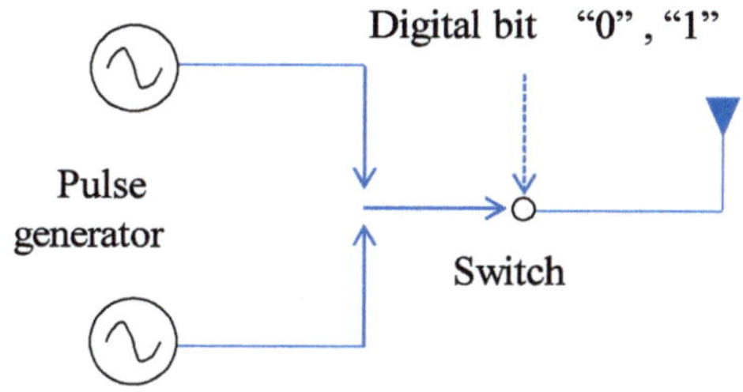

Fig. 1 Transmitter block diagram with FSK modulation scheme

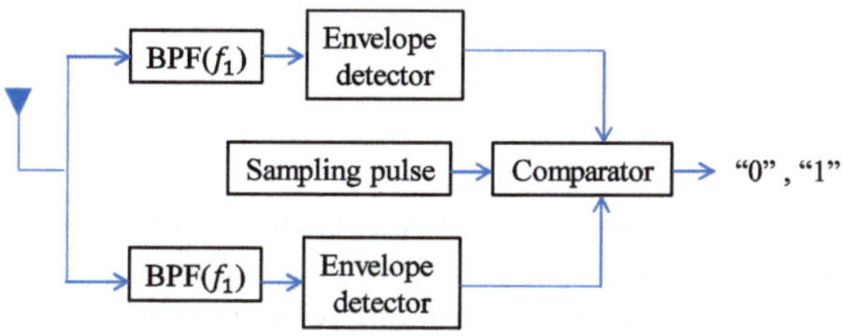

Fig. 2 Receiver block diagram with non-coherent detection

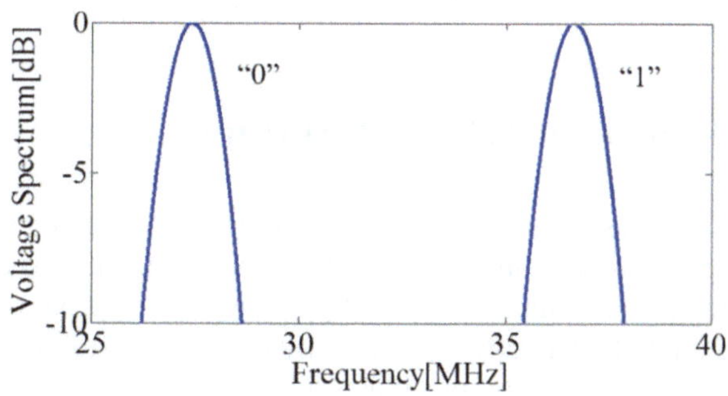

Fig. 3 Spectrum waveform of the transmitting FSK signals

cies of $f_1 = 27$ MHz and $f_2 = 37$ MHz, which are close to the resonant frequencies of after-mentioned antenna. Figure 3 shows the spectrum waveform of the transmitting FSK signals, in which center frequency of 27 MHz is assumed as bit "0" and 37 MHz as bit "1". The theoretical bit error rate with non-coherent detection in a FSK impulse radio system can be derived by

$$P_e = \frac{1}{2} e^{-\frac{\gamma}{2}}, \tag{2}$$

where γ represents the Signal-to-Noise Power Ratio (SNR) of the signals, and can be derived by

$$\gamma = \frac{E_b}{N_0} \times \frac{f_s}{B}. \tag{3}$$

$\frac{E_b}{N_0}$ represents the energy per bit per noise power spectral density, f_s and B represent the data rate and bandwidth of Band Pass Filter (BPF), respectively.

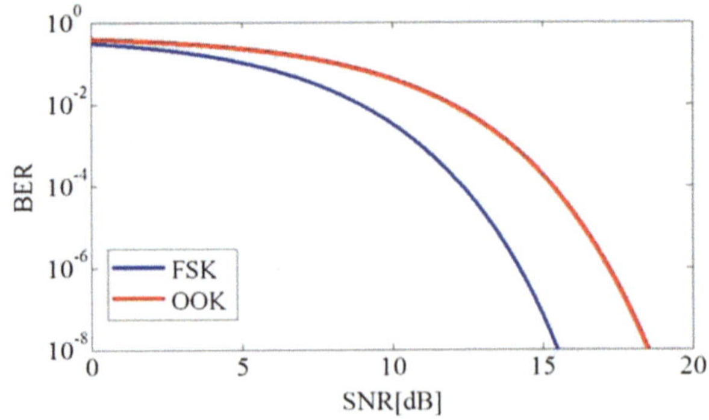

Fig. 4 BER performance versus SNR for FSK and OOK scheme with non-coherent detection

Figure 4 shows a comparison result of BER performance versus SNR in decibel. Since the bandwidth of FSK impulse radio scheme is twice of that in On-OFF Keying (OOK) scheme, 3 dB improvement at BER = 10^{-4} can be observed at the same data rate (f_s = 1 Mbps).

3 Implant Antenna Structure

Since the signal around 30 MHz in HBC band has a wavelength λ_0 of 10 m in free space, we paid attention to utilize flexible magnetic sheets to shorten the wavelength for antenna miniaturization [6]. A magnetic sheet has both a high relative permittivity ε_r and a high relative permeability μ_r. So we can expect a double effect on shortening the wavelength by manufacturing the antenna radiation elements on the magnetic sheet because the wavelength $\lambda \cong \lambda_0/\sqrt{\varepsilon_r \mu_r}$ in a magnetic material. The flexibility of the magnetic sheet also benefits to form a cylindrical pill shape. Figure 5 shows the structure of the implant transmit antenna, which is actually a monopole-type helical antenna with two shorted terminals.

The inner cylinder with a radius of a mm is composed of a copper-covered polyester cylinder. The polyester could sustain a sufficient weight without obvious deformation. The copper plane acts as the ground plane of the antenna. The flexible magnetic sheet noted as "layer 1" is covered on the copper-covered polyester cylinder with a thickness of 1 mm. Then the helical radiation element of copper foil with t-mm-thick is wrapped clockwise over "layer 1" with n turns and a spacing of s mm. Similarly, the flexible magnetic sheet with a thickness of 0.5 mm noted as "layer 2" is covered on the first layer helical copper foil and then wrapped anti-clockwise with the second layer copper foil. Another flexible magnetic sheet with a thickness of 0.5 mm is covered on the second layer helical copper foil at last. This antenna structure includes one copper plane, three layers of magnetic sheets, two layers of helical copper foil sandwiched between the magnetic sheets, in total. The copper plane and the first layer helical copper foil are connected by a copper wire at the bottom of the

Fig. 5 Implant antenna structure

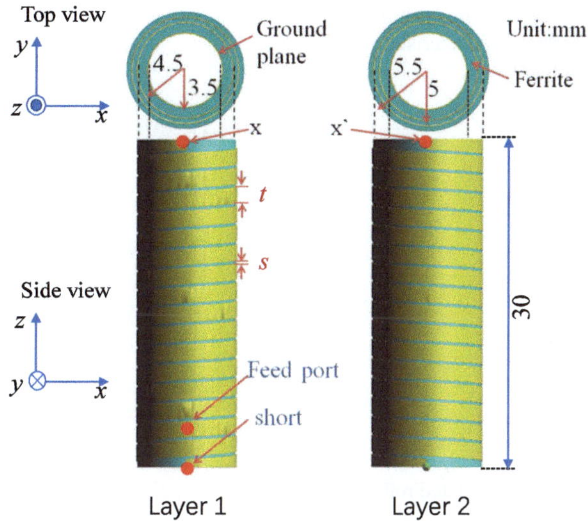

cylinder for shorting to form an invert-F type antenna structure, and the two layers of helical copper foil are connected also by a copper wire at the top of the cylinder. The feeding point is at a height of 4 mm from the bottom of the cylinder along the axis direction. The antenna dimensions were determined by using an electromagnetic field simulation tool based on Finite Difference Time Domain (FDTD) method to determine the parameters a, t, s and n for a commercially available magnetic sheet with $\mu_r = 20.7$, $\tan\delta_\mu = 0.12$, $\varepsilon_r = 13$, and $\tan\delta_\varepsilon = 0.17$, where $\tan\delta_\mu$ is the magnetic loss tangent and $\tan\delta_\varepsilon$ is the dielectric loss tangent. As a design result of numerical simulations, we determined the inner radius of the cylinder $a = 3.5$ mm, the thickness of the cooper foil t = 1.5 mm, the spacing between cooper wires s = 0.25 mm, and the number of turns n = 17. As a result, this antenna has a compact size of 1-cm diameter and 3-cm length.

4 Characteristics Evaluation

To evaluate the antenna characteristics inside human body, as can be seen in Fig. 6, we inserted the antenna into a biological-equivalent phantom model with a dimension of 28 cm × 16 cm × 28 cm in the electromagnetic field numerical simulations. The dielectric properties of the cubic phantom model were set to be $\epsilon_r = 56.05$ and conductivity σ = 0.52 S/m at 30 MHz, nearly 2/3 times muscle's values. An air layer of 1 mm was placed around the implant antenna in consideration of the capsule endoscope operating environment.

Figure 7 shows the simulation results of S_{11} of our designed monopole-type helical antenna with two shorted terminals in this study. It can be observed that this

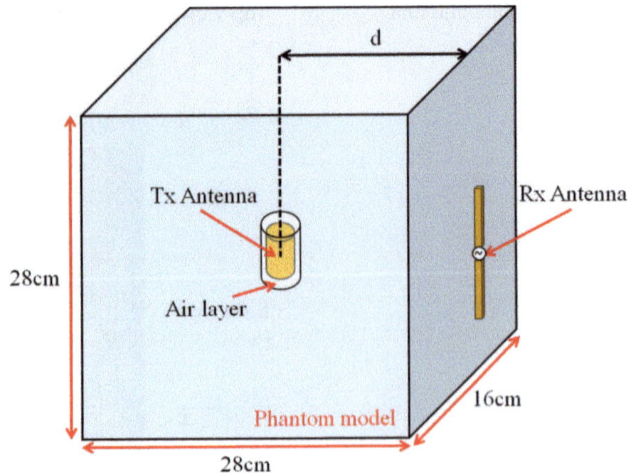

Fig. 6 Arrangement for antenna characteristic evaluation

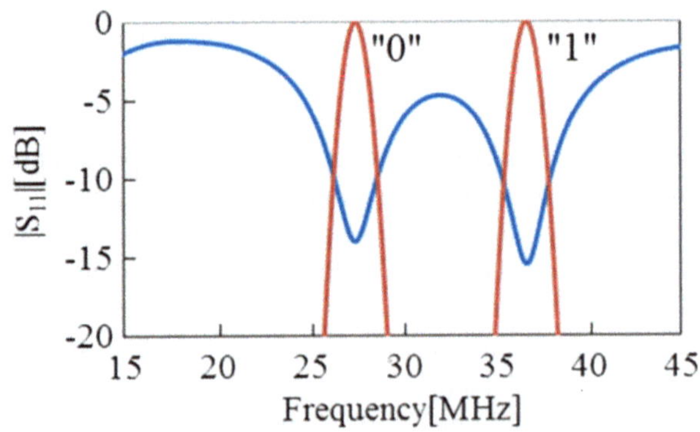

Fig. 7 Simulated reflection coefficient S_{11} in biological-equivalent phantom model

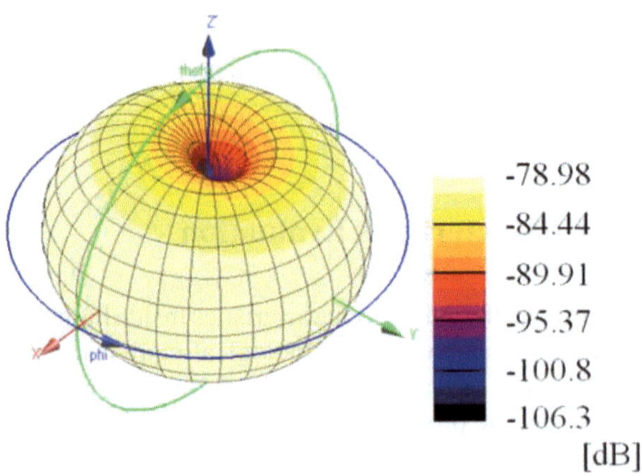

Fig. 8 Radiation pattern of the monopole-type helical antenna in biological-equivalent phantom model

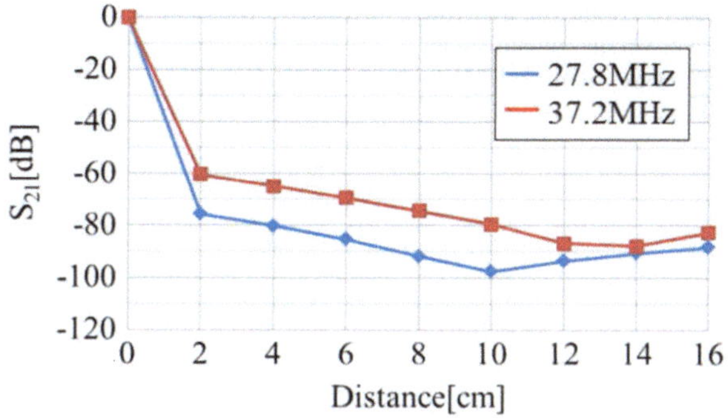

Fig. 9 Simulated transmission coefficient S_{21} as a function of transmission distance

antenna has double resonance point at 27.8 MHz and 37.2 MHz, respectively with reflection coefficient of −14.4 dB and −16 dB. The bandwidth of these two resonant frequencies at −10 dB is 2.32 MHz and 2.61 MHz, respectively. These results show good suitability for our proposed FSK-based impulse radio system for in-body communications. Figure 8 shows the 3D directivity of the antenna gain in the cubic biological-equivalent phantom model. The antenna exhibited good omni-directional radiation characteristics in the resonant frequency band, and small gain variation within 20 dB. However, due to the lossy characteristics of magnetic sheet and dielectric phantom model, the antenna gain in the maximum radiation direction was only −80 dB. Further improvement of the antenna gain should be a subject in future work for in-body communications.

Moreover, the transmission characteristic S_{21} was also calculated when considering a dipole antenna as a receive antenna with a parallel arrangement, as can be seen in Fig. 6. The transmit antenna was moved inside the phantom along a horizontal line to change the distance from the transmit antenna to the receive antenna. The receive antenna was fixed at the cubic phantom surface with a spacing of 5 mm to the phantom.

Figure 9 shows the results of S_{21} at the frequency of 27.8 and 37.2 MHz. From the calculated S_{21}, it is confirmed that the transmission loss is around 90 dB in a depth of 14 cm of the phantom, which can be demodulated in the receiver. This transmission loss contains not only the path loss in the biological-equivalent phantom but also the transmit and receive antenna gains. The required transmit power P_t was also derived by

$$P_{t,\mathrm{dBW}} = PL_{\mathrm{dB}} + P_{r,\mathrm{dBW}}, \tag{4}$$

where PL_{dB} represents the path loss, which can be obtained by the average value of S_{21} in FSK dominating bands corresponding to "0" and "1". P_r is the required receive power determined by receiver. Here, we took BER = 10^{-2} as a target transmission performance, in order to derive the required transmit power P_t. Figure 10 shows the relationship between required transmit power and transmission distance under the assumption of BER = 10^{-2}. The maximum required transmit power within a necessary in-body communication range was found to be −35 dBW (0.316 mW)

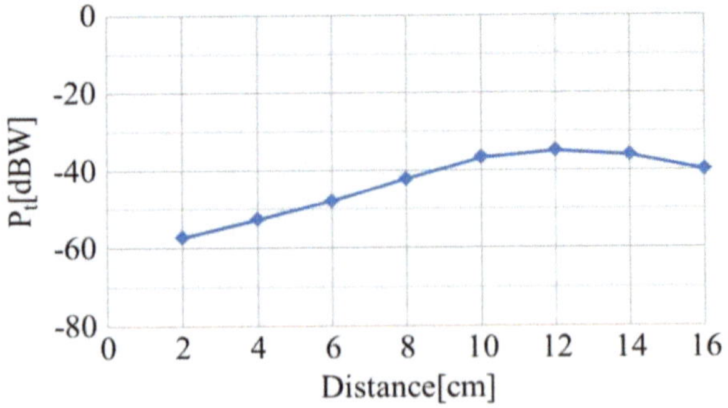

Fig. 10 Required transmit power versus transmission distance to achieve BER = 10^{-2}

at the distance of 12 cm, which shows sufficient possibility of using our designed antenna in capsule endoscope application.

5 Conclusions

In this study, we have first proposed a FSK-based impulse radio system in HBC band around 30 MHz for in-body communications. In order to design a small-size implant antenna suitable to the proposed FSK impulse radio scheme, we have adopted a monopole-type helical structure formed on flexible magnetic sheets based on its double shortening effect of wavelength. The designed antenna (1 cm × 3cm) has been evaluated in a biological-equivalent cubic phantom model by FDTD numerical simulations. It has been found to posses double-resonant frequencies in HBC band, which shows good applicability in FSK impulse radio scheme. The transmission characteristics and required transmit power have also been investigated to demonstrate the feasibility and usefulness of our proposed communication method and small-size implant antenna.

References

1. Wang, J., Wang, Q.: Body Area Communications. Wiley. IEEE (2012)
2. A future Created by "In-Body Hospital". NanoSky, vol. 1, Mar 2016
3. IEEE standard for local and metropolitan area network—part 15.6: wireless body area networks, Feb 2012. IEEE Std 802.15.6-2012
4. Shi, J., Anzai, D., Wang, J.: Channel modeling and performance analysis of diversity reception for implant UWB wireless link. IEICE Trans. Commun. **E95-B**(10), 3197–3205 (2012)
5. Anzai, D., Katsu, K., Chavez-Santiago, R., Wang, Q., Plettemeier, D., Wang, J., Balasingham, I.: Experimental evaluation of implant UWB-IR transmission with living animal for body area networks. IEEE Trans. Microwave Theory Tech. **62**(1), 183–192 (2014)
6. Wang, J., Liu, J., Suguri, K., Anzai, D.: An in-body impulse radio transceiver with implant antenna miniaturization at 30 MHz. IEEE Microwave Wirel. Compon. Lett. **25**(7), 484–486 (2015)

Motion-Based Gait Identification Using Spectro-temporal Transform and Convolutional Neural Networks

Omid Dehzangi, Mojtaba Taherisadr, Raghvendar ChangalVala and Priyanka Asnani

Abstract The wide range of usage and application of wearable sensors like as smart watches provide access to precious inertial sensor data that is usable in human identification based on their gait pattern. A large number of studies have been conduced on extracting high-level and various heuristic features out of inertial sensor data to identify discriminative gait signatures and distinguish the target individual from others. However, complexity of the collected data from inertial sensors, detachment between the predictive learning models and intuitive feature extraction module increase the error rate of manual feature extraction. We propose a new method for the task of human gait identification based on spectro-temporal two dimensional expansion of gait cycle. Then, we design a deep convolutional neural network learning in order to extract discriminative features from the two dimensional expanded gait cycles and also jointly optimize the identification model simultaneously. We propose a systematic approach for processing nonstationary motion signals with the application of human gait identification with 3 main elements: first gait cycle extraction, second spectro-temporal representation of gait cycle, and third deep convolutional learning. We collect motion signal from 5 inertial sensors placed at different locations including lower-back, chest, right knee, right ankle, and right hand wrist. We pre-process the acquired raw signals by motion signal processing and then we propose an efficient heuristic segmentation methodology and extract gait cycle from the segmented and processed data. Spectro-temporal two dimensional features are extracted by merging key instantaneous temporal and spectral descriptors in a gait cycle which is capable of characterizing the non-stationarities in each gait cycle inertial data. The two

O. Dehzangi
Rockefeller Neuroscience Institute, West Virginia University,
Morgantown, WV 26505, USA
e-mail: omid.dehzangi@hsc.wvu.edu
URL: http://www.wssplab.net

M. Taherisadr (✉) · R. ChangalVala · P. Asnani
University of Michigan, Dearborn, MI 48128, USA
e-mail: mojtabat@umich.edu

G. Fortino and Z. Wang (eds.), *Advances in Body Area Networks I*,
Internet of Things, https://doi.org/10.1007/978-3-030-02819-0_31

dimensional time-frequency distribution representation of the gait cycle extracted from acquired inertial sensor data from 10 subjects are fed as input to the designed and proposed 10 layers DCNN architecture. Based on our experimental analysis, 93.36% accuracy was achieved for subject identification task.

Keywords Spectro-temporal analysis · Motion analysis
Convolutional neaural network · Gait identification
Time frequency representation · Sensor fusion · Motion

1 Introduction

The action of gait refers to the manner in which an individual is stepping or walking. The analysis of human gait dates to early 60's when studies were focused on medical purposes for applications like early diagnosis of different diseases including:

(a) Musculoskeletal disorders such as spinal stenosis.
(b) ageing related disorders.
(c) neurological disorders such as Rett syndrome, Parkinson's or Cerebral Palsy.

Reliable and continuous monitoring of characteristics of gait over time may help in early diagnosis of many diseases and prevent their related complexities. Unlike palm veins, face, fingerprint, iris, or other existing biometric identifiers, identifying by gait can be less intrusive because it can be collected at a distance [1]. Furthermore, recent literature in medical studies revealed that human gait has 24 different components and gait pattern is capable to be unique if all movements are considered in the cycle. Consequently, gait cycle has the potential to be considered in several other applications such as biometric identification like as keyless vehicles, health monitoring [3], social security [2], etc.

Recently, wearable devices have been used more for activity recognition applications [4], activity level estimation [5], joint angle estimation [6], activity-based prompting [7], and sports training [8]. Recently, gait recognition using wearable devices has become an active research topic due to the widespread installation of sensors for measuring movements in smartphones, fitness trackers, and smartwatches [2, 9]. Most of the wearable devices use Micro Electro Mechanical Systems (MEMS) based inertial sensors. These inertial sensors (accelerometers, gyroscopes) are one of the most important members of MEMS family and are combined together as inertial measurement units (IMU). Most modern accelerometers are electromechanical devices that measure acceleration forces in one, two, or three orthogonal axes. Gyroscopes, also known as angular rate sensors, are devices that measures angular velocity in three directions. Due to their small-size, portability, and high processing power, inertial sensors are widely used for complex motion analysis. Hence gait recognition using wearable devices has become an effective Privacy Enhancing Technology (PET) [10].

Advent of MEMS based accelerometers and gyroscopes and wireless interfaces like Bluetooth and Wi-Fi have made the measurement setup for gait analysis data collection non-intrusive and ubiquitous. The inertial sensors have become an important part of ubiquitous smart devices and therefore integration of inertial sensors in smart devices has become a common practice. There are two facts that demonstrate inertial sensors are an important part for the gait analysis. First, there is a mass of people using smart devices on a daily basis. In 2012, In-stat reported that by the end of 2015, 65% of U.S. population will own a tablet or smartphone having inertial sensors integrated [11]. Second, with the latest achievements in the field of pervasive computing, limitations of inertial sensors such as cost, storage, computational power were overcome [12]. Owing to which inertial sensors are not only restricted to simple tasks such as tilt estimation but also for complex tasks such as advanced motion analysis, and activity recognition [9]. They have also been evaluated in medical applications, such as analysis of patient's health based on gait abnormalities [13], fall detection [14]. Thus, it is reasonable to identify gait using inertial data collected using smart phones.

Although the inertial sensor based wearable devices revolutionized the gait analysis, there are some intrinsic challenges with the system. Since, the wearable device is always worn casually, relative orientation between the sensors and the subject body cannot be fixed over different sessions of data acquisition [15]. As the coordinate system used by sensors is defined relative to the frame of the device, small orientation changes of sensor installation may make measurements quite different [15]. Moreover, noise interference and external environmental effects on complex sensor data makes feature extraction for gait analysis a difficult task. Hence, feature extraction always needs lots of handcraft work [10]. Therefore, we aim to exploit the additional information provided by expanding the motion signals to 2D spectro-temporal space via time-frequency (TF) analysis, and then, extracted features from the TF representation of the signals for the purpose of classifying them. Due to the non-stationarity of the motion signals as discussed above, TF and instantaneous frequency (IF) based methods seem naturally more suitable for motion data mining and pattern recognition. However, there are two major issues of: (1) selection of features from the 2D spectro-temporal expanded space due to high dimensionality of the space, and (2) selection of the appropriate classification model due to the fact that sensor data are collected sequentially and shallow models such as support vector machines (SVM), nearest neighbors (NN), decision tree (DT) might not be suitable choices to undertake the authentication task.

In this paper, we propose a gait recognition approach based on two complementary areas: time-frequency analysis and deep convolutional learning. We first conduct a window based zero crossing search for gait cycle extraction. Then, we extract 2D spectro-temporal representation of gait patterns per each gait cycle. We aim to capture subject dependent gait patterns in high resolution TF and IF spectral density of the sensor data. In order to effectively extract the gait signatures high dimensional expanded space and avoid cumbersome hand-craft feature extraction, we then employ deep convolutional neural networks (DCNN) as a feature extractor to extract high-level features by passing low-level features in the input space through the layers of the DCNN. We design a DCNN structure for predictive modeling of the 2D representation without needing to manually and subjectively extract features. DCNN performs discriminative joint feature space and predictive model optimization in an automatic fashion.

Raw motion data is collected from five inertial sensors placed at chest, lower-back, right hand wrist, right knee and right ankle. The 2D time frequency density image of the gait cycles extracted from inertial sensor data from a population of 10 subjects were fed as input to our proposed 10 layer DCNN architecture. In the last layer of the proposed DCNN architecture, a fully connected back-propagation logistic regression model will be trained to perform the human authentication task. Then, we investigate the fusion of multiple sensor nodes (i.e. five inertial sensors) and modalities (i.e. accelerometer and Gyroscope readings). We perform early (input level) and late (model level) fusion to increase the classification rate in the gait identification task. In this way, we generate the posterior probability scores from 10 DCNNs trained on the resulting factors of accelerometer and Gyroscope readings from five sensor nodes, separately. A discriminative performance measure is introduced to optimize the performance of the fusion of the 10 DCNNs. Our experimental results demonstrated a 97.06% subject authentication accuracy.

2 Methodology

The proposed system incorporates a gait cycle based identification to accommodate the temporal variations in the gait patterns, which might be very useful in the context of gait recognition. Our proposed system is unique in several ways:

(1) window based zero crossing search for gait cycle extraction.
(2) 2D spectro-temporal representation of the gait cycles.
(3) DCNN based feature extraction and predictive analysis.

In this work, we aim to automatically identify the target subject given their gait information. Assuming M target subject, given an unknown gait segment, a gait identification system gives the corresponding subject identity, φ, where $\varphi \in \{1, \ldots, M\}$. A set of reference models, learned in the training phase, is used to classify the input gait segments to one of the M classes.

2.1 Experimental Setup and Data Collection

A total of 10 subjects participated in the experimental procedure. A set of five inertial sensors were placed at various locations including chest, right wrist, right knee, right ankle and lower back. The motion sensor system employed for this study was Shimmer sensor platform, which is a wearable senor platform with wireless interface. It houses both the accelerometer and gyroscope inertial sensors. The data collection sessions were synchronized across all the sensors and labels using our in-lab designed Android app developed at the wearable sensing and signal processing (WSSP) lab, University of Michigan. Each subject was asked to walk the same route from a specific point the lab to outside the building.

2.2 Preprocessing

The raw accelerometer and gyroscope data (x, y and z vectors) collected during the experiment has various noise factors such as motion artifacts, step impacts, sensor orientation and location related noises along with the necessary gait information. To reduce the orientation related noise, resultant vectors of the triaxial sensor data (i.e. accelerometer and gyroscope recorded data) is computed using Eq. 1.

$$Mag(t) = \sqrt{x^2(t) + y^2(t) + z^2(t)} \tag{1}$$

Since typical gait data for normal walking has frequency components in the range of 0.5–2.5 Hz, a 10-th order *Butterworth* bandpass filter was used to extract the required frequency components from the resultant vectors of the IMUs. For Gyroscope data, we assume the first estimated value of direction vectors to be the same as the direction vectors measure by the accelerometer:

$$R_x Est(0) = R_x Acc(0),\ R_y Est(0) = R_y Acc(0),\ R_z Est(0) = R_z Acc(0), \tag{2}$$

In our algorithm, we assume the value of the accelerometer, when the sensor device is at rest (a first few seconds recordings before subjects started the gait paradigm), to be zero.

2.3 Gait Cycle Extraction

Since the data collection through all the sensors is time synchronized, we extracted the gait cycles from the sensor #4 (i.e. the ankle sensor) and used the same markers for other sensors. The cycle extraction process implements amplitude check and zero crossing check to extract noise free gait cycle data.

To approximate the gait cycle frequency, the resultant vectors of accelerometer data from sensor #4 R_{acc} are passed through a band pass filter of rage 0.5–1.5 Hz only for the purpose of cycle extraction. This eliminates the interference of any high frequency components while determining the gait cycle frequency. The frequency range of 0.5–1.5 Hz is chosen to include the average gait frequency is around 1 Hz. Frequency Analysis consists of finding the energy distribution as a function of the frequency index ω. Therefore, it is necessary to transform the signal to the frequency domain by means of the Fourier transformation:

$$X(\omega) = \int x(t)e^{-jt\omega}dt \tag{3}$$

where $x(t)$ is the time domain signal. A Fourier transform is performed on the resultant data and the dominant frequency component within the range of 0.5–1.5 Hz is selected as the gait cycle frequency of that subject f_{cycle}.

In order to remove irregular gait cycles f turns, the R_{acc} is checked against the amplitude threshold of a signal 0.05. If there are samples that have values below the threshold, those samples are neglected and samples with amplitude above the threshold are collected into small local windows. Gait cycles are extracted from these local windows (LW) following the process shown in the Algorithm 1.

Algorithm 1

$S_{cycleStart} = 0$
$\quad while(LW_{end} < length(LW))$
$\qquad S_{cycleEndTemp} = S_{cycleStart} + S_{cycleOffset}$
$\qquad S_{cycleEnd} = S_{cycleEndTemp} \pm S_{zeroCrossing}$
$\qquad S_{cycleStart} = S_{cycleEnd}$
$\qquad LW_{end} = S_{cycleStart}$
$\quad end$

To synchronize the cycles, zero crossing of the R_{acc} is taken as a reference point and the samples between a given zero crossing and the subsequent zero crossing at $S_{cycleOffset}$ samples away from previous zero crossing is considered as a gait cycle. $S_{cycleOffset}$ is calculated using Eq. 4, where f_s is the data sampling frequency 50 Hz in our empirical investigations.

$$S_{cycleOffset} = \frac{f_s}{f_{cycle}} \tag{4}$$

The end sample of this cycle $S_{cycleEndTemp}$ is approximated initially based on the f_{cycle} but a more precise $S_{cycleEnd}$ is calculated based on finding a zero-crossing sample near the initial approximation. The local window indices of the start and end of cycle data $S_{cycleStart}$ and $S_{cycleEnd}$ are then mapped to the global window of R_{acc}.

Cycle data collected across all subjects and sensors were observed for data consistency, i.e. the number of gait cycles extracted per sensor. Considering this number as a measure of data integrity, the ankle based accelerometer (sensor #4) captured nearly all of the gait cycles performed by different subjects consistently. Hence, it was selected as the reference sensor with discriminative features to represent the gait cycle. Gait cycles are extracted from other sensors based on the indices generated from the sensor 4 R_{acc}. Figure 1 shows the gait cycle extracted from Subject 1 for all the five sensors.

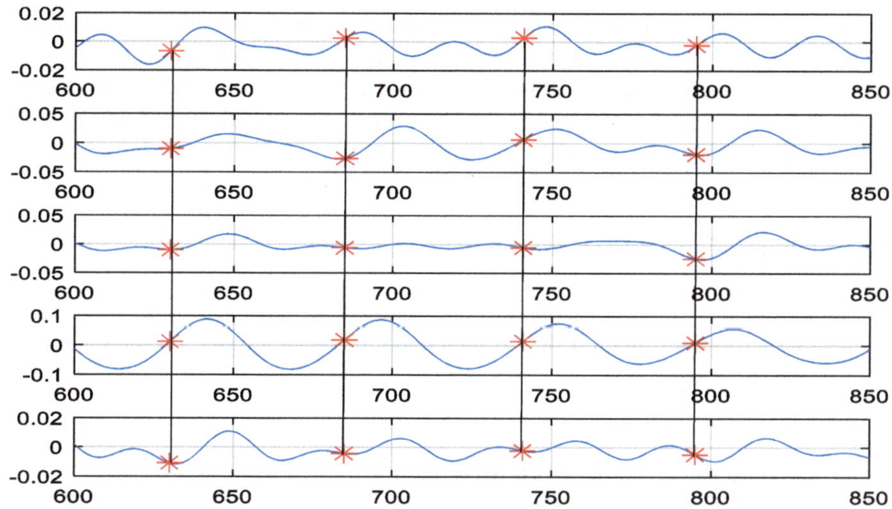

Fig. 1 Extracted cycle data sample plot for subject 1 across all the sensors

2.4 *Time Frequency Representation*

Extracting useful information from a given signal is an inherit concern for many signal processing tasks. Such information could be directly available from a given signal $x(t)$, or could be achieved after a proper transformation applied to the signal to have a new more informative space. An example of important information can be how the energy of the signal is distributed. This category of information has conventionally been divided into three main groups: frequency analysis, time analysis, and joint time and frequency analysis. Time domain analysis includes finding the energy distribution of the given signal $x(t)$ as a function of the index t (time) which often is referred to as the *instantaneous power* (*IP*). Frequency domain analysis includes considering the energy distribution of the signal as a function of the index w (frequency). Thus, it requires to first transform the given signal to the frequency space by applying Fourier transformation as discussed in Sect. 2.3. It represents the signal energy in the frequency space per unit frequency (J/Hz) at the frequency w, and then is referred to as the *powerspectrum* (*PS*). Joint time and frequency analysis merges the information provided by the *IP* and *PS* into one energy space. The purpose of merging these two domain is to map a one dimensional function of frequency or time into a two dimensional function of frequency and time.

Among the different approaches, *instantaneous frequency* (*IF*) has been considered more in the literature to represent and describe a non-stationary and multi-component signal simultaneously in frequency and time. In order to characterizing a multi-component signal, IF law could be assigned to every signal components. Various IF estimation approaches with the application in multi-component signals have been proposed [17]. These approaches initially localize and then extract components of the signal from the given signal's time-frequency representation and finally apply the IF estimation law. Advantage of these methods is not requiring any prior knowledge about the given signal except its components are separated in the time-frequency space. A typical implementation of a basic multi-component *IF* estimation method consists of the following two main steps.

2.4.1 Time-Frequency Transformation

The input signal is initially mapped to the time-frequency space by a proper time-frequency distribution method (TFD) which can be chosen based on the characteristics of the input signal.

2.4.2 Selectig the *IF* Estimation Approach

IF estimation approach first considers the resulted time-frequency distribution space ($S(T, F)$) as a two dimensional representation in which frequency and time are considered as its column and row coordinates, and also identifies the local extremums (with respect to frequency F) by applying the first and second derivative tests. Only local extremums which have values lower or higher than a predefined threshold are considered as acceptable peaks. Then, an algorithm specifically designed for detecting linked component in the $S(T, F)$ space counts the number of connected pixels and evaluates their connectivity to detect the linked components. This idea is based on the fact that the IF of a component of a signal where energy of signal concentrates will be detected in the $S(T, F)$ space as a ridge which describes the IF.

Selection of a proper time-frequency distribution for representing the given gait cycle is the first and main step in initializing any classification scheme in the time-frequency space. A proper approach is the one which is able to emphasize the non-stationarities of the signal that efficiently discriminate between various groups under consideration.

A thorough comparative review of the existing approaches shows that they can be categorized in 6 classes as following: modified-B distribution, separable kernel distribution, Gaussian kernel distribution, smoothed Wigner-Ville distribution, spectrogram, and Wigner-Ville distribution. Use approaches which reduces the interference like as smoothed Wigner-Ville distribution (SWVD) is capable of improving the performance of the prediction. The reason is that RIDs are capable of reducing the affections from the cross-terms, while they still provide an acceptable resolution. Therefore, SWVD is capable of generating the most smoothed patterns for gait cycles. In this study we choose SWVD as gait cycle representation approach.

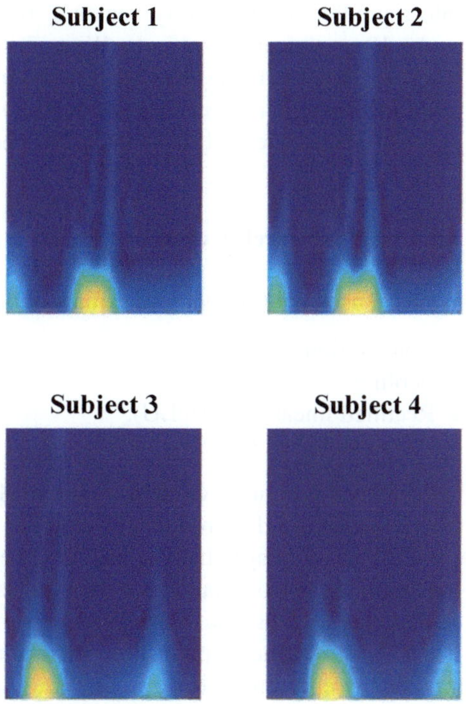

Fig. 2 TF representation of gait samples as CNN inputs for 4 subjects

2.5 Deep Convolutional Neural Networks (DCNN)

Figure 2 demonstrates an example TF representation of gait cycle for 10 different subjects. Taking a close look at the gait cycle data, it is evident that the temporal orientation and order of the data stand out as significant discriminative features. A Convolutional Neural network (CNN) was designed as deep structure for feature extraction and as the authentication model in this investigation. We intended to verify if multiple feature maps generated by CNN process would preserve the temporal aspect of the gait cycle data and provide higher level of discriminative feature space.

CNNs are most commonly used in pattern recognition. A simple CNN is a sequence of steps where each step transforms a volume of activations through a set of functions that are differentiable. They are made up of neurons with learnable weights and biases [19]. Though the weight vector optimization is similar to the conventional neural networks. CNNs are designed to deal specifically with 2D or 3D image data. A variety of combinations of linear and non-linear differentiable steps could be used to build a deep CNN and that determines the complexity of the system [20].

We have developed a non-parametric fully supervised DCNN model for motion-based gait authentication. The model takes a 3D input image x_i and transforms it into a prediction probability vector $\overline{y_i}$ for ten different classes which correspond to the 10 participating subjects. We train the model using N labeled images {x,y} where the label y_i is the class label of the input data. Training minimizes a *SoftMax* loss function with respect to network parameters such as weights and biases using a gradient descent method and network parameters are updated using back propagation.

2.5.1 DCNN Architecture

We have used the following four main building blocks in the DCNN model:

- Convolution.
- Pooling.
- Rectified linear unit (ReLU).
- Fully connected layer.

The convolution layer performs convolution of input with a set of predefined filters, which is a dot product between the filters and the region they are connected to in the input image. If we consider k kernels of spatial dimensions' h_k and w_k, then the filter tensor would have the dimension of $h_k \times w_k \times k \times k'$ which could be represented by tensor w. Considering a simple convolution layer with zero padding and a unit stride, the output y after performing convolution on an input layer of tensor x could be represented as

$$y_{i'j'k} = \sum_{i=0}^{h_k} \sum_{j=0}^{w_k} \sum_{k'=0}^{k} w_{ijk'k} \times x_{i+i',j+j',k'} \quad (5)$$

In Eq. 5, $x_{i+i',j+j',k'}$ represents the $(i+i',j+j',k')$ indexed element of the input tensor x.

The pooling layer applies a chosen operator and combines closely associated feature values. It is used to down sample the input image along the width and height. This layer does not require parameter learning. A simple implementation of max pooling can be represented as

$$y_{i'j'k} = \underset{0 \le i \le h_p,\ \ 0 \le j \le w_p}{Max} \{x_{i' * h_p + i,\ j' * w_p + j,\ k}\} \quad (6)$$

where x and y represent the $i'j'k$ indexed input and output layer and h_p, w_p are the pooling window dimensions.

The Rectified Linear Unit (ReLU) is a non-linear activation layer introduces the non-linearity when applied to the feature map. ReLU layer leaves the size of its input unchanged. A simple implementation of ReLU would be as below:

$$y_{i'j'k} = \max\{0, x_{i'j'k'}\}. \quad (7)$$

where x and y are input and output of corresponding tensors. Like pooling layer, ReLU doesn't need any parameter learning and it doesn't alter the dimensions of the input layer.

3 Results

We implemented a Matlab TF representation toolbox [21] to generate the SWVD TF representation of the data and use Matlab print function to output a 2D image with specific resolution and size. After experimenting with multiple image sizes, it was observed that higher image resolution did not necessarily mean better model performance. DCNN model was tested with different input image sizes. Figure 3 shows the prediction accuracies of the CNN models for different input image sizes. The image size was modified by changing the resolution of the figure generated by Matlab print function. It's observed that the model prediction accuracies are high for low image sizes. We selected the image size of 120 × 120 × 3 for our current analysis based on the observation in Fig. 3.

10-CV divides the total input data of N samples into ten equal parts. In every iteration one part is considered as test sample set and the rest of nine parts are considered as validation and training sample set. This set is further divided into 4:1 ratio of training and validation data samples. For an input data set of N samples

- Number of Test Samples: N/10
- Number of Validation samples: (N − N/10) * 1/5
- Number of Training samples: (N − N/10) * 4/5.

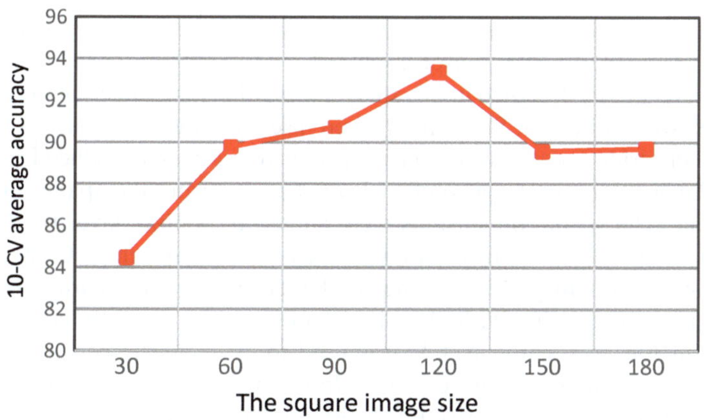

Fig. 3 The effect of the input image size selection on CNN 10-CV performance (size 30 means a 30 by 30 matrix)

Table 1 10-CV performance of first 3 sensors

Sensor	S1 Acc	S1 Gyro	S2 Acc	S2 Gyro	S3 Acc	S3 Gyro
10-CV accuracies (%)	86.67	80.01	86.01	82.69	91.83	87.75
	100	93.33	74.5	95.83	80.02	84.01
	93.75	80.01	84.02	80.39	88.01	92.01
	78.75	90.01	84.01	82.01	**98.03**	92.01
	80.01	86.67	79.59	83.33	79.59	**91.83**
	100	93.33	78.72	87.75	91.66	87.51
	86.67	80.01	73.46	79.59	86.27	90.19
	100	92.86	95.91	82.02	92.11	86.01
	92.85	93.33	88.01	90.01	90.01	**96.01**
	93.33	87.61	82.35	81.63	87.75	**100**
Average (%)	91.01	87.61	82.65	84.52	88.51	90.72

Data set has N = 4178 samples, which includes images from ten different subjects. Each iteration of the 10-fold cross validation would have the following division of data.

- Number of Test Samples: 417
- Number of Validation samples: 752
- Number of Training samples: 3008

Figure 2, shows a sample image set of all subjects for sensor four. The CNN model was trained using the training and validation set and tested independently with the testing set.

Subject prediction accuracies of the CNN models for different sensors are shown in Table 1. It can be observed that the gyroscope data of sensors is a better predictor than the accelerometer data in 4 out of 5 cases. For the gyroscope data the mean prediction accuracy for different sensors in decreasing order is sensor5 (lower back), sensor3 (right knee), sensor4 (right ankle), sensor1 (chest), sensor2 (right wrist). Similarly, for the accelerometer data, prediction accuracies of the sensors in descending order is sensor1, sensor3, sensor5, sensor2, sensor4.

For DCNN model designed in this study, the architecture guidelines as mentioned in [22, 23] were followed. Small filters were used and the image size was reduced using higher stride lengths where necessary. Padding was introduced in the convolution layers to prevent convolution layer from altering the spatial dimensions of the input. The spatial dimensions were altered in pooling layer by down sampling. Also, more ReLU activation layers were used across the DCNN after each convolution and pooling pair to bring in element wise non-linearity. The current model consists of 10 (CONV, RELU, POOL) layers and one fully connected layer as shown in Fig. 4. Finally, 10-CV was used to estimate the performance of the authentication model on the unseen data. In this work, we investigate aggregating the complementary discriminative information in all the synchronous IMUs. In this way, we conduct a sensor fusion at the input level to the DCNN model. The input level fusion is achieved by combining the 3D gait cycle images from five different

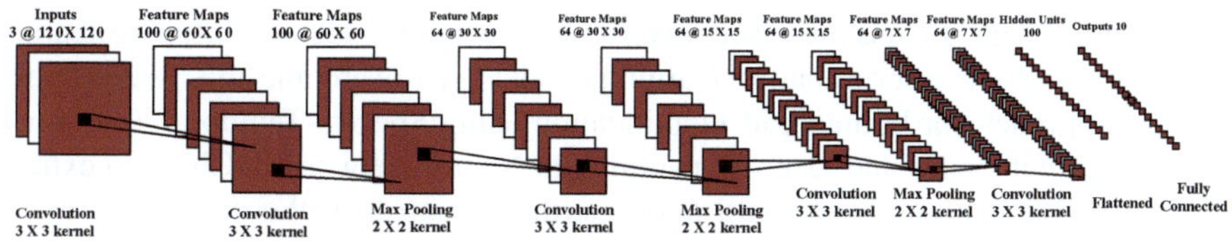

Fig. 4 CNN architecture

Table 2 Input image concatenation performance

	10-CV accuracies	
	Mean pooling (%)	Max pooling (%)
Subject#1	89.65	**91.14**
Subject#2	90.21	**96.84**
Subject#3	**89.78**	89.28
Subject#4	93.17	**96.77**
Subject#5	**91.29**	90.14
Subject#6	87.05	**93.19**
Subject#7	89.96	**90.45**
Subject#8	93.81	**97.84**
Subject#9	85.16	**89.92**
Subject#10	94.01	**98.01**
Average	90.41	93.36

sensors to form a $121 \times 121 \times 30 - \textit{dimensional}$ image. The average subject identification accuracy percentages of the DCNN model whose parameters are optimized using 10-CV method is shown in the Table 2. The image fusion at the input to the DCNN demonstrated good prediction accuracies as shown in Table 2, averaging at 93.88%.

The results suggests that using a simple input fusion of sensors increase the performance of the best individual sensor reading (i.e. 'S1 Acc') with 32% relative improvement. In order to compare the results of the input level fusion and S1 Acc, we conducted 10-CV test 10 times and run a statistical test with the null hypothesis that the performance accuracies of the fusion and 'S1 Acc' on different subjects comes from independent random samples from normal distributions with equal means, using the two-sample t-test. The test rejected the null hypothesis with the p-value = 0.023 which demonstrate significant improvement.

4 Conclusion and Future Work

In human gait recognition task, we aim to extract discriminative features and descriptors to identify our target subject from others. The manual feature extraction from inertial sensor recordings is prone to errors due to the complexity of data collected from inertial sensors and the disconnection between heuristic and intuitive feature extraction and the discriminative learning models. To overcome this, we propose a novel approach for human gait identification using Spectro-Temporal Transform of isolated gait cycles and Deep Convolutional Neural Network (DCNN) to extract discriminative features from multi-node inertial sensor data. We designed a novel methodology for processing non-stationary signals for the purpose of human gait identification. The proposed methodology comprises three main components of cycle extraction, Spectro-temporal 2D expansion and representation, and deep convolutional learning. Based on our experimental results, 93.88% subject identification accuracy was achieved by fusing the multi-node sensor readings.

As our future work, we will expand the conditions during the gait and investigate the variations and their impact on the identification performance. We will also investigate and design effective sensor data fusion and selection methods at the logical levels to model the dynamics of gait patterns by incorporating complementary information from multiple nodes and discover the key signature patterns and sensor locations.

References

1. El-Sheimy, N., Hou, H., Niu, X.: Analysis and modeling of inertial sensors using Allan variance. IEEE Trans. Instrum. Meas. **57**(1), 140–149 (2008)
2. Sprager, S., Juric, M.B.: Inertial sensor-based gait recognition: a review. Sensors **15**(9), 22089–22127 (2015)
3. Gafurov, D., Einar, S., Patrick, B.: Gait authentication and identification using wearable accelerometer sensor. In: 2007 IEEE Workshop on Automatic Identification Advanced Technologies. IEEE (2007)
4. Kim, E., Sumi, H., Diane, C.: Human activity recognition and pattern discovery. IEEE Pervasive Comput. **9**(1) (2010)
5. Mortazavi, B., et al.: Met calculations from on-body accelerometers for exergaming movements. In: 2013 IEEE International Conference on Body Sensor Networks (BSN). IEEE (2013)
6. Vikas, V., Crane, C.D.: Measurement of robot link joint parameters using multiple accelerometers and gyroscopes. ASME Paper No. DETC2013-12741 (2013)
7. Robertson, K., et al.: C-66 prompting technologies: is prompting during activity transition more effective than time-based prompting? Arch. Clin. Neuropsychol. **29**(6) (2014)
8. Ahmadi, A., et al.: Automatic activity classification and movement assessment during a sports training session using wearable inertial sensors. In: 2014 11th International Conference on Wearable and Implantable Body Sensor Networks (BSN). IEEE (2014)
9. Le Moing, J., Stengel, I.: The smartphone as a gait recognition device impact of selected parameters on gait recognition. In: 2015 International Conference on Information Systems Security and Privacy (ICISSP). IEEE (2015)
10. Gupta, M. (ed.): Handbook of Research on Social and Organizational Liabilities in Information Security. IGI Global (2008)
11. Vienne, A., et al.: Inertial sensors to assess gait quality in patients with neurological disorders: a systematic review of technical and analytical challenges. Front. Psychol. **8** (2017)
12. Chen, C., Jafari, R., Kehtarnavaz, N.: A survey of depth and inertial sensor fusion for human action recognition. Multimed. Tools Appl. **76**(3), 4405–4425 (2017)
13. Roberts, M.L., Zahay, D.: Internet Marketing: Integrating Online and Offline Strategies. Cengage Learning (2012)

14. Yamada, M., et al.: Objective assessment of abnormal gait in patients with rheumatoid arthritis using a smartphone. Rheumatol. Int. **32**(12), 3869–3874 (2012)
15. Sposaro, F., Tyson, G.: iFall: an Android application for fall monitoring and response. In: Annual International Conference of the IEEE Engineering in Medicine and Biology Society, EMBC 2009. IEEE (2009)
16. Tunca, C., et al.: Inertial sensor-based robust gait analysis in non-hospital settings for neurological disorders. Sensors **17**(4), 825 (2017)
17. Boashash, B.: Estimating and interpreting the instantaneous frequency of a signal. II. Algorithms and applications. Proceedings of the IEEE **80**(4), 540–568 (1992)
18. Boashash, B., Sucic, V.: Resolution measure criteria for the objective assessment of the performance of quadratic time-frequency distributions. IEEE Trans. Signal Process. **51**(5), 1253–1263 (2003)
19. Karpathy, A.: Cs231n: convolutional neural networks for visual recognition. Neural Netw. **1** (2016)
20. Zhong, Y., Deng, Y.: Sensor orientation invariant mobile gait biometrics. In: 2014 IEEE International Joint Conference on Biometrics (IJCB). IEEE (2014)
21. Auger, F., et al.: Time-frequency toolbox. CNRS France-Rice University 46 (1996)
22. Tong, M., Minghao, T.: LEACH-B: an improved LEACH protocol for wireless sensor network. In: 2010 6th International Conference on Wireless Communications Networking and Mobile Computing (WiCOM). IEEE (2010)
23. Rigamonti, R., Brown, M.A., Lepetit, V.: Are sparse representations really relevant for image classification?. In: 2011 IEEE Conference on Computer Vision and Pattern Recognition (CVPR). IEEE (2011)
24. Reynolds, D.A., Quatieri, T.F., Dunn, R.B.: Speaker verification using adapted Gaussian mixture models. Digit. Signal Process. **10**(1–3), 19–41 (2000)
25. Vapnik, V.: The Nature of Statistical Learning Theory. Springer (2013)
26. Bezdek, J.C., et al.: Fuzzy Models and Algorithms for Pattern Recognition and Image Processing, vol. 4. Springer (2006)
27. Woods, K., Kegelmeyer, W.P., Bowyer, K.: Combination of multiple classifiers using local accuracy estimates. IEEE Trans. Pattern Anal. Mach. Intell. **19**(4), 405–410 (1997)

Voluntary EMG-to-Force Estimation in Shoulder and Elbow During the Movement of Feeding Oneself

Jiateng Hou, Yingfei Sun, Lixin Sun, Bingyu Pan, Zhipei Huang and Jiankang Wu

Abstract Muscle force estimation opens up the possibility of objective evaluating human motion in both mechanical and physiological ways. This paper proposes an EMG-adjusted method to predict individual muscle force in the shoulder and elbow during a purposeful daily activity: feeding oneself. Two male subjects were asked to flex and extend their shoulders and elbows to simulate the movement of getting food from the pocket to the mouth. Three inertial sensors and six surface electromyography (sEMG) sensors were used to synchronously collect motion and sEMG data during the movement. A Hill-type musculotendon model was then employed to predict individual muscle force by the fusion of motion and adjusted sEMG data. The result shows that our method can predict individual muscle force accurately with the ability to cover subject-specific joint dynamics and neural control strategies in multi-joints movement.

Keywords Hill-type musculotendon model · Surface electromyography (sEMG) Shoulder · Elbow · Muscle force

J. Hou · Y. Sun · L. Sun · B. Pan · Z. Huang (✉) · J. Wu
School of Electronic Electrical and Communication Engineering, University of Chinese Academy of Sciences, Beijing 101408, China
e-mail: zhphuang@ucas.ac.cn

J. Hou
e-mail: houjiateng13@mails.ucas.ac.cn

Y. Sun
e-mail: yfsun@ucas.ac.cn

L. Sun
e-mail: sunlixin12@mails.ucas.ac.cn

B. Pan
e-mail: panbingyu14@mails.ucas.ac.cn

G. Fortino and Z. Wang (eds.), *Advances in Body Area Networks I*,
Internet of Things, https://doi.org/10.1007/978-3-030-02819-0_32

1 Introduction

Predicting individual muscle force during human movements is of great value in movement neuroscience and biomechanics research, especially in the neurorehabilitation field. Currently, the severity of impairment and disability in clinical settings is generally assessed by ordinal scales such as the Fugl-Meyer Assessment (FMA) [1], the Box and Block Test (BBT) [2], and Action Research Arm Test (ARAT) [3]. But they cannot account for neuromuscular mechanisms in patients and the results are often subjective. Thus, muscle force prediction, which opens up the possibility to examine the outcome of muscle deficiencies and to investigate causes of joint instability [4], should be introduced to quantitatively evaluate patient's impairment. However, direct measurement of muscle force is generally not feasible in a clinical setting, non-invasive methods based on musculoskeletal modeling should therefore be considered [5].

EMG-driven model, which has been developed aiming at many static or dynamic tasks for different anatomical sites such as elbow, shoulder, knee, ankle, jaw, lower back, and wrist [6], is the most frequently used method in non-invasive muscle force prediction. However, as the EMG signals of deep muscles cannot be measured using non-invasive methods and noise contamination from cross-talk and movement artifacts is very difficult to be removed, the ability of EMG-driven method in simulating musculoskeletal dynamics is limited.

To overcome the inherent drawbacks of the EMG-driven methods, we brought in an EMG-adjusted algorithm to help with the muscle force prediction. In our approach, we tuned experimental muscle excitations extracted from sEMG within a reasonable range to minimized joint moment tracking error. The reason why we chose to analyze the movement of feeding oneself is because it is one of the basic activities of daily living (ADL) and demands the shoulder and elbow working coordinately. The result indicates that our approach can predict individual muscle force accurately with the ability to reflect subject-specific joint dynamics and neural control strategies, making it possible for us to get a deeper understanding of human dynamic movement mechanism. In this paper, we first explicate how our approach works in biomechanical simulation. Then we assess the result of our method in muscle force prediction. Last, we give a discussion about some limitations in our experiment and our future work.

2 Methods

There are two main components in our approach as shown in Fig. 1. EMG-driven model takes joint angle information and adjusted muscle excitation as inputs to estimate individual muscle force. Static optimization embeds the EMG-adjusted algorithm to tune the experimental muscle excitation amplitudes.

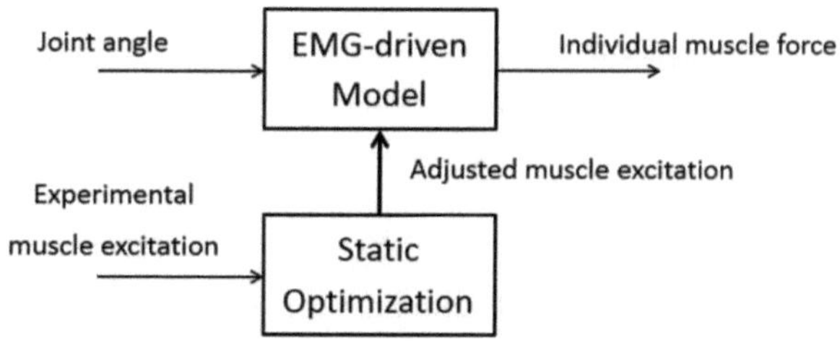

Fig. 1 The flowchart of our muscle force prediction method. It consists of two main parts: EMG-driven model and static optimization

2.1 Data Collection and Preprocessing

Two healthy male subjects (subject 1: age: 22 years, height: 174 cm, mass: 66 kg; subject 2: age: 23 years, height: 176 cm, mass: 60 kg) volunteered for this investigation and gave their informed, written consent.

In the initial position, the subject stood upright with the right arm dropped naturally to the body side. The right shoulder was orientated at about 0° of flexion, 0° of abduction, and neutral rotation. The right elbow was orientated at about 10° of flexion and neutral rotation. During the test, the subject slowly flexed his right shoulder from 0° to 90° with the elbow synchronously flexed from 10° to 90°, and then extended back to the initial position, to simulate the movement of getting food from the pocket to the mouth.

The motion data was collected (50 Hz) using three MPU-9150 (InvenSense Inc., San Jose, CA, USA) sensors attached to the lateral side of the right forearm, the lateral side of the right upper arm and the trunk. Simultaneously, sEMG signals including anterior deltoid (DELT_A), posterior deltoid (DELT_P), clavicular head of pectoralis major (PECM), latissimus dorsi (LAT), long head of biceps brachii (BIC), and long head of triceps brachii (TRI), were recorded (1925.93 Hz) by the Trigno wireless sEMG recording system (DELSYSINC, Massachusetts, USA). The electrodes were placed longitudinally with respect to the muscle fiber directions and the positions of each muscle according to the recommendations from the SENIAM (surface electromyography for the non-invasive assessment of muscles) conventions [7]. Traditional manual muscle test techniques [8] were adopted to obtain the maximum voluntary contraction (MVC) EMG values under isometric conditions. Experimental muscle excitations were calculated from raw EMG signals that were high-pass filtered (30 Hz), full-wave-rectified, and low-pass filtered (6 Hz) using a zero-lag fourth-order recursive Butterworth filter [9]. Experimental muscle excitations were then normalized with the peak MVC EMG values respectively.

Joint angle can be calculated by the integration of the angular rate using a gyroscope. However, the errors also accumulate when adopting integration, which results in a drift over a period of time. In this scenario, we employed a Kalman filter which fused angular rate, acceleration, and magnetic field to estimate joint angle. Thanks to the hierarchical structure of human body, joint angle can be easily represented by the adjacent sensors' orientations. The algorithm block diagram is shown in Fig. 2. More details are described as below.

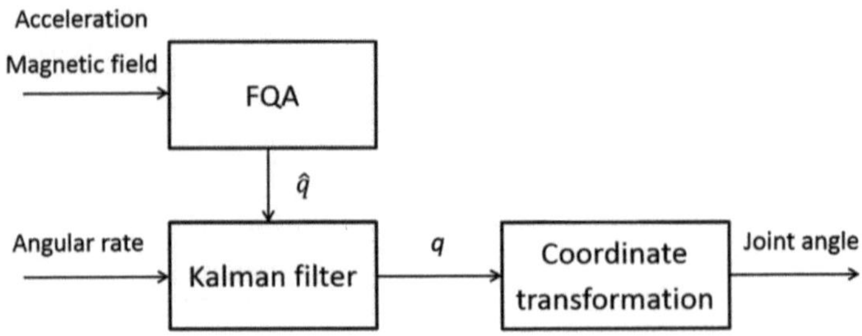

Fig. 2 Joint angle estimation algorithm block diagram. Joint angle can be calculated by the fusion of acceleration, magnetic field, and angular rate

First, we introduced factor quaternion algorithm (FQA) [10] to obtain sensor's raw quaternion $\hat{q}$. Here, quaternion is used to represent orientation because it has low computational complexity and does not suffer from singularity problem. FQA method has been successfully used in real-time human-body-motion-tracking applications for static or slow-moving tasks [10]. To make the algorithm more robust for normal speed moving tasks, we proposed a Kalman filter to fuse sensor's angular rate and raw quaternion for a more stable orientation estimation.

In the Kalman filter, the state vector is defined as $\left[w_x, w_y, w_z, q_0, q_1, q_2, q_3\right]^T$, where $w = \left[w_x, w_y, w_z\right]^T$ is the sensor's tri-axis angular rate, $q = \left[q_0, q_1, q_2, q_3\right]^T$ is the sensor's orientation in quaternion form. The state dynamic equations are given as follows:

$$w_k = w_{k-1} + n_w \tag{1}$$

$$q_k = q_{k-1} \otimes \Delta q + n_q \tag{2}$$

where

$$\Delta q = \left[\cos\left(\frac{\Delta\theta}{2}\right), \bar{e}\sin\left(\frac{\Delta\theta}{2}\right)\right]^T \tag{3}$$

$$q_k = q_{k-1} \otimes \Delta q + n_q \tag{4}$$

$$\bar{e} = \frac{w_{k-1}}{\|w_{k-1}\|} \tag{5}$$

subscript k represents the disperse time point, n_w and n_q represent state noise of angular rate and quaternion respectively, Δt stands for the sampling time interval.

The measurement model relates the measurement vector to the state vector. In this, we define measurement vector as $\left[Z_x^w, Z_y^w, Z_z^w, Z_0^q, Z_1^q, Z_2^q, Z_3^q\right]^T$, where $Z^w = \left[Z_x^w, Z_y^w, Z_z^w\right]^T$ is the sensor's angular rate measurement, $Z^q = \left[Z_0^q, Z_1^q, Z_2^q, Z_3^q\right]^T$ is the sensor's orientation measurement in quaternion form. The measurement equations are given as below:

$$Z_k^w = w_k + m_w \tag{6}$$

$$Z_k^q = q_k + m_q \tag{7}$$

where m_w and m_q represent measurement noise of angular rate and quaternion respectively. We then employed an extended Kalman filter (EKF) to deal with the non-linearity of the state relationship.

After the EKF, we can get sensor's orientation in quaternion form. To estimate joint angle, we employed a coordinate transformation using the adjacent sensors' orientations. Let's take elbow for example. Sensor A is attached to the forearm, sensor B is attached to the upper arm. Based on the assumption that during the movement, the sensors' and body's relative coordinate remain constant, the elbow joint angle can be calculated using the following equation:

$$q^J = (q^A)^{-1} \otimes q^B \tag{8}$$

where q^J, q^A, and q^B are the elbow joint angle, sensor A orientation and sensor B orientation in quaternion form respectively. Then we can easily calculate joint angle in Euler form through a quaternion to Euler transformation.

2.2 *EMG-Driven Model*

EMG-driven model is the core of our muscle force prediction. It consists of four main parts: muscle kinematics, muscle activation dynamics, muscle contraction dynamics, and joint dynamics.

2.2.1 Muscle Kinematics

In muscle kinematics, joint angle is transformed to muscle kinematics parameters such as musculotendon unit (MTU) lengths, moment arms and experimental joint moment based on an OpenSim [11] model. Also, the experimental joint moment can be calculated by the OpenSim Inverse Dynamics Tool [11]. The OpenSim model we used is shown in Fig. 3, which was developed by Saul et al. [12]. Generally speaking, OpenSim upper limb model characterizes the transformation from joint angle to muscle kinematics parameters, which is a necessary prerequisite to individual muscle force prediction [13].

2.2.2 Muscle Activation Dynamics

In muscle activation dynamics, muscle excitation is transformed to muscle activation to describe the time-delay and non-linearity in the EMG-to-force relationship.

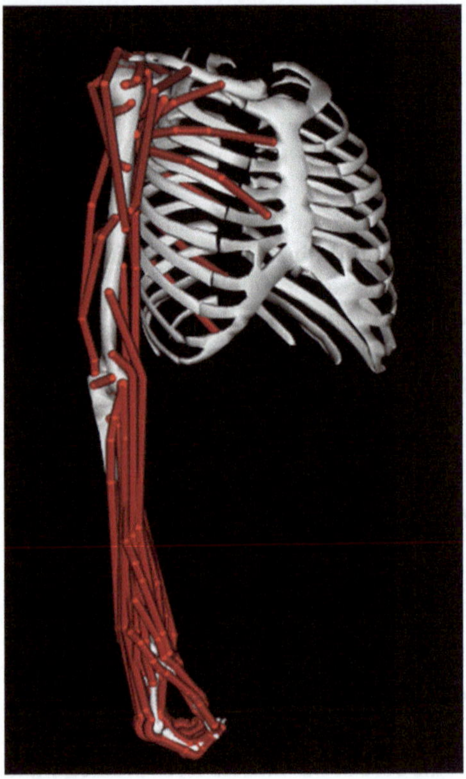

Fig. 3 OpenSim upper limb musculoskeletal model. It consists of 7 body segments and 32 muscles across the shoulder, elbow, forearm, and wrist

Accordingly, the transformation is formulated using a two-step process. First, muscle excitation is transformed into an intermediate variable, that is, neural activation:

$$u(t) = \alpha e(t-d) - (C_1 + C_2)u(t-1) - C_1 C_2 u(t-2) \tag{9}$$

where $e(t)$ is the muscle excitation at time t, $u(t)$ is neural activation at time t, α is the muscle gain, C_1 and C_2 are recursive coefficients, and d is the electromechanically delay. Then the shape of muscle activation is modeled using the following function:

$$a(t) = \frac{e^{Au(t)} - 1}{e^A - 1} \tag{10}$$

where $a(t)$ is the muscle activation at time t, $u(t)$ is the neural activation at time t, and A is the non-linear shape factor. The values of system parameters were adjusted to match the subject-specific physiological characteristics using a calibration process similar to previous study [13]. In our experiment, we extended our research to multi-joints and selected $DELT_A$, $DELT_P$, PECM, LAT, BIC and TRI as the six main action muscles. The selection of these muscles was in line with previous study [12, 14].

2.2.3 Muscle Contraction Dynamics

In muscle contraction dynamics, muscle activation and muscle kinematics parameters are transformed to muscle force using a Hill-type musculotendon model:

$$F^m = F^{max}\left[f_a\left(\hat{l}_m\right)f_v(\hat{v}_m)a + f_p\left(\hat{l}_m\right) + d_m\hat{v}_m\right]cos\varphi \tag{11}$$

where F^m is the muscle force, F^{max} is the maximum isometric muscle force, $\tilde{l}_m$ is the normalized muscle fiber length, $\tilde{v}_m$ is the normalized muscle fiber contraction velocity, a is the muscle activation, d_m is the muscle damping element, φ is the pennation angle. $f_a\left(\tilde{l}_m\right)$ is the active force-length relation that express the ability of muscle fibers to produce force at different lengths, $f_p\left(\tilde{l}_m\right)$ is the passive force-length relation that represents the force response of the fibers to strain, $f_v(\tilde{v}_m)$ accounts for the force contribution of the fiber contraction velocity.

2.2.4 Joint Dynamics

In joint dynamics, the force generated by each muscle is multiplied by the respective moment arm and then summed to yield the estimated joint moment:

$$\hat{\tau} = \sum_{i=1}^{N_{muscles}} \left(MA_i \times F_i^m\right) \tag{12}$$

where $\hat{\tau}$ is the estimated joint moment, $N_{muscles}$ is the number of muscles, MA_i is the moment arm of muscle i, F_i^m is the force of muscle i.

2.3 Static Optimization

As mentioned above, muscle force estimated by EMG-driven model is not reliable due to the uncertainties in sEMG signals. Thus, we employed a static optimization component to make the estimated joint moment more close to reality by adjusting experimental muscle excitation within a reasonable range:

$$minimize\ F_{obj} = \sum_{i=1}^{N_{DOFs}} \left|\tau_i - \hat{\tau}_i\right| \tag{13}$$

$$subject\ to \left|\frac{e_j - \hat{e}_j}{e_j}\right| < Th\ \forall j\ \in N_{muscles} \tag{14}$$

where τ_i is the experimental joint moment for degree of freedom (DOF) i, $\hat{\tau}_i$ is the estimated joint moment for DOF i, e_j is the experimental muscle excitation for muscle j, $\hat{e}_j$ is the adjusted muscle excitation for muscle j, N_{DOFs} is the number of DOFs, $N_{muscles}$ is the number of muscles, Th is the threshold constrained in the interval (0, 1).

In our experiment, two degree of freedoms, shoulder flexion-extension and elbow flexion-extension, were taken into account. Th is the only parameter we need to tune in static optimization. Larger Th value means muscle excitation can be adjusted in a wider range, which will improve the accuracy of joint moment estimation. According to previous study [15], we considered comparison results between experimental and estimated joint moment with $0 \leq NRMSD \leq 0.1$ and $0.95 \leq R^2 \leq 1.0$ are acceptable in our experiment, which provide us a criterion to set Th value. Here, $NRMSD$ represents the normalized root mean squared deviation with the form:

$$NRMSD = \frac{\sqrt{\frac{1}{N}\sum_{i=1}^{N}\left(\hat{X}_i - X_i\right)^2}}{\max\left(\hat{X}, X\right) - min\left(\hat{X}, X\right)} \tag{15}$$

where X and $\hat{X}$ referred to the two variables being compared, N referred to the number of points in the considered curves. R^2 represents the coefficient of determination.

3 Results and Discussion

Figures 4 and 5 demonstrate the joint moment estimation of shoulder and elbow for subject 1 and subject 2. Comparison between experimental and estimated joint moment is shown in Table 1. In our approach, we set $Th = 0.8$. The result showed that our method can predict joint moment accurately with lower $NRMSD$ and higher R^2 across joints and subjects compared with the EMG-driven model (see Table 1), revealing the fact that our method is able to predict muscle and joint dynamics more close to reality.

As stated previously, uncertainties in sEMG acquisition and processing limit the EMG-driven model's ability in muscle force and joint moment estimation. Figures 6 and 7 demonstrate the comparison between experimental and adjusted muscle excitations of subject 1 and subject 2. From the result we can see our adjustment mainly changed the amplitude of muscle excitation, the shape of adjusted excitation shows good agreement with the experimental one (average $NRMSD = 0.1407$ and $R^2 = 0.7598$ for subject 1 across all muscles; average $NRMSD = 0.1569$ and $R^2 = 0.6823$ for subject 2 across all muscles). In this way, adding a muscle excitation adjustment component in muscle force prediction is feasible to reduce the disadvan-

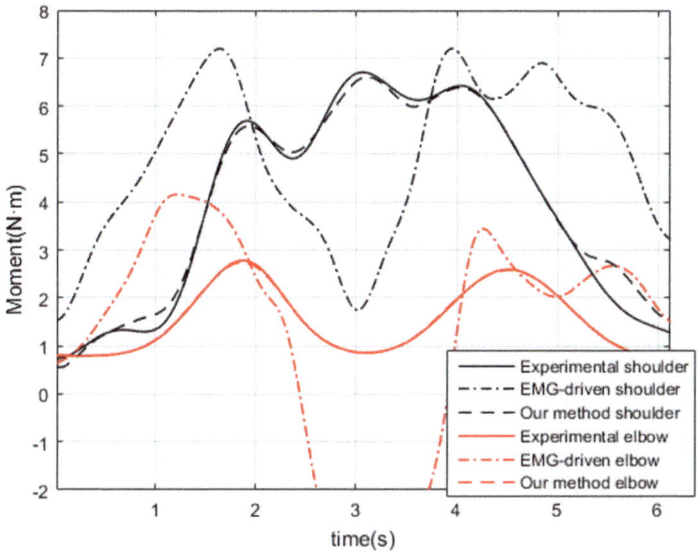

Fig. 4 Joint moment estimation of EMG-driven model and our method for subject 1

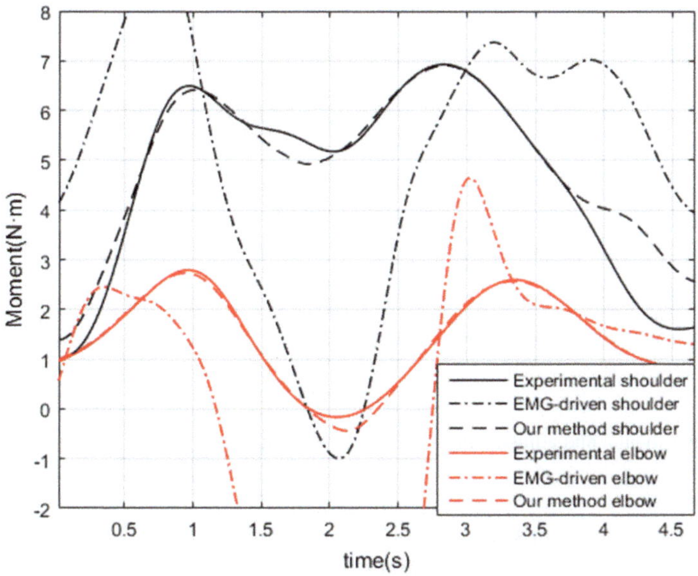

Fig. 5 Joint moment estimation of EMG-driven model and our method for subject 2

tage of uncertainties in sEMG signals while maintaining the subject-specific neural control strategy.

Table 1 Comparison between experimental and estimated joint moment using EMG-driven model and our method

	Subject 1		Subject 2	
	EMG-driven	Our method	EMG-driven	Our method
NRMSD_shoulder	0.4072	0.0405	0.3307	0.0939
NRMSD_elbow	0.2624	0.0064	0.2838	0.0255
R^2*_shoulder*	0.0183	0.9917	0.0041	0.9531
R^2*_elbow*	0.2157	0.9997	0.6290	0.9937

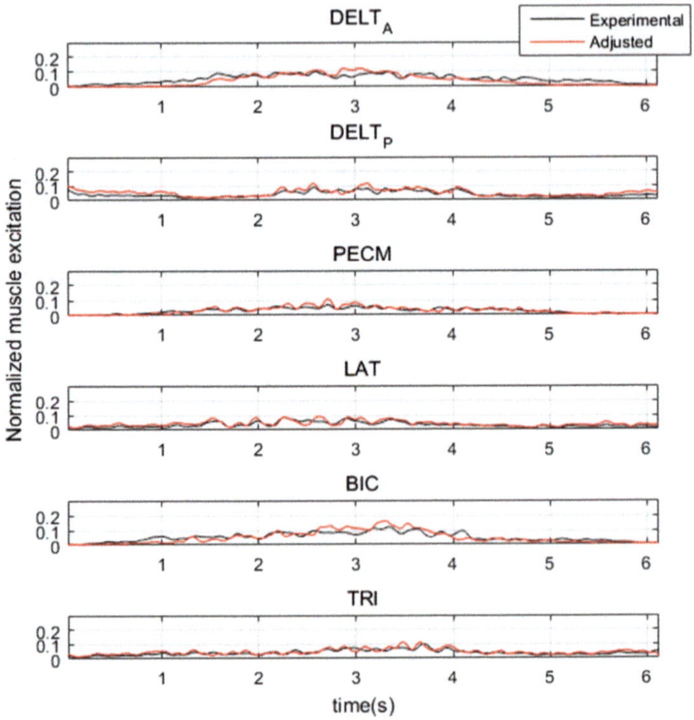

Fig. 6 Experimental and adjusted muscle excitations of subject 1

In general, our combination of EMG-driven model and static optimization is effective in muscle force and joint moment estimation during the movement of feeding oneself. However, some limitations of our study should be noted. First, normalizing sEMG is tricky because true maximum EMG values can be difficult to obtain [16]. We recorded maximal values for each muscle separately during muscle testing procedures under some certain isometric conditions. However, due to the changes of relative positions and orientations of surface electrodes with respect to the muscle volume across joint positions, it appears that a more reasonable approach is to normalize the EMG signals with respect to the maximum amplitude of the corre-

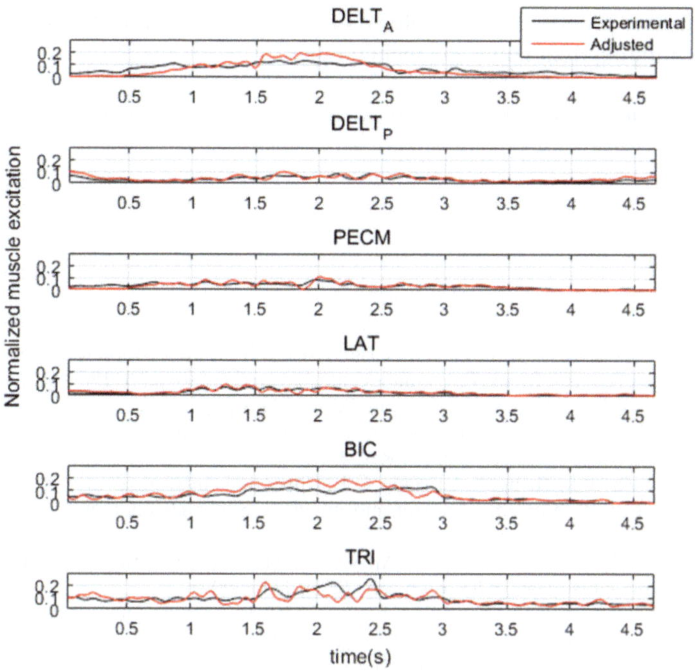

Fig. 7 Experimental and adjusted muscle excitations of subject 2

sponding joint positions [6]. Nevertheless, this approach is very time-consuming and can easily cause muscle fatigue which will affect the accuracy of maximum EMG values. Thus, how to obtain true maximum EMG values deserves further investigation. Second, the simulation task of our study was shoulder and elbow voluntary flexion-extension without any loading on the hand. Experiments should be further performed to facilitate the testing of different loading conditions or different movement velocities, which may provide us more information about muscle dynamics. Last, this work presented a study on two male subjects only. To be more rigorous, a general model validation across a larger number of individuals should be the subject of our future work.

4 Conclusions

In conclusion, this work proposed a novel EMG-adjusted muscle force predicting method for objective evaluating the movement of feeding oneself, which can predict joint dynamics accurately with the ability to reflect subject-specific neural control strategy. This may open up new avenues for obtaining a deeper understanding of neurological, physiological, and anatomical characteristics of human dynamic movement. The proposed methodology may also have implications in advanced neurorehabilitation treatments and technologies.

Acknowledgements This work was supported by the National Natural Science Foundation of China (No. 61431017).

References

1. Platz, T., Pinkowski, C., Van, W.F., Kim, I.H., Di, B.P., Johnson, G.: Reliability and validity of arm function assessment with standardized guidelines for the Fugl-Meyer test, Action Research Arm Test and Box and Block Test: a multicentre study. Clin. Rehabil. **19**(4), 404 (2005)
2. Box and Block Test. http://www.reha-stim.de/cms/index.php?id=122
3. Lyle, R.C.: A performance test for assessment of upper limb function in physical rehabilitation treatment and research. Int. J. Rehabil. Res. **4**(4), 483 (1981)
4. Zhang, L.L., Zhou, J., Zhang, X.A., Wang, C.T.: Upper limb musculo-skeletal model for biomechanical investigation of elbow flexion movement. J. Shanghai Jiaotong Univ. (Sci.) **16**(1), 61–64 (2011)
5. Erdemir, A., Mclean, S., Herzog, W., Aj, V.D.B.: Model-based estimation of muscle forces exerted during movements. Clin. Biomech. **22**(2), 131 (2007)
6. Koo, T.K.K., Mak, A.F.T.: Feasibility of using EMG driven neuromusculoskeletal model for prediction of dynamic movement of the elbow. J. Electromyogr. Kinesiol. Off. J. Int. Soc. Electrophysiol. Kinesiol **15**(1), 12 (2005)
7. Hermens, H.J., Freriks, B., Merletti, R., Stegeman, D., Blok, J., Rau, G.: European recommendations for surface electromyography (1999)
8. Gruen, J.P.: Handbook of Manual Muscle Testing. McGraw-Hill, Health Professions Division (1999)
9. Pizzolato, C., Lloyd, D.G., Sartori, M., Ceseracciu, E., Besier, T.F., Fregly, B.J., Reggiani, M.: CEINMS: a toolbox to investigate the influence of different neural control solutions on the prediction of muscle excitation and joint moments during dynamic motor tasks. J. Biomech. **48**(14), 3929–3936 (2015)
10. Yun, X., Bachmann, E.R., Mcghee, R.B.: A simplified quaternion-based algorithm for orientation estimation from earth gravity and magnetic field measurements. IEEE Trans. Instrum. Meas. **57**(3), 638–650 (2008)
11. Delp, S.L., Anderson, F.C., Arnold, A.S., Loan, P.: OpenSim: open-source software to create and analyze dynamic simulations of movement. IEEE Trans. Biomed. Eng. **54**(11), 1940–1950 (2007)
12. Saul, K.R., Hu, X., Goehler, C.M., Vidt, M.E., Daly, M., Velisar, A., Murry, W.M.: Benchmarking of dynamic simulation predictions in two software platforms using an upper limb musculoskeletal model. Comput. Methods Biomech. Biomed. Eng. **18**(13), 1445–1458 (2015)
13. Hou, J., Sun, Y., Sun, L., Pan, B., Huang, Z., Wu, J., Zhang, Z.: A pilot study of individual muscle force prediction during elbow flexion and extension in the neurorehabilitation field. Sensors **16**(12), 2018 (2016)
14. Gatti, C.J., Doro, L.C., Labgenderfer, J.E., Mell, A.G., Maratt, J.D., Carpenter, J.E., Hughes, R.E.: Evaluation of three methods for determining EMG-muscle force parameter estimates for the shoulder muscles. Clin. Biomech. **23**(2), 166–174 (2008)
15. Sartori, M., Gizzi, L., Lloyd, D.G., Farina, D.: A musculoskeletal model of human locomotion driven by a low dimensional set of impulsive excitation primitives. Front. Comput. Neurosci. **7**(26), 79 (2013)
16. Buchanan, T.S., Lloyd, D.G., Manal, K., Besier, T.F.: Neuromusculoskeletal modeling: estimation of muscle forces and joint moments and movements from measurements of neural command. J. Appl. Biomech. **20**(4), 367 (2004)

Muscle Synergistic Pattern and Kinematic Sensor Data Analysis During Upper-Limb Reaching in Stroke Patients

Bingyu Pan, Yingfei Sun, Zhipei Huang, Jiateng Hou, Jiankang Wu, Zhen Huang, Bin Xie and Yijun Liu

Abstract Quantitative and efficient measurement of motor impairment level is of vital importance in stroke rehabilitation. This paper investigates the muscle synergistic patterns and kinematic sensor data of upper limb reaching in stroke patients with different impairment level. Thirty-three stroke patients and nineteen healthy age-matched subjects serving as the control group were asked to do voluntary upward reaching. Inertial sensors and surface electromyography (sEMG) sensors were attached to subjects' upper limb to obtain the real-time joint angle through segment position by the inertial sensory data fusion and extract synergistic patterns from sEMG data by applying principal components analysis at the same time. The experimental results show that stroke patients not only have abnormal range of shoulder joint motion, which was correlated with the degree of clinical impairment level; but also have different muscle synergistic patterns at different impairment level, which can be used as a quantitative measurement of functional recovery status.

B. Pan · Y. Sun · Z. Huang (✉) · J. Hou · J. Wu
University of Chinese Academy of Sciences, Beijing, China
e-mail: zhphuang@ucas.ac.cn

B. Pan
e-mail: panbingyu14@mails.ucas.ac.cn

Y. Sun
e-mail: yfsun@ucas.ac.cn

J. Hou
e-mail: houjiateng13@mails.ucas.ac.cn

J. Wu
e-mail: jkwu@ucas.ac.cn

Z. Huang · B. Xie · Y. Liu
Rehabilitation Department, Peking University First Hospital, Beijing, China
e-mail: huangzhen6313@163.com

B. Xie
e-mail: beccasherry@163.com

Y. Liu
e-mail: lyx129@126.com

G. Fortino and Z. Wang (eds.), *Advances in Body Area Networks I*,
Internet of Things, https://doi.org/10.1007/978-3-030-02819-0_33

Keywords Internal sensors · Muscle synergistic pattern
Surface electromyography · Principal component analysis · Stroke rehabilitation

1 Introduction

Cerebral apoplexy, also known as stroke, is the leading cause of death and long-term disability around the world. Hemiplegia caused by stroke badly influence daily lives of stroke survivors. Luckily, human's central nervous system can compensate for lesions and recover under appropriate conditions according to the neuroplasticity theory [1]. Since training can change muscle patterns of stroke survivors [2], most of them rely on rehabilitation treatment to regain motion ability for daily life. Due to the high complexity of biomechanical characteristics and motor function of upper-limb, the rehabilitation of upper-extremities is more challenging compared with lower-limbs. Therefore, quantitative and efficient rehabilitation techniques of upper-limb to improve motor function after stroke is of vital importance.

Clinically, the motor impairment level is evaluated through a set of clinical assessment scales, for example, the Fugl-Meyer Assessment (FMA) [3] and modified Ashworth scale [4], which are all subjective and dependent on the therapist's experience. With the development of micro sensors, wearable devices are applied for quantitative assessment of stroke rehabilitation. Motion capture devices can provide quantitative and detailed description of movement of limbs [5, 6]. However, inertial sensors have inherent limitations in interpreting the mechanism of physiology and neurology. Moreover, neural deficits may be hided at kinematic level as similar movements may be produced through different neuromuscular mechanisms [7]. In contrast, surface electromyography (sEMG) analysis can make up such deficiency and a number of studies have been carried out. For instance, Barker et al. [8] and Hughes et al. [9] have found that stroke might change the activity of some muscles in affected side generating abnormal activation patterns to complete the movement. Cruz et al. [10] and Musampa et al. [11] illustrated that stoke survivors had abnormal muscle synergies with high level of agonist-antagonist co-contraction. What's more, evidence has proved that muscle synergy composition altered in stroke survivors from severe to mild impairment [12–14]. The extraction of muscle activation patterns using PCA has been reported in healthy individuals during postural control [15] and locomotion [16]. PCA provides a spatial and temporal pattern of muscle activity to describe synergistic patterns. By extracting the principal components during reaching movement revealed changes in motor control strategies that occurred after a stroke. Previous studies have investigated the muscle synergy alterations after stroke [12, 13], but they did not give the correlation between alterations and the impairment level. Overall, sEMG analysis can provide insight into the compensations of the non-paretic limb or adaptations of the paretic limb during movements in post-stroke, but it lacks of physical descriptions on movements.

Therefore, the combination of sEMG and kinematic system can provide both physical information and neuromuscular characteristics valuable for stroke rehabilitation. In this study, we adopted a novel inertial-sensor sEMG synchronous system to collect data from stroke patients and age-matched healthy subjects to give insights in neurorehabilitation. A set of inertia sensors and sEMG sensors were attached to upper limb segments of subjects to calculate real-time joint angle during the movement by fusion of inertial sensory data, meanwhile principal component analysis (PCA) was applied to sEMG signals to extract muscle synergistic patterns which can reflect the interaction of co-activated muscles. Understanding these muscle synergistic pattern alterations in stroke patients and the relationship with motor impairment level would be important to find out the mechanism behind motor impairment caused by stroke to improve stroke patients' recovery process.

2 Experimental and Computational Details

2.1 Experimental Protocol and Data Acquisition

Thirty-three stroke survivors and nineteen age-matched healthy people serving as control group were recruited from the Peking University First Hospital for this study. The general information of participants is shown in Table 1. Stroke participants should be diagnosed with stroke for the first time and the poststroke duration should be no longer than 6 months. Meanwhile, they should have the ability to lift their shoulders voluntarily without help. Moreover, patients should not have cognitive impairment that unable to collaborate, nor suffer from other musculoskeletal disease. Motor recovery of upper extremity for each patient was assessed by FM score, which was graded by professional rehabilitation therapist from Peking University First Hospital. This research has been approved by the Ethics Committee of Peking University First Hospital and all the participants gave informed consent before experimentations.

The kinematic data was recorded (50 Hz) by the wireless upper-limb motion capture system developed by University of Chinese Academy of Sciences, Sensor

Table 1 General information of participants

		Mean	SD	Range
Affected group (N = 33)				
Age (year)		60.06	11.41	[38, 83]
Sex (male/female)	25/8			
Side affected (left/right)	19/14			
Fugl-meyer (out of 52)		29.79	12.94	[9, 51]
Control group (N = 19)				
Age (year)		56.95	9.26	[37, 75]
Sex (male/female)	10/9			

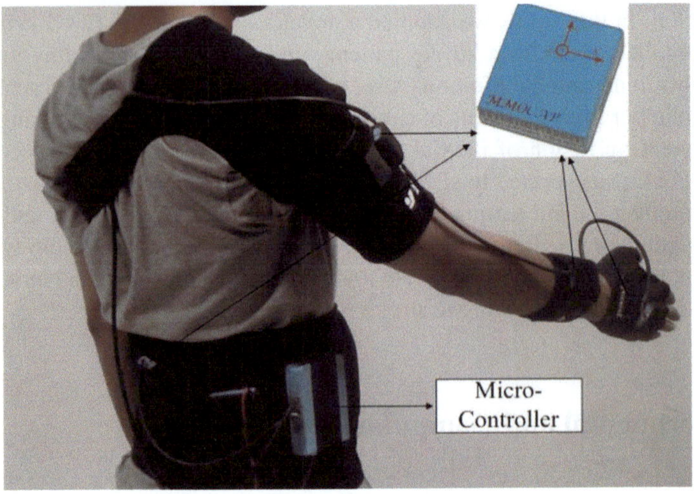

Fig. 1 Motion capture equipment of upper limb

Network and Application Joint Research Center [17]. The upper limb motion capture system consists of four inertial sensor measurement nodes attached to the trunk, upper arm, forearm and hand respectively (Fig. 1). The sensor chip in each measurement node is MPU-9150 (InvenSense Inc, USA), including a tri-axial accelerations, tri-axial gyroscope, and tri-axial magnetometer. By fusion of sensory data, we can calculate the real-time upper limb motion of a subject.

At the same time, sEMG activities of 7 muscles wrapping across shoulder, upper-arm, and forearm were recorded (1000 Hz) by ME6000 multi-channel bipolar sEMG recording system (Mega Electronics Ltd, Finland). Seven elbow and shoulder muscles were examined including pectoralis major (PECM), trapezius (TRA), anterior deltoid (DELA), medial deltoid (DELM), biceps brachii (BIC), triceps brachii (TRI) and brachioradialis (BRAC). Electrodes were placed along with corresponding muscles consistent with the recommendations of the Surface Electromyography for the Non-Invasive Assessment of Muscles (SENIAM) [18].

Getting objects is one of the basic functions of motion in rehabilitation. Both kinematic and sEMG data were collected when the stroke patients and the control subjects were asked to perform the movement of voluntary upward reaching with each arm. To start with, we placed an object for reaching at the point where subjects could reach by shoulder flexion of 90°. The subjects sat upright in front of an appropriate table with palm facing the thighs and perform voluntary upward reaching following instructions by keeping thumb up in case of pronation and then held on for 2 s at the highest point. Each subject repeated the task three times with an interval of three minutes to prevent fatigue.

For further analysis, the collected sEMG signals need to be preprocessed by the following steps: (1) filtered with a high-pass, window-based finite impulse response filter with the cutoff of 50 Hz; (2) rectified; (3) filtered by a low-pass window-based finite impulse response filter with the cutoff of 20 Hz; (4) integrated over 20 ms intervals so that it can be synchronous with the kinematic data. To avoid the extraction of sEMG activation patterns biased into the high-amplitude muscles only, the sEMG signals from each muscle of each arm were normalized within the task according to the task-specific interpretation of the data [19].

2.2 *Joint Angle*

Joint angle, measured by the range of individual joint's movement, directly demonstrates the completeness of the task. We used inertial sensors to monitor the movements of the upper extremity and calculated real-time joint angles during the task. The upper limb motion was obtained on the basis of the hierarchical biomechanical model. To calculate joint angles, we assumed that upper arm, forearm, and hand were all rigid bodies, rotating around their corresponding joints. In this case, shoulder joint angle is equal to the angle between upper limb vectors of start point to endpoint.

$$\theta = \cos^{-1}\left\langle P^{S}_{elbow}, P^{E}_{elbow}\right\rangle, \tag{1}$$

where P^{S}_{elbow} and P^{E}_{elbow} are the start point and endpoint of elbow joint, respectively. The joint position is represented by quaternion. Quaternion can be obtained by fusing accelerometer and magnetism sensor data:

$$p^{G}_{elbow} = P^{G}_{shoulder} + q^{GB} \otimes L^{B}_{upper} \otimes \left(q^{GB}\right)^{-1} \tag{2}$$

where P^{G}_{elbow} and $P^{G}_{shoulder}$ represent the positions of elbow joint and shoulder joint in global coordination system, q^{GB} is the quaternion of shoulder joint, and L^{B}_{upper} is the vector between shoulder joint and elbow joint in body coordinate system. In our hierarchical model, the shoulder joint, we can simplify Eq. (1) on the basis of Eq. (2). Then we get:

$$\theta = \cos^{-1}\left\langle \begin{matrix} q^{S}_{shoulder} \otimes L^{B}_{upper} \otimes \left(q^{S}_{shoulder}\right)^{-1} \\ q^{E}_{shoulder} \otimes L^{B}_{upper} \otimes \left(q^{E}_{shoulder}\right)^{-1} \end{matrix} \right\rangle, \tag{3}$$

in which $q^{S}_{shoulder}$ and $q^{E}_{shoulder}$ are the start point and endpoint of shoulder quaternion respectively.

2.3 Muscle Synergistic Patterns

Fixed patterns activated across a set of muscles were referred to as muscle synergy whose combinations were used to generate coordinated movements. The muscle activation patterns can interpret the interaction of co-activated muscles. Since they contain space and time information and very complex to deal with, we applied PCA algorithm to extract synergistic patterns of muscle activations. The algorithm of PCA has the virtue of ability to reduce dimension.

PCA is an algorithm that utilizes the covariance structure of a set of dependent variables to express the collective dataset into a new orthogonal space of the same dimensionality by calculating the eigenvalues and eigenvectors of the data covariance matrix, which can be compressed into fewer variables [20]. To extract principal components of muscle activation patterns, we first concatenated each row of the sEMG data matrix and represented it as a vector $EMGC_i$ [21]. Therefore, the mean vector of control group is

$$Avg = \frac{1}{N}\sum_{i=1}^{N} EMGC_i \tag{4}$$

where N is the number of subjects in the group and $EMGC_i$ is the sEMG vector of the ith subject in the control group. The deviation-from-mean vectors, which represent the difference of each signal to the mean vector, are expressed as:

$$Dev_i = EMGC_i - Avg. \tag{5}$$

Since the vector Dev_i is zero mean vector, the covariance matrix is given by:

$$\mathrm{Cov} = \frac{1}{N}\sum_{i=1}^{N} Dev_i Dev_i^{\tau} \tag{6}$$

The first K eigenvectors of matrix Cov in Eq. (6) formed the new feature space. Then we projected the vectors of both control group and stroke group onto the feature space to observe the altered of muscle synergistic patterns.

All statistical tests were made at alpha = 0.05. Statistical analyses were performed using MATLAB (Mathworks).

3 Results

Figure 2 shows the real-time shoulder joint angel during reaching task. We take two subjects from different impairment level whose FM scores are 13 and 44 respectively and one subject from control group for example. We can see from Fig. 2 that the

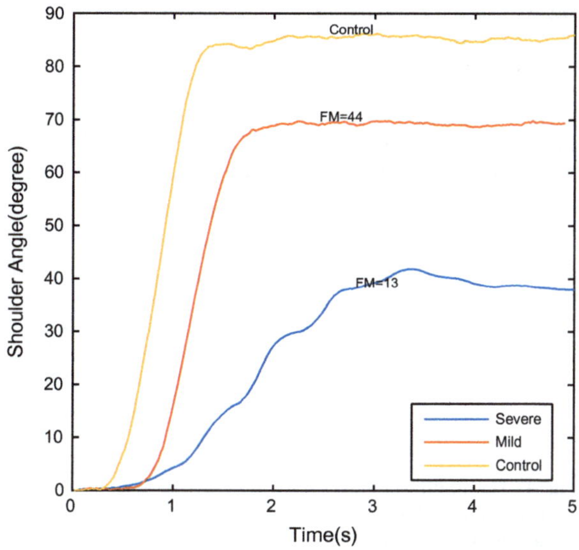

Fig. 2 Real-time shoulder angle of three subjects from severe, mild and control group separately

control subject has a normal range of shoulder angle and mild stroke patient has a limited shoulder angel.

The severe patient has the smallest joint angle and perturbed trajectory during the holding stage compared to the mild patient and control subject due to the serious hemiplegia. We calculated mean shoulder angle during the holding stage of reaching movement in patients and control group and summarizes as Fig. 3. The statistic difference of two group was assessed by t-test, and there is a significant difference between two groups ($p = 0.0026$). Furthermore, we computed the Pearson correlation between shoulder angle and the patients' FM scores and found a significant correlation between shoulder angle and FM score (Fig. 4, $r = 0.53$, $p = 0.0016$). Therefore, the maximum shoulder angle can be a useful index to evaluate the motor function of stroke patients.

After projecting the muscle activation pattern of stroke patients to the feature space, we obtained the first two principal components as synergistic patterns of poststroke. Figure 5 shows that the muscle synergistic patterns altered after stroke. We then categorized them into groups to examine the alterations of muscle activation patterns in stroke patients of different impairment level. Clustering was based on Euclidean distance.

The distances from the center of each group to control group are 16.65, 11.21 and 9.59, separately. Figure 6 illustrates the box-plot of different distance from the patterns of stroke group in three groups to the center of control template. The p value calculated by one-way analysis of variance (ANOVA) was 0.0008 indicating that differences between column means are significant. To further verify the assumption that the distance can be used to evaluate the stroke impairment level, we calculated the FM score and maximum shoulder joint angle during holding stage of each group

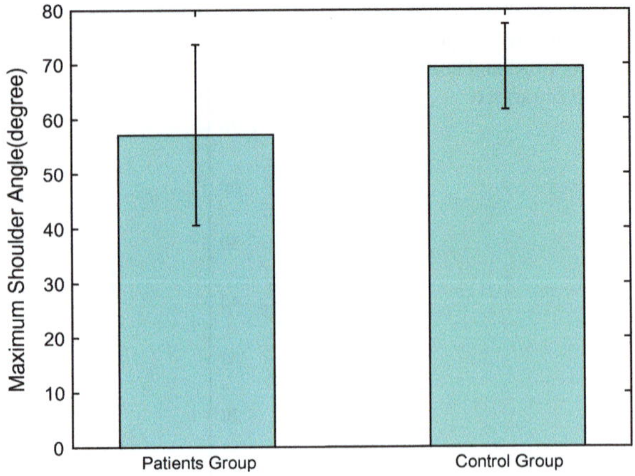

Fig. 3 Average and standard deviation of maximum shoulder angle in patients group and control group. There is a significant difference between two groups since p = 0.0026

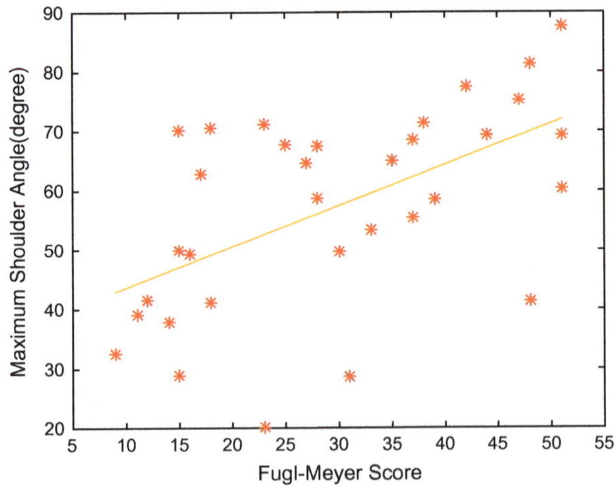

Fig. 4 Maximum shoulder angle and Fugl-Meyer score has a significant, positive correlation (r = 0.53, p = 0.0016). The diagonal line in the figure is the polynomial curve fitting between Fugl-Meyer score and shoulder angle

by one-way ANOVA. The results in Figs. 7 and 8 illustrate that shoulder angle of group 3 is significantly greater than the other two groups (F = 5.3, P = 0.0107), and the FM scores is the highest among three groups (F = 5.43, P = 0.0097), which means the mildest impairment. Therefore, from group 1 to group 3, the impairment is from severe to mild. This is consistent with what we get from Fig. 5. Therefore, we proved the hypothesis that muscle synergistic patterns altered after stroke and the distance can be used to evaluate the impairment level.

Voluntary upward reaching is a critical functional movement in daily lives. It's a task full of motor abundance as the number of joints available for movement and their coordination is unrestricted [22]. Healthy individuals reaching for objects placed within arm-reaching distance have smooth upper-limb trajectories and normal range of joint angle. Indeed stroke patients may have limited arm movement when reach objects placed within the reach of the arm by using a compensatory strategy

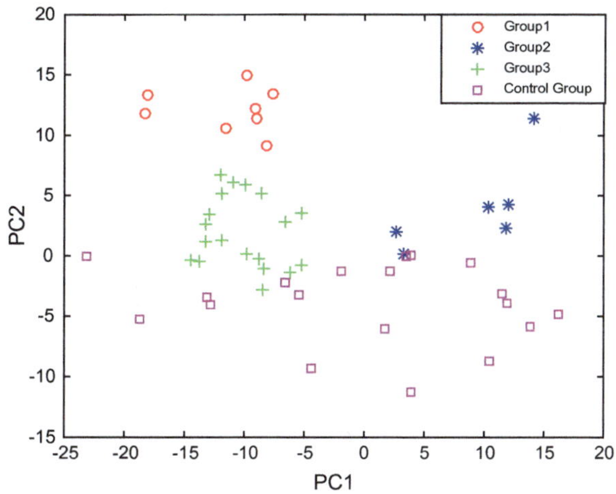

Fig. 5 Plot of muscle synergistic patterns of patients group and healthy group. The distance from the center of each group to the center of control group is 16.65, 11.21 and 9.59, respectively

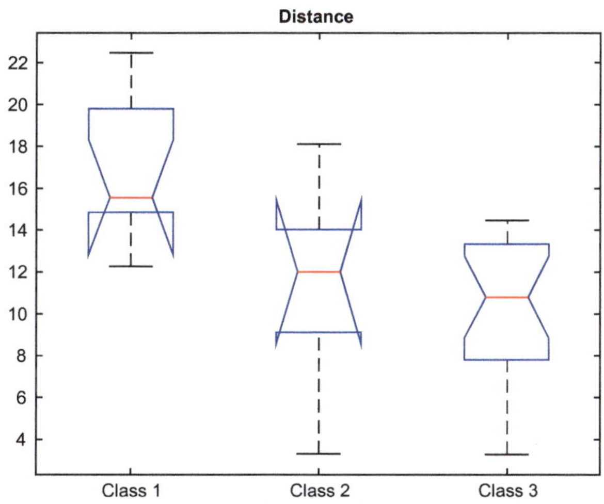

Fig. 6 Box-plot of Euclidean distance from the center of control template to the principal components of stroke group in three classes. The 3 horizontal bars representing each variable indicate, from bottom to top, the lower quartile, median, and upper quartile of Euclidean distance. Error bars indicate ±1.5* interquartile range

involving trunk recruitment [23]. In stroke patients, multi-joint reaching movements are characterized by decreased movement speed and increased movement variability and usually have smaller shoulder joint angel of affected arm due to hemiplegia, which can be observed in Fig. 2. It takes the longest time for the severe subject to get the maximum joint angle and the trajectory was fluctuant compared with moderately and mildly impaired subjects. The shoulder joint angel in stroke patients were significantly correlated with the level of arm motor impairment, which can be used as physiological markers of motor damage.

However, similar movements may be produced through different neuromuscular mechanisms due to compensatory strategies of neural system. The analysis of muscle activations may reflect differences in the flexibility of neural mechanisms in stroke

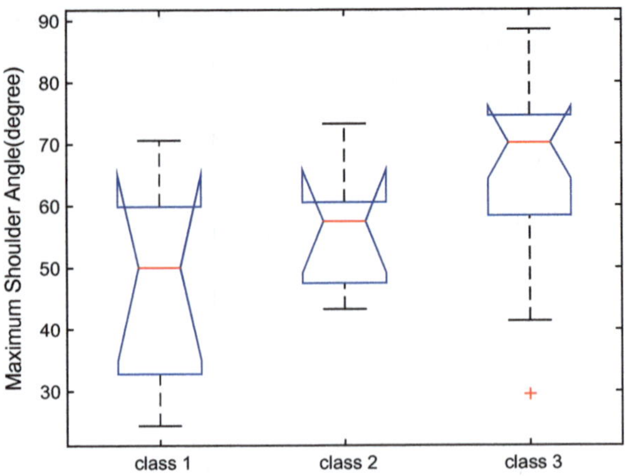

Fig. 7 Box-plot of maximum shoulder angle of three classes. Since F = 5.3, P = 0.0107, there is a significant difference in three groups

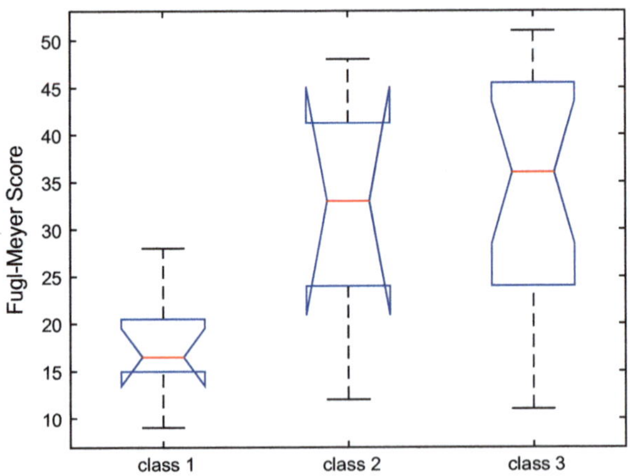

Fig. 8 Fugl-Meyer score of three groups. Since F = 5.43, P = 0.0097, there is a significant difference in three groups

patients. In this paper, we propose and research preliminarily on the shoulder joint angle and alterations of muscle synergistic patterns of upward reaching, which can help with the evaluation of stroke patients' motor function. The alterations of muscle synergistic patterns can be further used as physiological markers of motor cortical damage and become a new measure to reveal the biological mechanism of stroke.

There is still space of improvement in further research. We could take the different aetiology of stroke onset and time since stroke into consideration in further study since they may affect the synergistic patterns of sEMG. What's more, we can predict the impairment level of stroke patients quantitatively based on the results we obtained and provide more effective information to help with the rehabilitation treatment of stroke patients.

4 Conclusions

The goal of this paper is to explore alterations in kinematic data and muscle synergistic patterns with various groups of stroke patients and control group. We find that mean shoulder angle of holding stage during reaching has a significant correlation with FM score and can be used to evaluate stroke impairment level. What's more, patterns of control group were similar to each other, while stroke patients with different impairment level have different synergistic patterns. Our research throws a light on understanding of the mechanism underlying motor deficits and lay a foundation of quantitative evaluation of stroke rehabilitation, with which physicians can design proper rehabilitation measurement for stroke patients.

Acknowledgements This work was supported by National Natural Science Foundation of China, Grant No. 61431017 and 81272166.

References

1. Johansson, B.B.: Brain plasticity and stroke rehabilitation the Willis lecture. Stroke **31**, 223–230 (2000)
2. Kang, N., Idica, J., Amitoj, B., Cauraugh, J.H.: Motor recovery patterns in arm muscles: coupled bilateral training and neuromuscular stimulation. J. Neuroeng. Rehabil. **11**, 57 (2014)
3. Fugl-Meyer, A.R., Jääskö, L., Leyman, I., Olsson, S., Steglind, S.: The post-stroke hemiplegic patient. 1. A method for evaluation of physical performance. Scand. J. Rehabil. Med. **7**, 13–31 (1974)
4. Bohannon, R.W., Smith, M.B.: Interrater reliability of a modified Ashworth scale of muscle spasticity. Phys. Ther. **67**, 206–207 (1987)
5. Del Din, S., Patel, S., Cobelli, C., Bonato, P.: Estimating Fugl-Meyer clinical scores in stroke survivors using wearable sensors. In: 2011 Annual International Conference of the IEEE Engineering in Medicine and Biology Society, pp. 5839–5842. IEEE (2011)
6. Bai, L., Pepper, M.G., Yan, Y., Spurgeon, S.K., Sakel, M., Phillips, M.: Quantitative assessment of upper limb motion in neurorehabilitation utilizing inertial sensors. IEEE Trans. Neural Syst. Rehabil. Eng. **23**, 232–243 (2015)
7. Safavynia, S.A., Torres-Oviedo, G., Ting, L.H.: Muscle synergies: implications for clinical evaluation and rehabilitation of movement. Top. Spin. Cord Inj. Rehabil. **17**, 16–24 (2011)
8. Barker, R.N., Brauer, S., Carson, R.: Training-induced changes in the pattern of triceps to biceps activation during reaching tasks after chronic and severe stroke. Exp. Brain Res. **196**, 483–496 (2009)
9. Hughes, A.M., Freeman, C.T., Burridge, J.H., Chappell, P.H., Lewin, P.L., Rogers, E.: Shoulder and elbow muscle activity during fully supported trajectory tracking in people who have had a stroke. J. Electromyogr. Kinesiol. **20**, 465–476 (2010)
10. Cruz, E., Waldinger, H., Kamper, D.: Kinetic and kinematic workspaces of the index finger following stroke. Brain **128**, 1112–1121 (2005)
11. Musampa, N.K., Mathieu, P.A., Levin, M.F.: Relationship between stretch reflex thresholds and voluntary arm muscle activation in patients with spasticity. Exp. Brain Res. **181**, 579–593 (2007)
12. Roh, J., Rymer, W.Z., Perreault, E.J., Yoo, S.B., Beer, R.F.: Alterations in upper limb muscle synergy structure in chronic stroke survivors. J. Neurophysiol. **109**, 768–781 (2013)
13. Roh, J., Rymer, W.Z., Beer, R.F.: Evidence for altered upper extremity muscle synergies in chronic stroke survivors with mild and moderate impairment. Front. Hum. Neurosci. **9** (2015)

14. Berger, D.J., Ferrari, F., Esposito, A., Masciullo, M., Molinari, M., Lacquaniti, F., d'Avella, A.: Changes in muscle synergy organization after neurological lesions. In: Ibáñez, J., González-Vargas, J., Azorín, J.M., Akay, M., Pons, J.L. (eds.) Converging Clinical and Engineering Research on Neurorehabilitation II: Proceedings of the 3rd International Conference on NeuroRehabilitation (ICNR2016), 18–21 October 2016, Segovia, Spain, pp. 939–943. Springer International Publishing, Cham (2017)
15. Krishnamoorthy, V., Latash, M.L., Scholz, J.P., Zatsiorsky, V.M.: Muscle synergies during shifts of the center of pressure by standing persons. Exp. Brain Res. **152**, 281–292 (2003)
16. Suzuki, K., Nishida, Y., Mitsutomi, K.: Association between muscle synergy and stability during prolonged walking. J. Phys. Ther. Sci. **26**, 1637 (2014)
17. Huang, S., Luo, C., Ye, S., Liu, F., Xie, B., Wang, C., Yang, L., Huang, Z., Wu, J.: Motor impairment evaluation for upper limb in stroke patients on the basis of a microsensor. Int. J. Rehabil. Res. Internationale Zeitschrift fur Rehabilitationsforschung. Revue internationale de recherches de readaptation **35**, 161–169 (2012)
18. Hermens, H.J., Freriks, B., Merletti, R., Stegeman, D., Blok, J., Rau, G., Disselhorst-Klug, C., Hägg, G.: European recommendations for surface electromyography. Roessingh Res. Dev. **8**, 13–54 (1999)
19. Burden, A.: How should we normalize electromyograms obtained from healthy participants? What we have learned from over 25 years of research. J. Electromyogr. Kinesiol. Offic. J. Int. Soc. Electrophysiol. Kinesiol. **20**, 1023–1035 (2010)
20. Perez, M.A., Nussbaum, M.A.: Principal components analysis as an evaluation and classification tool for lower torso sEMG data. J. Biomech. **36**, 1225–1229 (2003)
21. Patil, A.M., Kolhe, S.R., Patil, P.M.: Face recognition by PCA technique. In: 2009 2nd International Conference on Emerging Trends in Engineering and Technology (ICETET), pp. 192–195 (2009)
22. McCrea, P.H., Eng, J.J., Hodgson, A.J.: Saturated muscle activation contributes to compensatory reaching strategies after stroke. J. Neurophysiol. **94**, 2999–3008 (2005)
23. Levin, M.F., Michaelsen, S.M., Cirstea, C.M., Roby-Brami, A.: Use of the trunk for reaching targets placed within and beyond the reach in adult hemiparesis. Exp. Brain Res. **143**, 171–180 (2002)

Inertial Sensor Based Human Activity Recognition via Reduced Kernel PCA

Donghui Wu, Huanlong Zhang, Cong Niu, Jing Ren and Wanwan Zhao

Abstract In the past decade, wearable inertial sensor based human activity recognition has attracted lots of attention from researchers in the world. High-dimensional feature set will increase the computation and memory cost. In this paper, kernel PCA has been utilized for dimensionality reduction to deal with inertial sensor based human activity recognition. However, kernel method may increase the computation and memory cost. Thus, reduced kernel method is proposed. The real dataset has been utilized to evaluate the proposed reduced kernel PCA (RKPCA) method. Experimental results demonstrate the efficacy of the proposed method, which achieves better results than traditional PCA method.

Keywords Human activity recognition · Inertial sensor · Kernel PCA
Reduced kernel method

1 Introduction

With the development of microelectronics, inertial sensor based human activity recognition (HAR) has attracted tremendous attention around the world. Human activity recognition can be monitored by wearable sensors placed on different locations and subjects. A major goal of wearable inertial sensor based health-related research is to long-term monitor daily activities of patients, which may provide effective information to medical staff to accurately diagnose diseases and design the care plan for the patients. The quality of healthcare for the elderly and children may be effectively improved by HAR [1–4].

D. Wu (✉) · C. Niu · J. Ren · W. Zhao
School of Building Environment Engineering, Zhengzhou University of Light Industry, Zhengzhou 450045, Henan, China
e-mail: w_donghui@163.com

H. Zhang
College of Electric and Information Engineering, Zhengzhou University of Light Industry, Zhengzhou 450045, Henan, China

G. Fortino and Z. Wang (eds.), *Advances in Body Area Networks I*,
Internet of Things, https://doi.org/10.1007/978-3-030-02819-0_34

Two main researches has been focused on by previous works about human activity recognition system: signal sensor based researches and multi-sensor mounted researches. Single sensor attached on different parts of subjects has been investigated by lots of systems [5–10]. A signal triaxial accelerometer placed on the chest has been utilized by Khan et al. [5] to recognize fifteen daily activities (such as sitting, lying, standing, running and so on). After extracting the features of AR-coefficients, signal-magnitude area (SMA), and tilt angle (TA), LDA and artificial-neural nets (ANNs) based hierarchical model has been utilized, and the experimental result achieves an average accurate of 97.9%. Cheng and Jhan [6] has introduced a triaxial accelerometer to detect the action fall, which is fixed on the left ankle, right ankle, chest and waist, respectively. After extracting the features of SMA, signal magnitude vector (SMV) and the raw signal data, the cascade-AdaBoost-SVM classifier is proposed for fall detection, and the accuracy rate of 98.23% and 98.48 with accelerometer around chest and waist has been achieved, respectively. Triaxial accelerometer worn on the wrist has been utilized by Sirichana et al. [7] to estimate oxygen uptake (VO2) and energy expenditure for 12 different daily physical activities, and the result can be achieved that during the physical activity<6 METs, the energy expenditure can be accurately predicted. Due to the low cost, such system has achieved a number of researches, and good recognition results has been achieved for basic activities.

Comparatively, multiple sensors mounted at different sites of subjects have also been researched by a number of systems [11–14]. Four sensor nodes placed at chest, right waist, left abdomen and left thigh, respectively, has been utilized by Gao et al. [15] to recognize eight scenario activities. The experimental results show that the proposed hierarchical classifier, which is based on the Decision Tree classifier and Naive Bayes classifier, can obviously reduce the energy cost with high accuracy. Multiple sensors have been utilized by Ertuğrul and Kaya [16] to identify 33 activities, and the sensors are placed on the left calf, right calf, right thigh, left thigh, right lower arm, left lower arm, right upper arm, left upper arm and back, respectively. In this paper, the optimal number of body-worn sensors has been discussed and the accuracy of 99.08% using the data from only 3Dgyro sensors has been achieved. Eight foot force sensors and smart phone GPS attached on the two insole positions have been utilized to recognize five different daily activities: walking, cycling, bus-passenger, car-passenger, and car-driver [17]. Though outstanding results is achieved by multiple sensors based systems, this approach is not feasible for the long-term activity monitoring that is because of the limitation of the number of sensors with energy cost and the cable connections.

Though most of aforementioned researches have investigated human activity recognition, with the development of measurement and data storage, it is strongly demanded effective feature extraction and dimensionality reduction methods. In this paper, we focused on a single node including accelerometer and gyroscope based human daily activity recognition and dimensionality reduction.

PCA as a classical and popular technique has been widely applied for data processing and dimensionality reduction [14]. However, for some complicated cases with particular nonlinear character, poor results will be achieved by PCA, that is because of its assumption that the signal data should be linear. In order to overcome this nonlinear problem, kernel PCA method is proposed in recent years. However, the large storage of the symmetric kernel matrix hiders the application of kernel PCA. In order to deal with this problem, reduced kernel method is utilized in this paper.

The organization of this paper is detailed as follows: after a brief review of the existing human activity recognition researches, the proposed reduced kernel based kernel PCA method is given in Sect. 2. In Sect. 3, the experimental results of the proposed method are analyzed. The conclusion is finally stated in Sect. 4.

2 Reduced Kernel PCA

Kernel PCA has been widely used for nonlinear systems, however, it is unsuitable for human activity recognition in mobile devices. Due to that the entire data should be constructed for kernel matric, the kernel matric is prohibitively expensive in storage especially when the size of samples increase dramatically. To overcome this limitation, the random sampling approach has be utilized to choose a small subset $x_b^j \in \{x^{(j)}\}_{i=1}^{L}$ from the M effective observations of the information matrix. The selected observation will be used for model training. According to the theorem presented in [18], the coefficients of the final model will be determined by the number of the data number.

For each principal component $\{\mu_j\}_{j=1,2,\cdots,\ell}$, the closest subset $\Psi(x_b^{(j)})$ to μ_j is chose by RKPCA from the vectors $\left\{\Psi(x_b^{(i)})\right\}_{i=1}^{L}$. In order to achieve this, the vectors $\left\{\Psi(x_b^{(i)})\right\}_{i=1}^{L}$ is projected on the component μ_j, and the retained observations $x_b^j \in \{x^{(j)}\}_{i=1}^{L}$ should satisfy the following model:

$$\begin{cases} \Psi(x_b^{(j)})_j = \max\limits_{i=1,2,\cdots,L} \widehat{\Psi}(x^{(i)})_j \\ \Psi(x_b^{(j)})_{i\neq j} < \xi \end{cases} \tag{1}$$

where $\widehat{\Psi}(x^{(i)})_j$ means the jth component in the projection of $\Psi(x^{(i)})_j$ on μ_j and ξ is a preset value.

Once the observations $\left\{x_b^{(j)}\right\}_{j=1,2,\cdots,\ell}$, which are corresponding to the ℓ principal component $\{\mu_j\}_{j=1,2,\cdots,\ell}$, are chosen, the vector $\Psi(x) \in \mathbb{R}^l$ is transformed into the vector $\widehat{\Psi}(x) \in \mathbb{R}^\ell$, which belongs to the space generated by $\left\{\Psi(x_b^{(i)})\right\}_{i=1}^{\ell}$. The RKPCA model can be described as following:

$$\widehat{T}_{reduced} = \sum_{j=1}^{\ell} \alpha_j \widehat{\Psi}(x)_j \tag{2}$$

where

$$\widehat{\Psi}(x)_j = \left\langle \Psi(x_b^{(j)}), \Psi(x) \right\rangle \quad \text{with} \quad j = 1, 2, \cdots, \ell. \tag{3}$$

According to the kernel method [19], the Eq. (2) can be modified as:

$$\widehat{T}_{reduced} = \sum_{j=1}^{\ell} \widetilde{\alpha}_j \Omega_j(x) \tag{4}$$

where

$$\Omega_j(x) = \Omega\left(x_b^{(j)}, x\right) \quad \text{for} \quad j = 1, 2, \cdots, \ell \tag{5}$$

If non recursive least square is utilized to calculate the parameters $\widetilde{\alpha}_j, j = 1, 2, \cdots, \ell$ in model (4). The solution is:

$$\widetilde{\alpha} = F^{-1} G \tag{6}$$

where

$$G = \begin{bmatrix} \sum_{i=1}^{L} \Omega_1(x^{(i)}) y^{(i)} \\ \vdots \\ \sum_{i=1}^{L} \Omega_\ell(x^{(i)}) y^{(i)} \end{bmatrix} \in \mathbb{R}^{\ell}$$

$$F = \begin{bmatrix} \sum_{i=1}^{L} \Omega_1(x^{(i)}) \Omega_1(x^{(i)}) & \cdots & \sum_{i=1}^{L} \Omega_1(x^{(i)}) \Omega_\ell(x^{(i)}) \\ \vdots & \vdots & \vdots \\ \sum_{i=1}^{L} \Omega\ell(x^{(i)}) \Omega_1(x^{(i)}) & \cdots & \sum_{i=1}^{L} \Omega_\ell(x^{(i)}) \Omega_\ell(x^{(i)}) \end{bmatrix} \in \mathbb{R}^{\ell \times \ell}$$

Avoiding the numerical instability of calculating the solution in Eq. (6), a minimization of the regularized least square method is introduced:

$$J_r(\widetilde{\alpha}) = \frac{1}{2}\sum_{i=1}^{L}\left(T_i - \sum_{j=1}^{\ell}\widetilde{\alpha}_j\Omega_j(x^{(i)})\right) + \frac{\beta}{2}\|\widetilde{\alpha}\|^2 \tag{7}$$

where β is a regularization parameter and the parameter $\widetilde{\alpha} = (\widetilde{\alpha}_1, \cdots, \widetilde{\alpha}_\ell)$.

The Eq. (7) can solved as:

$$\widetilde{\alpha} = (F + \beta E_\ell)G \tag{8}$$

where E_ℓ is a ℓ unit matrix.

The pseudo code of the RKPCA is shown in Algorithm 1.

Algorithm 1 RKPCA algorithm for human activity recognition

Given : Human activity training data D, kernel function Ω

Output: The parameter $\mathring{\alpha}$

1. Using the training data to construct the kernel matrix Ω ;
2. Calculating the eigenvalues μ wiping out zero value and the eigenvectors η of matrix Ω ;
3. Decrease the eigenvalues μ and correspondingly organize the eigenvectors.
4. Choose the 1 principle components, and select the observations $\left\{x_b^{(j)}\right\}_{j=1,2,L,1}$ satisfy the criterion (1);
5. Using equation (8) to calculate the out parameter.

3 Experiment and Results

3.1 *Data Sets and Processing*

In order to validate the effectiveness of the proposed RKPCA method, the wearable experimental platform [20, 21] has been utilized for evaluation. As the major component of the experiment platform, the inertial measurement units (IMU) is mainly composed of a three-axis accelerometer and a three-axis gyroscope. In the platform, the signal data is sent by using ZigBee protocol to the receiver node. Once reached the receiver node, the data packet is sent to the computer side by USB. The placements of IMUs are presented in Fig. 1a, and the corresponding data of activity "Rope skipping" from the five body parts: right wrist, left arm, waist, right ankle and left thigh is shown in Fig. 1b. The data-sampling rate is 50 Hz. Eight subjects take part in this experiment. The first column of Table 1 lists the activities performed by each participant.

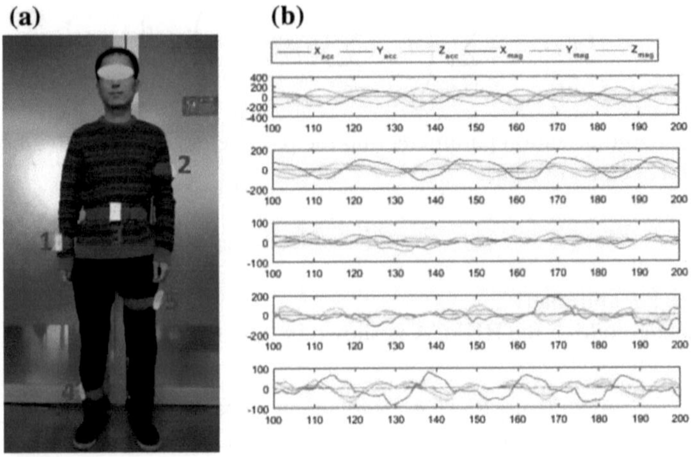

Fig. 1 The locations of IMUs and the inertial signal data of "Rope skipping" from five different body parts; **a** the fixed locations of sensor nodes; **b** The IMU signal data of "Rope skipping" from the five body parts

Table 1 The acquire features of different activities from the five body parts

Activity	Feature samples
Walk at the speed of 3 km/h	1400
Walk at the speed of 5 km/h	1400
Run at the speed of 6 km/h	1400
Run at the speed of 8 km/h	1300
Run at the speed of 12 km/h	1100
Go downhill	800
Climb the hill	970
Practice gymnastics	1200
Rope skipping	750
Riding a bike	1400

After preprocessing (e.g. removing the outliers), sliding window approach is introduced to split the signal into windows with the same length. 50 samples are chosen as the window size. 50% overlap is utilized to avoid information loss between the adjacent windows. The features mean, kurtosis, standard deviation and correlation between axes are extracted, respectively. All the features are normalized to [0, 1] [22]. In total, 24 features are extracted for each IMU data. The quantities of extracted feature samples in different actions are detailed in the second columns of Table 1.

3.2 *Experimental Results*

After signal preprocessing and feature extracting, SVM has been chosen to evaluate RKPCA method. In the experiment, the leave-one-out (LOO) strategy is utilized to train and classify activities. The validation is repeated five times with each union used exactly once for testing [22]. The reported final values are their average results.

All experiments are conducted on a PC with an Intel(R) Core(TM) i5, 2.55 GHz and 6 GB RAM. The algorithms are all implemented in Matlab 2010b.

3.2.1 Performance Comparison

In this section, the proposed RKPCA is compared with PCA and KPCA. Table 2 shows the accuracy of the dimension reduction approaches using SVM, and the cost time is presented in Table 3. The accuracy of PCA, KPCA and RKPCA approaches on the dataset from waist is presented Fig. 2, and the cost time is presented in Fig. 3. From Table 3 and Fig. 2, it can be seen that using RKPCA can achieve better results than PCA, which is similar to KPCA. From Table 3 and Fig. 3, it can be observed that RKPCA cost less time than KPCA, and much similar with PCA.

3.2.2 Performance of Different Placements

From Table 3 and Fig. 2, it can be found that best result is achieved by using the datasets acquired from all IMUs. The dataset from waist can much better identify the tested activities.

Table 2 The performance of dimensional deduction methods on five dataset

	PCA (%)	KPCA (%)	RKPCA (%)
Right wrist	81.9	88.4	88.3
Left arm	82.2	89.1	88.8
Waist	83.2	91.7	91.2
Right ankle	82.9	89.4	89.2
Left thigh	83.4	89.9	89.5
All data	88.1	95.1	94.6

Table 3 The cost time of dimensional deduction methods on five dataset

	PCA	KPCA	RKPCA
Right wrist	0.117	11.147	3.572
Left arm	0.131	11.908	3.854
Waist	0.127	11.270	3.403
Right ankle	0.114	11.134	3.419
Left thigh	0.122	11.624	3.428
All data	1.816	31.0975	5.9157

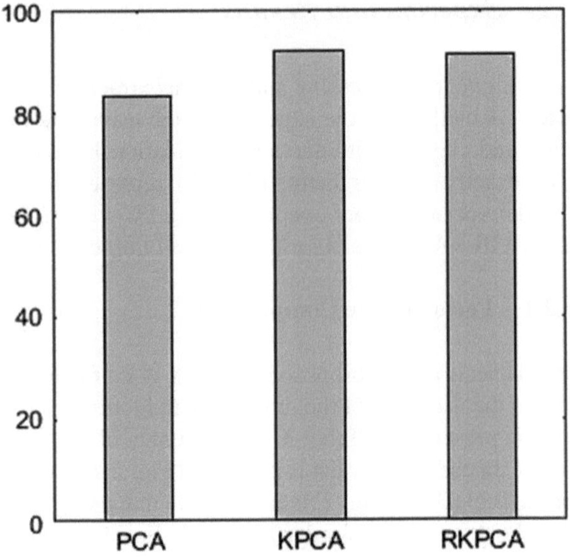

Fig. 2 The accuracy of dimensional deduction methods on the dataset from waist

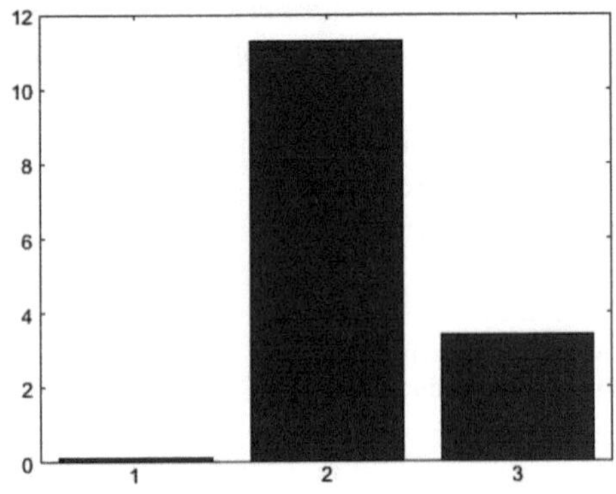

Fig. 3 The cost time of dimensional deduction methods on the dataset from waist

4 Conclusion

In order to improve the accuracy and effectively reduce the dimension, RKPCA method is proposed for inertial sensor based human activity recognition. The encouraging results demonstrate the efficacy and efficiency of the proposed method. The conclusion is also obtained that data from waist acquired in this experiment is much more discriminative than from other three body parts.

Acknowledgements This work was supported by Fundamental Research in cutting-edge technologies in the project of Henan province (162300410070), Key Science Technology Program of Henan Province, China (19A413013), Foundation of Henan Educational Committee (162300410070), and Ph.D. early development program of Zhengzhou University of Light Industry (13501050009).

References

1. Avci, A., Bosch, S., Marin-Perianu, M., Marin-Perianu R., Havinga, P.: Activity recognition using inertial sensing for healthcare, wellbeing and sports applications: a survey. In: Proceedings of 23rd International Conference on Architecture of Computing Systems (ARCS), Feb 2010, pp. 1–10 (2010)
2. Rashidi, P., Mihailidis, A.: A survey on ambient-assisted living tools for older adults. IEEE J. Biomed. Health Inf. **17**(3), 579–590 (2013)
3. Wang, Z., Zhao, H., Qiu, S., Gao, Q.: Stance-phase detection for ZUPT-aided foot-mounted pedestrian navigation system. IEEE/ASME Trans. Mechatron. **20**(6), 3170–3181 (2015)
4. Qiu, S., Wang, Z., Zhao, H., Liu, L., Jiang, Y.: Using body-worn sensors for preliminary rehabilitation assessment in stroke victims with gait impairment. IEEE Access **6**, 31249–31258 (2018)
5. Khan, M., Lee, Y.-K., Lee, S.Y., Kim, T.-S.: A triaxial accelerometer-based physical-activity recognition via augmented-signal features and a hierarchical recognizer. IEEE Trans. Inf. Technol. Biomed. **14**(5), 1166–1172 (2010)
6. Cheng, W.C., Jhan, D.M.: Triaxial accelerometer-based fall detection method using a self-constructing cascade-AdaBoost-SVM classifier. IEEE J. Biomed. Health Inf. **17**(2), 411–419 (2013)
7. Sirichana, W., Dolezal, B.A., Neufeld, E.V., et al.: Wrist-worn triaxial accelerometry predicts the energy expenditure of non-vigorous daily physical activities. J. Sci. Med. Sport (2017)
8. Gupta, P., Dallas, T.: Feature selection and activity recognition system using a single triaxial accelerometer. IEEE Trans. Biomed. Eng. **61**(6), 1780–1786 (2014)
9. Wu, D., Wang, Z., Chen, Y., et al.: Mixed-kernel based weighted extreme learning machine for inertial sensor based human activity recognition with imbalanced dataset. Neurocomputing **190**, 35–49 (2016)
10. Wang, Z., Wu, D., Chen, J., Ghoneim, A., Hossain, M.A.: A triaxial accelerometer-based human activity recognition via EEMD-based features and game-theory-based feature selection. IEEE Sens. J. **13**(9), 3198–3207 (2016)
11. Maurer, U., Smailagic, A., Siewiorek, D.P., Deisher, M.: Activity recognition and monitoring using multiple sensors on different body positions. In: Proceedings of International Workshop Wearable Implantable Body Sensor Networks (BSN), Apr 2006, pp. 113–116 (2006)
12. Ermers, M., Pärkkä, J., Mäntyjärvi, J., Korhonen, I.: Detection of daily activities and sports with wearable sensors in controlled and uncontrolled conditions. IEEE Trans. Inf Technol. Biomed. **12**(1), 20–26 (2008)
13. Qiu, S., Wang, Z., Zhao, H., et al.: Using distributed wearable sensors to measure and evaluate human lower limb motions. IEEE Trans. Instrum. Meas. **65**(4), 939–950 (2016)
14. Mukhopadhyay, S.C.: Wearable sensors for human activity monitoring: a review. IEEE Sens. J. **15**(3), 1321–1330 (2015)
15. Gao, L., Bourke, A.K., Nelson, J.: Activity recognition using dynamic multiple sensor fusion in body sensor networks. In: Conference on Proceedings of the IEEE Engineering in Medicine and Biology Society, p. 1077 (2012)
16. Ertuğrul, Ö.F., Kaya, Y.: Determining the optimal number of body-worn sensors for human activity recognition. Soft Comput., 1–8 (2016)
17. Zhang, Z., Poslad, S.: Improved use of foot force sensors and mobile phone GPS for mobility activity recognition. IEEE Sens. J. **14**(12), 4340–4347 (2014)
18. Taouali, O., Jaffel, I., Lahdhiri, H., et al.: New fault detection method based on reduced kernel principal component analysis (RKPCA). Int. J. Adv. Manuf. Technol. **85**(5), 1547–1552 (2016)
19. Aronszajn, N.: Theory of reproducing kernels. Trans. Am. Math. Soc. **68**(3), 337–404 (1950)
20. Chen, Y., Guo, M., Wang, Z.: An improved algorithm for human activity recognition using wearable sensors. In: 2016 Eighth International Conference on Advanced Computational Intelligence (ICACI), pp. 248–252. IEEE (2016)

21. Qiu, S., Wang, Z., Zhao, H., et al.: Inertial/magnetic sensors based pedestrian dead reckoning by means of multi-sensor fusion. Inf. Fus. **39**, 108–119 (2018)
22. Wang, Z., Jiang, M., Hu, Y., et al.: An incremental learning method based on probabilistic neural networks and adjustable fuzzy clustering for human activity recognition by using wearable sensors. IEEE Trans. Inf Technol. Biomed. **16**(4), 691–699 (2012)

Design Practice of Visual Art Forms Based on Targeted Selection of Microcosmic Appearance

Xiaonong Qian, Ying Wang, Caixia Du and Yuhui Yang

Abstract As a visual field that human eyes cannot directly observe, the microcosmic appearance is a rich element in the field of art and design, and it has great creative value. According to the design concept of microcosmic appearance of visual art form design, microscopic appearance imaging is obtained from the experiment. Furthermore, the microcosmic appearance of target selection is carried on the art transformation. As the elements of fashion design, diversification application of microcosmic appearance in clothing design. Thus, the visual effects of clothing and sustainable design concepts are reflected.

Keywords Microscopic appearance · Clothing design · Visual effect
Sustainable design

1 Introduction

The original intention of microscopic is "small", as opposed to "macroscopic", which is summarized in Xinhua Dictionary from two aspects. In natural science of particle generally refers to the space size less than 10–7 to 10–6 cm material system; in social science or the concept of broad sense, macroscopic refers to the observation from a large area and microscopic refers to the observation from a small side, and the latter is extended to the meaning of "local" [1]. The eyes of normal people all have a visual limit scale, and the healthy naked eye can distinguish objects more than 0.1 mm in diameter, while the objects or phenomena of smaller resolution distance are invisible to the human eye known as the "micro" [2]. Such as particles, cells and the corresponding particles movement, cells life activities, fields and waves, etc. In this paper, microscopic refers to the invisible category by the naked eye state, which is able to be visualized by the equipment with magnification imaging function.

X. Qian (✉) · Y. Wang · C. Du · Y. Yang
School of Fashion, Dalian Polytechnic University, Dalian 116034, China
e-mail: qianxiaonong@163.com; 974553936@qq.com

G. Fortino and Z. Wang (eds.), *Advances in Body Area Networks I*,
Internet of Things, https://doi.org/10.1007/978-3-030-02819-0_35

The "morphology" has a variety of information that is of great significance to the survival of the living body, such as its appearance and characteristics, as well as the circulation and transformation of material and energy [3]. A large number of visually perfect shapes and structures exist in the microscopic image, which manifest the image and beauty law of form in contrast with unity, contradiction and balance. Currently, the visual aesthetic trend of new graphic information is hided in the nature, organic characteristic and simplicity of the imaging metaphor, and people are more inclined to regain unexpected visual surprises by changing the usual sight [4]. The inherent form and structure of life are one of the unique manifestations of contemporary visual aesthetics. In this paper, the "morphology" will be explained word by word. "Form" as a noun, refers to the shape and structure of a targeted, which can be understood as the static state of the targeted. "Appearance" refers to the shape, appearance, state, expression, meaning of the dynamics of a targeted. From this, the meaning of the word "morphology" is obtained, that is, it includes the dynamic and static performance characteristics of the thing, its shape, appearance and manifestations.

The micro-scene is a brand-new perspective of human sight and a novel visual world opened up by scientific and technological progress. The inherent natural characteristics and life rhythms inherent in the micro-material are also closely related to the modern society's aesthetic awareness and ecological awareness [5].

The appropriate elements extracted from the microcosmic appearance, are suitable for the theme of clothing design source materials, can enhance the concept expression and practical efficacy of clothing. Besides, based on the aesthetic point of view, the extracted microcosmic appearance is carried on art transformation, and the new and different visual aesthetic feeling of clothing can be bestowed.

The abstract theme and concept of fashion design are closely related to the specific elements, and the two are reversible relations of reciprocal push. In other words, the abstract theme and concept of fashion design can be used to find suitable specific elements, and it can be derived from the inspiration of specific elements [6]. In this paper, the targeted selection of microcosmic appearance is implemented and the art transformation of fashion design is applied in this design thinking method.

2 The Concept and Path of Design

2.1 Design Concept

Based on the design concept of clothing with visual aesthetic function and sustainability function, a clothing that can maximize the wearer's clothing time is designed, which is an embodiment of the sustainable design principle, that is, the initial use cycle principle of the product is optimized.

So how extend the clothing time limit of clothing maximally? In this paper, the clothing seeking heterosexual, multipurpose and constant companionship of the

Table 1 New age group division of China in 2016

Age	Life stage	
0 to 6 years old	Childhood	Infancy: 0 to 3 months whole
		Toddler Period: 4 months whole to 2.5 years old
		Early childhood: 2.5 to 6 years old
7 to 17 years old	Juvenile	Enlightenment: 7 to 10 years old
		"Antagonistic period": 11 to 14 years old
		Growth period: 15 to 17 years old
18 to 40 years old	Youth	Adolescence: 18 to 28 years old
		Mature period: 29 to 40 years old
41 to 65 years old	Middle aged	Sturdy period: 41 to 48 years old
		Stable period: 49 to 55 years old
		Adjustable period: 56 to 65 years old
Over 66 years old	Elder	Early aging period: 67 to 72 years old
		Middle and old aged period: 73 to 84 years old
		Old-age period: 85 years old later

design concept is started. Furthermore, "seeking the different in nature "means that everyone has its own unique micro-material. When the characteristics of this microscopic material are implanted into clothing, clothing has the uniqueness and customization of "only this one belongs to you". "Multipurpose" refers to the clothing that not only has its own clothing function, has their own different applications and functions by different components of clothing. "Constant companionship" is the ideal design concept about clothing. The clothing which can meet the different needs of clothing products at different stages of the growing and maturing process is anticipated. The design of the clothing products, although not available to everyone in every age group, can meet the needs of a part of the audience for each age. According to the new age group division criteria of China in 2016, as shown in Table 1, the modular design of clothing is applied to the physiological or psychological needs of the wearers respectively, such as infants, young children, juvenile, youth, middle-aged and elder. The concept of sustainable design that "takes care of each other's life" is conveyed and the avoidance of disposable life behaviour is advocated, as shown in Fig. 1.

2.2 *Clothing Modular Design Path*

See (Fig. 1).

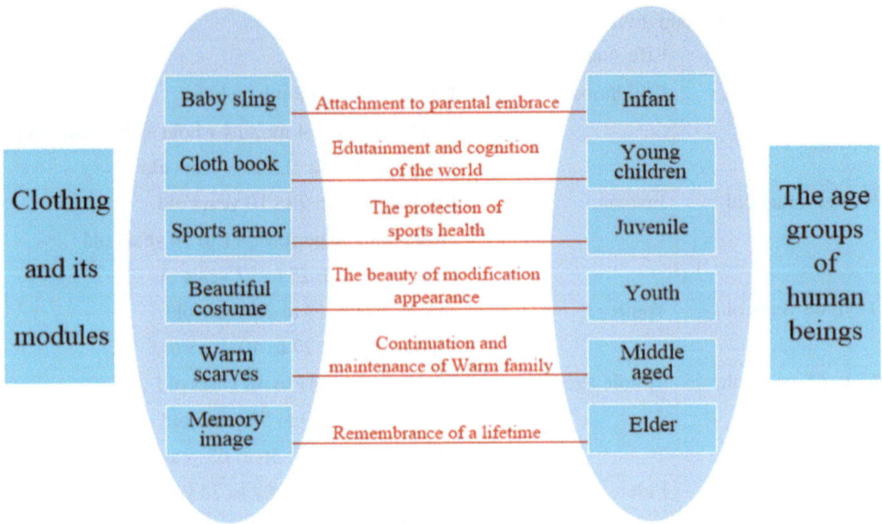

Fig. 1 The design path of clothing modularity

3 The Acquisition Experiment and Art Form Transformation of Microcosmic Appearance

3.1 Microcosmic Appearance Acquisition Experiment

According to the difference between the state and effect of micro-magnification, the different functional microscope needs to be used. Moreover, the observation sample needs to be treated differently in accordance with the different requirements of the observed samples by various microscopes. In this paper, the microcosmic appearance obtained by the special microscope, such as SZX12-131 type stereomicroscope, BX61 type fluorescence microscope and JSM-7800F type scanning electron microscope.

In experiments, the samples are observed firstly by the microscope in the style of casting net. The method of multiplied amplification is applied in the observation experimental operation of the sample, and step by step to observe the sample from the macro to the relative micro, and finally to the microscopic morphology features. The highest multiple of 90 times the magnification effect image is observed by stereomicroscope observation, as shown in Fig. 2, and human being's eyes naked state cannot see or overlooked details can be observed, such as epidermal villus, surface texture and other relatively microscopic image, and the presented image retains the primary colours of the sample. The operation is simple and convenient in the process of observation, since the sample can be directly putted on the objective table without making glass slides. It is necessary for observing with the fluorescence microscope to make a glass slide sample form for the observation sample, which can

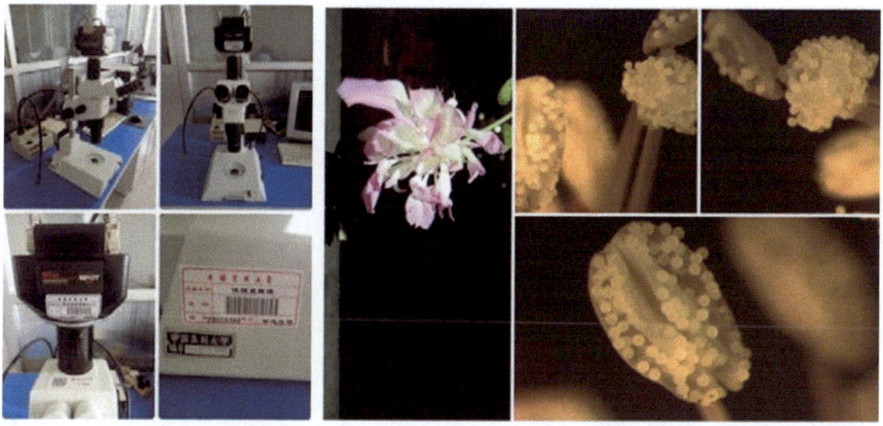

Fig. 2 Microscopic appearance observed by SZX12-131 stereoscopic microscope

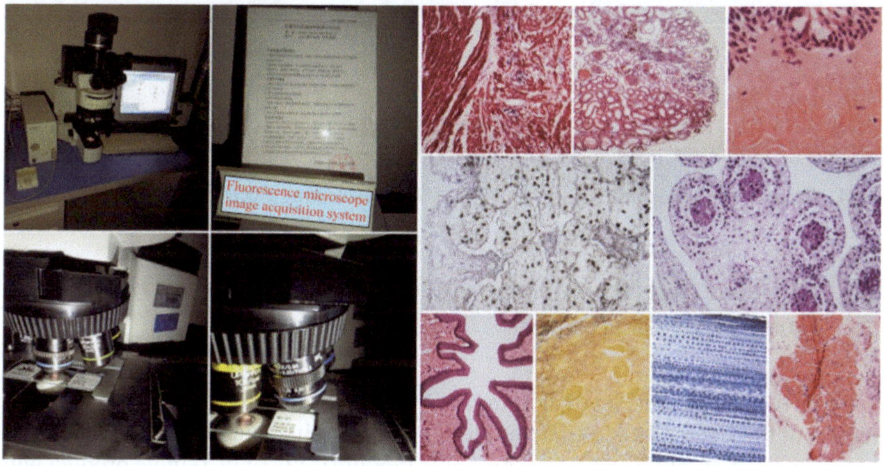

Fig. 3 Microscopic appearance observed by SZX12-131 stereoscopic microscope

present the microcosmic appearance characteristic of the material. However, since the ultraviolet ray is used as the light source and is treated by the fluorescence staining of the observed sample, the presented image is not the primary colour of material, as shown in Fig. 3. The requirement of scanning electron microscopy is higher for the observation samples and observation conditions, but the observation sample is magnified much higher than the above two optical microscope and the presented image with the very strong stereoscopic sense. However, the presented image only with black, white and gray colour, as shown in Fig. 4.

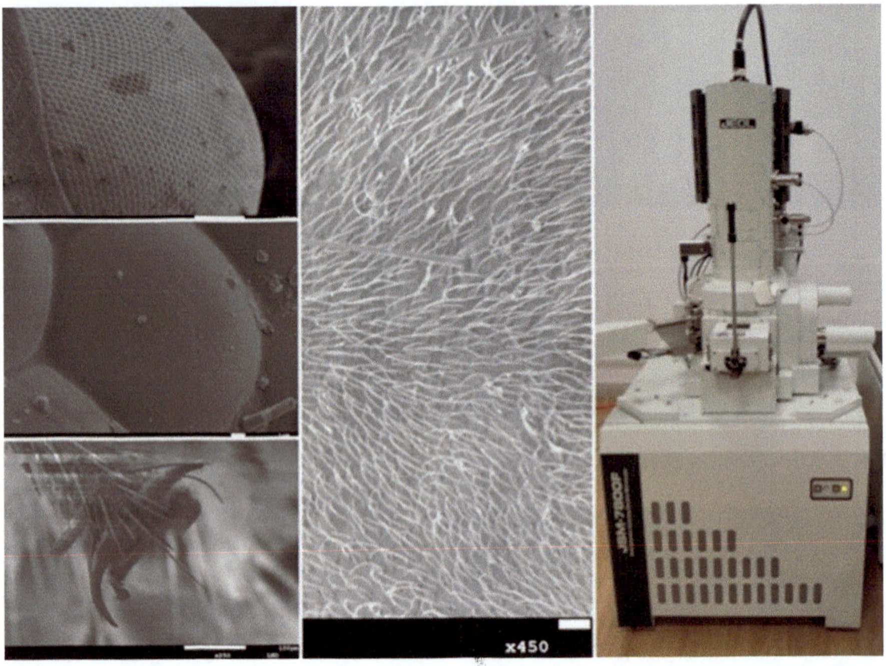

Fig. 4 Microscopic appearance observed by JSM-7800F scanning electron microscope

3.2 *Targeted Selection Microscopic Appearance and Art Form Transformation*

Based on the design concept of clothing, this kind of clothing designed has the unique element of the wearer. In addition, the clothing has the characteristics of being detachable and reconfigurable, and each detachable module is not only a single product with independent clothing functions. It can become the striking gorgeous clothing through the combination. Moreover, whether the clothing or its each module is incorporated the microscopic appearance features corresponding to its characteristics, such as the variability of microscopic appearance is applied in the deconstruction of clothing and structural design, the gorgeous appearance of microscopic appearance is applied in clothing fabric colour, pattern, texture design, as well as the unique natures and functions of micro-material are translated into the meaning symbol that is integrated into the fashion's innovative design, which also is the screening criteria of targeted selection microscopic appearance.

According to the purpose and significance of the clothing and its each module in the design idea, the microscopic appearance of targeted selection needs to have the protection function, communication and uniqueness. In the obtained and collected microscopic appearance from this experiment and research process, the human health immune cells which have the protection function, with different propagation routes

Fig. 5 Art transformation of immune cells

of fine pollen, and cell molecules with individual uniqueness are selected respectively for design transformation. In addition, in order to increase the visual richness and coordination beauty of clothing, some selected microscopic appearances play the role of visual collocation, which is mainly used in the aspects of clothing colour and fabric texture.

The design and transformation of microscopic appearance is analyzed by the selected three kinds of micro-material. The bionic method and abstract method are comprehensive applied to morphology transformation of human immune cells and pollen grains. The ontological features of the micro substance are observed and analyzed firstly. The overall feature of human immune cells is that many weirder synaptic structures are dispersed from a center body to all directions. The shape of the pollen grains is mostly the sphere and spindle, and the most significant feature is the pore structure of the surface and the protuberant tissue attached to the surface. The most obvious feature is enlarged exaggeratively and some complex but not important features are simplified abstractly by cognition of the material. The colours can be extracted and adjusted from the original material in the colouring, as well as colour recreation based on the need of overall design effect. The effect diagrams of the transformation of human body immune cells and pollen grains from ontology to art form are shown in Figs. 5 and 6.

The extraction and transformation of individual-unique cellular molecule, are very rich in technology content, have specialized complex operations that go far beyond the category of the art discipline which this paper belongs. This is done with the help of professionals such as medicine and biology. Some molecular structures of human body cannot be observed by microscope and can be visualized by chemical principles and biotechnology, which is confirmed by related scientific experiments. For example, the corn seed storage protein is transformed into the form of purity map, as shown in Fig. 7. In the book "Human Genome" by Carina Dennis, Richard Gallagher, the process of extracting and transforming cellular molecules into colour maps is described, with the meaning of artistic attributes From the perspective of

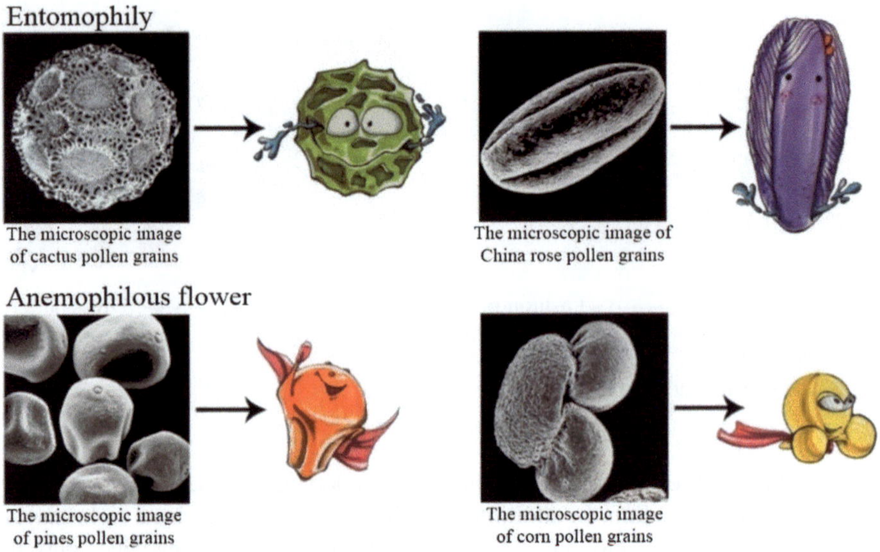

Fig. 6 Art transformation of pollen grains

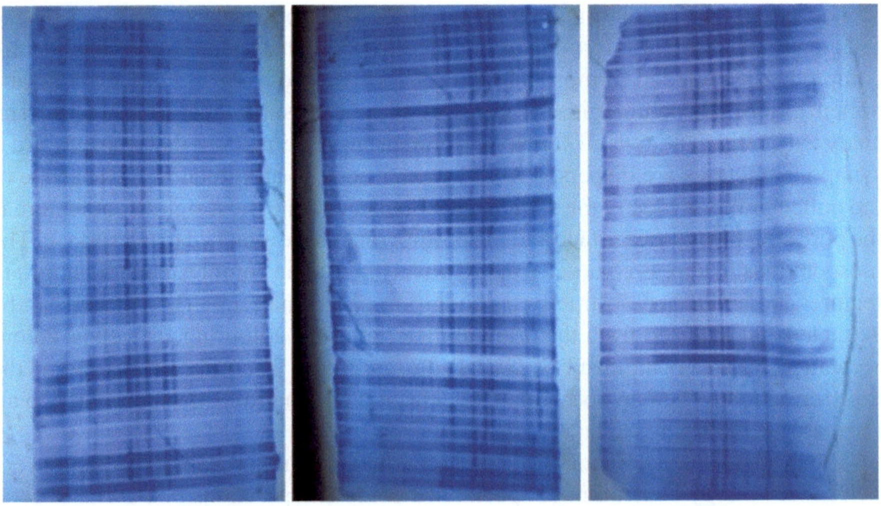

Fig. 7 Corn seed purity profiles

design vision, the colour map is redesigned to present the graphics in the art of design.., as shown in Fig. 8.

Through the above experiment, the ultimate goal is that the exquisite images discovered accidentally and the first-hand information are obtained in the experiment, and the art form of transformed visual effect is applied into clothing to show its uniqueness. In spite of the use of leading-edge equipment, the complicated research

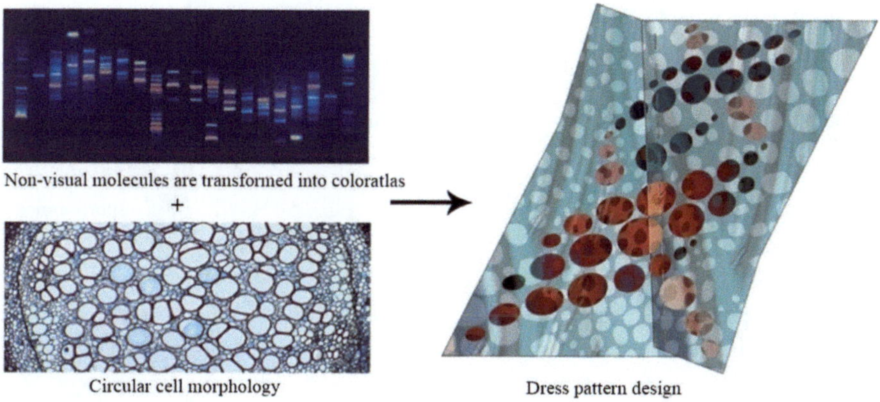

Fig. 8 Art transformation of clothing patterns

contents and the scientific research process of twists and turns, the ultimate goal is to return to the origin of "fashion design". All preparations are to provide rich resources and solid theoretical support for making more beautiful, valuable and forward-looking clothes.

4 Design and Application of Visual Art Forms with Target Select Micromorphology in Clothing

4.1 Clothing and the Composition Design of Its Each Module

4.1.1 Clothing Composition Design

As shown in (Fig. 9).

4.1.2 The Composition Design of Each Module

Module one: Baby sling (Fig. 10).
Module two: Cloth book (Fig. 11).
Module three: Sports armor (Figs. 12, 13).
Module four: Beautiful clothing (Fig. 14).
Module five: Warm scarf (Fig. 15).
Module six: Memory image + (Fig. 16).

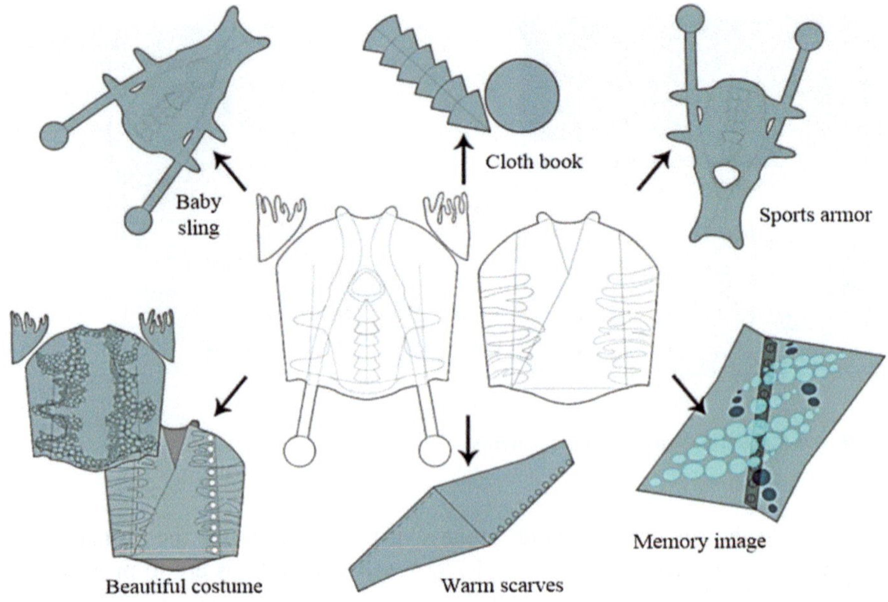

Fig. 9 Clothing composition design of each module

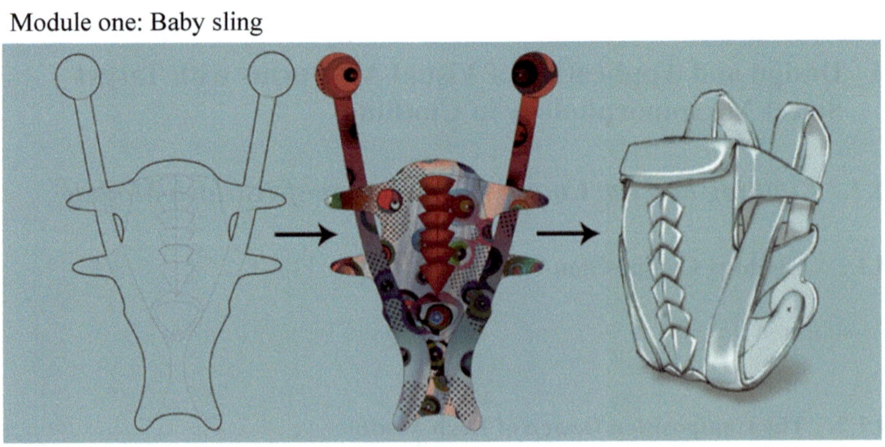

Fig. 10 Baby sling design composition diagram

4.2 The Visual Art Form Clothing Design of Targeted Selection Microscopic Appearance

As far as the overall design of clothing is concerned, the art form of target-selected microscopic appearance has been integrated from many aspects. In the clothing color, blue and green as the main tone rich has the futuristic sense of science and technology,

Module two: Cloth book

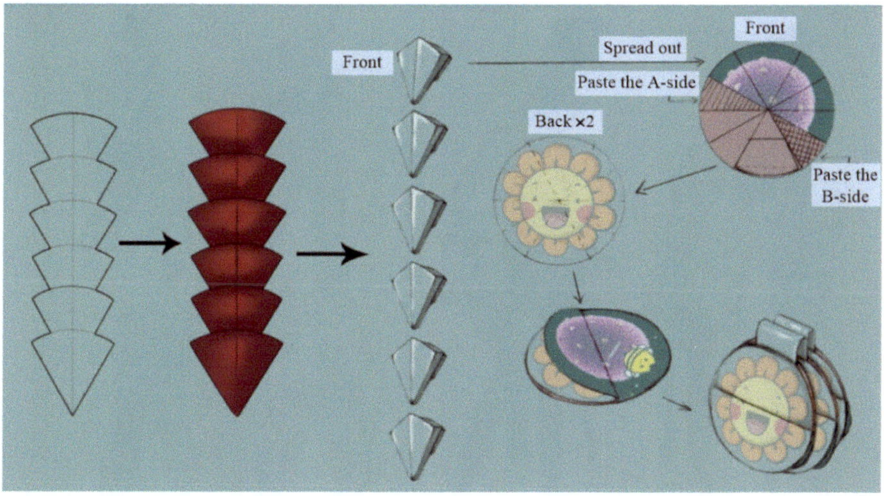

Fig. 11 Cloth book design composition diagram

Module three: Sports armor

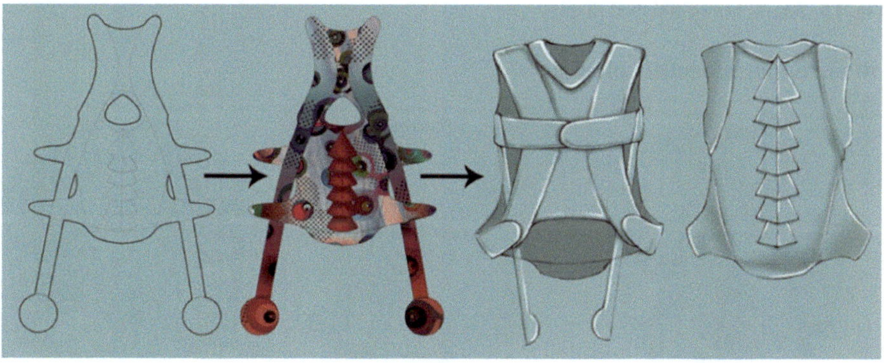

Fig. 12 Sports armor design composition diagram

the jumping colors with a variety of small area are added, such as both harmonious and not monotonous color scene of the micro-world mystery. In addition, the artistic transformation form of the targeted selection microscopic appearance is used as the clothing pattern. In fabric texture, the morphological features and hierarchical space of the microscopic appearance are referenced, which can achieve visual and tactile dual experience effect, the design effect diagram is shown in Fig. 17.

In addition, each single module of clothing has a unique design. Such as the baby sling for the infants and the sports armor designed for youth. The inspiration of its shape and pattern comes from the morphologic features of the human immune cells. The human immune cells are the guardians of human health function, which is

Fig. 13 Sports armor finished display

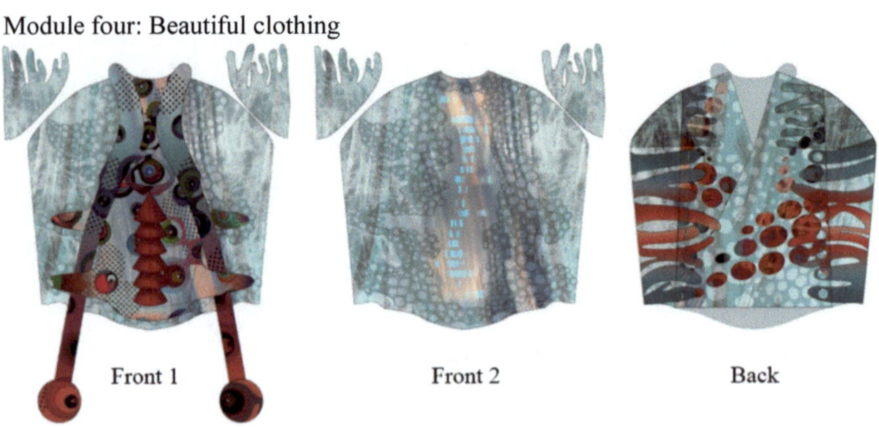

Fig. 14 Beautiful clothing design composition diagram

injected into meaning of peace and protection for two kinds of products. The designed cloth book is a changeable structure that can be split and combined for early childhood. Furthermore, it becomes an important part of sport armor with spine protection function after folding and combining, and it becomes the children's tear-proof book with pages after splitting and tiling. This versatile structure is designed with the inspiration come from microscopic variability of cells. The picture design in the

Module five: Warm scarf

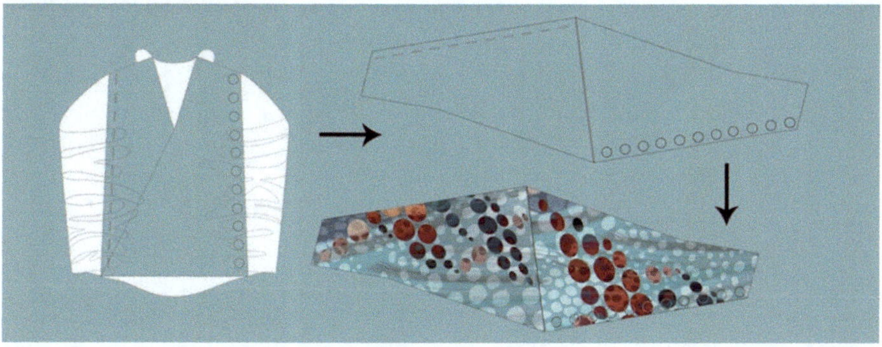

Fig. 15 Warm scarf design composition diagram

Module six: Memory image+

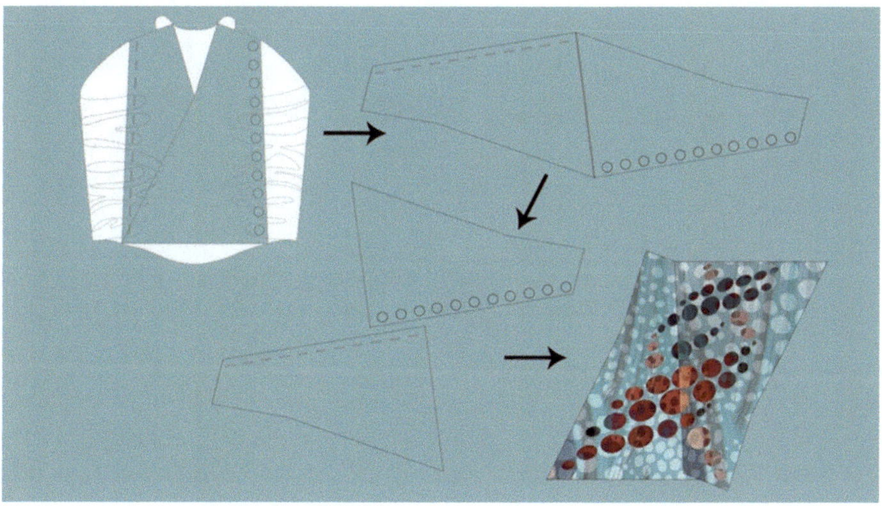

Fig. 16 Memory image design composition diagram

cloth book is based on the prototype of pollen that has transmissibility, which meaning that seeds of knowledge are imparted to children and satisfy their desire for the knowledge of world. Moreover, the cloth book and memory images designed for the elderly are used in a wide range of expressions. The vivid and vivid digital dynamic effect can be presented by the combination of digital technology and equipment. The design of the microscopic appearance and art two-dimensional code are integrated into the clothing or embed into clothing, thus, the multi-cherished cultural content of the wearer needs is shown from microscopic appearance art two-dimensional code through the smart phone identification. For children who is learning the knowledge of graphics, this method plays an important role in teaching through lively activities

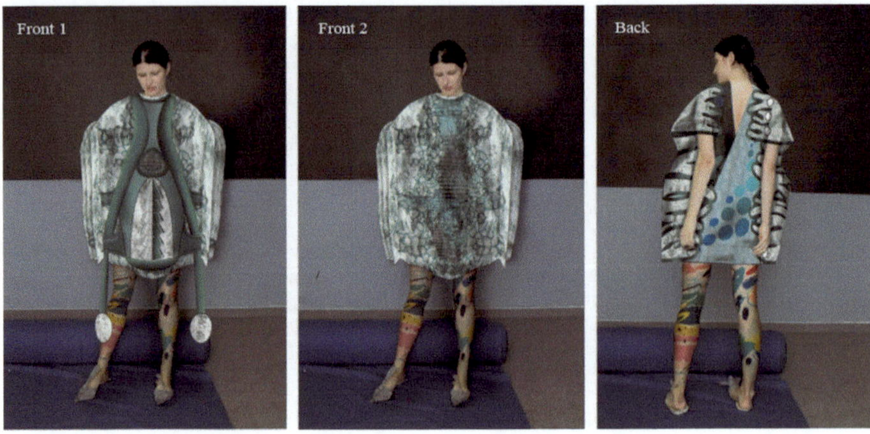

Fig. 17 Design effect drawing

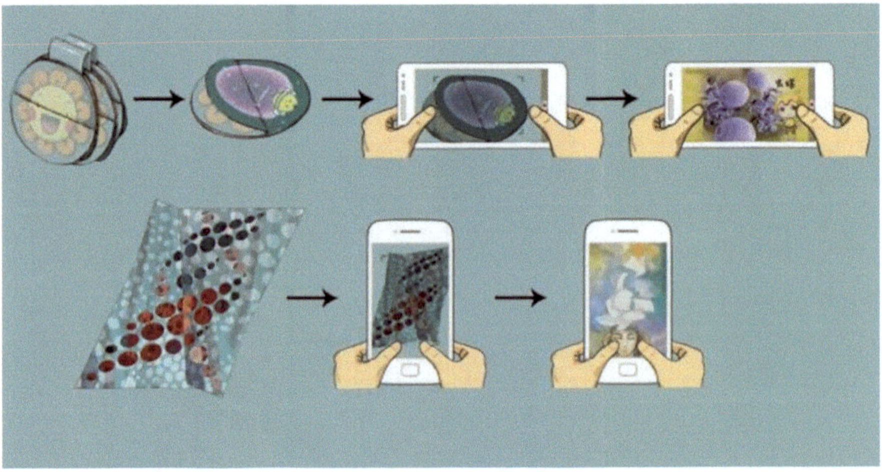

Fig. 18 The digital dynamic effects of cloth book and memory image

and its hidden image vivid memory will become a precious gift of life review meaning for the old man and his relatives, as shown in Fig. 18. Figure 19a–b is the finished product pattern that is integrated with the artistic visual design of microscopic appearance and two-dimensional code graphics into clothing or embedded wear.

Fig. 19 Clothing display

(c)

Fig. 19 (continued)

5 Conclusion

In this paper, the targeted selection microscopic appearance is transformed into artistic form which is applied in the clothing design through the experiment, so that the clothing has special effects. The artistic form of microscopic appearance used in clothing is designed, which pay attention to the value of visual decorative effect, and the efficacy of clothing sustainability is also excavated. In the aspect of observation sample of targeted selection microscopic appearance, the target element with special meaning and anisotropic characteristics are selected and the field of clothing design is extended. Thus, the morphology elements of the micro world are incorporated the process of clothing product design.

References

1. Wang, W.: Clothing function on the development of fashion design. J. Art Educ. 216–217 (2016)
2. Foreign dyeing and finishing technology. J. Print. Dye. 54–59 (2011)
3. Dennis, C., Gallagher, R.: The Human Genome. Science Press, M. Beijing (2003)
4. Jinguo, Z.: Unveil the mystery of the microscopic world—the scientific course of microscope invention. J. Sci. 24 hours (2008)
5. Yoda.: A flower world has beautiful—the development of micro-video works. J. Popular Arts (2011)
6. Haijun, S.: Nanotechnology—the invisible art. J. World Life (2012)

Practice Research on Chaos-Theory-Based Algorithmic Composition

Xiaonong Qian, Yiwen Sun, Caixia Du and Yuhui Yang

Abstract Chaotic music composition requires a series of "musical events" which constitute music to continuously take place, develop and evolve in time and space sequence orbit, the cross and integration of its certainty and uncertainty offer a new manner for music composition-Algorithmic Composition and extend to the domain of colorful music so as to form one kind of design thinking with randomness. In this paper, we summarize several Chaos composer algorithm design a set of installation art by using its randomness features. The results of the random combination of the music clips are visualized according to the emotional color of the interval, and realize the conversion between music and color. This research provides new ideas for the development of new media art.

Keywords Chaos theory · Algorithmic composition · Color music · Randomness

1 Introduction

1.1 Algorithmic Composition

Algorithm Composer is a kind of composer technique which was established in the late 20th century. It is a study that tries to use a formal process so that the composer can minimize the involvement of when the composer using a computer for music creation, which has far-reaching influence on the future music. In the meantime, the rapid development of computer technology brings about continuous change in the technique of Algorithmic Composition and constantly enriches the type of music [1]. The prime techniques that the domain of Algorithmic Composition adopts are the Markov transition diagram (or called Markov chain), random process, layering technology, Knowledge Base System, music grammar, artificial neural network technique and Genetic Algorithm.

X. Qian (✉) · Y. Sun · C. Du · Y. Yang
Dalian Polytechnic University, Dalian 116034, China
e-mail: qianxiaonong@163.com; 854883692@qq.com

G. Fortino and Z. Wang (eds.), *Advances in Body Area Networks I*,
Internet of Things, https://doi.org/10.1007/978-3-030-02819-0_36

What Algorithmic Composition emphasizes is one kind of absolute rational and rigorous. But every coin has two sides, the absolute rational mentioned here is not the absolute certainty of the result, it stresses the absolutely rational and procedural of the algorithmic process and can be expressed by the mathematical formula; Furthermore, not all final results of various musical key elements in the Algorithmic Composition technique are able to be predicted accurately, which is like contradictory and unified relativity theory, requiring certain results as well as uncertain results, and it is exactly similar to Chaos Theory.

1.2 Chaos Theory Design

Chaos Theory indicates under a certain physical model, if the given initial values have slight differences, the result obtained might be quite different compared to the initial values after going on the calculation within a certain period of time. Chaos theory reveals the dialectical relationship between the contradictory unity of complex systems, such as order and disorder, certainty and uncertainty, simplicity and complexity. Chaos Theory takes the nonlinear system as its own research content, hereby the nonlinear system refers to the line relationship without the direct ratio between two variables, for example, and the mathematical equations which describe the movements of complicated things are all non-linear. Therefore the world is non-linear in essence, marking humans' deepening and improving acquaintance on objective things; By understanding things from the simple relationship deep into the complicated relationship as well as always revealing some new facts, new features and new regular pattern is not only of significant value but also of important philosophical meaning on the development of science technology [2].

Music compositions are generated from the sequence orbit in the dynamic system of time and space by a series of "musical events" and carry out the simple equivalent superposition, progressively increase and progressively decrease, enlarging and shrinking in strict accordance with the ratio, these are all linear processes, represented by deterministic behavior; however, uncertain behavior indicates the debugging, some note might occur at any time and has influences on the next note more or less in the note series, all these belong to the non-linear process. Music compositions are one non-linear dynamic system, and music events in this system continue to generate, develop and evolve, the structure is similar to that of the sonata presentation section, the expansion section, and the reproduction section, during this process, the mode and status of coexistence in the same voice for certain behavior and uncertain behavior are formed, presenting the synthesis of order and disorder [3].

Chaotic composition has existed since ancient times, the algorithmic system started to use the rhythmic pattern to compose in a hymn of uniform rhythm during the period of fifteenth century. Followed by the Renaissance period as well as the Baroque age, the composition technique was improved ceaselessly, composers developed strict rules for the counterpoint of melodies. For instance, the creation of polyphony developed a given theme motivation almost completely by using the systematized processes such as inverted images and adding decorations to or simplifying the theme, the fugue is like the pure logic in music, as it were, most refined demonstrating the general order in symbolic coding. In the classical age, when Mozart created the Musical Dice Game, he applied the random manner to combine the module for the composition. In the work, he created 176 pieces of music and then the section number is arranged into two special matrix graphs. Decide the order of playing by the player rolling the dice. The desire to apply the algorithm to music may be as old as the music itself.

The concept of "randomness" in composition techniques is doomed to guide Algorithmic Composition creators to find more targeted composing skills in music design, and it has very practical reference value in whole new digital music creation and certainly become one of core composing skills in the future artificial intelligent composition.

2 Chaos-Theory-Based Composition Practice

From the perspective of chaotic thinking, it can be considered that music works are fractal in spatial sequence, which develop to fine structure and form partial and whole self-similarity. Some classical works entirely are derived on the basis of constant repetition of the same themes, but these themes contain some unexpected elements, rather than being rigid repetition. To take one example, the Third Movement of Brahms Symphony No. 1, the repetition when the same melody is played slightly different brings fresh ears every time, which is extremely fascinating.

Musical works turns out to be as chaotic movement in time sequence, gradually transition from order to disorder, this chaotic movement appears to be random, but as a matter of fact, there is own inherent and in-depth regularity in it, which refers to Chaos sequence [4]. The randomness is an important contrast factor that grows with the absolute rationality of the composition of the algorithm, which is a necessary condition for the real improvement of the algorithm composition technique. Whether it is four harmony or registration method needs to be in accordance with a large number of rules to limit, essentially, the composing of classical music is recklessly elaborate under comply with the rules of all composing. However, if the computers can be told about those rules in computer language, since the computer is always right about everything, When ensuring that two or more tones (chords), a combination of two or more tones (harmony) do not violate the rules, then other tones may allow the computer to composite music by random selection, which may even be possible to create harmonies that human composers fail to even think of it in this way [5].

2.1 Using Interval Relation to Harmonious Music Note Sequences

Interval refers to, in the music system, the pitch relationship between the two tones, or pitch difference between the two [6]. "Chord intervals" are two or more notes that occur simultaneously. Concord Interval sounds sweet, confluent, while discordant interval sounds awkward, even harsh. In auditory neuroscience and psychology of music, the former is called harmonic, the latter is called disharmony, and the regularity summarization is shown in Table 1.

Using the interval relation to compose music at random, the first thing needed to be done is that the note sequence is generated randomly in the piano window of FL Studio; and then, based on the interval relation, with reference to the interval table and circle of fifths degree, we need to delete or remove illegal (inconsonant) notes, and sustain Octave, Major sixth, Minor sixth, Perfect fifth, Perfect forth, Major third, and Minor third; Make chord, rhythm and pitch standardized, match the note distribution with the appropriate chords, remove the upbeat on the notes to the downbeat, adjust the notes to the appropriate pitch. It is the sequence of notes (that is, the melody of the song) that is based on the interval relationship. In addition, classical music is generally eight or four bars of a phrase, the visual shape of the melody is arched, that is, by compositing it from the low to the high and then gradually to the low again, which can form a more beautiful phrase, and it also can dig some of composition laws under the circumstances of not violating relevant rules, and all of which can be input into the computer in the form of computer language, so as to achieve the goal of composing randomly.

Table 1 Consonant and dissonant chords of Western music

Name	Consonance
Perfect 5th	Perfectly consonant
Perfect 4th	
Major 6th	Imperfectly consonant
Major 3rd	
Minor 6th	
Minor 3rd	
Minor 7th	Dissonant
Tri-Tone	
Major 7th	
Major 2nd	
Minor 2nd	

2.2 Gain Harmony Interval Sequence According to Chord Progression

Chord progression is the connection form of chord accompaniment, is the using of chord function, consistent with the main music nature, which varies with the melody [7]. Some classic chords are enduring, such as I-V-I, I-IV-I, I-IV-VI, I-VI-II-VI and others common are used in the 18th century, which are still widely used by later generations. Therefore, input the classic chord into the computer and establish database for use at any time, has become an important creative resource for modern composition.

It first we need to determine the style and range according to their own ideas when compose random according to the chord progression, for example, choice major C if you want to create a cheerful song, and choice minor A for sad song. Secondly, determine the vocal range of the performer, choose C3 for male and C4 for female. And then according to the determined tone and range, invoke the chord from the database, display in the piano window of FL Studio and randomly select the chord, and then delete the extra note. Finally, modify the chord according to the lyrics to determine the melody and rhythm to complete the creation. Figure 1 shows the melody produced by the author touching the keys randomly in calm mood according to the direction of the C-Am-F-G chord.

Randomness is to achieve some humanity characteristics and reduce subjective consciousness in the composition, rather than meaningless disorderly random, which is similar with "fuzzy logic" design of artificial intelligence. As a concept of mathematical algorithm, "randomness" is mostly a description of mathematical operations or an allusion to things that have not been revealed, but so far, no true positioning in the field of music creation has been made. Music is subordinate to art and requires dramatic tension. Dramatic refers to the visual reaction of the inner mind of the assumed artistic image in a scene, and generally includes the phenomena of chance, coincidence, sudden change and the like. The development of artistic plot needs such a sudden and unexpected story. Music art also needs such randomness to produce various unexpected results. Resulting in a variety of unexpected results, breaking the habit of people's expectations of the music development context, given the dramatic musical driving force for the creation of music to provide endless inspiration.

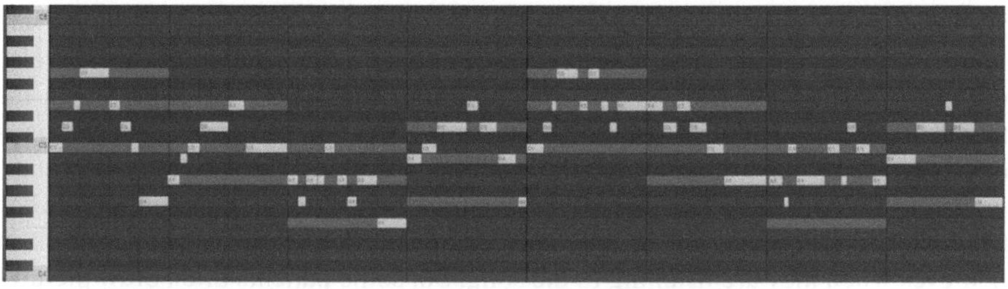

Fig. 1 Gain harmony interval sequence according to chord progression

3 Research on the Color Transformation of Algorithm Composing

Color music is a brand-new integrated art. It combines the music, optics, electronics, mechanical engineering, laser and other disciplines and the development of the formation. With the development of new media technology and the advent of the visual industrial era, "all information visualization" is the goal of development. At present, the art forms of Color Music are mostly based on the artist's subjective understanding of the musical works, so as to design the electro-acoustic effects in conjunction with the music delivery so as to enhance the theme and emotional expression of the music [8].

3.1 Flowing Color

Just as painting cannot separate with color, music art cannot separate with tone, and there is a natural connection between tone and color. From the angle of physical, they are all a kind of fluctuation, but they just have different properties and frequency ranges. The sound waves people can hear are from about 16 cycles per second to about 20,000 cycles per second, the light waves people can see are from about 4150 thousand billion cycles per second to 7800 thousand billion cycles per second [9]. The use of different timbres in musical compositions is very similar to the use of different colors in fine art. Tone and color can both give people the feelings of clear, bright, warm, bleak and so on. Many musicians compare the music with the colors and relate them each other. For example, the famous composer Scriabin believes that the tone of tonality is obvious and offers specific ideas [10].

It can be seen form Scriabin's understanding of the tonality color that with the rise and fall of tone mark, its color sense shifts from red to purple in spectral order, which happens to coincide with the natural relationship between the two, showing that people's subjective feeling and phenomenon can also show some objective regularity.

3.2 Conversion of Music to Color

No matter from what point of view, the relationship between music and color are indirectly compared through subjective feelings, and the final conclusion is also affected by the intervention of individual culture, cultivation and learning. In the natural state, there is not no way to feel the intuitive color of music. Synesthesia is a rare psychiatric disorder in which the numbers, words, and music are color, shape, texture and mood [11]. The melody will appear in the mind synchronously when some patients are seeing the image, and the unreal picture will appear in front of their eyes when they are listening to the song. So, some patients often draw pictures of what appears in front of them when they hear a variety of songs. These works are refreshing.

Compared with ordinary people, Synesthesia patients have a much lower degree of subjective participation in the conversion process among different art forms. As a result, a set of installation named "consciousness stream of Huarong road" (Fig. 2) is created according to the direct process of "entering music—producing pictures—outputting images". Huarong Road is an ancient Chinese folk puzzle game. Ten quadrilaterals are arranged on a 20-grid board. By moving each quadrilateral, the largest square can be moved from the initial position to the lowest center of the board, but crosses are not allowed (Fig. 3). Huarong Road is selected as the carrier of the color music conversion device. First, 10 accelerometer modules are placed in 10 quadrilaterals to sense the movement track and send the data to the Max/MSP application in the computer through a Bluetooth wireless signal, so that 10 quadrilaterals are mapped with 10 simple music pieces in similar style that have been created in

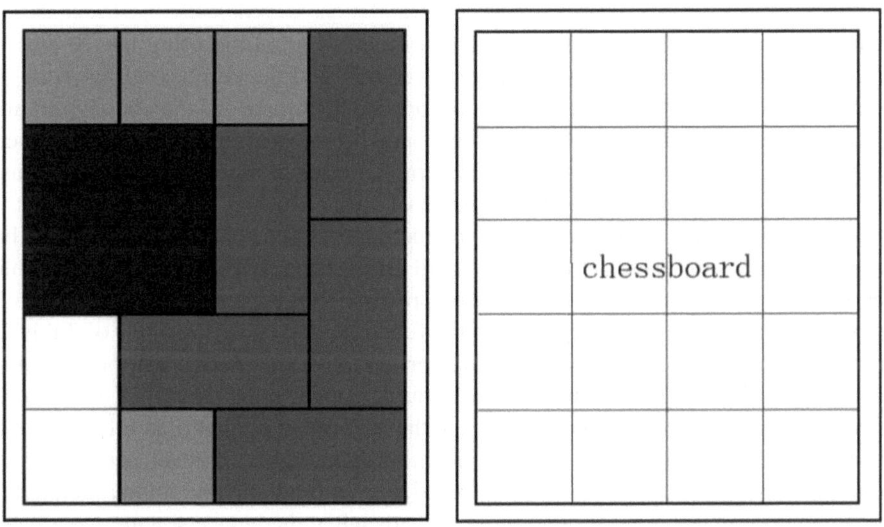

Fig. 2 Huarong road

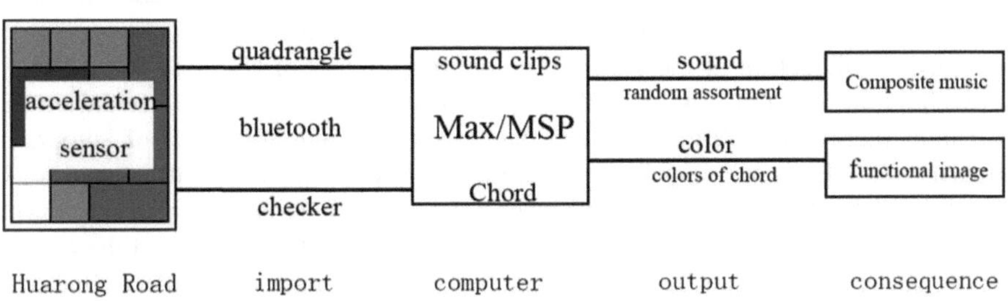

Fig. 3 Color music installation process

Table 2 The music and color comparison table of Scriabin

Major tonal	Color
C	Red
G	Orange, Rose
D	Yellow, brilliant
A	Green
E	Light blue
B	Light blue
#F	Azure blue
bD	Purple
bA	Fuchsia
bE	Iron color, Shiny
bB	Iron color, Shiny
F	Red

the computer. Ten pieces of music are controlled in real time using the 10 pieces of data, a quadrilateral is moved in any direction, and the corresponding piece of music is played in the computer. In addition, the 20 squares in the chessboard are mapped to several groups of Chord progression preselected in the computer. When a quadrilateral is moved in any direction, simple pieces of music and the chords just below it is compounded into complex pieces of music to achieve the purpose of automatic composition, enabling real-time interaction between people and music. At the same time, the simple music piece and chord progression will produce unpredictable chords during the continuous combination process. Then represent these chords with digital and substituting the digital into the function, and using the color of Table 2, the image of function is drawn by computer to realize the transformation of music and image. Music piece produce images simultaneously in the process of continuous compounding, and the acceleration data in the movement of quadrilaterals affect the characteristics of lightness, heaviness, slowness and urgency of music and images in real time and enhance the immersive experience of dramatic and human-computer interaction of music and images. The song ends when the largest square is removed from the board below [12].

The “randomness” of composing techniques is embodied in the randomness of music pieces combination and the randomness of real-time data processing, that is, the random allocation of the data in the moving process of Huarong Road Block, as well as the random influence of the parameters of the acceleration data on the music elements. In addition, because there are more than 200 kinds of layouts in Huarong Road, each layout has an optimal solution, and the solution method is regular and the quantity is limited, a certain solution can get a deterministic regularity through algorithms, thus reflecting the controllability of the chaos. And the regular combination of the quadrilateral and the chessboard changes cyclically in line with the chaotic fractal structure of the music. Finally, excluding the subjective influence, the function image is drawn according to the composite musicality tonality. This random design also reflects the taste and aesthetic value of the music itself.

4 Summary

As a fresh musical vocabulary at the end of the last century, "randomness" appeared with the emergence of chaos theory, which is gradually attached importance to by contemporary composers, and all kinds of experimental works are coming out. Chaotic composition is characterized by its contradiction and unity, and the dramatic tension is very strong. With the rapid development of computer technology, it is bound to bring the constant changes of algorithm composition technology, making the types of music constantly rich. And color music has broken through the limitations of the screen. It can not only be an extension of music, but also enable people to gain immersive experience to better understand works, enhancing the interaction. The development of random composition technology will blur the boundary between the artist and the audience. Everyone can create music, which is conducive to the spread of music and promote the development of industry.

References

1. Peng, D.U., Zhou, C.L., Zhi-Qiang, H.E.: An auto composing method for motive of inventions based on genetic algorithm. Comput. Technol. Dev. (2007)
2. Lefresne, B.J: Applications of Chaos Theory and Fractal Geometry in the Music of Gyorgy Ligeti, University of Ottawa (2005)
3. Milicevic, M.: The impact of fractals, chaos, and complexity theory on computer music composition. In: Proceedings of the 1996 International Computer Music Conference. The International Computer Music Association, pp. 473–476 (1996)
4. Jia, C.: Application of interactive genetic algorithm in computer aided innovation of music composition. Comput. Eng. Appl. **43**(3), 198–201 (2007)
5. Yin, F.: Zhou changle: research progress of algorithmic composition. J. Softw. **17**(2), 209–215 (2006)
6. Zhang, Y.: Huang, T.: Three changes in the process of the transposition of interval. Beauty and The Times, **8**, 50–51 (2002)
7. Hu, Y.: The functional principle and connection method of changing chord in pop songs. Art Eval. **19**, 61–63 (2016)
8. Zhai, M.: Discussion on the development of color music. Pop. Lit. **24**, 16–17 (2011)
9. Wang, L.: Theory of color music. J. Huaihua College **1**, 87–88 (1993)
10. Perle, G.: Scriabin's self-analyses. Music Anal. **3**(2), 101–122 (1984)
11. Simner, J., Ward, J.: Synaesthesia: the taste of words on the tip of the tongue. Nature **444**(7118), 438 (2006)
12. Williams, D., Kirke, A., Miranda, E.R., et al: Investigating affect in algorithmic composition systems. Psychol. Music **43**(6) (2014)

MIX
Papier aus verantwortungsvollen Quellen
Paper from responsible sources
FSC® C105338

FSC
www.fsc.org

If you have any concerns about our products,
you can contact us on
ProductSafety@springernature.com

In case Publisher is established outside the EU,
the EU authorized representative is:
Springer Nature Customer Service Center GmbH
Europaplatz 3, 69115 Heidelberg, Germany

Printed by Libri Plureos GmbH
in Hamburg, Germany